Algal Biorefinery: An Integrated Approach

Debabrata Das
Editor

Algal Biorefinery: An Integrated Approach

Springer

CP

Editor
Debabrata Das
Department of Biotechnology
Indian Institute of Technology
Kharagpur, India

Co-published by Springer International Publishing, Cham, Switzerland, with Capital Publishing Company, New Delhi, India.

Sold and distributed in North, Central and South America by Springer, 233 Spring Street, New York 10013, USA.

In all other countries, except SAARC countries—Afghanistan, Bangladesh, Bhutan, India, Maldives, Nepal, Pakistan and Sri Lanka— sold and distributed by Springer, Haberstrasse 7, D-69126 Heidelberg, Germany.

In SAARC countries—Afghanistan, Bangladesh, Bhutan, India, Maldives, Nepal, Pakistan and Sri Lanka—printed book sold and distributed by Capital Publishing Company, 7/28, Mahaveer Street, Ansari Road, Daryaganj, New Delhi 110 002, India.

ISBN 978-3-319-22812-9 ISBN 978-3-319-22813-6 (eBook)
DOI 10.1007/978-3-319-22813-6

Library of Congress Control Number: 2015957995

Springer Cham Heidelberg New York Dordrecht London

Printed on acid-free paper

Springer International Publishing AG Switzerland is part of Springer Science+Business Media (www.springer.com)

Foreword

Algae was intensively studied for biofuel production from the late 1970s through the mid-1990s; however, in 1996 as the oil prices dipped to $20/barrel, algal biofuel research was abandoned as it seemed like a distant possibility against the low cost fossil fuels. However, by 2009, as oil prices reached $100 per barrel, amid rising concerns about domestic energy security and greenhouse gas emissions, the voyage in algal-based biofuels was re-embarked in a major way.

Microalgae are single-cell, photosynthetic organisms known for their rapid growth and high energy content. Some algal strains have a rapid doubling speed of several times per day. In some cases, more than half of that mass consists of lipids or triacylglycerides—the same material found in vegetable oils. These bio-oils can be used to produce advanced biofuels such as biodiesel, ethanol, green gasoline, biohydrogen and green jet fuel.

Algal biofuels are not economical to produce using the technology available today. Based on conservative estimates, algal biofuels produced in large volumes with current technology would cost more than the extraction of regular vegetable oil. Lowering this cost will require coordinated R&D efforts over the next decade. Although the technical challenges are significant, the broad public benefit of successfully commercializing algal biofuels warrants placing a high priority on the needed research. Particular attention must be paid to the engineering of sustainable microalgal systems. Against this background, the release of a book based on integrated algal biofuels appears well coordinated.

Using algal biomass for extracting one product might not be commercially feasible, thus the concept of biorefinery has surfaced. Under this concept, from algal biomass various products could be harnessed thereby making the process economically viable. Biochemical compounds such as pigments, PHB and starch produced by algae could be used in various fields. Many antioxidant, pharmacy-relevant compounds have been identified in algae like carotenoids, phycobillin proteins etc.

However, the commercial application and potential of algal products encompass a far wider perimeter than what has been explored.

In recent times, researchers have shown interest towards the bioremediation property of algae. The ability to trap heavy metals and toxic compounds in wastewater makes algae a suitable candidate for bioremediation. The challenges and prospect of algal bioremediation have been comprehensively encored in this book. Readers would find a detailed review of heavy metal remediation and engineering challenges of implementation of such technology. This book thus represents all aspects of algal biorefinery concept.

This book is a concerted effort to bring all the major areas of the emerging technologies in algal biology together under one cover. These include: biomass production, cultivation, harvesting, extraction, biorefinery, feedstock conversion into fuels, bioproducts, and scale up of the technology.

I strongly recommend this book to energy scientists, engineers and student enthusiasts who are interested in algal-based biofuel research as well as to industrial concerns that are seeking inexpensive algal-based technologies. This book thus serves as an ideal platform to discuss the wider scope of algal research. It could be an ideal *vade mecum* for students, researchers environmentalist etc. who want to know the prospect of algal technologies towards sustainable development of mankind.

Uppsala University
Uppsala, Sweden

Peter Lindblad

Preface

A book should serve as the ax for the frozen sea within us. (*Franz Kafka*)

Rapid industrialization and anthropological activity has dented the environment to a great extent. Outcome of such activities lead to global weather changes. In recent times, the world economics has shown a positive correlation between per capita energy consumption and per capita growth of gross domestic production (GDP). The major contributor towards the energy demand for past century was mainly based on fossil fuels such as coal, natural gas, petroleum, etc. The emission of greenhouse gases are responsible for the increase of earth's temperature thus leading to the global warming. The reservoirs of fossil fuels are shrinking rapidly and might be available for one more century only. An urgent necessity for future fuel has gained interest in recent times. Future fuel should be renewable in nature, carbon neutral and easily accessible.

Atmospheric carbon dioxide has been captured by photoautotrophic life to produce biomass. Terrestrial phototrophic organisms such as plants, trees, herbs and shrubs are advanced organisms and efficiently capture CO_2 as their biomass. They are also an integral part of the food web of the ecosystem. The terrestrial photoautotrophic biomass could be used as feedstock for fuels and biochemical products but it can supply a fraction of the total energy need. Moreover, destruction of natural habitats associated with terrestrial photoautotrophic biomass could pose a serious environmental and ecological threat. On comparison with terrestrial photoautotrophic organisms, aquatic photoautotrophs are less developed. But they have huge potential as feedstock for fuel and biochemical products. The faster growth rate, non-stringent nutritional demand and ease of cultivation are the salient features of algal biomass production. They can capture atmospheric CO_2 and can convert them to biomass. Use of such biomass could offset the dependency on conventional fossil fuels. Cultivation of algae could counter the debate regarding "food vs fuel" as it doesn't require fertile land or any other seasonal requirements. Many technologies have been developed to realize the potential of algae.

A biorefinery integrates biomass conversion processes and equipment to produce fuels, power, heat, and value-added chemicals from biomass. For realizing the commercial potential of algae, a biorefinery concept has been envisioned that could help to extract maximum benefits out of algal biomass. A refinery concept promotes harvesting of multiple products from the feedstock so as to make the process economically attractive. A functional biorefinery operation should encompass efficient technologies for production, extraction, collection of feedstock, transportation of feedstock and products, life cycle analysis, favourable policies, etc. For last few decades, algal biomass has been explored for various products such as fuel, pigments, pharmaceuticals, bioremediation, etc. To meet the huge demand of algal biomass, a greater emphasis has been given on large scale production of algal biomass in closed or open photobioreactors. Different nutritional conditions for algal growth have been explored like photoautotrophic, heterotrophic, mixotrophic and oleaginous. Open raceway ponds were also explored for studying the effect of seasonal changes on algal productivity. The present book deals with different aspects of algal production systems. These provide a background of the state-of-the-art technologies towards algal cultivation, CO_2 sequestration, and large scale application of algal cultivation systems. Suitability of algal biomass as feedstock for biofuels has been discussed in details. The present book also highlights the potential chemicals extracted from algae that could be used in pharmaceutical industries. Moreover, importance of algal biomass as food supplement to counter deficiency of vital nutrients is also discussed elaborately. The bioremediation ability of algae is highlighted in few chapters where it is proposed that heavy metal contamination could be mitigated through algae cultivation. Moreover, wastewater can also be used for algal cultivation thereby helping in overall reduction in chemical oxygen demand. Thus an effective wastewater management concomitant with energy production could be achieved via algal cultivation.

This book is aimed at a wide audience, mainly undergraduates, postgraduates, energy researchers, scientists in industries and organizations, energy specialists, policy makers, energy specialists, research faculty and others who wish to know the Algal Biorefinery as also wish to get abreast with the latest developments. Each chapter in the book begins with the fundamental explanation for general readers and ends in in-depth scientific details suitable for expert readers. Algal bioengineering laboratories may find this book a ready reference for their routine use.

We hope this book will be useful to our readers!

Kharagpur, India Debabrata Das

Contents

Acronyms

PO-$(OH)_3$	phosphoric acid
μm	micrometre
$^1O_2$	singlet oxygen
2OG	2-oxaloglycerate
3,6-AG	3,6-anhydro-L-galactose
3PGA	3-phosphoglycerate
AA	ascorbate
ACC	acetyl CoA carboxylase
ACC	1-aminocyclopropane, 1-carboxylate
AcCoA	acetyl co-enzyme A
acetyl-CoA	acetyl coenzyme A
ACP	acyl carrier protein
AD	anaerobic digestion
Adh	alcohol dehydrogenase
ADP	adenosine diphosphate
ADP	adenosine dinucleotide phosphate
AEM	anion exchange membrane
AgNP	silver nanoparticle
AHS	acetohydroxy acid synthase
AIDS	acquired immunodeficiency syndrome
Al	aluminium
ALI	acute lung injury
AOM	algogenic organic matter
AOX	adsorbable organic halogens
APC	allophycocyanin
APX	ascorbate peroxidise
ARA	acetylene reduction assay
ARTP	atmospheric and room temperature plasma
As(III)	arsenite
As(V)	arsenate
ASC	ascorbate

ASTM	American Society for Testing and Materials
ASW	artificial sea water
ATP	adenosine triphosphate
ATP	adenosine trinucleotide phosphate
Ba^{2+}	barium ion
BBM	Bold's basal medium
Bcl-2	B-cell lymphocytic-leukaemia proto-oncogene 2
BEA	beta polymorph A
BES	bioelectrochemical system
BG-11	blue green algae 11 media
BGA	blue green algae
BGY	billion gallons per year
BMP	biomethane production
BNF	biological nitrogen fixation
BOD	biological oxygen demand
BTL	biomass to liquid
C	carbon
C:N	carbon:nitrogen
CA	carbonic anhydrase
Ca	calcium
Ca^{2+}	calcium ion
$CaCO_3$	calcium carbonate
cAMP	cyclic adenosine monophosphate
CaO	calcium oxide
CARPT	computer automated radioactive particle tracking
CAT	chloramphenicol acetyl-transferase
CAT	catalase
CBB	cyanobacterial biofilmed biofertilizers
CCM	carbon concentrating mechanisms
CCM	CO_2 concentrating mechanisms
Cd	cadmium
Cd^{2+}	cadmium ion
CD4	cluster of differentiation 4
CD59	cluster of differentiation 59
CD8	cluster of differentiation 8
CE	Coulombic efficiency
CFD	computational fluid dynamics
cGMP	cyclic guanosine monophosphate
CH_4	methane
Chl *a*	chlorophyll *a*
CHO	chinese hamster ovary
CIP	clean-in-place
CO	$(NH_2)_2$ urea
CO_2	carbon dioxide
Co^{2+}	cobalt ion

CoA	co-enzyme A
COD	chemical oxygen demand
$COOH^-$	carboxyl ion
COX-2	cyclooxygenase 2
C-PC	C-phycocyanin
Cr	chromium
CrI	crystalline index
CTAB	cetyl trimethyl ammonium bromide
Cu^{2+}	cuprous ion
CV	cyclic voltammetry
CVDs	cardiovascular diseases
DAF	dissolved-air flotation
DCMU	3-(3,4-dichlorophenyl)-1,1-dimethylurea
DCW	dry cell weight
DET	direct electron transfer
DF	dietary fibre
DGAT	diacylglycerol acyltransferase
DHA	docosahexaenoic acid
DHAP	dihydroxyacetone-phosphate
DIC	dissolved inorganic carbon
DMA	dimethyl arsenate
DMAE	dimethylarsenoyl ethanol
DMAP	dimethylallyl diphosphate
DMMA	dimethyl arsenic acid
DMSO	dimethyl sulfoxide
DNA	deoxyribonucleic acid
DO	dissolved oxygen
DOC	dissolved organic carbon
DP	degree of polymerization
DPPH	1,1-diphenyl-2-picrylhydrazyl
DW	dry weight
DXP	1-deoxyxylulose-5-phosphate
EAB	electrochemically active bacteria
EC	electrical conductivity
ECF	electro-coagulation–flotation
EDTA	ethylene diamine triacetic acid
EF-Ts	elongation factor Ts
ELISA	enzyme linked immune sorbant assay
EMB	effective microbial biofilms
EN	European standards
ENR	enoyl-Acp reductase
EPA	eicosapentaenoic acid
EPS	extracellular polysaccharides
ETC	electron transport chain
exDNA	extrachromosomal deoxyribonucleic acid

FACoA	fatty-acylCo-enzyme A
FAEE	fatty acid ethyl esters
FAME	fatty acid methyl ester
FAO	Food and Agriculture Organization
FAU-X	faujasite framework X
FBR	fluidized bed reactor
Fd	ferredoxin
FDA	fluorescein diacetate
FDA	Food and Drug Administration
Fe	iron
FFA	functional fatty acid
FFAs	free fatty acids
FNR	ferredoxin nadp$^+$ reductase
FNR	ferredoxin NADPH reductase
FNR	ferredoxin-NADP$^+$-reductase
FOG	fats oils and grease
FTS	Fischer–Tropsch synthesis
FYM	farmyard manure
G3P	glyceraldehyde-3-phosphate
G3P	glucose-3-phosphate
GAP	glyceraldehyde-3-phosphate
GC	guanidine cytosine
GHG	green house gas
GICON	German Engineering and Consulting Company
GK	glucokinase
GLA	γ-linoleic acid
GlgC	glucose 1-phosphate adenyltransferase
GMO	genetically modified organisms
GOT	glutamic oxaloacetate transaminase
GPT	glutamate pyruvate transaminase
GPX	glutathione peroxidise
GR	glutathione reductase
GSH	glutathione
GSP	glycosylated serum protein
H_2O	water
H_2O_2	hydrogen peroxide
H_2S	hydrogen sulphide
H_2SO_4	sulphuric acid
HCl	hydrochloric acid
HCO_3	bicarbonate
HDL-C	high-density lipoprotein cholesterol
HEPI	Health Enhancement Products, Inc.
Hg (II)	mercuric ion
hIAPP	human islet amyloid polypeptide
HISTAR	Hydraulically Integrated Serial Turbidostat Algal Reactor

HIV	human immunodeficiency virus
HMBPP	1-hydroxy-2-methyl-2-butenyl-4-pyrophosphate
HOCl	hypochlorous acid
HPLC	high performance liquid chromatography
HRAP	high rate algal ponds
HRT	hydraulic retention time
HupL	hydrogen uptake, large subunit
HupS	hydrogen uptake, small subunit
I/R	ischemia-reperfusion
IAA	indole-3-acetic acid
IBA	indole-3-butyric acid
IC	inferior colliculus
ICAM-1	intercellular cell-adhesion molecule 1
ICP-MS	inductively coupled plasma mass spectrometry
IEM	ion exchange membrane
IgA	immunoglobulin A
IL-1β	interleukin-1β
iNOS	inducible nitric oxide synthase
INR	Indian rupee
IPP	Isopentyl diphosphate
IREP	integrate a renewable energy park
K	potassium
KAR	3-ketoacyl-Acp reductase
KAS	3-ketoacyl-Acp synthase
kDa	kilodaltons
K_m	affinity constant
LCA	life cycle assessment
LDA	laser Doppler anemometry
LDL	low density lipoprotein
LDL-C	low density lipoprotein cholesterol
LDPE	low density polyethylene
LED	light emitting diode
LHC	light harvesting centre
LPA	lyso-phosphatidic acid
LPAAT	lyso-phosphatidic acid acyltransferase
LPS	lipopolysaccharide
LSF	liquid seaweed fertilizer
MA	methyl arsonate
MAT	malonyl-Coa:Acp transferase
MDA	malondialdehyde
MDHA	mono-dehydro-ascorbate
MEP	methylerythritol pathway
MFC	microbial fuel cell
Mg	magnesium
Mg^{2+}	magnesium ion

MgO	magnesium oxide
MLDP	major lipid droplet protein
MMAA	monomethyl arsenic acid
MMC	mitomycin-C
Mn	manganese
Mn^{2+}	manganese ion
MOG	malonyl-CoA-oxaloacetate-glyoxylate
MPN	most probable number
mRNA	messenger ribonucleic acid
MSTR	membrane-sparged helical tubular photobioreactor
MVA	mevalonic acid
N	nitrogenous compounds
N	nitrogen
N:P:K	nitrogen:phosphorous:potassium
N_2	nitrogen
Na_2CO_3	sodium carbonate
Na_2SO_4	sodium sulphate
NaCl	sodium chloride
NAD(P)H	nicotinamide adenine dinucleotide phosphate
NADH	nicotinamide adenine dinucleotide
NADPH	nicotinamide adenine dinucleotide phosphate
NADPH	nicotinamide adenine dinucleotide phosphate-reduced form
NaOMe	sodium methoxide
NASH	non-alcoholic steatohepatitis
NER	net energy requirement
NEWS	nutrient export from watersheds
NF- κ B	nuclear factor-kappa B
NGO	non governmental organisation
NH_3	ammonia
NH_4^+	ammonium ion
Ni	nickel
Ni^{2+}	nickel ion
nm	nanometer
NMVFR	non-mixed vertical flow reactor
NO	nitrous oxide
NO	nitric oxide
NO_2^-	nitrite
NO_3^-	nitrate
NOx	nitrogen oxides
NPQ	non-photochemical quenching
NR2B	N-methyl D-aspartate receptor subunit 2B
O_2	superoxide anion
O_2	oxygen
O_2^-	superoxide radical
OA	oxaloacetate

OCV	open circuit voltage
OH•	hydroxyl radical
OLR	organic loading rate
$ONOO^-$	peroxynitrite
ONOOH	peroxynitrous acid
OPP	oxidative pentose phosphate
ORFs	open reading frames
ORP	oxidation reduction potential
P	phosphorous
P	phosphates
PAH's	polycyclic aromatic hydrocarbons
PAM	pulsed amplitude modulated fluorescence technique
PAP	phosphatidic acid phosphatase
PAR	photosynthetically active radiation
Pb	lead
Pb^{2+}	lead ion
PBP	phycobili proteins
PBR	photobioreactor
PBRs	photobioreactors
PBS	Phycobilisome
PC	phycocyanin
PC	plastocyanin
PCB	phycocyanobilin
PCR	polymerase chain reaction
PD	power density
PDA	polydopamine
Pdc	pyruvate decarboxylase
PE	phycoerythrin
PEF	pulse electric field
PEP	phosphoenolpyruvate
PepC	phosphoenolpyruvate carboxylase
PGPR	plant growth promoting rhizobacteria
PHB	poly-3-hydroxybutyrate
PIV	particle image velocimetry
PKC	protein kinase-C
PKU	phenylketonuria
PMFC	photosynthetic microbial fuel cell
PO_4	phosphate
PPB	polyphosphate
ppm	parts per million
PQ	plastoquinone
Prx	peroxiredoxin
PS	photosynthetic apparatus
PSI	photosystem I
PSII	photosystem II

PUFA	poly unsaturated fatty acid
PUFAs	polyunsaturated fatty acids
PVA	polyvinyl alcohol
PVC	poly vinyl chloride
$PVDF^-$	polyvinylidene fluoride
RCC	reinforced cement concrete
RNA	ribonucleic acid
RNAi	RNA interference
RNAi	ribonucleic acid interference
RNAse P	ribonuclease P
ROO•	peroxyl radical
ROS	reactive oxygen species
R-PC	R-phycocyanin
R-PE	R-phycoerythrin
rRNA	ribosomal ribonucleic acid
RuBisCO	ribulose 1,5-bisphosphate carboxylase
RuBP	ribulose-1,5-biphosphate
S/V	surface–to–volume ratios
SARS	severe acute respiratory syndrome
$SC\text{-}CO_2$	supercritical carbon dioxide
SCP	single-cell proteins
SDS	sodium dodecylsulfate
Se	selenium
Se-PC	Se-enriched phycocyanin
SHGs	self help groups
SLF	seaweed liquid fertilizer
SO_4^-	sulphate ion
SOD	superoxide dismutase
SOx	sulfur oxides
SPC	*Spirulina platensis* concentrate
SPRTC	Spirulina Production Research and Training Centre
Sr^{2+}	strontium ion
SRT	solids retention time
STKs	serine/threonine kinases
TAG	triacylglyceride
TAG	triacylglycerol
TAGs	triacylglycerides
TAIGEM-EB	Taiwan general equilibrium model energy
T-AOC	total antioxidative capability
TAP	tris acetate phosphate
TBT	tributyltin
TC	total cholesterol
TDS	total dissolved solids
TETRA	tetramethylarsonium ion
TFF	tangential flow filtration

TG	triglyceride
THF	tetrahydrofuran
TMAO	trimethylarsine oxide
tmRNA	transfer messenger ribonucleic acid
TNF-α	tumour necrosis factor-α
TOC	total organic carbon
tRNA	transfer ribonucleic acid
Trx	thioredoxin
TSS	total suspended solids
USD	US Dollar
USY	ultrastable Y
UV	ultraviolet
VASP	vasodilator-stimulated phosphoprotein
VFAs	volatile fatty acids
VS	volatile solids
WHO	World Health Organization
WIS	water insoluble solids
WSP	waste stabilization pond
XRD	X-ray diffraction
Zn	zinc
ΔΨm	mitochondrial membrane potential

About the Editor

Debabrata Das pursued his doctoral studies from Indian Institute of Technology (IIT) Delhi. He is a Senior Professor at IIT Kharagpur. He has pioneered the promising R&D of Bioenergy production processes by applying fermentation technology. He is actively involved in the research of hydrogen biotechnology for a period of last sixteen years. His commendable contributions towards development of a commercially competitive and environmentally benign bioprocess began with the isolation and characterization of high-yielding bacterial strain *Klebsiella pneumoniae* IIT-BT 08 (previously known as *Enterobacter cloacae* IIT-BT 08), which, as of today, is known to be the highest producer of hydrogen by fermentation. He has conducted basic scientific research on the standardization of physico-chemical parameters in terms of maximum productivity of hydrogen by fermentation and made significant contribution towards enhancement of hydrogen yield by redirection of biochemical pathways. Apart from pure substrates use of several other industrial wastewaters such as distillery effluent, starchy wastewater, deoiled cake of several agricultural seeds like groundnut and coconut and cheese whey were also explored successfully as feedstock for hydrogen fermentation.

Prof. Das has been also involved in CO_2 sequestration, biodiesel, phycobillin protein etc. production from microalgae research work for the last seven years. He has been leading the Indian Group in the Indo-Danish sponsored research project of Department of Biotechnology, Government of India on "High rate algal biomass production for food, biochemicals and biofuels". He organized one International Conference on "Algal Biorefinery: a Potential Source of Food, Feed, Biochemicals, Biofuels and Biofertilizers (ICAB 2013)" at IIT Kharagpur, India during January 10–12, 2013. He also explored the potentiality of microalgae in the operation of Microbial Fuel Cells (MFCs). The major aim of the research work was to synchronize the bioremediation of wastewater with clean energy generation. He has also been associated as MNRE Renewable Energy Chair Professor at IIT Kharagpur. He has the Thomson Reuters ISI h-index of 27 (Google Scholar h-index of 35) for his research work. He has about 120 research publications in the peer reviewed journals and contributed more than 15 chapters in the books published by International publishers. He is the author of the book entitled "Biohydrogen Production: Fundamentals and Technology Advances" published by CRC Press. He has two Indian patents.

Prof. Das is the Editor-in-Chief of *American Journal of Biomass and Bioenergy*. He is a member of the editorial board of *International Journal of Hydrogen Energy*; *Biotechnology for Biofuels*; and *Indian Journal of Biotechnology*. He has successfully completed six pilot plant studies in different locations in India. He is involved in several national and international sponsored research projects like NSF, USA; DAAD, Germany; etc. He has been leading a Technology Mission Project of Ministry of New and Renewable Energy (MNRE), Government of India for the installation of several pilot plants on biohydrogen production process in different locations in India for its commercial exploitations. He has been awarded IAHE Akira Mitsue award 2008 and Malaviya Memorial award 2013 for Senior Faculty for his contributions in hydrogen research. He is a Fellow of Indian National Academy of Engineering 2015; Institute of Engineers (India) 2012, Biotechnology Research Society of India 2011; and West Bengal Academy of Science and Technology, 2004.

Chapter 1
Introduction

Debabrata Das

1.1 Background

The primitive earth consisted of an atmosphere filled with CO_2 which could not sustain any life forms. The advent of life on earth was possible due to *Cyanobacterium* and algae. They sequestered the atmospheric CO_2 via photosynthesis and in turn released molecular oxygen. This led to a rapid decrease in CO_2 levels as a result of which life started evolving on earth. Time has come once again that these humble organisms save us from the threat of global warming. The conversion of solar energy to chemical energy is the major attribute of all photosynthetic organisms including algae. The energy gets stored in the cells as oils, carbohydrates and proteins. The conversion of solar energy to chemical energy largely depends on the photosynthetic efficiency of the organism. Algae are the most photosynthetically efficient organisms on earth which makes them a prospective feedstock for different purposes. The oil production by microalgae is much higher as compared to any other oil crop. Algal biotechnology could provide a plausible solution to many problems ranging from green house gas emission reduction to treatment of wastewater and generation of value added products. Algae are diverse in nature and can be found from unicellular to complex and differentiated levels. They generally inhabit damp surroundings and are abundant in terrestrial as well as freshwater and marine environments. Similar to all photosynthetic organisms, algae require sunlight, carbon dioxide and water for their growth. The affinity of CO_2 for microalgae is very high which makes them a prospect for CO_2 mitigation. The productivity of oil per acre is the highest in microalgae and it can surpass any other oil crop for biodiesel production. Moreover, it does not require arable land for growth and can be sustained on wastewater (Demirbas, 2009a).

D. Das (✉)
Department of Biotechnology, Indian Institute of Technology,
Kharagpur 721302, India
e-mail: ddas.iitkgp@gmail.com

D. Das (ed.), *Algal Biorefinery: An Integrated Approach*,
DOI 10.1007/978-3-319-22813-6_1

There are four main classes of algae: diatoms, green algae, blue green algae and golden algae. They can be filamentous or are found as phytoplanktons. Rapid increase in the population of algae leads to algal blooms in major water bodies (Demirbas, 2009b). Sustainability is the keyword for the future. To achieve an economically stable and environmentally sustainable future, the fuel production processes need to go green. A step forward towards a renewable and sustainable future can be the use of algae in a biorefinery concept. The ever rising of production costs and low yields of microalgal biomass provide a barrier to realising this dream. These problems may be solved if we can utilize microalgae in the biorefinery concept. It works on similar lines with the petroleum biorefinery concept. The algal biomass derived after cultivation contains different elements such as proteins, carbohydrates and lipids and the whole biomass itself can be used to derive various value added products. The integration of different techniques of production is the necessity of the hour in order to tackle the price involved in algal biomass production. The two major approaches of algal biorefinery are: energy generation and bioremediation.

1.1.1 Definition of Algal Biorefinery

The integration of different biomass conversion processes to produce energy and value added chemicals into a single facility is called a biorefinery. In a broad definition, it converts all kinds of biomass (all organic residues, energy crops, and aquatic biomass) into numerous products (fuels, chemicals, power and heat, materials, and food and feed) (Fig. 1.1). The conversion of biomass to different products is sustainable because the processes produce minimal wastes to the environment. This concept is similar to the crude oil refinery where various products are produced at different stages of petroleum refining. The biorefinery concept presents a conceptual model for future biofuel generation along with production of high value added products. This in turn reduces the cost of liquid fuel production with maximum utilization of the biomass. Looking into the future, we need more efficient biorefineries to operate where there is maximum utilization of the heat released from the process as well as utilization of biomass to the fullest extent. The heat requirements for the biorefinery may be obtained from the recirculation of heat generated from the process (WI, 2007).

Similar to a petroleum refinery, the biomass is used as a raw material for production of varied products. Different conversion processes (physical, chemical, biological and thermal) are used either individually or in combination to provide products for economic purposes. The products obtained after conversion are fractionated into various separate products or may undergo further processing to obtain value added products. The waste products obtained after each step of treatment is either used for different purposes or is recycled into the production chain to be used as raw materials for the process. The various uses of products obtained from algal biorefineries include transport fuels, therapeutics, food additives or as biofertilizers (Fig. 1.2).

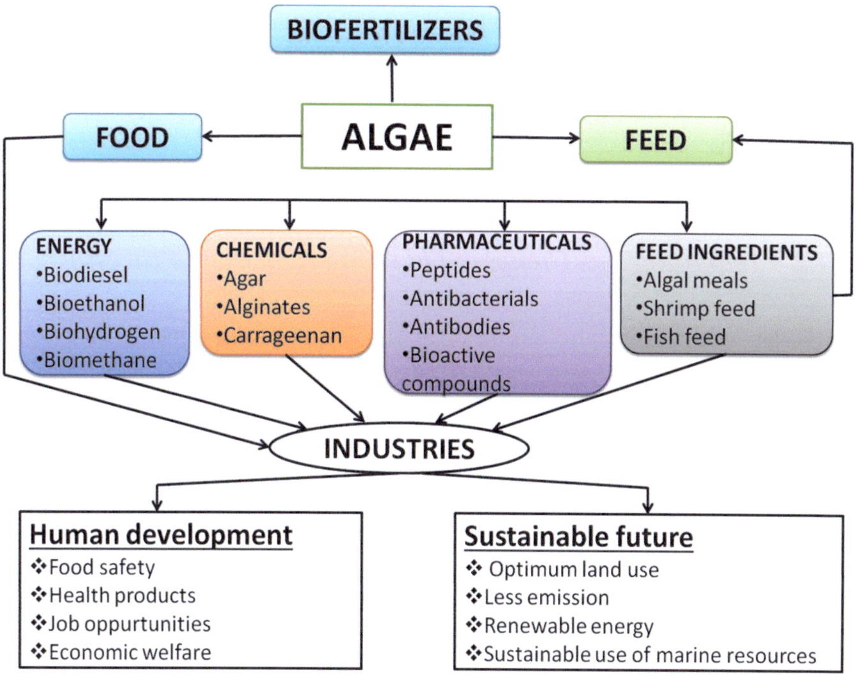

Fig. 1.1 Various uses of algae in a biorefinery concept.

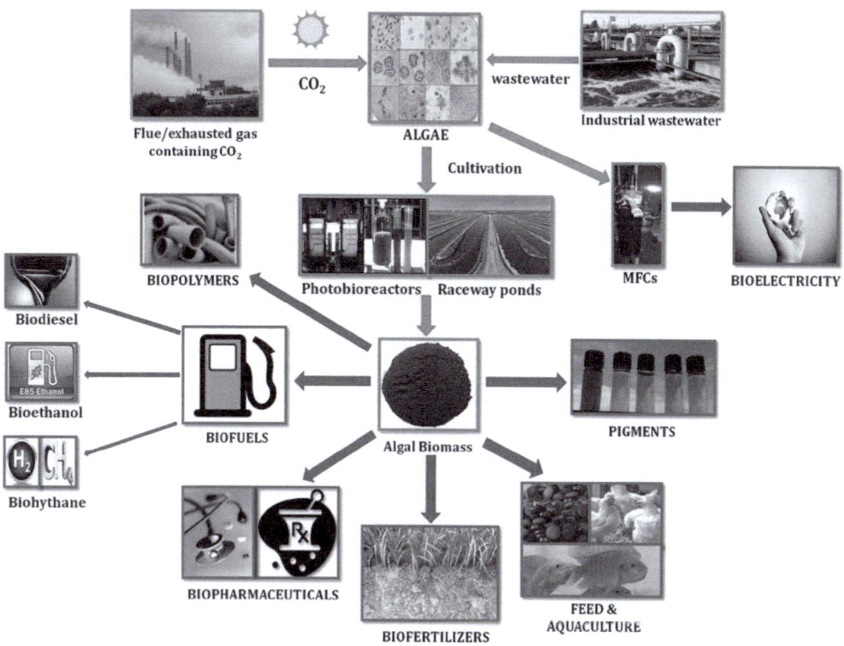

Fig. 1.2 Visualization of algal biorefinery concept.

The process can be made more economical when the raw materials used are basically waste products. This in turn serves the dual purpose of energy generation as well as bioremediation. The biorefineries may also be integrated with the existing power generation infrastructure to lower the production costs and reduction of carbon footprint.

1.1.2 Importance as Compared to Other Biological Processes

Algae are generally photosynthetic organisms which inhabit marine as well as freshwater environments. They have an advantage over land-based plants because of their simple photosynthetic apparatus and proximity to water, CO_2 and other nutrients. This accounts for their higher photosynthetic efficiency. With cell structures dedicated to photosynthesis (no root, stems, etc.), they are favourable candidates for aquaculture too (John et al., 2011). They appear to be the only possible solution to the feedstock issue for generating renewable energy (Chisti, 2007; Schenk et al., 2008). The environmental impact of microalgal biofuels is significantly less as compared to conventional plant biofuels. It also does not go into a conflict with the world's food supply. They have fuel properties similar to conventional fuels which give them an edge over lignocellulosic biofuels (Miao and Wu, 2004). They have also very high inherent lipid content and can be grown in varied climatic zones (Clarens et al., 2010). Algal species are widespread and divergent in nature. The variety of species provides an advantage for selection. The algal species can be selected according to the product required and can be manipulated for higher yields. A wide spectrum of products ranging from energy sources (biodiesel, bioethanol, jet fuel, etc.) to nutritional compounds and biofertilizers, recombinant proteins, pigments, medicines, pharmaceuticals and vaccines can be produced from algae (Pulz, 2004; Pienkos and Darzins, 2009).

The main advantages of microalgae are given below (Campbell, 1997; Chisti, 2007; Huntley and Redalje, 2007; Schenk et al., 2008; Li et al., 2008; Rodolfi et al., 2009; Khan et al., 2009):

- A higher photosynthetic efficiency (approximately 3–8 % against 0.5 % for terrestrial plants), leading to higher biomass yields per hectare and higher growth rates (e.g. 1–3 doublings/day).
- A higher CO_2 sequestration ability as compared to terrestrial plants.
- It can grow in liquid medium. It reduces the use of freshwater for growth because it can grow in wastewater as well as saline and brackish water.
- Algae may be used for the bioremediation because it can grow on a variety of wastewaters (agricultural, municipal, industrial wastewater).
- There is no competition with arable land for food production as it can utilize marginal and non-arable lands for growth.
- Algal biomass can be harvested almost all the year round with no seasonal breaks.

- Induction of desired products (lipids, proteins, carbohydrates) is possible by adjusting different cultivation conditions.
- The methods of algal culturing are very simple, easy to operate and can be scaled up for higher biomass productivity.
- Less use of fertilizers and pesticides leading to less pollution.
- The emission from algae-based biofuels contains less amount of NO_x (Li et al., 2008) causing minimum damage to the environment.
- They produce a large amount of storage products which can be used for biofuel production: protons and electrons (for biohydrogen), sugars and starch (for bio-ethanol), oils (for biodiesel) and biomass (for BTL and biomethane).
- The value added co-products are also obtained (proteins, polysaccharides, bio-fertilizers, pigments, etc.).

1.2 Global Scenario

The conversion of light energy into chemical energy is the driving force for all the reactions which ultimately leads to the formation of feedstock required for the formation of different biofuels: synthesis of protons and electrons (for bio-H_2), sugars and starch (for bio-ethanol), oils (for biodiesel) and biomass (for BTL products and biomethane) (Hankamer et al., 2007; Costa and Morais, 2011). The most acceptable and technically feasible biofuels in the international market are biodiesel and bio-ethanol. They score over other biofuels because no engine modification is required for their use and can be produced by using existing technologies. The overall solar energy conversion efficiency determines net energy yield/ha and this in turn determines land requirements for fossil fuel displacement. In the case of ethanol from sugarcane, the overall solar energy conversion energy efficiency is currently ~0.16 % (Kheshgi et al., 2000) and in the case of biodiesel from palm oil ~0.15 % (Reijnders and Huijbregts, 2009). These percentages are much higher than those for transport biofuels from European wheat and rapeseed (Reijnders, 2009). Both biodiesel and bioethanol are being produced in increasing amounts as renewable biofuels, but their production in large quantities is not sustainable (Chisti, 2007, 2008a, b). The increasing debate of food vs fuel has put algae in the forefront. Microalgae can be a solution to this problem because it does not use arable land for production of biofuels. But higher production and maintenance costs have deterred the growth of microalgae as feedstock for biofuels. This problem can be solved by using the microalgae in a biorefinery concept.

Recent studies have intensively focussed on biorefineries (Taylor, 2008), and stipulations that biomass-based feedstocks can also produce industrially important chemicals (van Haveren et al., 2008). In the coming decades, production of biomass-based bulk chemicals will come to the forefront (Willems, 2009). The already proposed plans include coproduction of animal feed along with production of biofuels. Methods are to be devised on similar lines with an oil refinery using biomass as initial feedstock. Mineral oil being a highly concentrated material can be utilized for

Table 1.1 Types of photobioreactors with their optimal features (Dasgupta et al., 2010)

Type of photo bioreactor	S/V ratio	Agitation system	Temperature control	Gas exchange	Advantages	Disadvantages
Tubular reactors						
Vertical tubular	Small	Airlift, bubble column		Open gas exchange at headspace	Good mixing, efficient CO_2 supply and O_2 removal	Scale up is limited, major light is reflected due to angle
Horizontal tubular	Large	Recirculation with diaphragm/ mechanical pumps	Shading, overlapping, water spraying	Injection into feed, and dedicated degassing units	Adequate angle towards sunlight	High shear due to pumps, risks of O_2 buildup, biofouling, separate gas exchange unit required
Helical tubular	Large	Centrifugal pumps	Heat exchanger	-do-	High S/V, easy scale up by increasing the number of units	O_2 buildup, separate gas exchange, pumps exert more shear, cell debris accumulate inside
α-shaped reactor	Large	Airlift	-do-	Injection in the vertical units and degassed at top	High unidirectional flow rate with low air flow rate, high S/V	Foam formation due to high cell density

Flat plate reactors						
Flat panel bubbled at bottom	Medium	Bubbling at bottom or from sides, recirculation	Heat exchange coils	Bubbling	Open gas transfer avoids O_2 buildup	Shear due to entrainment of cells till bubbles burst
Flat panel pivoted at centre	Medium	Pulsating motion	Heat exchange coils	Degasser	Good mixing, low shear	Scale up is difficult
Floating type bioreactor	Medium	Sea saw motion	No cooling required	-do-	Low energy for operation, good agitation, can be installed on lakes and sea floor	
Fermentor type with internal/external lighting	Small	Impellers	Heat exchange coils	By sparger	High degree of control of various parameters	Light conversion efficiency is less
Torus shaped reactor	Medium	Marine Impeller	Cooling fans	CO_2 inlet after impeller, outlet at top	Good mixing conditions owing to shape avoiding dead zones	
Annular triple jacketed with lighting from innermost chamber	Medium	Magnetic stirrer	Outer water jacket	Open gas exchange	Good S/V and temperature control, open gas exchange	Scaling up is difficult, biofouling
Induced diffused PBR	Large	Not required	–	–	Greater thickness achievable due to	Material costs

production of various coproducts. The oil biorefineries also use the concept of integration of heat energy to minimize energy input. But biomass is a locally available product which tempts the producers to decentralize the facility and go for smaller biorefineries as compared to large oil refineries. The small biorefineries in cooperation with each other produce different value added products for the market (Willems, 2009). Furthermore, the biorefinery approach should not be the same as the oil refinery approach because of the present-day constraints on fossil fuel inputs and GHG emissions from industrial units into the atmosphere. In addition, in an emission-constrained industrial environment, we might not have a long time frame to maintain the same level of processing refinements that oil refineries have attained over many decades. A much more feasible and practical approach to this issue would be the integration of many sources of renewable energy into biofuel production and processing so that total fossil fuel input and subsequent emissions are minimized.

1.3 Classification of Algae and Their Uses

Algae are thallophytes (plants lacking roots, stems and leaves) that have chlorophyll-*a* as their primary photosynthetic pigment, and that lack the sterile cover of cells around the reproductive cells. They have defined tissues containing specialized cells. However, the degree of specialization or differentiation of cell types is much less than for terrestrial vascular plants (Wiencke and Bischof, 2012).

1.3.1 Microalgae

Microalgae are microorganisms (prokaryotic or eukaryotic), which are able to accumulate biomass by photosynthetic process assimilating sunlight, water and carbon dioxide. The time of microalgal cultivation varies from 24 h to several days (Mata et al., 2009) and doubling time can be within few hours during their exponential growth period. They live in various ecosystems and can be found not only in water but also in soil environments. There are 50,000 microalgal species and only approximately 30,000 species have been studied (Mata et al., 2009). These organisms represent exciting possibilities as promising sources of a diverse range of metabolites of immense medical and industrial significance (Spolaore et al., 2006) and can convert solar energy into biomolecules including carbohydrates, proteins, lipids and triglycerides (Varfolomeev et al., 2010). Their ability to grow rapidly and adapt to extreme environments and ecologies make them suitable models for not only understanding the metabolic and evolutionary processes, but also miniature factories for producing useful value added products.

1.3.2 *Cyanobacteria*

Cyanobacteria [or bluegreen algae (BGA)] are a prokaryotic microorganism which can accumulate biomass by photosynthetic process like the microalgae. Cyanobacteria are commonly found in paddy fields because of presence of favourable environment such as high temperature required by rice, nutrient management, reducing conditions in soil and ability of these organisms to withstand desiccation (Mitra, 1951). Watanabe and Yamamoto (1971) observed the presence of BGA in 71 % of Japanese soils. Relatively lower occurrence of BGA has also been reported in Australian paddy fields (Bunt, 1961). The BGA accounted for 33 % on average of the 2,213 samples and some reports showed up to 50 % of BGA in some of the southern and eastern states of India (Venkataraman, 1979; Kaushik, 1995). More than 125 strains of N_2 fixing free living BGA such as *Anabaena, Nostoc, Aulosira, Calothrix, Tolypothrix, Aphanothece, Cylindrospermum* and *Gloeotrichia* are common in flooded rice ecosystem. The survey of paddy fields revealed a remarkable difference in the cyanobacteria growing in soil and those in flood water above the soil surface (Rother and Whitton, 1989). Thirty-eight soil samples from 11 districts of Dhaka (Bangladesh) analyzed for blue green algal flora recorded a total of 84 strains, 50 % of which were reported to be heterocystous diazotrophic forms belonging predominantly to *Fischerella, Nostoc* and *Calothrix* (Khan et al., 1994).

1.3.3 *Macroalgae*

Macroalgae are eukaryotic photosynthesizing organisms that are somewhat more differentiated than microalgae, however less than plants. The ocean covers two thirds of the world, and the upper layer of the ocean is inhabited by vegetation dominated by these evolutionarily primitive plants like macroalgae (Lobban and Harrison, 1997; Wiencke and Bischof, 2012). Marine macroalgae have been identified as a group of organisms of vital importance for ecosystem function within coastal ecosystems. They form enormous underwater forests of considerable size with a structure on rocky coasts similar to terrestrial forests, and provide a very diverse habitat and breeding area for an uncountable number of organisms (e.g. fish and crustaceans). Large masses of macroalgae can either bloom in some coastal regions and may end as beach cast in some localities, and support meiofauna (small animals) species. The beach cast can also occur under normal circumstances due to storms, and macroalgae are tossed and drift to shore. These macroalgal-blooms constitute often a great nuisance for humans (Wiencke and Bischof, 2012).

The term "macroalgae" (also known as seaweeds) includes macroscopic, multicellular marine green, brown and red algae. Each of these groups has a microscopic and even unicellular, representative. Furthermore, all macroalgae are at some stage of their lifecycle unicellular as spores or zygotes, and may be temporarily planktonic. Evolutionarily the seaweeds are quite diverse and divided into four traditional divisions (or phyla)—Cyanophyta, Rhodophyta, Phaeophyta and Chlorophyta. They assigned to two or more kingdoms, depending on the systematists (Eubacteria, Plantae/Protista), and a proposed new kingdom, Chromista (for the brown algae). These divisions of food chain, producing organic material from sunlight, carbon dioxide and water (Wiencke and Bischof, 2012).

Macroalgae act as ecological engineers mainly on rocky coasts of the oceans. The muddy and sandy areas have fewer macroalgae, because most of the species cannot anchor there. The production of macrophytes (macroalgae and seagrasses (vascular water plant)) amounts to 5–10 % of the total oceanic production, even though they only cover a minute area of the world's oceans (Lobban and Harrison, 1997; Wiencke and Bischof, 2012). Carbon assimilation of kelps (large brown algae of the order Laminariales) is with 1.8 kg carbon m^{-2} $year^{-1}$ similar to that of dense terrestrial forests (Wiencke and Bischof, 2012).

1.4 Overview of the Biochemical Pathways

1.4.1 Lipids and Their Biosynthesis

The *de novo* synthesis of fatty acids occurs in chloroplast of algae. TAGs can be synthesized in the cell using two pathways, one acyl CoA dependent pathway (*de novo* fatty acid biosynthesis pathway) and another is acyl CoA independent pathway. The *de novo* fatty acid biosynthesis pathway starts with carboxylation reaction with formation of 3-carbon compound, malonyl CoA (Fig. 1.3). In this step, acetyl CoA irreversibly combines with CO_2 with the help of acetyl CoA carboxylase (ACCs) using ATP. Acetyl-CoA used for the reaction can be derived from either plastidial or cytosolic glycolysis, or directly from dihydroxyacetone phosphate (Greenwell et al., 2010). Phosphoenolpyruvate (PEP) forms in cytosolic glycolysis in oil seed plants, and PEP is transported to plastid from cytosol where it is converted to acetyl CoA via pyruvate (Hu et al., 2008). In green algae, however, acetyl CoA forms in both cytosol and chloroplast. Over expression of ACCs was proposed to push the substrate malonyl CoA for synthesis of fatty acids, because this expression is the first committed step in fatty acids biosynthesis pathway (Fig. 1.3). However, this only increased lipid synthesis by 5 % in the seed of higher plants (Yu et al., 2011). Similarly, while ACCs was over expressed 2 or 3 fold, there was no increase in the oil content in the diatom *Cyclotella cryptic* (Sheehan et al. 1998). Further, malonyl-CoA:ACP transferase (MAT) helps in transfer of malonyl moiety to acyl carrier protein (ACP) forming malonyl-ACP. Acetyl CoA combines with malonyl-ACP and undergoes condensation, reduction, dehydration and again reduction reaction, each catalyzed by respective enzymes such as 3- ketoacyl-ACP synthase (KAS), 3-ketoacyl-ACP reductase (KAR), 3-hydroxyacyl-ACP dehydrase (HD) and enoyl-ACP reductase (ENR) resulting into elongation of two carbon atoms per cycle to the precursor acyl-ACP moiety chain (Greenwell et al., 2010).

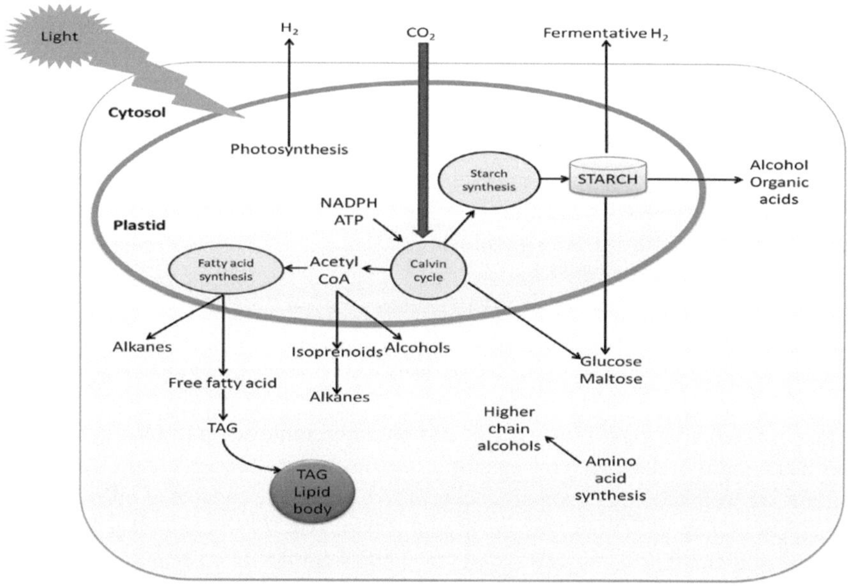

Fig. 1.3 Overview of the biochemical pathways in microalgae.

Saturated 16 and 18 carbon chain fatty acid are formed at the end of the cycle. The conversion of acetyl CoA to 16-carbon compound palmitic acid requires 7 ATP and 14 NAD(P)H (Eq. 1.1; Greenwell et al., 2010).

$$8\ \text{acetyl CoA} + 7\ \text{ATP} + 14\text{NAD(P)H} \rightarrow CH_3(CH_2)_{14}COOH \qquad (1.1)$$

Double bond is introduced by stearoyl ACP desaturase. Elongation reaction terminates in action of one of the followings: first, removal of acyl group from ACP by the action of acyl-ACP thioesterase (FAT) and second, direct transfer of ACP to G3P backbone by acyltransferases and make them available for attachment into glycerol-3-phosphate backbone using Kennedy pathway for the formation of TAGs. This involves stepwise addition of acyl group into the glycerol-3-phosphate backbone. Overexpression of glycerol-3-phosphate resulted into 40 % increase in oil content of the algae (Yu et al., 2011). FFAs join G3P backbone one by one. Transfer of first FFA chain to position one of G3P is catalyzed by GPAT resulting into formation of lyso-phosphatidic acid (LPA). Second FFA transfers to LPA using lyso-phosphatidic acid acyltransferase (LPAAT) and formation of phosphatidic acid (PA) takes place. This is followed by dephosphorylation of PA leading to formation of DAG using enzyme phosphatidic acid phosphatase (PAP). In the last committed step of TAGs synthesis, position three of DAG is filled with further addition of one FFA with the help of diacylglycerol acyltransferase (DGAT; Chen et al., 2009; Hu et al., 2008).

Malic enzyme (a NADPH-generating enzyme) was reported to play a key role in enhancing lipid synthesis in fungi and also found to control lipid synthesis pathway in algae (Gong and Jiang, 2011). Different strategies for lipid biosynthesis and altering fatty acid profile are: (i) increasing the supply of reducing sugar, (ii) overexpression of thioesterase to decrease feedback inhibition resulting from increased acyl-ACP concentration, (iii) introduction of plant-derived thioesterases to optimize the chain length of fatty acids and (iv) elimination of enzymes responsible for fatty acid β-oxidation pathway (Gong and Jiang, 2011).

1.4.2 Biohydrogen from Microalgae

The conversion of algal biomass (oil and/or starch removal) into bio-H_2 by dark-fermentation is orchestrated by anaerobic organisms. *Enterobacter* and *Clostridium* strains are well known as good producers of bio-H_2 that are capable of utilizing various types of carbon sources (Angenent et al., 2004; Das, 2009; Cantrell et al., 2008). On the other hand, cyanobacteria and green algae are the only organisms currently known to be capable of both oxygenic photosynthesis and bio-H_2 production. In cyanobacteria, hydrogen is produced by a light-dependent reaction catalyzed by nitrogenase or in dark anaerobic conditions by hydrogenase (Rao and Hall, 1996; Hansel and Lindblad, 1998), while in green algae, hydrogen is produced photosynthetically by the ability to harness the solar energy resource, to drive H_2 production, from H_2O (Melis et al., 2000; Ghirardi et al., 2000; Melis and Happe, 2001; Ran et al., 2006; Yang et al., 2010). To date, H_2 production has been observed in only 30 genera of green algae (Boichenko and Hoffmann, 1994) highlighting the potential to find new H_2-producing eukaryotic phototrophs with higher H_2 producing capacities.

For photobiological H_2 production, cyanobacteria, formerly called "blue green algae" and "nitrogen-fixing" bacteria, are among the ideal candidates, because they have the simplest nutritional requirements. They can grow using air, water and mineral salts, with light as their only source of energy (Tamagnini et al., 2007; Lindblad et al., 2002). In fact, cyanobacteria (mainly its mutants) are considered the highest biological producer at low cost, since they require only air (N_2 or CO_2), water and mineral salts, using light as the only energy source. H_2 production by cyanobacteria requires two enzymes: the nitrogenase(s) and the bi-directional hydrogenase. In N_2-fixing strains, the net H_2 production is the result of H_2 evolution by nitrogenase and H_2 consumption mainly catalysed by an uptake hydrogenase. Consequently, the production of mutants deficient in H_2 uptake activity is necessary. Moreover, the nitrogenase has a high ATP requirement and this lowers considerably its potential solar energy conversion efficiency. On the other hand, the bi-directional hydrogenase requires much less metabolic energy, but it is extremely sensitive to oxygen (Das and Veziroğlu, 2001; Schütz et al., 2004).

1.4.3 Bioethanol Production from Microalgae

Microalgal bioethanol can be produced through two distinct processes: via dark fermentation or yeast fermentation. The dark fermentation of microalgae consists of the anaerobic production of bioethanol by the microalgae itself through the consumption of intracellular starch. The yeast fermentation process is well established industrially and to achieve higher yields, it is necessary to screen strains with high starch and other sugar contents and induce accumulation of intracellular starch. Some microalgae have high starch content and, therefore, a high potential for bioethanol production (Schenk et al., 2008; Hankamer et al., 2007). However, only limited research (Huntley and Redalje, 2007; Rosenberg et al., 2008; Subhadra and Edwards, 2010) has been reported on the same (Douskova et al., 2008). It has been estimated that approximately 46,760 to 140,290 L of ethanol ha^{-1} yr^{-1} can be produced from microalgae (Cheryl, 2010). This yield is several orders of magnitude higher than yields obtained for other feedstocks.

Matsumoto et al. (2003) have screened several strains of marine microalgae with high carbohydrate content, and identified a total of 76 strains with a carbohydrate content ranging from 40 to 53 %. Hirano et al. (1997) conducted an experiment with *C. vulgaris* microalga (37 % w/w starch content) through fermentation and yielded a 65 % ethanol-conversion when compared with the theoretical conversion rate from starch. Ueda et al. (1996) found that microalgae, such as *Chlorella*, *Dunaliella*, *Chlamydomonas*, *Scenedesmus* and *Spirulina*, contain large amounts (>50 %) of starch and glycogen which are useful as raw materials for ethanol production. Microalgae can assimilate cellulose that can be fermented to bioethanol (Chen et al., 2009). The microalgae *Chlorococum* sp. has also been studied as a feedstock for ethanol production (Harun et al., 2010). The production of bioethanol from the fermentation of microalgal biomass presents itself some advantages because it can use leftover microalgae from other processes (e.g. oil extraction) or intact biomass. It occurs in an aqueous medium, therefore, there is no need to spend energy drying the biomass and the biomass necessary can be concentrated by simply settling. The chemical cell disruption techniques can simultaneously breakdown complex sugars necessary for yeast fermentation and the yeast fermentation technology is well established industrially.

1.4.4 Biomethane Production by Anaerobic Digestion

Organic material such as crop biomass or liquid manure can be used to produce biogas via anaerobic digestion and fermentation. Mixtures of bacteria are used to hydrolyze and break down the organic biopolymers (i.e. carbohydrates, lipids and proteins) into monomers, which are then converted into a methane-rich gas via fermentation (typically 50–75 % CH_4). Carbon dioxide is the second main component found in biogas (approximately 25–50 %) and like other interfering impurities, has to be removed before the methane is used for electricity generation. Microalgae biomass is a source of a vast array of components that can be anaerobically digested to produce biogas. The use of this conversion technology eliminates several of the key obstacles that are responsible for the current high costs associated with algal biofuels, including drying, extraction, and fuel conversion, and as such may be a cost-effective methodology. Several studies have been carried out that demonstrate the potential of this approach. According to Sialve et al. (2009), the methane content of the biogas from microalgae is 7 to 13 % higher when compared with the biogas from maize. For biogas production, the microalgae species should have a high degree of degradation and low amount of indigestible residues (Mussgnug et al., 2010). The substrates should be concentrated but drying process should be avoided, as it results in a general decrease in the biogas production potential in around 20 %. This result represents a good one as it saves energy and time. However, to avoid transportation of the wet biomass, the algal production facility and the biogas fermentation plant should be as close as possible (Mussgnug et al., 2010). According to Das (1985), wastewater grown algal biomass played very important role for the improvement of the biomethanation process. Anaerobic digestion well explored in the past, will probably re-emerge in the coming years either as a mandatory step to support large-scale microalgal cultures or as a standalone bioenergy-producing process (Sialve et al., 2009). This technology could be very effective for situations such as integrated wastewater treatment, where algae are grown under uncontrolled conditions using strains that are not optimized for lipid production.

1.4.5 Fuel Gas or Syngas by Gasification

This combustible gas mixture is the product of the gasification of biomass, at high temperature (~800 to 900 °C) by the partial oxidation of biomass with air, oxygen and/or steam (Wang et al., 2008). Gasification of the algal biomass may provide an extremely flexible way to produce different liquid fuels, primarily through Fischer–Tropsch synthesis (FTS) or mixed alcohol synthesis of the resulting syngas. FTS is also a relatively mature technology, where the syngas components (CO, CO_2, H_2O, H_2 and impurities) are cleaned and upgraded to usable liquid fuels through a water–gas shift and CO hydrogenation (Okabe et al., 2009). The bio-oil yields for the microalgae are 5–25 % w/w lower than the yields of bio-crude, and depending on the biochemical composition. The yields of bio-crude follow the trend lipids > proteins > carbohydrates (Biller and Ross, 2011). Conversion of bio-syngas has several advantages as compared to other methods. First and foremost, it is possible to create a wide variety of fuels with acceptable and known properties. In addition, bio-syngas is a versatile feedstock, and it can be used to produce a number of products, making the process more flexible. Another advantage is the possibility to integrate an algal feedstock into an existing thermochemical infrastructure and wet biomass could be processed (Clark and Deswarte, 2008). It may be possible to feed algae into a coal gasification plant to reduce the capital investment required and improve the process efficiency through economy of scale. In addition, because FTS is an exothermic process, it should be possible to use some of the heat for drying the algae during a harvesting/dewatering process (for other applications) with a regenerative heat exchanger. Another interesting approach would be the study of the feasibility using the oxygen generated by algae for the use in the gasifier to reduce or eliminate the need for a tar reformer.

1.4.6 Pigments from Microalgae

The principles of photosynthesis are similar in higher plants and algae, but the algae stand in contrast to the higher plants in regard to the diversity of pigmentation among marine algae and the diversity of the light regime in the oceans. A broad region of light called the photosynthetically active radiation (PAR, 350–700 nm) is the absorption region of both the chlorophylls and other light harvesting pigments having different absorption peaks (Lobban and Harrison, 1997).

Chlorophyll-*a* is the pigment responsible for photosynthesis, but also chlorophyll-b, c_1 and c_2 are present in macroalgae. Accessory pigments such as carotenoids (β-carotene, lutein, fucoxanthin, siphonaxanthin, violaxanthin, antheraxanthin, zeaxanthin) and phycobilliproteins red phycoerythrin (absorb in green region: 495–570 nm) and blue phycocyanin (absorb in the green-yellow region: 550–630 nm) can help with the harvesting of photons from light of other wavelengths.

A large array of natural products of economic potential is produced by cyanobacteria. These also represent an attractive source of natural pigments such as phycocyanin and phycoerythrin and allo-phycocyanin, and carotenoids. Among them, phycocyanin and phycoerythrin are commercially valuable. Phycocyanin (PC), the light harvesting pigment in cyanobacteria, gives these organisms their bluish colour and that is why are referred as blue green algae. Phycocyanin is water soluble, strongly fluorescent and has antioxidant property. PC and related PBPs of cyanobacteria have application in food and agriculture, cosmetics, biotechnology, diagnostics and pharmaceuticals.

PBPs as natural colorants are important over synthetic colours as they are environment friendly, non-toxic and non-carcinogenic. Dainippon Ink & Chemicals (Sakura, Japan) has developed a product called "Lina blue" (PC extract from *Spirulina platensis*) which is used in chewing gum, ice sherbets, popsicles, candies, soft drinks, dairy products and wasabi. PC is considered more versatile than gardenia and indigo, showing a bright blue colour in jelly gum and coated soft candies, despite its lower stability to heat and light (Jespersen et al., 2005). Besides this, there are number of other companies commercializing different products based on PCs like – C-phycocyanin from Cyanotech; PhycoLink® biotinylated C-phycocyanin from PROzyme; PhycoPro™ C-phycocyanin from Europa Bioproducts Ltd.; C-phycocyanin from Sigma Aldrich, C-phycocyanin from Fisher Scientific, etc. (Chakdar and Pabbi, 2012). Use of phycobilins in cosmetics like lipstick, eyeliners, etc., are also gaining importance. Many properties like high molar absorbance coefficients, high fluorescence quantum yield, large Stokes shift, high oligomer stability and high photostability make PBPs particularly PC and PE very powerful and highly sensitive fluorescent reagents.

Purified native phycobiliproteins and their subunits fluoresce strongly and have been widely used as external labels for cell sorting and analysis and a wide range of other fluorescence based assays (Glazer and Stryer, 1984). The stabilized phycobilisomes designated PBXL-3L was accessed as a fluorochrome for flow cytometric immuno-detection of surface antigens on immune cells (Telford et al., 2001). A number of commercial products based on C-phycoerythrin are also available in the market e.g. Lightning Link® C-PE from Innova Biosciences Ltd.; C-phycoerythrin and lightning link C-PE antibody labeling kit from Novus Biologicals; biotin Cr-PE (C-Phycoerythrin) from AssayPro; and Stretavidin C-phycoerythrin from Sigma Aldrich. PE is also a very important reagent in proteomics and genomics and form the basis of the detection system in affymetrix chips (DNA microarrays). Phycoerythrin labeled streptavidin is added after complete binding and produces a strong signal from array elements containing the biotin-labeled DNA or protein probes (De Rosa et al., 2003). *In vivo* fluorescence from PC has been used for online monitoring of growth in cyanobacterial cultures (Sode et al., 1991) and detection of toxic cyanobacteria in drinking water (Izydorczyk et al., 2005).

1.4.7 Algae as Food and Feed

Microalgae namely *Spirulina, Chlorella*, *Scenedesmus*, *Dunaliella salina*, and *Aphanizomenon flos-aquae* have found an application for food industry. *Chlorella pyrenoidosa* can be very significant for human health, because this biomass contains proteins (50–65 %), lipids (5–10 %), hydrocarbons (10–20 %), antioxidants, vitamin C (200–500 mg kg^{-1}) and vitamin A (120–300 mg kg^{-1}) (Sheng et al., 2008). The accumulated microalgae starch can be hydrolysed with the formation of organic acids (Rodjaroen et al., 2007). The high lipid, carbohydrates and proteins of many microalgal species have driven research in a wide spectrum of uses for human consumption. They are used as tablets, capsules and are also added in sauces, candies and beverages (Yamaguchi, 1997). Analyses of gross chemical composition of these algal extracts indicates an increase in antioxidative property with an increasing content of unsaturated fatty acid (Tokusoglu and Ünal, 2003; Becker, 2007). *Aphanizomenon flos-aquae* is used in food individually or with other nutrients and natural products. The average protein quality of most of the algae examined is equal, sometimes even superior to that of conventional plant proteins (Becker, 2007). Proteins of *Dunaliella* can be used in baking industry (Finney, 1984), and the biomass can be used for animal and fish feed (Dufossé et al., 2005).

Many species of microalgae have higher protein content, protein efficiency ratio, apparent biological value, true digestibility value of protein, amino acid content, proportion and availability of amino acids in their protein profile which give them advantages over other conventional food proteins (Becker, 2007). Japanese scientists found *Chlorella* as a good source of essential amino acid except methionine. Aspartic and glutamic acids are found to be large fraction of amino acid group in seaweeds, and in brown seaweed like *Fucus* sp. These two amino acids are occupying 22–44 % of total amino acid family (Munda, 1977). In green seaweed the total share of aspartic and glutamic acid goes up to 32 % in *Ulva rigida* and *Ulva rotundata* (Fleurence et al., 1995). Algal diets, especially extract of microalgae, comprise ω-3 fatty acid for infant. Fatty acid family of ω-3 and ω-6 are abundant in marine algal species, and *Crythecodinium cohnii* contains 40–50 % DHA, but lacks EPA and other long chain PUFAs (Jiang et al., 1999). EPA was abundantly found in *Porphyridium purpureum*, *Phaeodactylum tricornutum*, *Isochrysis galbana*, *Nannochloropsis* sp. and *Nitzschia laevis* (Zittelli et al., 1999).

The algal biomass may also contain high levels of vitamins such as A, B_1, B_2, B_6, B_{12}, C, E, nioctinate, folic acid, pantothenic acid and β-carotene and minerals (Becker, 2007). Mineral contents of seaweeds (8–40 %) are sufficient to meet recommended dietary allowances on daily basis (Ruperez, 2002). Depositions of minerals are lower in red and green algae as compared to brown algae or pheophytes (Holdt and Kraan, 2011). Fresh water *Chlorella* contains 6.30 % w/w minerals whereas a marine *Isochrisis* have higher ash content of 16.08 % w/w (Tokusoglu and Unal, 2003). Most of the algae are rich in iodine, potassium, iron, magnesium and calcium (Poulíčková et al., 2008), and *Laminaria japonica* was found to be an excellent source of iodine (Holdt and Kraan, 2011). Microalgae have the potential to become a novel source of bioactive molecules. They may also have potent probiotic compounds that enhance health (Kay and Barton, 1991).

1.4.8 Algal Pharmaceuticals

Carotenoids possess many properties, important for pharmaceutical industry. Antioxidants such as β-carotene and asthaxanthin are sources of pro-vitamin A. Asthaxanthin, widely produced from *Haematococcus pluvialis*, is being used for its possible roles in human health for protection from UV-light, enhancing immune system and acting against inflammation and tumour formation (Guedes et al., 2011; Brennan and Owende, 2010). Phycocyanin which is mostly obtained from *Spirulina* has potential applications in diagnostics purpose in the field of medicine and biotechnology. For example, the strong and highly sensitive fluorescent properties of phycocyanin can be exploited for labelling antibodies, receptors and other biological molecule in immunolabelling experiments (Bermejo Roman et al., 2002; Brennan and Owende, 2010). Similarly, microalgae have special carbohydrates binding proteins within protein bodies of the cells called lectins, and these were found highly specific for altered complex oligosaccharides, glycoprotein or glycolipids during disease (Skjånes et al., 2013). Several brown algae such as *Wakame*, *Kombu* and *Mozuki* were found rich in fucoxanthin (Kanazawa, 2012). Fucoxanthin and the metabolite, fucoxanthinol exhibit diverse and significant biofunctions, such as cancer preventing action, anti-obesity effect, improvement of lipid metabolism, and antioxidant potency. Heparinoids is a sulphated heteropolysaccharides known for antithrombotic action including anti-inflammatory action, but is prone to contamination. However, heparinoids extracted from red marine algae *Hypnea musciformis* was found to have antithrombotic action, and since it was obtained from marine source, the heparinoid was less prone to contamination by prions and viruses (Alves et al., 2012). Algal extracts were also found effective in controlling pathogenic microbial activity. For example, hydroalcoholic extracts of filamentous green algae, *Cladophora glomerata* had antimicrobial activities on different Gram negative and Gram positive bacteria including *Staphylococcus aureus, Bacillus subtilis, Pseudomonas aeruginosa, Salmonella typhimurium* and *Proteus mirabilis* (Soltani et al., 2011). Similarly, n-hexane extracts of *Sargassum polycystum* and *C. agardh* exhibited promising bacteriostatic agents against many Gram negative like *Escherichia coli, P. aeruginosa* and Gram positive bacteria such as *Staphylococcus aureus, B. cereus* (Chiao-Wei et al., 2011).

Production of specific recombinant proteins such as vaccines, antibodies, hormones and enzymes using algal machinery have many advantages such as low production costs, ease of scalability, absence of human pathogens and the ability to fold and assemble complex proteins accurately (Rosales-Mendoza et al., 2012). In addition, it offers an attractive alternative to traditional mammalian-based expression systems, as both the plastid and nuclear genomes are easily and quickly transformed. Therefore, chloroplast could be used as factory for the production of antibiotics (Tran et al., 2009), reporter protein, bioactive mammalian protein and other pharmaceutical important protein. For example, Franklin and Mayfield (2005) expressed human IgA antibody directed against herpes simplex virus in unicellular eukaryotic green algae, *Chlamydomonas reinhardtii.*

1.4.9 Natural Colours

Natural colours from microalgae are used in food, pharmaceutical, cosmetics and textile industries. The use of colours from microalgae is becoming more popular as they are non-toxic with no bad influence on the environment and are renewable sources of raw material. Nevertheless, it should be mentioned that natural colours may not be stable; hence, there are certain difficulties in their applicability (Dufossé et al., 2005). Carotenoids are used as colouring matters for food stuffs, like in production of orange juice, as feed additives and for production of cosmetics (Del Campo et al., 2000). Phycocyanin is used as a dye in food industry (ice creams, candies, nonalcoholic beverages and dietary foodstuffs), cosmetics and pharmaceuticals. Carotenoids are also widely used in clinical and immunology research laboratories because the effective molecular absorption, high fluorescence ability and photo stability make these pigments powerful and sensitive fluorescent reagents. β-carotene is converted into vitamin A, which enables the immune system to function in human organism (Dufossé et al., 2005). The biomass of *D. salina* can be used to make preparations which are antihypersensitive, broncholytic and analgesic drugs (Villar et al., 1992). Furthermore, *D. salina* powder is used in medioprophylactic diet (Törnwall, 2004) and this microalga contains oxidized carotenoids (xanthophylls) having anticancer properties (Roodenberg, 2000).

1.4.10 Polymers

The major advantage of algal bioplastics is the complete biodegradability properties making it suitable for manufacture of wide range of materials like consumable and disposable plastic products, agricultural plastic products and containers for horticultural planting, etc. (Sharma et al., 2011). Gelling polysaccharides like agar, agarose and carrageenans are some of the unique sulphated galactans produced from red macroalgae (*Rhodophyta*) (Usov, 2011). For example, agar obtained from the red alga, *Hydropuntia cornea,* can be blended with polyvinyl alcohol in different ratio to produce biodegradable films, which was found having wide applications in the biodegradable packaging industry (Santana et al., 2010). Cellulose hydrogels obtained from *Cladophora* green macroalgae were used for the polymer energy storage device and manufacture of ion exchange membranes (Mihranyan et al., 2010). The green macroalga, *Ulva armoricana,* could be used for the production of hydrophilic, eco-compatible polymer, such as polyvinyl alcohol (PVA) as continuous matrix (Chiellini et al., 2008). In addition, extracellular polymeric substance of green microalga *Penium margaritaceum* was found suitable for biofilm formation (Domozych et al., 2005). Poly-3-hydroxybutyrate (PHB) is biodegradable polyester with thermoplastic properties and commonly found in the bacteria. Currently, bacterial PHB pathway of *Ralstonia eutropha* H16 was inserted into the microalga *Phaeodactylum tricornutum* and found capable of producing PHB (Hempel et al., 2011). *Scenedesmus communis* and *Botryococcus braunii* are microalgae known for the extraction of algaenans, a resistant biopolymer in the cell wall of green algae (Allard et al., 1998). *Spirulina* biomass has potential to be applied as extracellular matrix containing very thin fibres for stem cell culture and treatment of spinal cord injury (de Morais et al., 2010).

1.4.11 Biofertilizers

The agricultural importance of bluegreen algae (BGA) lies in their capacity to metabolize the molecular nitrogen, liberation of part of fixed nitrogen and growth promoting substances as extra metabolites, solubilising the insoluble phosphates, addition of organic matter and improving the physical and chemical nature of soil. Heterocysts of BGA are the nitrogen fixing stations of heterocystous algae. Non-heterocystous forms of BGA fix nitrogen anaerobically. Nitrogen fixed by BGA may become available to plants (e.g. rice) only after its release extracellularly into the surroundings, either as extracellular products or by mineralization of their intracellular contents through microbial decomposition after death (Srinivasan, 1978). Nitrogen (N) fixation by BGA and its release in the soil water system may become more useful for crop production during vegetative growth stage of rice crops than at the later stages (Roger et al., 1993). Recovery of cyanobacterial fixed N by rice varied 13–50 % depending upon the nature of inoculum, method of application and the absence of soil fauna in inoculated soil (Tirol et al., 1982). Field studies on the indigenous BGA in presence of urea super granules and urea (87 kg N ha^{-1}), ammonium sulphate (58 kg N ha^{-1}), SSP (30 kg ha^{-1}), potash (20 kg ha^{-1}) and Zn (10 kg ha^{-1}) showed that surface application of nitrogenous fertilizers inhibited nitrogen fixation, whereas deep placement of urea increased 70 % nitrogen fixing activity as compared to control. In addition, BGA was also encouraged by surface broadcast which increased the pH of flood water and loss of nitrogen by ammonia volatilization. The BGA nitrogen fixation has a 'switch on' mechanism which is activated when the level of combined nitrogen falls below a threshold level (~40 ppm). In addition, the cyanobacteria metabolize atmospheric carbon dioxide during photosynthesis. Application of chemical fertilizers at the recommended level or lower levels stimulated growth of diazotrophic cyanobacterial population and nitrogenase activity in a paddy field whereas higher fertilizer levels proved to be inhibitory (Jha et al., 2001). The rice-mustard-mung crop rotation was observed to be more suitable for cyanobacterial nitrogen fixation than rice-wheat-maize rotation. The low fertility coupled with rice-mustard-mung rotation were found to be best suited for promoting nitrogen fixation by cyanobacteria during rice cultivation (Jha et al., 2001).

1.5 Development of Photobioreactors

Most microalgae grow photoautotrophically i.e. they require light and CO_2 as energy and carbon sources respectively. Some species, however, are capable of growing heterotrophically by utilizing external carbon sources such as glucose and acetate for carbon and energy. Microalgae cultivation using sunlight energy can be carried out in open or covered ponds or closed photobioreactors, based on tubular, flat plate, or other designs. The different configurations for photobioreactors are given in Fig. 1.4.

Due to some operating challenges like overheating, fouling and gas exchange limitations, it is difficult to scale up much beyond approximately 100 m^2 for an individual growth unit. Currently there are three types of industrial reactors used for algal culture: (1) photobioreactors, (2) open ponds, and (3) closed and hybrid systems (Fig. 1.5). Photobioreactors are different types of tanks or closed systems for cultivation of algae. Open-pond systems are shallow ponds in which algae are cultivated. Nutrients can be provided through runoff water from nearby land areas or by channelling the water from sewage/water treatment plants. Technical and biological limitations of these open systems have given rise to the development of enclosed photo bioreactors. A few open systems are presented for which particularly reliable results are available. Emphasis is then put on closed systems, which have been con-

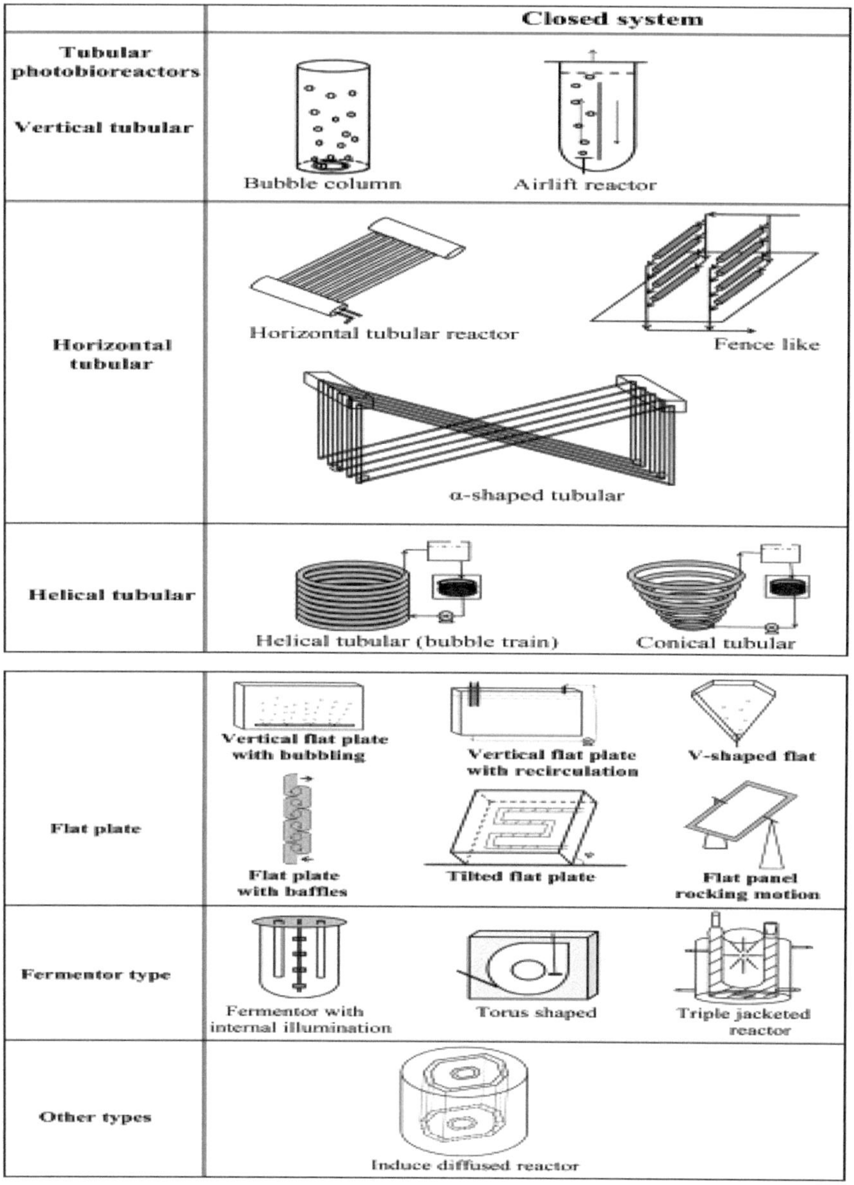

Fig. 1.4 Schematic representation of different bioreactors for biomass production (Dasgupta et al., 2010).

sidered to be capital intensive and are justified only when a fine chemical is to be produced. Closed systems are much more expensive than ponds. However, closed systems require much less light and agricultural land to grow algae. High oil species of microalgae cultured in growth optimized conditions of photobioreactors have the potential to yield 19,000 to 57,000 L of microalgal oil per acre per year. The yield

Fig. 1.5 Utilization of algal biomass for biohydrogen and biodiesel production.

of oil from algae is over 200 times the yield from the best-performing plant/vegetable oils (Chisti, 2007).

Large-scale production of microalgal biomass generally uses continuous culture during daylight. In this method of operation, fresh culture medium is fed at a constant rate and the same quantity of microalgal broth is withdrawn continuously (Molina Grima et al., 1999). Feeding ceases during the night, but the mixing of broth must continue to prevent settling of the biomass (Molina Grima et al., 1999). As much as 25 % of the biomass produced during daylight may be lost during the night due to respiration. The extent of this loss depends on the light level under which the biomass was grown, the growth temperature, and the temperature at night (Chisti, 2007). Algal cultures consist of a single or several specific strains optimized for producing the desired product. Water, necessary nutrients and CO_2 are provided in a controlled way, while oxygen has to be removed (Carlsson et al., 2007). Algae receive sunlight either directly through the transparent container walls or via light fibers or tubes that channel it from sunlight collectors. A good amount of developmental work to optimize different photobioreactor systems for algae cultivation has been carried out by Janssen et al. (2003), Choi et al. (2006), Carvalho et al. (2006) and Hankamer et al. (2007). Unlike open ponds, bioreactors can ensure a single alga species is grown without interference or competition (Campbell, 2008). Open ponds are a very efficient and cost-effective method of cultivating algae, but they could become contaminated with unwanted species very quickly. PBR are excellent for maintaining axenic cultures, but set up costs are generally ten times higher than for open ponds. A combination of both systems is probably the most logical choice for cost-effective cultivation of high yielding strains for biofuels. Inoculation has always been a part of algal aquaculture. Open ponds are inoculated with a desired strain that was invariably cultivated in a bioreactor, whether it was as simple as a plastic bag or a high-tech fibre optic bioreactor. Importantly, the size of the inoculum needs to be large enough for the desired species to establish in the open system before an unwanted species. Although sooner or later contaminating species will end up dominating an open system (if they do not require extreme conditions) and it will have to be cleaned and re-inoculated. Therefore, to minimize contamination issues, cleaning or flushing the ponds should be part of the aquaculture routine, and as such, open ponds can be considered as batch cultures.

The growing, harvesting and processing of any feedstock, including algal biomass requires considerable energy. The use of fossil-based energy sources for these actions would reduce the net carbon gain in a life-cycle assessment for this new fuel pathway (Subhadra and Edwards, 2010). However, if algal energy production uses non-fossil, renewable energy sources such as wind and solar energy, the process will show a substantial net carbon gain. A green house-based algal production may need heat to sustain high productivity in winter. Greenhouses with solar panels to harvest solar energy or greenhouses to operate with the heat from geothermal would significantly contribute to the sustainability issue.

1.6 Recent Developments

The term biorefinery was coined to describe the production of a wide range of chemicals and biofuels from biomasses through the integration of bioprocessing and appropriate low environmental impact chemical technologies in a cost-effective and environmentally sustainable manner (Li et al., 2008). The microalgal biomass biorefinery concept is not new. However, it assisted in making biofuel production economically viable. An algal biorefinery could potentially integrate several different conversion technologies to produce biofuels including biodiesel, green diesel, green gasoline, aviation fuel, ethanol and methane, as well as valuable co-products, such as fats, polyunsaturated fatty acids, natural dyes, sugars, pigments (mainly β-carotene and astaxanthin), antioxidants and polyunsaturated fatty acids (EPA, DHA). Conceptually, the biorefinery would involve sequentially the cultivation of microalgae in a microalgal farming facility (CO_2 mitigation), extracting bioreactive products from harvested algal biomass, thermal processing (pyrolysis, liquefaction or gasification), extracting high value chemicals from the resulting liquid, vapour and/or solid phases, and reforming/upgrading biofuels for different applications (Li et al., 2008). After oil and/or starch removal from the microalgal biomass (for biodiesel and/or ethanol production, respectively), the leftover biomass can be processed into methane or livestock feed, used as organic fertilizer due to its high N:P ratio, or simply burned for energy cogeneration (electricity and heat) (Wang et al., 2008).

The high-value bioactive compounds could be used in nutritional supplements, food/feed additives, aquaculture, cosmetics, pharmaceuticals, biofertilizers, edible vaccines through genetic recombination (Chisti, 2006; Rosenberg et al., 2008) and pollution prevention. Microalgae play an imperative role in bioremediation and wastewater treatment. They can eliminate heavy metals, uranium and other pollutants from wastewater, and they can degrade carcinogenic polyaromatic hydrocarbons and other organics. Furthermore, algae are accountable for at least 50 % of the photosynthetic biomass production in our planet and they are great sources of biofuels (Chisti, 2006). Nevertheless, several authors (e.g. van Harmelen and Oonk, 2006; van Beilen, 2010; Park et al., 2011), indicate that only if the algal biomass is a by-product of wastewater treatment systems, GHG abatement and/or the production of high-value compounds, such as astaxanthin, β-carotene, EPA and DHA, commercially viable energy production from algal biomass might be feasible. Mussgnug et al. (2010) proposed another approach for the microalgae producers of H_2. The cell's response to the induction of H_2 production cycle is the strong increase of starch and lipids (high fermentative potential compounds), which results in an increase in the biogas production (second step), after a first step of H_2 production, showing a synergistic effect in a biorefinery concept.

A microalgae biomass biorefinery was proposed by © PetroAlgae company (Greenwell et al., 2010) where little or no waste products are present and allows for residual energy capture, recycling of unused nutrients, and water purification and recycling. This system would mean low environmental impact and maximization of the value of products from the system (Greenwell et al., 2010). Subhadra and Edwards (2010) stated that it would be better to integrate a renewable energy park (IREP) where the facilities are centralized, instead of a single central facility, such as giant petroleum refineries operated by a single firm. Major firms can be a part of IREPs and might play an important role in the development of this concept. However, other small-scale renewable energy (wind, solar, geothermal and biomass) firms, working as a consortium, may also be an integral component of IREPs. These firms together can cross-feed power, heat, raw materials and products with the shared goal of minimizing emissions of pollutants to the atmosphere and optimizing the utilization of natural resources such as land, water and fossil fuels, and fossil agricultural chemicals. The integration of established prototype carbon capture devices, which feed algal cultures, should also be examined.

Several novel green technologies such as geothermal heat pumps (Dickinson et al., 2009), dual fuel (bivalent) ground source heat pumps (Ozgener and Hepbasil, 2007), solar-assisted heat pump systems (Benli and Durmus, 2009), solar wind turbine (which harvest wind and sun energy in one element) have been receiving increased attention because of their potential to reduce primary energy consumption and thus reduce GHG emission. Further, newer energy conservation and utilization concepts such as bioheat from wood (Ohlrogge et al., 2009), bioelectricity from biomass (deB Richter et al., 2009) and hybrid hydrogen–carbon process for the production of liquid hydrocarbon fuels (Agrawal et al., 2007) can also be envisioned into the broader design concept of IREPs. Together, these technologies and concepts can maximize the ecological and environmental benefits of energy production from IREPs. The green electricity from these IREPs may flow into the existing grid. OriginOil (2010) proposed an integrated system called Optimized Algae Production System where the first step is a low-pressure Quantum Fracturing™. It works by breaking up carbon dioxide and other nutrients into micron-sized bubbles and infusing them into the growth vessel. The growth occurs in OriginOil's Helix BioReactor™, which features a rotating vertical shaft of low energy lights (highly-efficient LEDs) arranged in a helix or spiral pattern and tuned precisely to the waves and frequencies for optimal algae growth. They claim that in the Cascading Production™ and Single-Step Extraction™, the oil and biomass are separated without having to dewater the algae, and the continuous process is called Live Extraction™. After extraction, the water is recycled back into the system. The harvested oil is packaged for refining and distribution, and the algae mass is devoted to various 'green' applications such as fuel, animal feed, fertilizers, chemicals, health products and construction materials (OriginOil, 2010).

1.7 Economic Overview

The road to cost effective biofuel and bioproduct manufacture from algae can be realised by a biorefinery-based strategy and advances in genetic and metabolic engineering (Chisti, 2007). In the coming decade, genetic and metabolic engineering will play a big role in deciding the future of microalgal biotechnology (Dunahay et al., 1996). The various advantages of algae over other terrestrial crops make them an interesting perspective for the future. Sustainable and cost effective solutions to the energy security problem can be made possible through algal means. The various value added chemicals add to the plethora of products which we can derive from microalgal biomass. To make the products available for the market, the processes need to be modified for better conversion efficiencies resulting in reduction of costs. Unlike terrestrial crops, algae can be grown almost throughout the year with little seasonal variations. The appropriate temperature range (293 to 303 K) for cultivation of microalgae prevails in the equatorial region of the earth. The adaptability of algal cultures for other extreme climatic conditions increases the cultivation area to other areas of the earth. Inland and coastal cultures are possible due to the properties of algae utilizing saline as well as freshwater sources for growth. The competition for land resources does not exist with algal cultivation as they thrive on non-arable lands for growth and production. Minimal water and nutrient requirements are required for algal biomass production. In addition, we obtain the property of bioremediation by microalgae which adds to all the advantage. They may be rightly projected as the crop of the future. The developing nations of the world suffer from a wide range of malnutrition-related diseases. Algae can be the solution to all these woes predating the world and making it a better place to live in.

The current focus on algae has attracted many industries for commercialization of the processes of conversion. This has led to a scarcity of data related to the costs incurred in production and downstream processing. Reports by Schenk et al. (2008) suggested that sustainable biofuel production requires biomass at less than \$300 US ton^{-1} dry weight. A high lipid content with average biomass yields of 20 $g\ m^{-2}\ d^{-1}$ in open ponds may lead to production costs of \$340 US ton^{-1} (Solix_Biofuels, 2010). The present status for production costs suggests that for average microalgal biomass production and harvesting (5–60 ton ha^{-1} yr^{-1}), a minimum of \$5–\$15 ton^{-1} is required. The final costs for open and closed systems stand to \$8–15 ton^{-1} and \$30–70 ton^{-1} respectively (van Beilen, 2010). High density cell cultures are possible in photobioreactors. But high costs incurred for construction and maintenance are detrimental to their use. Large scale bioreactors were demonstrated by Willson and Cloud at a very low cost of \$15 US m^{-2} using disposable plastics as construction material (Schenk et al., 2008). A high cell density (~10 g L^{-1}) with low power consumption (less than 20 kW ha^{-1}) was achieved by Proviron (2010) for an investment below \$25 m^{-2}. Capital costs for photobioreactors range from \$1.1 million to \$1.75 million ha^{-1} which is much higher as compared to open pond systems (\$125,000 ha^{-1}). Economic feasibility of microalgal biomass production is not possible even when the assumptions are most favourable. Considering into fact the high revenues for fuels (\$125 ton^{-1} algae) and GHG (\$65 ton^{-1}), the production cost still leads to \$210 ton^{-1} (Carlsson et al., 2007).

Production of algal biomass is largely performed in open ponds. But they suffer from several disadvantages. The need for large areas of land, water loss by evaporation and susceptibility of algae to contamination places the open ponds for high maintenance costs. The typical theoretical productivity is around 0.025 kg m^{-2} d^{-1} (82 ton ha^{-1} yr^{-1}) and the maximum biomass concentration 1 g L^{-1} (Chisti, 2010). But due to the constraints typical commercial yields decrease to around 10–30 ton ha^{-1} yr^{-1} (van Beilen, 2010). Closed photobioreactors provide higher biomass productivity [1.535 kg m^{-2} d^{-1} (~158 ton ha^{-1} yr^{-1})] with higher algae cell densities (4 g L^{-1}) (Chisti, 2010). An annual productivity of 120 ton ha^{-1} was stated by Subitec (2010). Production of 365 ton ha^{-1} $year^{-1}$ were made by commercial bioreactor supplier Algae Link for one of their systems. On the other hand, Green Fuel Technologies Corporation (USA), who has several large-scale pilot plants operating focus on CO_2 capture from industrial emitters, indicates productivities of ~250–300 ton ha^{-1} yr^{-1} (Singh et al., 2011). Various reports such as Ono and Cuello (2006) doubted that the economic feasibility would be difficult to achieve because to obtain a target CO_2 mitigation price of \$30 ton^{-1} CO_2 at 40 % biological conversion efficiency, the allowable net cost should be less than \$2.52 m^{-2} year at low-light intensity (average US location). The average price required for production of biodiesel from algae have been provided in Table 1.2.

The encouragement from the various governments is required in the form of subsidies and tax cuts for realization of an algal biorefinery-based economy in addi-

Table 1.2 Cost incurred for production of algal biodiesel

Cultivation systems	Cost of biodiesel (\$ L^{-1})	Reference
Open pond (400 ha)	0.50–0.82	Benemann and Oswald (1996)
Open pond (1950.58 ha)	2.73	Davis et al. (2011)
PBR (1950.58 ha)	5.70	
Open pond (333.3 ha)	1.68	Delrue et al. (2012)
PBR	2.80	
Raceway + PBR	2.69	

tion to high biomass yields (Gallagher, 2011). Algal biomass production is a technically feasible as well as environment friendly process. In order to make it feasible, the methods for harvesting and dewatering as well as supply of CO_2 and downstream processing need to be cost effective and energy sufficient (Mata et al., 2010). Taxation on high carbon emitting industries may significantly decrease the costs incurred for CO_2 supply (Gallagher, 2011). According to Chisti (2007), for algal diesel to potentially replace fossil fuels, it must be priced as follows:

$$C_{\text{algaloil(per L)}} \leq 6.9 \times 10^{-3} \times C_{\text{petroleum(per L)}}$$

The cost of crude oil would have to exceed \$100 per barrel to make a high return scenario plausible. Despite many economists feel that this is unlikely to occur in the near-term period (3–5 years) due to present economic conditions and the low global demand for oil, the geologists and petroleum engineers are predicting that global oil production rates will soon peak and then begin to decline, resulting in a steady increase in oil prices (Gallagher, 2011). It is widely accepted that microalgal biomass could assist in fossil fuels in the near future, if commercial production will intertwine the following requisites:

- Highly productive microalgae that could be cultivated on a large scale using wastewater as a nutrient supply, and waste CO_2 as carbon source.
- Harvesting, dewatering and extraction of algal biomass could be developed at a low cost.
- Production of biofuels could be combined with that of higher value co-products.

If these issues are resolved following long-term R&D and concerted efforts by the public and private sectors in addition to large investments in the area, microalgae cultures might become an economically viable, renewable and carbon-neutral source of transportation biofuels, which do not jeopardize our forests and food supply. According to Lee (2011), microalgae will undoubtedly become an important feedstock for diesel in the future. This author, adopting the Taiwan General Equilibrium Model Energy for Biofuels (TAIGEM-EB), estimated for 2040 that the share of petroleum of the total production energy will reduce to 19.24 %, while algal biodiesel will reach 19.24 %. CO_2 emissions will further have a reduction of 21.7 %.

1.8 Conclusion

Algal biofuels have a tremendous potential for contributing to environmental, social and economic sustainability. Algal biofuel production should integrate other environmentally sustainable technologies, such as CO_2 sequestration, emissions cleanup from industrial and agricultural wastes and the purification of water and should be done in conjunction with the production of valuable co-products. Many biofuels can be produced from "green coal", such as liquid fuels (e.g. biodiesel, green diesel, jet fuel and bioethanol) and gas fuels (e.g. biogas, syngas and bio-hydrogen) which can be used in engines and turbines and as feedstock for refineries. However, several drawbacks of microalgal biomass production should be solved, such as more efficient production, harvesting, dewatering, drying and extraction (if applicable). Thus, fuel-only algal systems are not plausible, at least not in the foreseeable future, and additional revenues are required. Microalgal production should be assisted by alternative energies, for mixing the culture, illumination, dewatering and processing, and drying. Both thermochemical liquefaction and pyrolysis appear to be feasible methods for the conversion of algal biomass to biofuels, after the extraction of oils from algae. Anaerobic digestion shows potential to reduce external energy demand and to recycle a part of the mineral fertilizers, avoiding eutrophication, especially when coupled with a wastewater treatment system. Bioethanol production from algae through fermentation could be another interesting alternative, due to the fact that it requires less energy consumption and is a simplified process when compared to biodiesel production. To make algae-to-energy systems a practical reality, considerable research should continue such as genetic improvements, biorefinery and microbial fuel cell concepts, and the integration of alternative energies into wastewater and CO_2 treatment systems.

References

Agrawal, R., Singh, N.R., Ribeiro, F.H. and Delgass, W.N. (2007). Sustainable fuel for the transportation sector. *Proceedings to the National Academy of Sciences USA*, 104, 4828–4833.

Allard, B., Templier, J. and Largeau, C. (1998). An improved method for the isolation of artifact-free algaenans from microalgae. *Organic Geochemistry*, 28(9–10), 543–548.

Alves, M.G.d.C.F., Nobre, L.T.D.B., Monteiro, N.d.K.V., Moura, G.E.D.d.D., Dore, C.M.P.G., de Medeiros, V.P. and Leite, E.L. (2012). Effects of heparinoids from algae on hemostasis and their action on the cycle cell. *Biomed Prev Nutr.* 2, 163–168.

Angenent, L.T., Karim, K., Al-Dahhan, M.H., Wrenn, B.A. and Rosa, D.E. (2004). Production of bioenergy and biochemicals from industrial and agricultural wastewater. *Trends in Biotechnology*, 22, 477–485.

Becker, W. (2007). Microalgae in Human and Animal Nutrition. Handbook of Microalgal Culture: Biotechnology and Applied Phycology (ed A. Richmond), Blackwell Publishing Ltd, Oxford, UK. pp 312–351.

Benemann, J.R. and Oswald, W.J. (1996). Systems and economic analysis of microalgae ponds for conversion of CO_2 to biomass. DOE/PC/93204 – T5.

Benli, H. and Durmus, A. (2009). Evaluation of ground-source heat pump combined latent heat storage system performance in greenhouse heating. *Energy Building*, 41, 220–228.

Bermejo Roman, R., Alvarez-Pez, J.M., Acien Fernandez, F.G. and Molina Grima, E. (2002). Recovery of pure B-phycoerythrin from the microalga *Porphyridium cruentum. Journal of Biotechnology*, 93(1): 73–85.

Biller, P. and Ross, A.B. (2011). Potential yields and properties of oil from the hydrothermal liquefaction of microalgae with different biochemical content. *Bioresour Technol*, 102, 215–225.

Boichenko, V.A. and Hoffmann, P. (1994). Photosynthetic hydrogen-production in prokaryotes and eukaryotes—occurrence, mechanism, and functions. *Photosynthetica*, 30, 527–552.

Brennan, L., & Owende, P. (2010). Biofuels from microalgae—a review of technologies for production, processing, and extractions of biofuels and co-products. *Renewable and sustainable energy reviews, 14*(2), 557–577.

Bunt, J.S. (1961). Nitrogen fixing blue-green algae in Australian soils. *Nature*, 192, 479–480.
Campbell, C.J. (1997). The coming oil crisis. Multi-science Publishing Company and Petroconsultants, S.A Essex.
Campbell, M. N. (2008). Biodiesel: algae as a renewable source for liquid fuel. *Guelph Engineering Journal*, *1*(1), 2–7.
Cantrell, K.B., Ducey, T., Ro, K.S. and Hunt, P.G. (2008). Livestock waste-to-bioenergy generation opportunities. *Bioresour Technol*, 99, 7941–7953.
Carlsson, A.S., van Beilen, J.B., Moller, R. and Clayton, D. (2007). Micro- and macro-algae utility for industrial applications. *In:* Dianna, B. (Ed). Outputs from the EPOBIO project. CPL Press, UK.
Carvalho, A.P., Meireles, L.A. and Malcata, F.X. (2006). Microalgal reactors: A review of enclosed system designs and performances. *Biotechnol Progress*, 22, 1490–1506.
Chakdar, H. and Pabbi, S. (2012). Extraction and purification of phycoerythrin from *Anabaena variabilis* (CCC421). *Phykos.*, 42(1): 25–31.
Chen, P., Min, M., Chen, Y., Wang, L., Li, Y., Chen, Q., Wang, C., Wan, Y., Wang, X., Cheng, Y., Deng, S., Hennessy, K., Lin, X., Liu, Y., Wang, Y., Martinez, B. and Ruan, R. (2009). Review of the biological and engineering aspects of algae to biofuels approach. *Int J Agri Biological Eng*, 2(4), 1–24.
Cheryl (2010). Algae becoming the new biofuel of choice. Available online 2008. http://duelingfuels.com/biofuels/non-food-biofuels/algae-biofuel.php#more-115N.
Chiao-Wei, C., Siew-Ling, H. and Ching-Lee, W. (2011). Antibacterial activity of *Sargassum polycystum* C. Agardh and *Padina australis* Hauck (Phaeophyceae). *African J Biotechnol.*, 10(64), 14125–14131.
Chiellini, E., Cinelli, P., Ilieva, V.I. and Martera, M. (2008). Biodegradable thermoplastic composites based on polyvinyl alcohol and algae. *Biomacromolecules*, 9(3), 1007–1013.
Chisti, Y. (2006). MICROALGAE AS SUSTAINABLE CELL FACTORIES. *Environmental Engineering & Management Journal (EEMJ)*, *5*(3).
Chisti, Y. (2007). Biodiesel from microalgae. *Biotechnology Advances*, 25, 294–306.
Chisti, Y. (2008a). Biodiesel from microalgae beats bioethanol. *Trends in Biotechnol*, 26, 126–131.
Chisti, Y. (2008b). Response to Reijnders: Do biofuels from microalgae beat biofuels from terrestrial plants. *Trends in Biotechnol*, 26, 351–352.
Chisti, Y. (2010). Fuels from microalgae. *Biofuels*, *1*(2), 233–235.
Choi, S.K., Lee, J.Y., Kwon, D.Y. and Cho, K.J. (2006). Settling characteristics of problem algae in the water treatment process. *Water Sc Technol*, 53, 113–119.
Clarens, A.F., Resurreccion, E., White, M. and Colosi, A. (2010). Environmental life cycle comparison of algae to other bioenergy feedstocks. *Env Sc Technol*, 44, 1813–1819.
Clark, J. and Deswarte, F. (2008). Introduction to chemicals from biomass. Wiley Series in Renewable Resources. ISBN978-0-470-05805.
Costa, J.A.V. and Morais, M.G. (2011). The role of biochemical engineering in the production of biofuels from microalgae. *Bioresour Technol*, 102, 2–9.
Das, D. (1985). Ph.D. Thesis on "Optimization of methane production from agricultural residues". IIT Delhi, India.
Das, D. (2009). Advances in biological hydrogen production processes: An approach towards commercialization. *Int J Hydrogen Energy*, 34, 7349–7357.
Das, D. and Veziroğlu, T.N. (2001). Hydrogen production by biological processes: A survey of literature. *Int J Hydrogen Energy*, 26 (1), 13–28.
Dasgupta, C.N., Gilbert, J.J., Lindblad, P., Heidorn, T., Borgvang, S.A., Skjanes, K. and Das, D. (2010). Recent trends on the development of photobiological processes and photobioreactors for the improvement of hydrogen production. *Int J Hydrogen Energy*, 35(19), 10218-10238.
Davis, R., Aden, A. and Pienkos, P.T. (2011). Techno-economic analysis of autotrophic microalgae for fuel production. *Appl Energy*, 88(10), 3524–3531.
de Morais, M.G., Stillings, C., Dersch, R., Rudisile, M., Pranke, P., Costa, J.A.V. and Wendorff, J. (2010). Preparation of nanofibers containing the microalga *Spirulina* (Arthrospira). *Bioresour Technol*, 101(8), 2872–2876.
De Rosa, S.C., Brenchley, J.M. and Roederer, M. (2003). Beyond six colours: A new era in flow cytometry. *Nat Med.*, 9, 112–117.

deB Richter, Jr, D., Jenkins, J.H., Karakash, J.T., Knight, J., McCreery, L.R. and Nemestothy, K.P. (2009). Wood energy in America. *Science*, 323, 1432–1433.
Del Campo, J. A., Moreno, J., Rodrı́guez, H., Vargas, M. A., Rivas, J., & Guerrero, M. G. (2000). Carotenoid content of chlorophycean microalgae: factors determining lutein accumulation in Muriellopsis sp.(Chlorophyta). *Journal of Biotechnology*, *76*(1), 51–59.
Delrue, F., Setier, P.A., Sahut, C., Cournac, L., Roubaud, A., Peltier, G. and Froment, A.K. (2012). An economic, sustainability, and energetic model of biodiesel production from microalgae. *Bioresour Technol*, 111, 191–200.
Demirbas, A.H. (2009a). Inexpensive oil and fats feedstocks for production of biodiesel. *Energy Education Sc Technol A*, 23, 1–13.
Demirbas, A. (2009b). Production of biodiesel from algae oils. *Energy Sources A*, 31, 163–168.
Dickinson, J., Jackson, T., Matthews, M. and Cripps, A. (2009). The economic and environmental optimisation of integrating ground source energy systems into buildings. *Energy*, 34, 2215–2222.
Domozych, D.S., Kort, S., Benton, S. and Yu, T. (2005). The extracellular polymeric substance of the green alga *Penium margaritaceum* and its role in biofilm formation. *Biofilms*, 2(2), 129–144.
Douskova, I., Doucha, J., Machat, J., Novak, P., Umysova, D., Vitova, M. and Zachleder, V. (2008). Microalgae as a means for converting flue gas CO_2 into biomass with a high content of starch. Bioenergy: challenges and opportunities. International conference and exhibition on bioenergy. Guimarães, Portugal, April 6th–9th.
Dufossé, L., Galaup, P., Yaron, A., Arad, S. M., Blanc, P., Murthy, K. N. C., & Ravishankar, G. A. (2005). Microorganisms and microalgae as sources of pigments for food use: a scientific oddity or an industrial reality?. Trends in Food Science & Technology, 16(9), 389–406.
Dunahay, T.G., Jarvis, E.E., Dais, S.S. and Roessler, P.G. (1996). Manipulation of microalgal lipid production using genetic engineering. *Appl Biochem Biotechnol*, 57–58, 223–231.
Finney, K. F., Pomeranz, Y., & Bruinsma, B. L. (1984). Use of algae Dunaliella as a protein supplement in bread. *Cereal chemistry (USA).*
Fleurence, J., Coeur, C.L., Mabeau, S., Maurice, M., & Landrein, A. (1995). Comparison of different extractive procedures for proteins from the edible seaweeds *Ulva rigida* and *Ulva rotundata*. Journal of Applied Phycology 7, 577–582.
Franklin, S.E. and Mayfield, S.P. (2005). Recent developments in the production of human therapeutic proteins in eukaryotic algae. *Expert Opinion on Biological Therapy*, 5(2), 225–235.
Gallagher, B.J. (2011). The economics of producing biodiesel from algae. *Renewable Energy*, 36, 158–162.
Ghirardi, M.L., Kosourov, S., Tsygankov, A. and Seibert, M. (2000). Two-phase photobiological algal H_2-production system. *In:* Proceedings of the 2000 U.S. DOE hydrogen program review. National Renewable Energy Laboratory, Golden, Colorado, pp 1–13.
Glazer, A.N. and Stryer, L. (1984). Phycofluor probes. *Trends Biochem Sci*. 9, 423–427.
Gong, Y. and Jiang, M. (2011). Biodiesel production with microalgae as feedstock: from strains to biodiesel. *Biotechnol Letters*, 33, 1269–1284.
Greenwell, H.C., Laurens, L.M.L., Shields, R.J., Lovitt, R.W. and Flynn, K.J. (2010). Placing microalgae on the biofuels priority list: A review of the technological challenges. *Journal of the Royal Society Interface*, 7, 703–726.
Guedes, A.C., Amaro, H.M. and Malcata, F.X. (2011). Microalgae as sources of high added-value compounds: A brief review of recent work. *Biotechnol Progress*, 27(3), 597–613.
Hankamer, B., Lehr, F., Rupprecht, J., Mussgnug, J.H., Posten, C. and Kruse, O. (2007). Photosynthetic biomass and H_2 production by green algae: From bioengineering to bioreactor scale-up. *Physiology Plant*, 131, 10–21.
Hansel, A. and Lindblad, P. (1998). Towards optimization of cyanobacteria as biotechnologically relevant producers of molecular hydrogen, a clean and renewable energy source. *Appl Microbial Biotechnol*, 50, 153–160.
Harun, R., Danquah, M.K. and Forde, G.M. (2010b). Microbial biomass as a fermentation feedstock for bioethanol production. *J Chem Technol Biotechnol*, 85, 199–203.
Hempel, F., Bozarth, A.S., Lindenkamp, N., Klingl, A., Zauner, S., Linne, U., Steinbüchel, A. and Maier, U.G. (2011). Microalgae as bioreactors for bioplastic production. *Microbial Cell Factories*. 10(81), 1–6.

Hirano, A., Ryohei, U., Shin, H. and Yasuyuki, O. (1997). CO_2 fixation and ethanol production with microalgal photosynthesis and intracellular anaerobic fermentation. *Energy*, 22, 137–142.
Holdt, S.L., Kraan, S., (2011). Bioactive compounds in seaweed: functional food applications and legislation, J Appl Phycol, 23:543–597.
Hu, Q., Sommerfeld, M., Jarvis, E., Ghirardi, M., Posewitz, M., Seibert, M. and Darzins, A. (2008). Microalgal triacylglycerols as feedstocks for biofuel production: Perspectives and advances. *Plant J*, 54, 621–639.
Huntley, M.E. and Redalje, D.G. (2007). CO_2 mitigation and renewable oil from photosynthetic microbes: A new appraisal. *Mitigation and Adaption Strategies for Global Change*, 12, 573–608.
Izydorczyk, K., Tarczynska, M., Jurczak, T., Mrowczynski, J. and Zalewski, M. (2005). Measurement of phycocyanin fluorescence as an online early warning system for cyanobacteria in reservoir intake water. *Environ Tox*., 20, 425–430.
Janssen, M., Tramper, J., Mur, L. R., & Wijffels, R. H. (2003). Enclosed outdoor photobioreactors: light regime, photosynthetic efficiency, scale-up, and future prospects. *Biotechnology and bioengineering*, *81*(2), 193–210.
Jespersen, L., Strømdahl, L. D., Olsen, K., & Skibsted, L. H. (2005). Heat and light stability of three natural blue colorants for use in confectionery and beverages. *European Food Research and Technology*, *220*(3–4), 261–266. (Lone et al., 2005)
Jha, M.N., Prasad, A.N., Sharma, S.G. and Bharti, R.C. (2001). Effects of fertilization rate and crop rotation on diazotrophic cyanobacteria. *World J Microbiol Biotechnol*, 17, 463.
Jiang, Y., Chen, F., & Liang, S.Z. (1999). Production potential of docosahexaenoic acid by the heterotrophic marine dinoflagellate *Crypthecodinium cohnii*. Process Biochemistry 34, 633–637.
John, R.P., Anisha, G.S., Nampoothiri, K.M. and Pandey, A. (2011). Micro and macroalgal biomass: A renewable source for bioethanol. *Bioresour Technol*, 102(1), 186–193.
Kanazawa, K. (2012). High bioavailability and diverse biofunctions of fucoxanthin in brown algae. *Nippon Shokuhin Kagaku Kogaku Kaishi*, 59(2), 49–55.
Kaushik, B.D. (1995). Cyanobacterial response to salinity and amelioration technology. *In:* Rice management biotechnology (ed. S. Kannaiyan). pp. 323–329, Associated Publishing Co., New Delhi.
Kay, R. A., & Barton, L. L. (1991). Microalgae as food and supplement. *Critical reviews in food science & nutrition*, *30*(6), 555–573.
Khan, S.A., Rashmi Hussain, M.Z., Prasad, S. and Banerjee, U.C. (2009). Prospects of biodiesel production from microalgae in India. *Renewable and Sustainable Energy Reviews*, 13, 2361–2372.
Khan, Z.U.M., Tahmida Begum, Z.N., Mandal, R. and Hossain, M.Z. (1994). Cyanobacteria in rice soils. *World Journal of Microbiology and Biotechnology*, 10, 296–298.
Kheshgi, H.S., Prince, R.C. and Marland, G. (2000). The potential of biomass fuels in the context of global climate change. *Annual Reviews in Energy and Environment*, 25, 199–244.
Lee, D.H. (2011). Algal biodiesel economy and competition among bio-fuels. *Bioresource Technology*, 102, 43–49.
Li, Y., Horsman, M., Wu, N., Lan, C.Q. and Dubois-Calero, N. (2008). Biofuels from microalgae. *Biotechnology Progress*, 24, 815–820.
Lindblad, P., Christensson, K., Lindberg, P., Fedorov, A., Pinto, F. and Tsygankov, A. (2002). Photoproduction of H_2 by wildtype *Anabaena* sp. PCC 7120 and a hydrogen uptake deficient mutant: From laboratory experiments to outdoor culture. *Int J Hydrogen Energy*, 27, 1271–1281.
Lobban, C.S. and Harrison, P.J. (1997). Seaweed ecology and physiology, IX. Cambridge University Press.
Mata, T.M., Martins, A.A. and Caetano, N.S. (2009). Microalgae for biodiesel production and other applications: A review. *Renewable and Sustainable Energy Reviews*, 14, 217–232.
Mata, T. M., Martins, A. A., & Caetano, N. S. (2010). Microalgae for biodiesel production and other applications: a review. *Renewable and sustainable energy reviews*, *14*(1), 217–232.
Matsumoto, M., Hiroko, Y., Nobukazu, S., Hiroshi, O. and Tadashi, M. (2003). Saccharification of marine microalgae using marine bacteria for ethanol production. *Applied Biochemistry and Biotechnology*, 105, 247–254.

Melis, A. and Happe, T. (2001). Hydrogen production. Green algae as a source of energy. *Plant Physiology*, 127, 740–748.
Melis, A., Zhang, L., Forestier, M., Ghirardi, M.L. and Seibert, M. (2000). Sustained photobiological hydrogen gas production upon reversible inactivation of oxygen evolution in the green alga *Chlamydomonas reinhardtii. Plant Physiology*, 122, 127–136.
Miao, X. and Wu, Q. (2004). High yield bio-oil production from fast pyrolysis by metabolic controlling of *Chlorella protothecoides. Journal of Biotechnology*, 110, 85–93.
Mihranyan, A., Nyström, G., Razaq, A., Nyholm, L. and Strømme, M. (2010). A battery from algae cellulose. PULPAPER 2010 Conference - Implementing the New Rise: Winning Change. pp. 135–142.
Mitra, A.K. (1951). The algal flora of certain Indian soils. *Indian Journal of Agricultural Sciences*, 21, 375.
Molina Grima, E., Fernandez, F. and Camacho, F. (1999). Photobioreactors: Light regime, mass transfer and scale up. *Journal of Biotechnology*, 70, 231–247.
Munda, I. M. (1977). Differences in amino acid composition of estuarine and marine fucoids. *Aquatic Botany*, *3*, 273–280.
Mussgnug, J.H., Klassen, V., Schlüter, A. and Kruse, O. (2010). Microalgae as substrates for fermentative biogas production in a combined biorefinery concept. *Journal of Biotechnology*, 150, 51–56.
Ohlrogge, J., Allen, D., Berguson, B., DellaPenna, D., Shachar-Hill, Y. and Stymne, S. (2009). Driving on biomass. *Science*, 324, 1019–1020.
Okabe, K., Murata, K., Nakanishi, M., Ogi, T., Nurunnabi, M. and Liu, Y. (2009). Fischer–Tropsch synthesis over Ru catalysts by using syngas derived from woody biomass. *Catalysis Letters*, 128, 171–176.
Ono, E., & Cuello, J. L. (2006). Feasibility assessment of microalgal carbon dioxide sequestration technology with photobioreactor and solar collector. *Biosystems engineering*, *95*(4), 597–606.
OriginOil (2010). http://www.originoil.com November 2010.
Ozgener, O. and Hepbasil, A. (2007). A review on the energy and exergy analysis of solar assisted heat pump systems. *Renewable and Sustainable Energy Reviews*, 11, 482–496.
Park, J.B.K., Craggs, R.J. and Shilton, A.N. (2011). Wastewater treatment high rate algal ponds for biofuel production. *Bioresource Technology*, 102, 35–42.
Pienkos, P.T. and Darzins, A. (2009). The promise and challenges of micro-algal derived biofuels. *Biofuels Bioproducts and Biorefineries*, 3, 431–440.
Poulíčková, A., Špačková, J., Kelly, M. G., Duchoslav, M., & Mann, D. G. (2008). Ecological variation within *Sellaphora species* complexes (Bacillariophyceae): specialists or generalists?. Hydrobiologia, 614(1), 373–386.
Proviron (2010). http://www.proviron.com/algae.
Pulz, O. (2004). Valuable products from biotechnology of microalgae. *Applied Microbiology and Biotechnology*, 65, 635–648.
Ran, C.Q., Chen, Z.A., Zhang, W., Yu, X.J. and Jin, M.F. (2006). Characterization of photobiological hydrogen production by several marine green algae. *Wuhan Ligong Daxue Xuebao*, 28(suppl 2), 258–263.
Rao, K.K. and Hall, D.O. (1996). Hydrogen production by cyanobacteria: Potential, problems and prospects. *Journal of Marine Biotechnology*, 4, 10–15.
Reijnders, L. (2009). Microalgal and terrestrial transport biofuels to displace fossil fuels. *Energies*, *2*(1), 48–56.
Reijnders, L. and Huijbregts, M.A.J. (2009). Energy Balance: Cumulative fossil fuel demand and solar energy conversion efficiency of transport biofuels. *In:* Transport biofuels: A seed to wheel perspective. Springer, London. pp. 49–74.
Rodolfi, L., Zitelli, G.C., Bassi, N., Padovani, G., Biondi, N., Bonini, G. and Tredici, M.R. (2009). Microalgae for oil: Strain selection, induction of lipid synthesis and outdoor mass cultivation in a low-cost photobioreactor. *Biotechnology and Bioengineering*, 102, 100–112.
Rodjaroen, S., Juntawong, N., Mahakhant, A., & Miyamoto, K. (2007). High biomass production and starch accumulation in native green algal strains and cyanobacterial strains of Thailand. Kasetsart J (Nat Sci), 41, 570–5.

Roger, P.A., Zimmerman, W.J. and Lumpkin, T.A. (1993). Microbiological management of wetland rice fields. *In:* Soil Microbiological Ecology—Application in Agricultural and Microbiological Management (ed. B. Metting). Marcel Dekker, New York. pp. 417–455.

Rosales-Mendoza, S., Paz-Maldonado, L.M.T. and Soria-Guerra, R.E. (2012). *Chlamydomonas reinhardtii* as a viable platform for the production of recombinant proteins: Current status and perspectives. *Plant Cell Reports*, 31(3), 479–494.

Rosenberg, J.N., Oyler, G.A., Wilkinson, L. and Betenbaugh, M.J. (2008). A green light for engineered algae: Redirecting metabolism to fuel a biotechnology revolution. *Biotechnology*, 19, 430–436.

Rother, J.A. and Whitton, B.A. (1989). Nitrogenase activity of blue green algae on seasonally flooded soils in Bangladesh. *Plant Soil*, 113, 47–52.

Ruperez, P. (2002). Mineral content of edible marine seaweeds. Food Chemistry 79: 23–26.

Santana, G. C. S., Martins, P. F., Da Silva, N. D. L., Batistella, C. B., Maciel Filho, R., & Maciel, M. W. (2010). Simulation and cost estimate for biodiesel production using castor oil. *Chemical Engineering Research and Design*, *88*(5), 626–632.

Schenk, P.M., Skye, R., Thomas-Hall, Stephens, E., Marx, U.C., Mussgnug, J.H., Posten, C., Kruse, O. and Hankamer, B. (2008). Second generation biofuels: High-efficiency microalgae for biodiesel production. *Bioenergy Research*, 1, 20–43.

Schütz, K., Happe, T., Troshina, O., Lindblad, P., Leitão, E., Oliveira, P. and Tamagnini, P. (2004). Cyanobacterial H_2-production—A comparative analysis. *Planta*, 218, 350–359.

Sharma, S., Hunt, R., Cao, J. and Zeller, M.A. (2011). Bioplastics from algae biomass (Conference Paper) Society of Plastics Engineers—Global Plastics Environmental Conference 2011.

Sheehan, J., Dunahay, T., Benemann, J. and Roessler, P. (1998). A Look Back at the U.S. Department of Energy's Aquatic Species Program–Biodiesel from Algae. Edited by: DOE: National Renewable Energy Laboratory.

Sheng, G., Schluchter, W. M., & Bryant, D. A. (2008). Biogenesis of Phycobiliproteins I. cpcS-I AND cpcU mutants of the cyanobacterium Synechococcus sp. PCC 7002 define a heterodimeric phyococyanobilin lyase specific for β-phycocyanin and allophycocyanin subunits. Journal of Biological Chemistry, 283(12), 7503–7512.

Sialve, B., Bernet, N. and Bernard, O. (2009). Anaerobic digestion of microalgae as a necessary step to make microalgal biodiesel sustainable. *Biotechnology Advances*, 27, 409–416.

Singh, A., Nigam, P.S. and Murphy, J.D. (2011). Renewable fuels from algae: An answer to debatable land based fuels. *Bioresource Technology*, 102, 10–16.

Skjånes, K., Rebours, C. and Lindblad, P. (2013). Potential for green microalgae to produce hydrogen, pharmaceuticals and other high value products in a combined process. *Crit Rev Biotechnol.*, 33(2), 172–215.

Sode, K.J., Horikoshi, K., Takeyama, J., Nakamura, N. and Matsunga, T. (1991). On line monitoring of marine cyanobacterial cultivation based on phycocyanin fluorescence. *J Biotechnol.*, 21, 209–218.

Solix_Biofuels (2010). http://www.solixbiofuels.com.

Soltani, S., Saadatmand, S., Khavarinejad, R. and Nejadsattari, T. (2011). Antioxidant and antibacterial activities of *Cladophora glomerata* (L.) Kütz. in Caspian Sea Coast, Iran. *Afri J Biotechnol.*, 10(39), 7684–7689.

Spolaore, P., Cassan, C.J., Duran, E. and Isambert, A. (2006). Commercial Applications of Microalgae. *J Bioscience and Bioeng*, 101, 87–96.

Srinivasan, S. (1978). Algae multiplication fertilizer practices. *International Rice Research Newsletter*, 1 l(4), 35–36.

Subhadra, B. and Edwards, M. (2010). An integrated renewable energy park approach for algal biofuel production in United States. *Energy Policy*, 38, 4897–4902.

Subitec (2010). http://www.subitec.com.

Tamagnini, P., Leitão, E., Oliveira, P., Ferreira, D., Pinto, F., Harris, D. and Heidorn, T. (2007). Cyanobacterial hydrogenases. Diversity, Regulation and Applications. *FEMS Microbiology Reviews*, 31, 692–720.

Taylor, G. (2008). Biofuels and the biorefinery concept. *Energy policy*, *36*(12), 4406–4409.

Telford, W.G., Moss, M.W., Moreseman, J.P. and Thomas Allnutt, F.C. (2001). Cyanobacterial stabilized phycobilisomes as fluorochromes for extracellular antigen detection by flow cytometry. *J Immunol Meth.*, 254, 13–30.

Tirol, A.C., Roger, P.A. and Watanabe, I. (1982). Fate of nitrogen from blue green alga in a flooded rice soil. *Soil Science and Plant Nutrition*, 28, 559–569.

Tokusoglu, O., & Unal, M.K. (2003). Biomass Nutrient Profiles of Three Microalgae: *Spirulina platensis, Chlorella vulgaris*, and *Isochrisis galbana*. Journal of Food Sciences 68, 1144–1148.

Törnwall, M. E., Virtamo, J., Korhonen, P. A., Virtanen, M. J., Taylor, P. R., Albanes, D., & Huttunen, J. K. (2004). Effect of α-tocopherol and β-carotene supplementation on coronary heart disease during the 6-year post-trial follow-up in the ATBC study. *European heart journal*, *25*(13), 1171–1178.

Tran, M. et al. (2009). Synthesis and Assembly of a Full-Length Human Monoclonal Antibody in Algal Chloroplasts. *Biotechnol Bioeng.*, 104, 663–673.

Ueda, R., Hirayama, S., Sugata, K. and Nakayama, H. (1996). Process for the production of ethanol from microalgae. US Patent 5,578,472.

Usov, A.I. (2011). Polysaccharides of the red algae. *Advances in Carbohydrate Chemistry and Biochemistry*, 65, 115–217.

Varfolomeev, S. D., Efremenko, E. N., & Krylova, L. P. (2010). Biofuels. *Russian Chemical Reviews*, *79*(6), 491.

van Beilen, J.B. (2010). Why microalgal biofuels won't save the internal combustion machine. *Biofuels Bioproducts Biorefineries*, 4, 41–52.

van Harmelen, T. and Oonk, H. (2006). Microalgae biofixation processes: Applications and potential contributions to greenhouse gas mitigation options. TNO Built Environmental Geosciences, Apeldoorn.

van Haveren, J., Scott, E. L., & Sanders, J. (2008). Bulk chemicals from biomass. *Biofuels, Bioproducts and Biorefining*, *2*(1), 41–57.

Venkataraman, G.S. (1979). Algal inoculation in rice fields. *In:* Nitrogen and rice. International Rice Research Institute, Los Banos, Philippines. Pp 312–321.

Villar, R., Laguna, M. R., Calleja, J. M., & Cadavid, I. (1992). Effects of Phaeodactylum tricornutum and Dunaliella tertiolecta extracts on the central nervous system. *Planta medica*, *58*(5), 405–409.

Watanabe, A., & Yamamoto, Y. (1971). Algal nitrogen fixation in the tropics. *Plant and Soil*, *35*(1), 403–413.

Wang, B., Li, Wu, N. and Lan, C.Q. (2008). CO_2 bio-mitigation using microalgae. *Applied Microbiology and Biotechnology*, 79, 707–718.

WI (Worldwatch Institute) (2007). Biofuels for Transport: Global Potential and Implications for Energy and Agriculture. Earthscan, London.

Wiencke, C. and Bischof, K. (eds) (2012). Seaweed Biology. Novel insights into ecophysiology, ecology and utilization. Ecological Studies Vol 219, Springer Publishers.

Willems, P. A. (2009). The biofuels landscape through the lens of industrial chemistry. *Science*, *325*(5941), 707–708

Yamaguchi, K., 1997. Recent advances in microalgal bio-science in Japan, with special reference to utilization of biomass and metabolites: a review. J. appl. Phycol. 8: 487–502.

Yang, Z., Guo, R., Xu, X., Fan, Xa. and Li, X. (2010). Enhanced hydrogen production from lipid-extracted microalgal biomass residues through pretreatment. *International Journal of Hydrogen Energy*, 35, 9618–9623.

Yu, W. L., Ansari, W., Schoepp, N. G., Hannon, M. J., Mayfield, S. P., & Burkart, M. D. (2011). Modifications of the metabolic pathways of lipid and triacylglycerol production in microalgae. *Microb Cell Fact*, *10*(11).

Zittelli G C, Lavista F, Bastianini A, Rodolfi L, Vincenzini M, Tredici MR 1999. Production of eicosapentaenoic acid by Nannochloropsis sp. cultures in outdoor tubular photobioreactors. J Biotechnol 70:299–312.

Chapter 2
CO_2 Sequestration Through Algal Biomass Production

Kanhaiya Kumar, Sanjiv Kumar Mishra, Gang-Guk Choi, and Ji-Won Yang

2.1 Introduction

The world is facing the threat of global warming. This is associated with the modernization and increasing dependency of human beings on the fossil fuel. Fossil fuel are necessary for providing the increasing demand on energy. However, their combustion produces greenhouse gases such as carbon dioxide, methane, ozone, NO_x, water vapour etc. (Kumar et al., 2011). CO_2 is also continuously being added into the earth's atmosphere through several natural sources such as volcanic eruptions, combustion of organic matters, autotrophic and heterotrophic respiration (Kumar and Das, 2014; Sharma et al., 2011). However, robust natural mechanisms of CO_2 capture could maintain the balance of CO_2 in the earth's atmosphere. Global carbon cycle is disturbed mainly due to anthropogenic emissions of CO_2 because of human activity (Kumar and Das, 2014). Coal is the major contributor of CO_2, which is in the range of 14–17 % depending upon its quality. Therefore, coal-based industries such as cement, steel and thermal power plants pollute the earth's environment to a greater extent.

The greenhouse gases trap the solar heat and prevent it from dissipating into the space. This increases the earth's temperature. The ecosystem of earth is affected by even a little increase in the global mean temperature. Melting glaciers, sea level rise, unpredicted rainfall and drought are some examples of global warming. Besides these, global warming increases the soil microbe's respiration, which further adds CO_2 into the earth's atmosphere (Bardgett et al., 2008). CO_2 is the major greenhouse gas of the flue gas (Kumar and Das, 2014). Keeling curve indicates a sharp rise in the global CO_2 concentration after 1950 (Kumar et al., 2011). In July, 2014, CO_2 concentration in the earth's atmosphere was 399.00 ppm as measured at Mauna Loa

K. Kumar (✉) • S.K. Mishra • G.-G. Choi • J.-W. Yang
Department of Chemical and Biomolecular Engineering, KAIST,
291 Daehak-ro, Yuseong-gu, Daejeon 305-701, Republic of Korea
e-mail: k_kumar222@kaist.ac.kr

D. Das (ed.), *Algal Biorefinery: An Integrated Approach*,
DOI 10.1007/978-3-319-22813-6_2

Observatory, Hawaii, USA (http://www.esrl.noaa.gov/gmd/ccgg/trends/). This is significantly higher than the range of the CO_2 concentration (180–300 ppm) measured over the history of earth. Similarly, CO_2 concentration was found to increase at the rate of 2.05 ppm per year in 2013 compared to 0.94 ppm per year in 1959 (http://www.esrl.noaa.gov/gmd/ccgg/trends/).

Thus, CO_2 sequestration is imperative to combat the challenges of global warming. Several natural processes continuously remove CO_2 from the atmosphere. For example, large mass of oceanic water dissolves atmospheric CO_2 and forms weak acids, which eventually forms bicarbonate due to reaction between carbonate anions and water (Kumar and Das, 2014; Falkowski et al., 2000). Similarly, large quantity of CO_2 is continuously sent to the bottom of the ocean due to differential solubility of CO_2 in cold and hot saline water. Besides these natural processes of CO_2 capture, CO_2 can be sequestered artificially, either by physical or biological methods. Physical methods involve natural mineral carbonation, scrubbing and geological (depleted oil and gas reservoirs) and oceanic injection (deep aquifers) etc. (Mirjafari et al., 2007; Allen et al., 2005). These methods are expensive as it also involves the collection, transportation and storage of CO_2 gas. In addition, it poses potential threat to the environment due to possibility of the accidental leakage of CO_2 (Bachu, 2000). Biological methods are another alternative, which mainly involves the photosynthesis process by oceanic phytoplanktons, terrestrial plants, algae and use of carbonic anhydrase (CA) enzyme. In addition, some non-photosynthetic bacteria also utilize CO_2 for their metabolic activity (Kumar and Das, 2014).

As an aggressive move, CO_2 sequestration through algal biomass production is a promising option. CO_2 sequestered algal biomass can be further utilized for the biofuel productions (biodiesel, bioethanol, biohydrogen) and several other value added products of industrial importance (Kumar et al., 2011; Loubiere et al., 2009). Microalgae can be cultivated in non-agriculture land or even in brown fields. Contrary to energy crops such as soybeans, algae can use various water sources ranging from wastewater to saline water. Furthermore, waste containing different volatile fatty acids can also be used for the supplementation of nutrient required for the algal growth. The photosynthetic efficiency of microalgae is nearly 10 to 50 times higher than that of terrestrial plants (Kumar and Das, 2014; Li et al., 2008). This results into higher specific growth rate and biomass productivity. Moreover, their cultivation is relatively easier and biomass can be easily managed during its processing for the extraction of value-added products. In addition, genetic modification targeting desired products is easy. Terrestrial plants are effective in sequestering CO_2 at low CO_2 concentration. Contrary to this, microalgae are efficient in growing at a relatively higher CO_2 concentration such as flue gas. Algal biomass can be directly used as food or food supplements as these are rich in protein, carbohydrates, essential fatty acids, antioxidants and several other biomolecules beneficial for the body. In fact, *Nostoc* sp. was used as food nearly 2000 years back during Chinese famine (Kumar and Das, 2014; Spolaore et al., 2006).

Thus, the present study aimed to summarize the different bottlenecks and the present state of art in CO_2 sequestration process through algal biomass production.

It also highlights the mechanisms of CO_2 sequestration process. The microalgal cultivation systems such as open ponds and photobioreactors are discussed in details.

2.2 Microbiology

Microalgae and cyanobacteria are distributed throughout the biosphere with an immense range of genetic diversity. Up until now, nearly 35,000 species have been described in the literature (Sydney et al., 2014). However, their actual number may be much higher. They can be found as unicells, colonies and extended filaments, growing even under extreme climatic conditions from aquatic to terrestrial places (Kumar et al., 2011). Their uniqueness is due to the presence of chlorophyll and their CO_2 fixing ability through photosynthesis in a single cell.

It is always desired to isolate a robust strain of algae, which can withstand the fluctuating physico-chemical conditions such as temperature, pH, light and CO_2 concentration and can grow efficiently in the open cultivation system. The low chances of contamination, the high CO_2 fixation rate and the product formation rate are the other desired characteristics of the algae. In the closed photobioreactor such as airlift photobioreactor, shear stress varies in different sections. Therefore, algal species having tolerance to high shear stress are other desired characteristics of the robust strain. Some of well known microalgal species used for the large scale production are described below.

Chlorella spp. are simple, non-motile and spherical unicellular eukaryotic green microalgae that have a thick cell wall (100–200 nm). *Chlorella* sp. measure between 2 and 10 μm. *Chlorella* sp. have a high concentration of chlorophyll and photosynthetic ability compared to higher plants. The optimal CO_2 concentration for the growth of *Chlorella* sp. was found in the range of 5–6 % (Kumar and Das, 2012; de Morais and Costa, 2007). They are known to have a high specific growth rate (μ): 0.08 h^{-1} for *Chlorella vulgaris* and 0.11 h^{-1} for *Chlorella sorokiniana* (Cordero et al., 2011). *Chlorella* sp. are known as an attractive source of food due to high concentration of protein (nearly 50 %) and essential nutrients. *Scenedesmus* spp. are freshwater, non-motile microalgae. Various biotic and abiotic factors determine the extent of aggregation of these microalgae. For example, they are unicellular in axenic condition and known to form colonies in xenic system of the natural system (Geng et al., 2014). *Botryococcus* spp. are found in colony held together within a lipid biofilm matrix. They are characterized by their slow growth rate and accumulation of the high amount of hydrocarbons (up to 50 %, w/w), which is released outside the cells.

Chlorococcum spp. are spherical unicellular marine green microalgae with cells of diameter about 10 μm. They have a high tolerance to CO_2 concentrations and can be grown to a very large cell density (Iwasaki et al., 1998). *Dunaliella* spp. are biflagellated, motile, unicellular marine green microalgae. They are mostly round in shape of diameter 9 to 11 μm. They have a high tolerance for salts, temperature and light (Segovia et al., 2003). *Dunaliella* lacks the cell walls, which make them an extremely fragile microorganism. They reproduce by binary fission with no evidence of cell lysis, encystment or spore formation (Segovia et al., 2003). *Dunaliella* is well known for the production of wide varieties of commercial products such as β-carotene (up to 14 % w/w). β-carotene can be easily extracted due to the easy rupture of their cell wall. *Nannochloropsis* spp. are small, non-motile spherical shaped and mostly marine in nature. They have only chlorophyll-*a* and completely lack chlorophyll-*b* and *c*, which make them distinct from other related microalgae. *N. salina*, *N. Gaditana* and *N. oceanic* are some of the well known species of *Nannochloropsis*. Along with CO_2 sequestration, they have wide industrial applications. They are oleaginous strains, rich in polyunsaturated fatty acids and other value-added products such as astaxanthin, zeaxanthin and canthaxanthin (Lubian et al., 2000). In addition, they are choice of researchers because of easier genetic manipulation aimed for the genetic improvement. *Haematococcus* sp. are mobile, single-celled green algae (Chlorophyta), capable of synthesizing high amount of astaxanthin in response to environmental conditions, ranging from 1 to 5 % of dry cell weight (Wan et al., 2014). They have two stages of growth: growth stage (green cell) followed by astaxanthin accumulation stage (red cell). Therefore, *Haematococcus* cell changes its resistance to shear stress along the growth cycle. They have a slow growth rate (nearly 1.20 div d^{-1}), which make them prone to contamination (González-López et al., 2012).

Cyanobacteria are Gram-negative bacteria having a thin peptidoglycan layer. Cyanobacteria can be single cellular or multicellular. Filamentous colonies of cyanobacteria can have different type of cells such as vegetative cells, akinetes and heterocysts, which have a uniqueness to carry out complete oxygenic photosynthesis, storing food reserves and nitrogen fixing ability, respectively (Kumar et al., 2011). These are the only prokaryotic microorganism having the ability to reduce both atmospheric CO_2 and nitrogen into carbohydrates and ammonia, respectively. Similar to microalgae and plants, cyanobacteria carry out oxygenic photosynthesis, involving both PSII and PSI. In anaerobic conditions, they use only PSI, causing anoxygenic photosynthesis. Nitrogen is fixed by the enzymes nitrogenase in the heterocysts of cyanobacteria. Carbohydrates and ammonia are utilized for the growth of cyanobacteria. *Synechocystis* is unicellular cyanobacteria. *Anabaena* and *Aphanothece* are linear multi-cellular, nitrogen fixing cyanobacteria. *Spirulina* sp. are freshwater, long, thin spiral thread like cyanobacteria, largely cultivated for the mass scale production. *Spirulina* has filaments (or trichomes) of helical shape in liquid medium and spiral shape in solid medium, which is the characteristics of the genus. Their large size (up to 1 mm in length and 12 μm in diameter) makes them easier in harvesting. Chlorophyll and phycocyanin are their main photosynthetic pigments. *Spirulina* sp. usually grow efficiently at alkaline pH with optimal pH in the range of 8.5 to 11. They are known for the food or food supplements as *Spirulina* cells are rich in protein, polyunsaturated fatty acids (PUFAs), vitamins, antioxidants and several important minerals.

2.3 Mechanism of CO_2 Fixation

2.3.1 Photosynthetic Pathway

Photosynthesis is the process in which CO_2 is fixed into carbohydrates using the light energy and water. Algae use this process to prepare their food for growth and survival. Chloroplast is the site of photosynthesis in the microalgae. Whereas, in cyanobacteria, photosynthetic apparatus is found in the cytoplasm. It has two stages: light dependent and light independent. Chlorophyll and light harvesting complex harness the light energy and conserve it in the form of energy currencies such as ATP and NADPH formed in the light dependent stage. In the light independent stage, ATP and NADPH are consumed during the synthesis of carbohydrates. Photosynthetic apparatus consisting of protein complexes, electron carriers and lipid molecules are located in or around the thylakoid membranes. Thylakoid membrane has two reaction centers called PSII and PSI, connected with several electron carriers such as plastoquinone, cytochrome b_6f complex and plastocyanin (PC) in the order of increasing redox potential. This arrangement of electron carriers allows the flow of electron from negative redox potential to positive redox potential. PSII and PSI are specialized to absorb the light of wavelength 680 nm and 700 nm, respectively. Electron carriers are arranged in the order of increasing redox potential. The water split at PSII into protons, electrons and oxygen molecule with the help of light of 680 nm (Eq. 2.1). Protons generated by splitting of water are accumulated into the lumen of the thylakoid membrane and oxygen escapes from the cell. A proton gradient is generated across the thylakoid membranes due to accumulation of the protons in the lumen. These protons eventually escape from the lumen into the stroma generating ATP with the help of ATP synthase. On the other hand, electron gets excited at the PSII and travels to PSI by the linear electron transport chain. At PSI, it is further excited with the help of light of 700 nm. Antenna chlorophyll molecules help in harnessing the light at PSI. Excited electron reduces Ferredoxin (Fd), which is loosely bounded to the thylakoid membrane from outside (Kumar and Das, 2014). Reduced Fd transfers its electron to the ferredoxin $NADP^+$ reductase (FNR), which catalyses the formation of NADPH using $NADP^+$ and H^+ coming from the lumen.

It should be noted that electron carriers transfer only one electron at a time with the help of one photon at each of the reaction center. Eventually the light energy is stored in the form of NAD(P)H and utilized in the light independent phase during CO_2 fixation using Calvin cycle. The production of two moles of NAD(P)H requires eight photons and two moles of water as shown in Eq. 2.2.

$$2H_2O \rightarrow 4H^+ + 4e^- + O_2 \quad (2.1)$$

$$2H_2O + 2NADP^+ + 8\text{ photons} \rightarrow O_2 + 2NAD(P)H + 2H^+ \quad (2.2)$$

2.3.2 *Calvin Cycle*

CO_2 is reduced as carbohydrates into the algal cells during the light independent phase through the process called Calvin cycle. Calvin cycle takes place in the stroma of chloroplast and all the participating proteins are found residing outside the thylakoid membrane in the aqueous phase (Kumar and Das, 2014; Whitmarsh and Govindjee, 1999). Calvin cycle has broadly three steps: carboxylation, reduction and regeneration. Several complex reactions are involved in the Calvin cycle. However, only major reactions are shown in the diagram (Fig. 2.1).

Ribulose-1,5-bisphosphate (RuBP), a 5-carbon compound, is the starting compound of the Calvin cycle (Calvin and Benson, 1948). Ribulose 1,5-bisphosphate carboxylase(RuBisCO) catalyzed the carboxylation reaction between RuBP and CO_2 forming two molecules of glycerate 3-phosphate. In the reduction reaction, glycerate 3-phosphate is reduced to glyceraldehyde 3-phosphate (Eq. 2.3). It is followed by regeneration reaction, where five molecules of glyceraldehyde 3-phosphate is needed to regenerate one molecule of RuBP to start the Calvin cycle again and remaining one molecule is channeled for the synthesis of cellular biosynthetic materials for immediate energy source. Carbohydrates such as sucrose is transported to

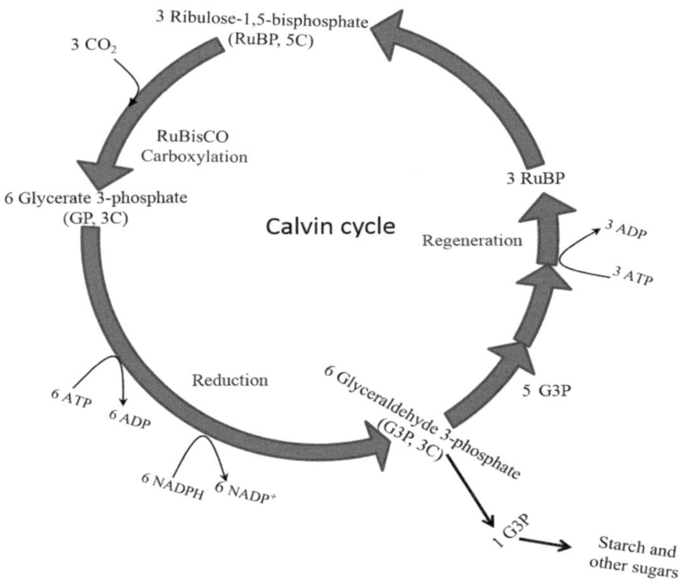

Fig. 2.1 Schematic diagram of the Calvin cycle. Only major intermediates have been shown. (Modified from Kumar and Das, 2014)

the cytosol, where these are stored in the form of glycogen and starch in cyanobacteria and green algae, respectively (Kumar and Das, 2014).

$$\text{Ribulose 1,5bisphosphate} + CO_2 + H_2O \rightarrow 2 \text{ glyceraldehyde } 3-\text{phosphate} \quad (2.3)$$

Calvin cycle is the most energy intensive processes among all the known pathways of CO_2 capture (Fast and Papoutsakis, 2012). It requires seven molecules of ATP and four molecules of NAD(P)H to transform two molecules of CO_2 into acetyl–CoA through the Calvin cycle as shown in Eq. 2.4 (Kumar and Das, 2014; Fast and Papoutsakis, 2012).

$$2\ CO_2 + 7\ \text{ATP} + 4\ \text{NAD(P)H} + \text{HS}-\text{CoA} \rightarrow \text{Acetyl}-\text{CoA} + 7\text{ADP} + 4\text{NAD(P)}^{+} \quad (2.4)$$

One reduced CO_2 molecule stores energy equivalent to 1400 kJ, which is produced at the expense of two moles of NAD(P)H. A synthesis of two moles of NAD(P)H in the light dependent phase requires eight moles of photons containing total energy of 3840 kJ mol^{-1}. In this way, maximum theoretical efficiency of the process can be calculated as 36.5 % (Kumar and Das, 2014).

2.3.3 *CO_2 Concentrating Mechanisms*

Cells of microalgae and cyanobacteria are simple in structure, due to which they cannot stop the diffusion of CO_2 to outside (Kumar and Das, 2014). They have low affinity for CO_2. In addition, CO_2 diffuses very slowly (nearly 10,000 times) in the aqueous solution compared to that in air (Kumar and Das, 2014). These factors limit the CO_2 availability to the RuBisCO to discharge carboxylase activity of fixing CO_2 into cellular components. In fact, under normal atmospheric condition, RuBisCO is only half saturated with the CO_2. Most of the algal cells have developed carbon concentrating mechanisms (CCMs) to cope with the low partial pressure of CO_2 outside the cell. The purpose of CCMs is to enhance the local partial pressure of CO_2 (up to 1000 times higher than the outside culture) near the RuBisCO (carboxysome in cyanobacteria and pyrenoid in microalgae). The affinity constant (K_m) of cyanobacteria and microalgae for CO_2 are generally ≥200 μM and ~20 μM, respectively. This may be the reason of need to store large amount of bicarbonate within the cells of cyanobacteria (100 fold) compared to microalgae (20 fold).

CO_2 is available inside the liquid culture in the form of gaseous dissolved CO_2 and different other chemical forms such as carbonic acids (H_2CO_3), bicarbonate (HCO_3^-), carbonate (CO_3^{2-}) depending upon the pH of the culture. Most of algal species were found to assimilate both CO_2 or HCO_3^-. However, only HCO_3^- is delivered into the cytosol regardless of the chemical species assimilated by the whole cells. This may have several reasons. For example, intracellular pH of cytoplasm is near 8, hence HCO_3^- predominant inside the cytoplasm. HCO_3^-, being a charged species, is unable to diffuse out the lipid bilayer of the cell membrane. CCMs have been extensively studied in the cyanobacteria. DIC is transported across the membranes by several mechanisms: transport of HCO_3^- into the cytosol by the ABC type transporter utilizing ATP, transport of HCO_3^- into the cytosol with the help of HCO_3^-/Na^+symporter or Na^+/H^+antiporter, transport of CO_2 by NADH dehydrogenase having constitutively low or inducible high affinity for CO_2. Besides these, HCO_3^- can have direct access to the cytoplasm; however its transfer rate across the membrane does not contribute significantly in the DIC accumulation compared to its active transport (Kumar and Das, 2014). The HCO_3^- transported to cytosol is further delivered into the carboxysome/pyrenoids. The pyrenoid is a densely packed compartment in the chloroplast containing RuBisCO. Both the carboxysome and pyrenoids have limited permeability of CO_2 and act as a store house for CO_2. Cyanobacteria have several carboxysomes (5 to 20 in number) into each cell depending upon species and growth conditions. RuBisCO enzymes have affinity only for gaseous CO_2. Therefore, inorganic carbon must be converted back into gaseous CO_2. Carbonic anhydrase (CA) does this job and catalyses the interconversion of inorganic carbon back to gaseous CO_2 as shown in Eq. 2.5. Under low partial pressure of CO_2 outside the algal cells, CA over-expresses to maintain high local CO_2 concentration. High concentration of local CO_2 activates the carboxylase activity of RuBisCO and fix the CO_2 more efficiently. CCMs have been less studied in the microalgae compared to cyanobacteria. Contrary to HCO_3^- in cyanobacteria, CO_2 is the predominant species entering into the microalgae such as *Chlamydomonas reinhardtii* (Spalding, 2008). The uptake of CO_2 into the whole cell and through the chloroplast of microalgae is due to diffusion and mediated transfer, respectively (Kumar and Das, 2014).

$$HCO_3^- + H^+ \xleftrightarrow{CA} CO_2 + H_2O \tag{2.5}$$

2.3.4 CO_2 Fixation Through Algal Biomass Production

The CO_2 is fixed during algal growth through several processes such as biomass formation, mineralization (transformation of gaseous CO_2 into chemical species such as bicarbonates, carbonates) and production of extracellular products such as polysaccharides, volatile organic compounds, organohalogens, hormones etc. (Sydney et al., 2014). Among them, biomass formation is the major factor, which was found to account for 70–88 % of total CO_2 fixation during algal cultivation (Sydney et al., 2014). CO_2 fixed during other processes are wasted or easily lost to the environment. Therefore, most of the researchers quantify the CO_2 fixation taking only biomass formation into account. The CO_2 fixation ability of microalgae and cyanobacteria varies (Table 2.1).

Table 2.1 Biomass productivity and CO_2 fixation rate of different microalgae

Algal species	Photobioreactor (volume)	Biomass productivity (mg L^{-1} d^{-1})	CO_2 fixation (mg L^{-1} d^{-1})	References
Anabaena sp. ATCC 33047	Bubble column	310	1450	Lopez et al., 2009
Aphanothecemicroscopica	Bubble column (3.2 L)	301	562	Jacob-Lopes et al., 2009
Spirulina platensis	BioFlowfermentor (11 L)	156	319	Sydney et al., 2010
Spirulina sp.	Vertical tubular (1.8 L)	280	–	de Morais and Costa, 2007
Synechocystis aquaticlis	Vertical flat plate (24 L)	78	128	Zhang et al., 2001
Botryococcusbraunii	BioFlowfermentor (11 L)	207	497	Sydney et al., 2010
	–	27	–	Yoo et al., 2010
Chlorella vulgaris	BioFlowfermentor (11 L)	129	252	Sydney et al., 2010
		298	624	Yun et al., 1997
	–	105	–	Yoo et al., 2010
Chlorella sorokiniana	Airlift reactor (1.4 L)	338	619	Kumar and Das, 2012
Chlorococcum littorale	Vertical tubular (20 L)	530	900	Kurano et al., 1995
	Vertical tubular (15 L)	120	200	Yang et al., 2013
Scenedesmusobliquus	–	218	–	Yoo et al., 2010
	Erlenmeyer flask (0.2 L)	142	253	Basu et al., 2013
	Vertical tubular (1.8 L)	160	–	de Morais and Costa, 2007
Dunaliellatertiolecta	BioFlowfermentor (11 L)	143	272	Sydney et al., 2010
Nannochloropsisoculata	Cylindrical (0.8 L)	480	902	Chiu et al., 2009

The elemental composition of biomass of algal species such as *Chlorella* sp. changes depending upon experimental conditions such as the concentration of CO_2 used for growing algae. C, H, N and S were reported in the range of 46.1 %–50.1 %, 6.1 %–7.74 %, 6.7 %–8.52 % and <0.5 % (w/w), respectively (Kumar et al., 2014; Rizzo et al., 2013; Friis et al., 1998). The molecular formula of the algal biomass of microalgae can be determined from the relative percent of C, H and N present in the algal biomass and can be represented in the form of $CH_xN_yO_z$. The corresponding molecular weight of algal biomass of *C. sorokiniana* was in the range of 24–26 g (Kumar et al., 2014). For example, 24.04 g per mole of biomass of *C. sorokiniana* was obtained, when the air was the sole carbon source. Material balance can be written in the form of Eq. 2.6 (Kumar et al., 2014).

$$CO_2 + 0.92\ H_2O + 0.14\ NO_3^- \rightarrow CH_{1.84}N_{0.14}O_{0.51} + 1.415\ O_2 \tag{2.6}$$

Assuming 50 % carbon content in the algal cells, 1.83 kg of CO_2 from the air is fixed for the production of one kg of algal biomass along with the release of nearly 1.9 kg of oxygen (Kumar et al., 2014; Kumar and Das, 2012).

2.4 Bottlenecks in the Algal Biomass Production

2.4.1 Temperature

Temperature is one of the most important factors influencing all the biological systems including algae. Each of the microalgal species has an optimal temperature at which they grow most suitably. The temperature affects the cellular enzymatic reactions, including the cytosolic pH. The optimal temperature of most of the algal species varies in the range of 25–35 °C. Algal growth is more sensitive to high temperature compared to low temperature. In fact, a little increase over the optimum temperature may severely affect the algal growth. The low temperature reduces the cellular activities such as limiting the electron transport chain (ETC), activity of CA etc. (Kumar et al., 2014). The high temperature affects the CO_2 fixation due to inhibitory effects on the cellular physiology and denaturing of essential proteins/enzymes. For example, photosystem II is very thermolabile and prone to denature at higher temperature. Besides these, an increase in photorespiration, a decrease in RuBisCO affinity for CO_2 and a decrease in CO_2 to O_2 solubility ratio are the other vital phenomena negatively affecting the CO_2 fixation at higher temperature (Kumar et al., 2014; Kumar et al., 2011).

The target of most of the CO_2 capture using microalgae is from the flue gas, generally having higher temperature. In addition, outdoor cultivation of algae is influenced by the local climatic temperature, which is higher for the countries closer to the equator and lower for the countries located far away from the equator. Further, the temperature exposed to algal cells depends upon the time of light exposure and the season variations. The use of thermophilic algae is an alternative for the fixation of CO_2 coming from the flue gas having high temperature.

2.4.2 *pH*

Most of the algae has neutral or slightly alkaline cytosolic pH as cellular enzymes are pH sensitive and become inactive at acidic conditions (Kumar et al., 2014). Therefore, the optimal pH of most of the algal species has been found in the range of 7 to 9. The extreme pH negatively affects the growth and CO_2 fixation by disrupting cellular processes. The introduction of CO_2 and other acidic gases such as Sox and NOx (generally present in the flue gas) in the culture medium decreases the pH. CO_2 dissolves in water by forming carbonic acids. The availability of carbonaceous species (aqueous CO_2, carbonic acids, bicarbonate and carbonate) to the algal cells depends upon the pH of the culture as the equilibrium shifts among them on change in the culture pH.

Along with the growth of the algae, pH generally increases. This may be due to consumption of CO_2 and other volatile fatty acids (if present in the medium). The rise in pH along the growth is higher at the inlet gas stream with low CO_2 concentration. This may be because of the release of hydroxide ions outside the cell at the expense of capture of H^+ ions inside the thylakoid membranes during the activation of CCMs (Kumar and Das, 2012; Jacob-Lopes et al., 2008). The presence of nitrogen source in the medium also influences the pH of the culture broth. For example, the consumption of nitrate from the medium helps in the increase of alkalinity via OH^- production (Kumar et al., 2014; Hulatt and Thomas, 2011). Contrary to this, uptake of NH_4^+ by the microalgae leads to H^+ production (Goldman and Brewer, 1980).

2.4.3 *Light*

Algae fix the CO_2 in the presence of light during photosynthesis. Therefore, light is crucial in CO_2 capture through algal biomass production. Algal cells in the culture can experience broadly three different phases of light such as light limitation, light saturation and light inhibition (Kumar et al., 2013). The light utilization efficiency and biomass productivity proportionally increases with the light intensity till the cells become light saturated. The light saturation value of algae is in the range of 80–200 μ molphotos m^{-2} s^{-1} (Kumar et al., 2014).

At low light intensity, photosynthesis becomes limiting. Lee and Pirt (1981) reported an increase in the cellular maintenance requirements during long time exposure of dark, which causes reduction in the biomass productivity. At high light intensity, photosynthesis is disrupted due to photoinhibition. The excess light energy is dissipated as fluorescence or heat through non-photochemical quenching (NPQ)

(Kumar et al., 2014; Yamakawa and Itoh, 2013). The photoinhibition disrupts the synthesis and degradation cycle (D1/D2 proteins) of light harvesting complex at PSII, eventually leading to inhibition of photosynthesis (Kumar et al., 2011). Along with the quantity of light, quality of light (light of a specific wavelength or light regime) also affects the algal biomass production. For example, in a study by Kim et al. (2014), the red light regime was found favoring the CO_2 fixation in *N. gaditana* followed by blue and white light regime. The pigmentation of algal cells exponentially reduces the light intensity inside the culture broth. The reduction of the size of the antenna molecules of algal cells (antenna size mutant) by molecular tools was found an effective way to decrease the pigment content and thus increasing the extent of light penetration inside the algal culture. This will eventually lead to better light energy utilization causing higher CO_2 fixation and biomass productivity. Proper mixing and efficient photobioreactor design are the other strategies for the effective distribution of light energy inside the algal culture and to improve the light utilization efficiency (Kumar et al., 2011).

2.4.4 Mixing

The importance of CO_2 for algal cultivation can be understood by the fact that the cost of the carbon accounts for 8–27 % of the total production cost of the algal biomass (Li et al., 2013). The bottlenecks of CO_2 transfer to the algal cells is due to higher resistance to the mass transfer of CO_2. RuBisCO is the crucial enzyme for the Calvin cycle used in CO_2 fixation. However, RuBisCO has low affinity for CO_2 compared with O_2 (Kumar and Das, 2013; Kumar et al., 2011). Moreover, CO_2 gas faces resistance to transfer from the gas phase to the liquid phase and from the liquid phase to inside the algal cells. Therefore, adequate supply of CO_2 in the culture is necessary. It has been estimated that CO_2 concentration greater than 65 μmol L^{-1} at a pH of 8.5 is necessary for the optimal productivity of majority of microalgae (Weissman et al., 1988). However, the dissolved CO_2 concentration in water in equilibrium with air is nearly 10 μmol L^{-1} at 25 °C.

2.4.5 Substrate Inhibition

An optimal CO_2 concentration in the inlet gas stream increases the specific growth rate and enhances the biomass productivity. The optimal CO_2 concentration was found in the range of 2 to 10 % in most of the microalgae. However, ability to tolerate CO_2 can be as high as 100 %. High CO_2 concentration causes substrate inhibition and environmental stress causing reduction in the algal growth. In addition to the environmental stress, high CO_2 concentration decreases the pH of the culture. Algae has higher CO_2 tolerance capability compared to terrestrial plants and therefore more effective in CO_2 sequestration. CO_2 tolerance ability of algal cells can be enhanced by the method of adaptation in gradually higher CO_2 concentration. According to Miyairi (1995) drop in pH is less at higher temperature due to decrease in the solubility of CO_2 in the culture.

2.4.6 *Product Inhibition*

2.4.6.1 Biomass Concentration

Cell growth causes exponential decrease in the light penetration inside the photobioreactor due to self-shading effect by the microalgal cells. This is similar to the product inhibition. At fixed light intensity, algal cells may be in a light inhibition zone at low cell density and light limiting phase at higher cell density. The biomass concentration in the range of 1–2 g L^{-1} was found optimum taking both biomass productivity and self-shading effect into consideration (Hu et al., 1996; Zhang et al., 2001). The continuous or semi-continuous method of algal cultivation may be an effective way to maintain optimum cell density.

2.4.6.2 Oxygen Accumulation

Accumulation of high concentration of oxygen leads to photooxidative damage to the algal cells. It has been reported that maximum tolerable dissolved oxygen concentration should not increase more than 400 % (>35 mg L^{-1}) of air saturation value (Kumar et al., 2011; Carvalho et al., 2006; Lee and Lee, 2003). RuBisCO enzyme has affinity for oxygen as well as carbon dioxide. In fact, affinity constant (K_m) for oxygenase activity is significantly lower than that of carboxylase activity. Therefore, RuBisCO participate in the CO_2 fixation only when the CO_2 concentration in the compartment is significantly higher. In case of oxygenase reaction, algal cells negatively affect the CO_2 fixation. For example, glycolate 2-phosphate is the end product when oxygenase dependent reaction of RuBisCO is active (Eq. 2.7). A significant amount of cellular energy is wasted during the formation of glycolate 2-phosphate, which is not required by algal cells. Further, glycolate 2-phosphate breaks down into glycine, which forms serine by combining with another glycine molecule resulting in a loss of CO_{2-} (Kumar and Das, 2014; Zeng et al., 2011). The loss of CO_2 may be up to 50 % of the algal biomass (Giordano et al., 2005). Furthermore, RuBP synthesis is affected because of loss of fixed carbon.

$$\text{Ribulose 1,5bisphosphate} + O_2 \rightarrow \text{Glycerate 3} - \text{phosphate} + \text{Glycolate 2} - \text{phosphate} \tag{2.7}$$

Shapes and sizes of the photobioreactor greatly influence the rise in dissolved oxygen (DO). For example, an increase in DO is a severe problem for the horizontal tubular photobioreactor but for the photobioreactor such as airlift, it is not a critical issue due to having an open gas disengagement zone (Kumar and Das, 2014; Kumar et al., 2011; Lee and Lee, 2003).

2.5 Algal Biomass Cultivation System

2.5.1 Open Ponds

Open ponds have shallow configuration, which allow maximum light penetration inside the culture. Open ponds can be categorized into four types: shallow big ponds, tanks, circular ponds and raceway ponds (Rao et al., 2012). Among them, raceway pond is considered as an attractive cultivation system. This is because of ease of mixing and better biomass productivity. In fact, nearly 95 % of the total world algal biomass facility is based on the raceway ponds (Mendoza et al., 2013). The large scale production of microalgae was started by Oswald and Golueke (1960) in the 1950s in a raceway pond using *Chlorella* sp.

Algal cultivation in open ponds has several advantages such as ease in operational handling, practical way of scaling up microalgal production facility, low initial investment cost, free solar radiation, low energy requirement for mixing etc. (Jorquera et al., 2010; Mendoza et al., 2013). The product extracted from the algal biomass obtained from the open ponds are comparatively cheaper than the photobioreactor. For example, the cost of biodiesel obtained from raceway ponds was calculated as 12.73 USD per gallon, which was nearly 2.5 times lower to that of photobioreactor (Richardson et al., 2012). Open ponds also have several drawbacks: areal biomass productivity is in the range of 0.12–0.48 g L^{-1} d^{-1}, which is nearly 2 to 8 times lower than the photobioreactor (Ketheesan and Nirmalakhandan, 2012; Brennan and Owende, 2010). Algal culture is susceptible to contamination, which decreases the biomass productivity and makes algal biomass unfit for the extraction of high value product. Only few selected algal species such as *Chlorella, Haematococcus, Dunaliella, Nannochloropsis, Spirulina,* etc. were found successful in growing in the open system. Further, the shallow depth of the open ponds allows low residence time to the introduced CO_2 gas, which in turn reduces CO_2 transformation efficiency into cellular constituents. In fact, nearly 80–90 % of the sparged gas is lost to the atmosphere (Richmond, 2004; Weissman et al., 1988). Besides these, open system has disadvantages in controlling the physico-chemical parameters such as availability of light, agitation, pH, temperature and nutrient concentrations (Kumar et al., 2011). The fluctuation in temperature and light availability during diurnal cycles and seasonal variations are other challenges with the open systems (Brennan and Owende, 2010).

Open ponds are generally light and carbon limited. The pond depth of the culture influence the light utilization efficiency, temperature variation in the culture, mixing (especially vertical mixing) and associated power consumption by the paddle-wheel. An increase in the pond depth enhances the total areal productivity of the algal biomass. For example, nearly 134–200 % increase in the areal productivity was obtained, when pond depth was 40 cm compared to 20 cm deep raceway ponds

(Sutherland et al., 2014). Similarly, mixing is very important for algal biomass productivity as it determines the nutrient and light availability to the algal cells. It has been found that optimum mixing enhances the biomass productivity by nearly 10 fold (Hreiz et al., 2014). The vertical mixing in the raceway ponds is more important as it limits the culture depth and hence the total areal biomass productivity. The vertical mixing causes cells to move from bottom (dark zone) to surface (light zone) of the open ponds. Therefore, it ensures the frequency by which each of the algal cells experience solar light exposure. Further, vertical mixing also prevents cells from either photolimitation or photoinhibition (Chiaramonti et al., 2013).

2.5.2 *General Raceway Ponds in Operation*

2.5.2.1 The Paddle-Wheel Driven Raceway Ponds

Paddle-wheels are mechanically simple to construct, require little maintenance and generate gentle mixing. These are traditionally used in the raceway ponds. In the raceway ponds, clearance zone between paddle wheel and the bottom floor is kept small (Chiaramonti et al., 2013). In this way, paddle-wheel acts as a positive displacement pump due to no backflow of the liquid. A wide recirculation flow of liquid can be visualized long after the bend of the raceway ponds showing the effectiveness of the raceway ponds (Chiaramonti et al., 2013). However, paddle-wheel driven raceway ponds have disadvantage of high power consumption and low vertical mixing (Xu et al., 2014). Paddle-wheel driven raceway pond in operation is shown in Fig. 2.2.

Fig. 2.2 Raceway pond for the large scale microalgae cultivation (200 m^3) at HighTech Algae Center, Hadong, South Korea.

2.5.2.2 Sump-Assisted Raceway Ponds

Most of the design of the raceway ponds is focused on the enhancement of CO_2-liquid contact time. For example, a sump was incorporated in the raceway ponds and gas bubbles were introduced at the bottom of the sump (Mendoza et al., 2013). Sump is a long tubular dig, at the bottom of which gas bubbles are introduced. In this way, CO_2 bubbles have to diffuse out into the atmosphere by passing through a long water column. This increases the CO_2-liquid contact time and thus better CO_2 utilization efficiency and minimum CO_2 loss into the atmosphere (Craggs et al., 2012). Areal biomass productivity in the sump-assisted raceway pond was found 17 g m^{-2} d^{-1} with CO_2 capture of nearly 66 % from the inlet gas stream (deGodos et al. 2014).

2.5.2.3 Airlift/Split Sump-Assisted Raceway Ponds

The sump-assisted raceway pond was further modified by introducing baffles in the sump. Baffle partitioned the sump into two halves. In this way, this acts as an airlift reactor with one side as riser and the other side as downcomer. The CO_2-air gas mixture is introduced from the riser. The density difference between the two portions of the split sump was the driving force for the liquid movement in the raceway ponds. For further enhancing the CO_2-liquid contact time, normal air was used to circulate and generate culture velocity to the liquid and CO_2 was introduced in the counter-current direction at half the height of the downcomer (deGodos et al., 2014). The airlift driven raceway pond was compared with the traditional paddle-wheel driven raceway ponds. Ketheesan and Nirmalakhandan (2012) claimed the reduction of nearly 80 % energy consumption in airlift-driven raceway ponds at same culture velocity.

2.5.3 *Photobioreactors*

Microalgae are cultivated in photobioreactor to overcome the bottlenecks of the open ponds. Better control of process parameters such as pH, temperature, mixing, light penetration etc., maintaining axenic culture and necessity to obtain higher biomass productivity are some of the advantages of growing algae in the photobioreactor. Photobioreactors are the closed system, similar to bioreactors with additional provision of supply of light into the photobioreacotr. This requires a high surface area to volume ratio (S/V ratio) and transparent surface of the photobioreactor. Necessity to consider these two criteria make photobioreactors difficult to scale-up. Photobioreactors are generally made up of transparent materials such as glass, plexiglass or polycarbonate (Kumar et al., 2011). Closed surface of the photobioreactor helps in introducing the gaseous CO_2 in a controlled way and with increased CO_2 gas-liquid contact time. Higher residence time of the CO_2 into the photobioreactor enhances its mass transfer and correspondingly CO_2 sequestration efficiency and algal biomass productivity. Photobioreactors designed for CO_2 sequestration take advantage of using gaseous CO_2 for the supply of carbon source as well as ensuring sufficient mixing. However, in some photobioreactors such as stirred tank, agitation is provided by mechanical devices as well as sparging with CO_2-rich gas. It is always desirable to introduce the CO_2-rich gas mixture in order to reduce energy consumption. Mixing with CO_2 rich gas stream also helps in the removal of oxygen generated during the algal growth.

2.5.3.1 Vertical Tubular Photobioreactor

Vertical tubular photobioreactors are most commonly used for the CO_2 sequestration. Height to diameter ratio is generally kept greater than two (Kumar et al., 2011). The CO_2-rich gas stream is passed at the bottom of the cylindrical vessel through sparger. Long liquid column in the cylindrical vessel increases the CO_2-liquid contact time. Very tiny gas bubbles can be produced by using sparger with suitable pore size, which further enhances the volumetric mass transfer of gaseous CO_2. Advantages of vertical tubular photobioreactors are in low infrastructure cost, ease in the design and handle, absence of moving parts, relatively homogeneous culture environment, higher S/V ratio and efficient removal of unused gas mixture and photosynthetically produced O_2 (Kumar et al., 2011). Working volume of this type of reactor is generally 80 % of the total volume. Based on the placement of the light source, vertical tubular photobioreactors can be called as externally or internally illuminated photobioreactor.

Airlift and bubble column photobioreactors are the well known examples of vertical tubular photobioreactor. Airlift and bubble column are different from each other due to difference in mixing (Kumar and Das, 2014). Airlift photobioreactors have additionally draft tube placed at the center of the bubble column. In this way, airlift reactor has two different zones: one light zone (annular area) and another dark zone (central area). The CO_2-rich gas stream can be passed either through a central tube or through the annular space. The zone of airlift, where sparger is placed, is named as riser whereas the other one is called downcomer (Kumar and Das, 2013; Kumar and Das, 2014). Riser is also called gassed zone, where liquid has less density compared to downcomer. The density difference between the two sides drives the liquid culture in axial circular and oriented motion along the length of the photobioreactor. The direction of movement of the culture broth is from riser to downcomer. It is to be noted that maintaining the liquid level above the draft tube is necessary in order to experience the axial movement of liquid. Algal cells periodically move through the light and dark zone due to elliptical movement of the culture along the length of the reactor (Kumar and Das, 2013; Kumar et al., 2011). This gives flashing light effect to algal cells (Barbosa et al., 2003). The liquid moves in zig-zag and haphazardly on removing the draft tube (bubble column), leaving many dead zones in the photobioreactor. Vertical tubular photobioreactors have the disadvantages of poor control of temperature and low surface area for receiving the sunlight.

2.5.3.2 Horizontal Tubular Photobioreactor

Horizontal tubular reactors are long tubular tubes placed horizontally generally made up of transparent polypropylene acrylic or polyvinylchloride pipes with small internal diameter. The CO_2-rich gas stream is introduced from one side of the tube through the dedicated gas exchange unit. The longer length of pipes gives higher residence time to the CO_2 gas bubbles. This type of reactor is mostly used for algal cultivation in the outdoor. Parallel tubes are oriented in such a way to receive a maximum of solar light. Disadvantages of this type of photobioreactor are poor gas disengagement, wall growth and cleaning problem. This becomes more severe as the length of the tubes increases. Oxygen produced during the photosynthesis get trapped inside the photobioreactor. Lee and Lee (2003) observed an increase in photosynthetically produced DO concentration up to 400 % in horizontal tubular photobioreactor. Temperature is controlled by sparging the cold water over the surface of the tubes or overlapping tubes of cold water and photobioreactor or putting the light harvesting unit in the pool of cold water (Kumar and Das, 2014).

2.5.3.3 Helical Tubular Photobioreactor

Helical tubular photobioreactor is a flexible, transparent and a coiled shape structure placed vertically. It has separate gas engagement unit attached at the top of the coil. The CO_2-rich gas stream is passed from the bottom opening of the helical reactor. The long helical unit gives a high S/V ratio. In addition, CO_2 gas has high residence time inside the reactor, which increases the CO_2 sequestration efficiency. The light source can be placed co-axially at the center of the photobioreactor to minimize the loss of light energy and better biomass productivity. Helical photobioreactor can be scaled-up by increasing the number of coils in it. However, accumulation of oxygen increases proportionally with the increase in the coil of helical tubular photobioreactor (Kumar and Das, 2014). It has the advantage of requiring a small land footprint. Helical tubular photobioreactor suffers from many disadvantages such as fouling inside the tube, shear stress during the introducing of the gas, trapping of photosynthetically produced oxygen, most importantly design and handling of the photobioreactor. Helical reactors were further modified and transformed into a cone shape for better solar light utilization efficiency (Kumar and Das, 2014; Kumar et al., 2011; Watanabe and Hall, 1996).

2.5.3.4 Flat Panel Photobioreactor

The flat panel reactor has a cuboidal shape with flat surface at two sides for receiving the light. Most of the studies are focused on the flat panel photobioreactor. It has a high S/V ratio and better illuminated light path length. The agitation is carried out by the CO_2-rich gas stream, which is introduced from the bottom through tubes with fine holes. Some researchers placed the tubes upto half of the length of the photobioreactor to induce spiral mixing pattern. Flat panel photobioreactors have been further modified by putting draft tubes inside it. Similar to airlift photobioreacter, flat panel photobioreactor can have the riser and downcomer. The simple arrangement of flat plate reactor makes it easy to construct with any desired light path length. However, wall growth, difficulty in cleaning and low biomass productivity per unit space are some of the problems with this type of reactor. An increase in the hydrostatic pressure with the increase in the volume of the reactor during scale-up is another problem associated with this type of arrangement.

2.6 Conclusions

Large scale microalgal cultivation has potential to sequester significant amounts of CO_2 through biomass production and thus contributing in reducing the global warming effect. Several physico-chemical parameters influence the algal biomass productivity. Therefore, it is important to optimize these parameters and grow the microalgae at optimal environmental conditions. All the cultivation systems have their own merits and demerits. The raceway ponds are more suited for the cheap and large scale algal cultivation. Closed photobioreactor is used for growing the algae in controlled physico-chemical conditions, which will result into better biomass productivity. Various shapes and design of the closed photobioreactor have an influence on the algal biomass productivity. In addition, longer residence time of CO_2 in the closed photobioreactor can improve the CO_2 sequestration efficiency of microalgae. However, the cost effective biomass production in large scale is yet to achieve in the closed photobioreactor.

Acknowledgements This work was supported by the Advanced Biomass R&D Center (ABC) of Korea Grant funded by the Ministry of Science, ICT and Future Planning (ABC-2010-0029728) and the New & Renewable Energy of the Korea Institute of Energy Technology Evaluation and Planning (KETEP) Grant funded by the Korea government Ministry of Knowledge Economy (2012T100201665).

References

Allen, D.E., Strazisar, B.R., Soong, Y. and Hedges, S.W. (2005). Modeling carbon dioxide sequestration in saline aquifers: Significance of elevated pressures and salinities. *Fuel Process Technol.*, 86(14), 1569–1580.

Bachu, S. (2000). Sequestration of CO_2 in geological media: Criteria and approach for site selection in response to climate change. *Energy Convers. Mgmt.*, 41, 953–970.

Barbosa, M.J., Janssen, M., Ham, N., Tramper, J. and Wijffels, R.H. (2003). Microalgae cultivation in airlift reactors: Modeling biomass yield and growth rate as a function of mixing frequency. *Biotechnol. Bioeng.*, 82(2), 170–179.

Bardgett, R.D., Freeman, C. and Ostle, N.J. (2008). Microbial contributions to climate change through carbon cycle feedbacks. *The ISME Journal*, 2, 805–814.

Basu, S., Roy, A.S., Mohanty, K. and Ghoshal, A.K. (2013). Enhanced CO_2 sequestration by a novel microalga: *Scenedesmus obliquus* SA1 isolated from bio-diversity hotspot region of Assam, India. *Bioresour. Technol.*, 143, 369–377.

Brennan, L. and Owende, P. (2010). Biofuels from microalgae – A review of technologies for production, processing and extractions of biofuels and co-products. *Renew. Sustain. Energy Rev.*, 14, 217–232.

Calvin, M. and Benson, A.A. (1948). The path of carbon in photosynthesis. *Science*, 107, 476–480.

Carvalho, A.P., Luis, A., Meireles, A. and Malcata, F.X. (2006). Microalgal reactors: A review of enclosed system designs and performances. *Biotechnol. Prog.*, 22, 1490–1506.

Chiaramonti, D., Prussi, M., Casini, D., Tredici, M.R., Rodolfi, L., Bassi, N., Zittelli, G.C. and Bondioli, P. (2013). Review of energy balance in raceway ponds for microalgae cultivation: Re-thinking a traditional system is possible. *Appl. Energy*, 102, 101–111.

Chiu, S.Y., Kao, C.Y., Tsai, M.T., Ong, S.C., Chen, C.H. and Lin, C.S. (2009). Lipid accumulation and CO_2 utilization of *Nannochloropsis oculata* in response to CO_2 aeration. *Bioresour. Technol.*, 100(2), 833–838.

Cordero, B.F., Obraztsova, I., Couso, I., Leon, R., Vargas, M.A. and Rodrigue, H. (2011). Enhancement of lutein production in *Chlorella sorokiniana* (Chlorophyta) by improvement of culture conditions and random mutagenesis. *Mar. Drugs.*, 9, 1607–1624.

Craggs, R., Sutherland, D. and Campbell, H. (2012). Hectare-scale demonstration of high rate algal ponds for enhanced wastewater treatment and biofuel production. *J. Appl. Phycol.*, 24, 329–337.

Dayananda, C., Sarada, R., Kumar, V. and Ravishankar, G.A. (2007). Isolation and characterization of hydrocarbon producing green alga *Bortyococcus braunii* from Indian freshwater bodies. *Electron. J. Biotechnol.*, 10, 78–91.

deGodos, I., Mendoza, J.L., Acién, F.G., Molina, E., Banks, C.J., Heaven, S. and Rogalla, F. (2014). Evaluation of carbon dioxide mass transfer in raceway reactors for microalgae culture using flue gases. *Bioresour. Technol.*, 153, 307–314.

de Morais, M.G. and Costa, J.A.V. (2007). Carbon dioxide fixation by *Chlorella kessleri*, *C. vulgaris*, *Scenedesmus obliquus* and *Spirulina* sp. cultivated in flasks and vertical tubular photobioreactors. *Biotechnol. Lett.*, 29, 1349–1352.

Falkowski, P., Scholes, R.J., Boyle, E., Canadell, J., Canfield, D., Elser, J., Gruber, N., Hibbard, K., Hogberg, P., Linder, S., Mackenzie, F.T., Moore III, B., Pedersen,T., Rosenthal, Y., Seitzinger, S., Smetacek, V. and Steffen, W. (2000). The Global Carbon Cycle: A Test of Our Knowledge of Earth as a System. *Science*, 290, 291–296.

Fast, A.G. and Papoutsakis, E.T. (2012). Stoichiometric and energetic analyses of non-photosynthetic CO_2-fixation pathways to support synthetic biology strategies for production of fuels and chemicals. *Curr. Opin. Chem. Eng.*, 1, 1–16.

Friis, J.C., Holm, C. and Sorensen, B.H. (1998). Evaluation of elemental composition of algal biomass as toxical endpoint. *Chemosphere.*, 37(13), 2665–2676.

Geng, L., Zhang, J., Qin, B. and Yang, Z. (2014). No difference in colony formation of *Scenedesmus obliquus* exposed to lakes with different nutrient levels. *Biochem. Sys. Ecol.*, 57, 178–182.

Giordano, M., Beardall, J. and Raven, J.A. (2005). Mechanisms in algae: Mechanisms, environmental modulation and evolution. *Annu. Rev. Plant Biol.*, 56, 99–131.

Goldman, J. and Brewer, P. (1980). Effect of nitrogen source and growth rate on phytoplankton-mediated changes in alkalinity. *Limnol. Oceanogr.*, 25, 352–357.

González-López, C. V., Acién Fernández, F. G., Fernández-Sevilla, J. M., Sánchez Fernández, J. F., & Molina Grima, E. (2012). Development of a process for efficient use of CO2 from flue gases in the production of photosynthetic microorganisms. *Biotechnology and bioengineering*, *109*(7), 1637–1650.

Hreiz, R., Sialve, B., Morchain, J., Escudié, R., Steyer, J.-P. and Guiraud, P. (2014). Experimental and numerical investigation of hydrodynamics in raceway reactors used for algaculture. *Chem. Eng. J.*, 250, 230–239.

Hu, Q., Guterman, H. and Richmond, A. (1996). A flat inclined modular photobioreactor for outdoor mass cultivation of photoautotrophs. *Biotechnol. Bioeng.*, 51, 51–60.

Hulatt, C.J. and Thomas, D.N. (2011). Productivity, carbon dioxide uptake and net energy return of microalgal bubble column photobioreactors. *Bioresour. Technol.*, 102, 5775–5787.

Iwasaki, I., Hu, Q., Kurano, N. and Miyachi, S. (1998). Effect of extremely high-CO_2 stress on energy distribution between photosystem I and photosystem II in a 'high-CO_2' tolerant green alga, *Chlorococcum littorale* and the intolerant green alga *Stichococcus bacillaris*. *J Photochem. Photobiol. B-Biol.*, 44(3), 184–190.

Jacob-Lopes, E., Revah, S., Hernandez, S., Shirai, K. and Franco, T.T. (2009). Development of operational strategies to remove carbon dioxide in photobioreactors. *Chem. Eng. J.*, 153(1–3), 120–126.

Jacob-Lopes, E., Scoparo, C.H.G. and Franco, T.T. (2008). Rates of CO_2 removal by *Aphanothece microscopic* Nageli in tubular photobioreactors. *Chem. Eng. Prog.*, 47, 1365–1373.

Jorquera, O., Kiperstok, A., Sales, E.A., Embiruçu, M. and Ghirardi, M.L. (2010). Comparative energy life-cycle analyses of microalgal biomass production in open ponds and photobioreactors. *Bioresour. Technol.*, 101, 1406–1413.

Ketheesan, B. and Nirmalakhandan, N. (2012). Feasibility of microalgal cultivation in a pilot-scale airlift-driven raceway reactor. *Bioresour. Technol.*, 108, 196–202.

Kim, C.W., Sung, M-G., Nam, K., Moon, M., Kwon, J-H. and Yang, J-W. (2014). Effect of monochromatic illumination on lipid accumulation of *Nannochloropsis gaditana* under continuous cultivation. *Bioresour. Technol.*, 159, 30–35.

Kumar, K. and Das, D. (2012). Growth characteristics of *Chlorella sorokiniana* in airlift and bubble column photobioreactors. *Bioresour. Technol.*, 116, 307–313.

Kumar, K. and Das, D. (2013). CO_2 sequestration and hydrogen production using cyanobacteria and green algae. *In:* Razeghifard, R. (ed.), Natural and Artificial Photosynthesis Solar power as an Energy Source. John Wiley and Sons, Inc., pp. 173–215.

Kumar, K. and Das, D. (2014). Carbon Dioxide Sequestration by Biological Processes. *In:* Bhanage, B.M. and Arai, M. (eds), Transformation and Utilization of Carbon Dioxide. Springer, pp. 303–334.

Kumar, K., Dasgupta, C.N. and Das, D. (2014). Cell growth kinetics of *Chlorella sorokiniana* and nutritional values of its biomass. *Bioresour. Technol.*, 167, 358–366.

Kumar, K., Dasgupta, C.N., Nayak, BK., Lindblad, P. and Das, D. (2011). Development of suitable photobioreactors for CO_2 sequestration addressing global warming using green algae and cyanobacteria. *Bioresour. Technol.*, 102(8), 4945–4953.

Kumar, K., Sirasale, A. and Das, D. (2013). Use of image analysis tool for the development of light distribution pattern inside the photobioreactor for the algal cultivation. *Bioresour. Technol.*, 143, 88–95.

Kurano, N., Ikemoto, H., Miyashita, H., Hasegawa, T., Hata, H. and Miyachi, S. (1995). Fixation and Utilization of Carbon-Dioxide by Microalgal Photosynthesis. *Energy Convers. Mgmt.*, 36(6–9), 689–692.

Lee, J.S. and Lee, J.P. (2003). Review of advances in biological CO_2 mitigation technology. Biotechnol. *Bioprocess Eng*, 8, 354–359.

Lee, Y.K. and Pirt, S.J. (1981). Energetics of photosynthetic algal growth: Influence of intermittent illumination in short (40s) cycles. *J. Gen. Microbiol.*, 24, 43–52.

Li, S., Luo, S. and Guo, R. (2013). Efficiency of CO_2 fixation by microalgae in a closed raceway pond. *Bioresour. Technol.*, 136, 267–272.

Li, Y., Horsman, M., Wu, N., Lan, C.Q. and Dubois-Calero, N. (2008). Biofuels from microalgae. *Biotechnol. Prog.*, 24, 815–820.

Lopez, C.V.G., Fernandez, F.G.A., Sevilla, J.M.F., Fernandez, J.F.S., Garcia, M.C.C. and Grima, E.M. (2009). Utilization of the cyanobacteria *Anabaena* sp ATCC 33047 in CO_2 removal processes. *Bioresour. Technol.*, 100(23), 5904–5910.

Loubiere, K., Olivo, E., Bougaran, G., Pruvost, J., Robert, R. and Legrand, J. (2009). A new photobioreactor for continuous microalgal production in hatcheries based on external-loop airlift and swirling flow. *Biotech. Bioeng.*, 102(1), 132–147.

Lubian, L.M., Montero, O., Moreno-Garrido, I., Huertas, I.E., Sobrino, C., Gonzalez-del Valle, M. and Pares, G. (2000). *Nannochloropsis* (Eustigmatophyceae) as source of commercially valuable pigments. *J. Appl. Phycol.*, 12(3–5), 249–255.

Mendoza, J.L., Granados, M.R., de Godos, I., Acién, F.G., Molina, E., Banks, C. and Heaven, S. (2013). Fluid-dynamic characterization of real-scale raceway reactors for microalgae production. *Biomass Bioenerg.*, 54, 267–275.

Mirjafari, P., Asghari, K. and Mahinpey, N. (2007). Investigating the Application of Enzyme Carbonic Anhydrase for CO_2 Sequestration Purposes. *Ind. Eng. Chem. Res.*, 46, 921–926.

Miyairi, S. (1995). CO_2 assimilation in a thermophilic cyanobacterium. *Energy Convers. Mgmt.*, 36(6–9), 763–766.

Oswald, W.J. and Golueke, C.G. (1960). Biological transformation of solar energy. *Adv. Appl. Microbiol.*, 11, 223–242.

Rao, A.R., Ravishankar, G.A. and Sarada, R.(2012). Cultivation of green alga *Botryococcus braunii* in raceway, circular ponds under outdoor conditions and its growth, hydrocarbon production. *Bioresour. Technol.*, 123, 528–533.

Richardson, J.W., Johnson, M.D. and Outlaw, J.L. (2012). Economic comparison of open pond raceways to photo bio-reactors for profitable production of algae for transportation fuels in the Southwest. *Algal Research.*, 1, 93–100.

Richmond, A. (2004). Principles for attaining maximal microalgal productivity in photobioreactors: An overview. *Hydrobiologia.*, 512, 33–37.

Rizzo, A.M., Prussi, M., Bettucci, L., Libelli, I.M. and Chiaramonti, D. (2013). Characterization of microalga *Chlorella* as a fuel and its thermogravimetric behavior. *Appl. Energy*, 102, 24–31.

Scragg, A.H., Illman, A.M., Carden, A. and Shales, S.W. (2002). Growth of microalgae with increased calorific values in a tubular bioreactor. *Biomass Bioenerg.*, 23(1), 67–73.

Segovia, M., Haramaty, L., Berges, J.A. and Falkowski, P.G. (2003). Cell death in the unicellular chlorophyte *Dunaliella tertiolecta*. A hypothesis on the evolution of apoptosis in higher plants and metazoans. *Plant Physiol.*, 132(1), 99–105.

Sharma, A., Bhattacharya, A. and Shrivastava, A. (2011). Biomimetic CO_2 sequestration using purified carbonic anhydrase from indigenous bacterial strains immobilized on biopolymeric materials. *Enzyme Microb. Technol.*, 48, 416–426.

Spalding, M.H. (2008). Microalgal carbon dioxide concentrating mechanisms: *Chlamydomonas* inorganic carbon transporters. *J. Exp. Biol.*, 59(7), 1463–1473.

Spolaore, P., Joannis-Cassan, C., Duran, E. and Isambert, A. (2006). Commercial applications of microalgae. *J. Biosci. Bioeng.*, 101(2), 87–96.

Sutherland, D.L., Turnbull, M.H. and Craggs, R.J. (2014). Increased pond depth improves algal productivity and nutrient removal in wastewater treatment high rate algal ponds. *Water Research*, 53, 271–281.

Sydney, E.B., Sturm, W., de Carvalho, J.C., Soccol, V.T., Larroche, C., Pandey, A. and Soccol, C.R. (2010). Potential carbon dioxide fixation by industrially important microalgae. *Bioresour. Technol.*, 88(10), 3291–3294.

Sydney, E.B., Novak, A.C., de Carcalho, J.C. and Soccol, C.R. (2014). Respirometric balance and carbon fixation of industrially important algae. *In:* A. Paney, D.-J. Lee, Y. Chisti and C.R. Soccol (eds.), Biofuels from algae. Elsevier, MA, USA, pp. 67–84.

Wan, M., Zhang, J., Hou, D., Fan, J., Li, Y., Huang, J. and Jun Wang, J. (2014). The effect of temperature on cell growth and astaxanthin accumulation of *Haematococcus pluvialis* during a light–dark cyclic cultivation. *Bioresour. Technol.*, 167, 276–283.

Wang, B., Li, Y.Q., Wu, N. and Lan, C.Q. (2008). CO_2 bio-mitigation using microalgae. *Appl. Microbiol. Biotechnol.*, 79(5), 707–718.

Watanabe, Y. and Hall, D.O. (1996). Photosynthetic production of the filamentous Cyanobacterium *Spirulina platensis* in a cone-shaped helical tubular photo-bioreactor. *Appl. Microbiol. Biotechnol.*, 44, 693–698.

Weissman, J.C., Goebel, R.P. and Benemann, J.R. (1988). Photobioreactor design: mixing, carbon utilization and oxygen accumulation. *Biotechnol. Bioeng.*, 31, 336–344.

Whitmarsh, J. and Govindjee (1999). The photosynthetic process. *In:* Singhal, G.S., Renger, G., Sopory, S.K., Irrgang, K.-D., Govindjee (eds), Concepts in Photobiology: Photosynthesis and Photomorphogenesis. Narosa Publishers, New Delhi and Kluwer Academic, Dordrecht, pp 11–51.

Xu, B., Li, P. and Waller, P. (2014). Study of the flow mixing in a novel ARID raceway for algae production. *Renew. Energy.*, 62, 249–257.

Yamakawa, H. and Itoh, S. (2013). Dissipation of excess excitation energy by drought-induced nonphotochemical quenching in two species of drought-tolerant moss: Desiccation-induced acceleration of photosystem II fluorescence decay. *Biochemistry.*, 52, 4451–4459.

Yang, X., Xiang, W., Zhang, F., Wu, H., He, H. and Fan, J. (2013). Adaptability of oleaginous microalgae *Chlorococcum alkaliphilus* MC-1 cultivated with flue gas. Sheng Wu Gong Cheng Xue Bao., 29, 370–381.

Yoo, C., Jun, S.Y., Lee, J.Y., Ahn, C.Y. and Oh, H.M. (2010). Selection of microalgae for lipid production under high levels carbon dioxide. *Bioresour. Technol.*, 101 Suppl 1, S71–74.

Yun, Y.S., Lee, S.B., Park, J.M., Lee, C.I. and Yang, J.W. (1997). Carbon dioxide fixation by algal cultivation using wastewater nutrients. *J. Chem. Technol. Biotechnol.*, 69(4), 451–455.

Zeng, X., Danquah, M.K., Chen, X.D. and Lu, Y. (2011). Microalgae bioengineering: From CO_2 fixation to biofuel production. *Renew. Sust. Energ Rev.*, 15, 3252–3260.

Zhang, K., Miyachi, S. and Kurano, N. (2001). Evaluation of a vertical flat-plate photobioreactor for outdoor biomass production and carbon dioxide bio-fixation: Effects of reactor dimensions, irradiation and cell concentration on the biomass productivity and irradiation utilization efficiency. *Appl. Microbiol. Biotechnol.*, 55(4), 428–433.

Chapter 3
Growth Characteristics of Different Algal Species

Sanjeev Mishra and Kaustubha Mohanty

3.1 Introduction

Algae are large and diverse group of simple, typically autotrophic organisms, ranging from unicellular to multi-cellular forms (Singh and Gu, 2010). Productivity of these photosynthetic microorganisms which converts CO_2 into carbon-rich lipids is only a step or two away from biofuel which in turn is produced by several chemical, biochemical and thermochemical processes (Wijffels and Barbosa, 2010; Kirrolia et al., 2013; Beneroso et al., 2013). Globally algal biofuel has been considered as 3rd and 4th generation biofuel based on its potential over 1st and 2nd generation crop based biofuels. Numerous scientists have discovered various applications of algal biomass apart from biofuel applications for the production of value added products to reduce its production cost towards bio-refinery approach (Rawat et al., 2013). Wijffels and Barbosa (2010) reported in *Science* about the broad prospect of microalgae over terrestrial crop based biofuel. In their report they mentioned how a 50-year-old concept came into focus during the oil crisis of 1970s. Since then over millions of algal species have been isolated, identified and studied towards its potential for biofuel and value added products. Table 3.1 represents microscopic view of some potential algal strains which has been studied as model organism at lab-scale and pilot-scale. Recent studies suggest that green algae are promising species bearing a substantial potential to obtain various products in a biorefinery concept (Suali and Sarbatly, 2012). The algal oil can be transesterified to fatty acid methyl ester (FAME) and non-lipid components of algal biomass such as carbohydrates and proteins can be used for the production of bioethanol, biobutanol, neutraceuticals and

S. Mishra
Centre for Energy, Indian Institute of Technology Guwahati, Guwahati 781039, India

K. Mohanty (✉)
Department of Chemical Engineering, Indian Institute of Technology Guwahati, Guwahati 781039, India
e-mail: kmohanty@iitg.ernet.in

D. Das (ed.), *Algal Biorefinery: An Integrated Approach*,
DOI 10.1007/978-3-319-22813-6_3

Table 3.1 Microscopic view of potential algae

Microorganisms	Microscopic picture	Microorganisms	Microscopic picture
Scenedesmous sp.	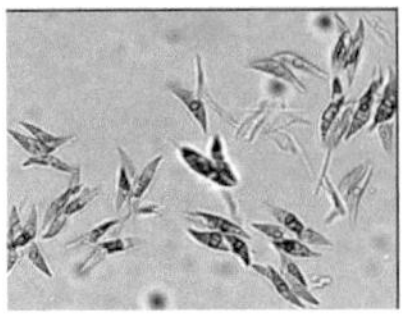	*Nitzschia* sp.	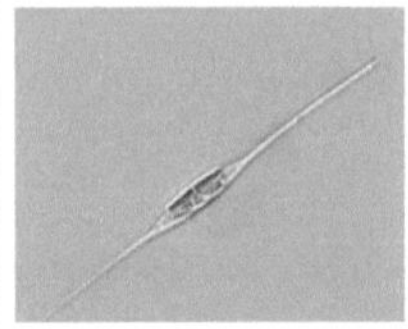
Chlorella sp.	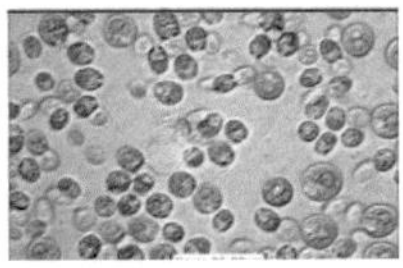	*Chloromonas* sp.	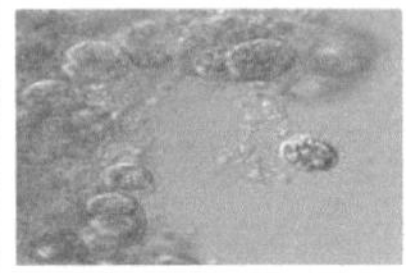
Nannochloropsis sp.	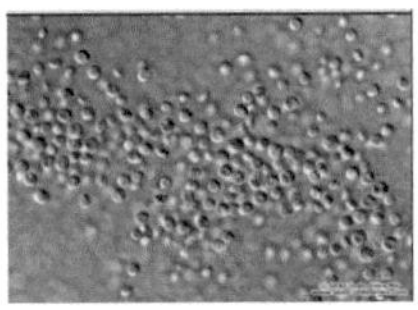	*Botryococcus* sp.	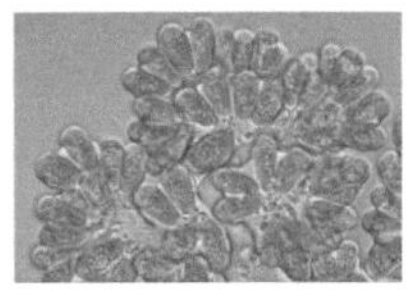
Dunaliella sp.	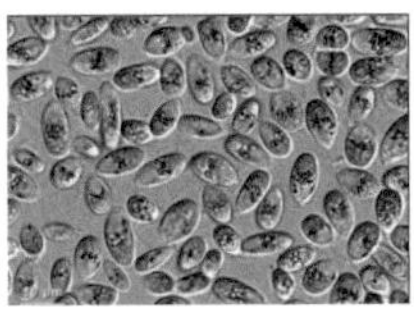	*Chlamydomonas* sp.	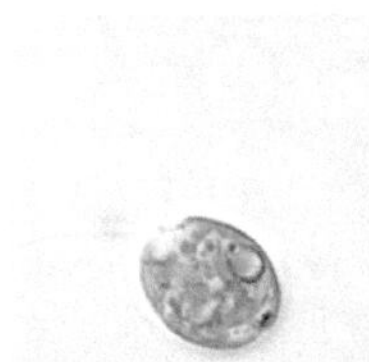
Micractinium sp.	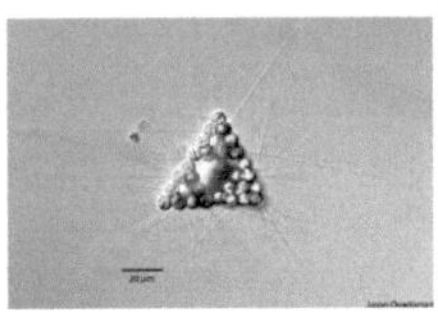	*Spirulina* sp.	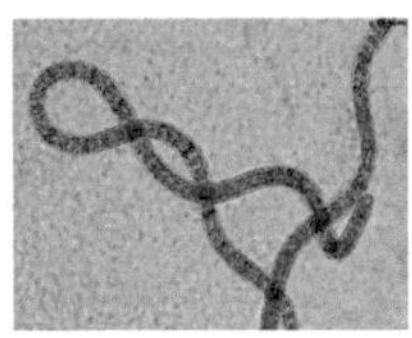
Ourococcus sp.	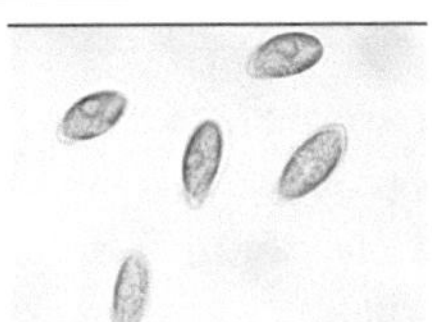	*Anabaena* sp.	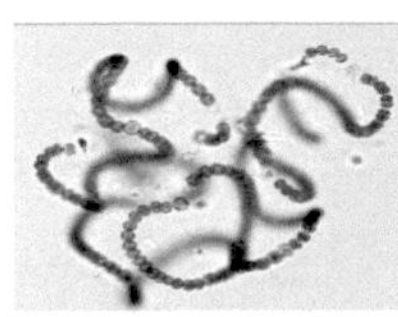
Pavlova sp.	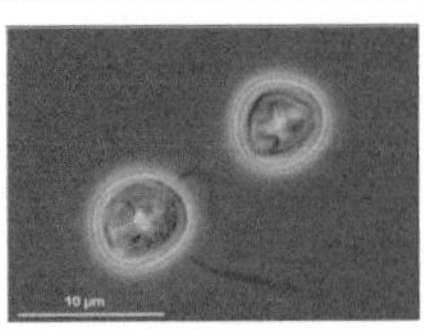	*Tetraselmis* sp.	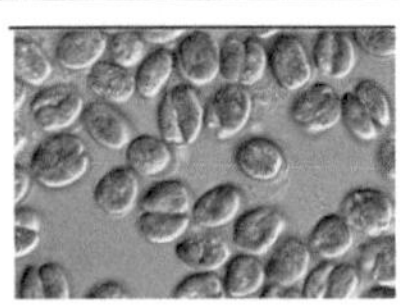

animal feed (Kirrolia et al., 2013). Moreover, the residue biomass cake can be used further to produce liquid fuel using the process of pyrolysis (Beneroso et al., 2013).

Bearing substantial feature towards biorefinery approach, triumphing significant biomass through cost effective process has always been a pronounced challenge (Acién et al., 2012). Amongst the several challenges, isolating and identifying potential algal strains having fast growth rate, high biomass yield, and significant lipid productivity are vital. In identifying potential algal strains, growth characteristics play significant role which further depends on growth medium, growth parameters, cultivation system, CO_2 concentration, algal growth rate, biomass productivity, and many more (Roleda et al., 2013). The current chapter elaborates the growth characteristics of different algal species studied in recent time and their feasibility towards algal biorefinery approach.

3.2 Potential Algal Strains

Photosynthetic algae represent an exceptionally diverse but highly specialized group of organisms of which microalgae exists with 1 to 10 million (upper limit) species in the nature (Norton et al., 1996). However, only few have shown the potential to be grown successfully in lab-scale and pilot-scale using photobioreactors and raceway pondss towards production of biofuel and value added products (Wu et al., 2012). Reports suggest that green algae have high prospective towards mass cultivation due to their high growth rate and lipid content which meets the present requirement of algal biorefinery concept (Singh and Gu, 2010). Potential algal strains can be obtained from the repositories present worldwide, but indigenous isolated algal strains will have an inherent adaptability that may be the competitive edge required for mass cultivation systems. In addition, they are able to adapt quickly to the changes in the local environment and climate (Abou-Shanab et al., 2011; Rawat et al., 2013). Indigenous algal strains can be isolated from different aquatic habitats such as pond, lake, river, etc. as well as salt water or brackish water lakes and oceans. Further different microorganism isolation techniques are used to isolate pure algal strains and identified morphologically and genetically. To screen model algal strain for future experiment and mass cultivation studies, those isolate strains undergo several lab-scale experiments such as growth rate estimation, photosynthetic efficiency, nutrient recovery, biomass productivity, lipid yield, lipid profiling etc. (Singh and Gu, 2010).

In recent time, many reports have been published on identifying potential algal strains and their feasibility towards biofuel production. Abou-Shanab et al. (2011) isolated 33 numbers of microalgae strains, from which screened eight strains and studied their potential towards biodiesel production. In evidence to their characteristics study they concluded that *Scenedesmous obliquus* YSR01 was the most potential strain towards biodiesel production. Chaichalerm et al. (2012) isolated six microalgae strains enriched in four different growth mediums for the study of their biomass yield, lipid content, and lipid productivity in batch mode. Based on their study, *Chlorococcum humicola* was identified as a potential strain. In contrast to isolating potential strains which can grow in artificial culture medium, there are several strains isolated having the potential to grow in wastewater. These algal strains were proficient towards removing nitrogen and phosphorus as growth nutrients while accumulating lipids in effluents from different domestic and industrial wastewater. Sydney et al. (2011) conducted a screening experiment on 20 algal strains and found positive growth on 13 strains towards nutrient removal and lipid accumulation in secondary domestic wastewater effluents. They concluded that *Botryococcus braunii* showed highest nutrient removal of 79.63 % on 14 days of cultivation period. Cai et al. (2013) in their review article on "Nutrient recovery from wastewater streams by microalgae: Status and prospects" stated, *Chlorella* sp. has been widely studied and shown to be effective in nutrient removal and high value biomass production. It has also been claimed that some species of *Chlorella* are heterotrophic or mixotrophic and can consume organic forms of carbon in addition to inorganic nutrients as part of their metabolic process. This can be an advantage when using wastewater streams containing carbon residues, such as digested dairy manure (Wang et al., 2010).

3.3 Algal Growth Characteristics

Algae are photosynthetic organisms whose growth characteristics are subjected to light intensity, growth medium (nitrogen, phosphorous, potassium and micronutrients) and carbon dioxide supply (Wang et al., 2010). Since long time many scientists have developed different growth conditions and optimized several growth medium to enhance algae growth rate towards high value biomass yield. Bold's basal medium (BBM), BG-11, TAP and artificial sea water (ASW) are the most common nutrient mediums which have been widely used as the algal (green algae and cyanobacteria) growth medium (He et al., 2013; Basu et al., 2014; Kumar et al., 2014; Bellou and Aggelis, 2012). In recent times, wastewater containing nitrogen and phosphorous as major nutrients has also been widely studied replacing lab-grade nutrient medium as low cost cultivation process coupled with wastewater treatment (Cai et al., 2013). Moreover, recent studies suggest that there are several stress effects of nutrients, light, temperature, CO_2 and pH which affect the growth rate and biochemical properties in algal biomass (Suali and Sarbatly, 2012). These stress conditions can be applied during batch mode (single/two-step) or continuous mode (Roleda et al., 2013). As algal cultivation needs to be studied towards biorefinery approach, achieving feasible biomass productivity during mass cultivation has always been a challenge. Acién et al. (2012) in their report suggested using low cost cultivation system in addition to flue gases as carbon source and wastewater as growth medium can help achieving substantial biomass yield up to 4.86 g L^{-1} (Basu et al., 2014).

3.3.1 Methodology to Estimate Algal Growth Rate

Determining algal growth rate is essential in identifying potential algal strains during the screening process and validates its productivity towards mass cultivation prospect. There are several methods through which algal growth rate can be estimated of which specific growth rate and biomass productivity is essential.

3.3.1.1 Spectroscopic Growth Rate Observation

Algae are photosynthetic organisms which contain different pigments. Chlorophyll '*a*' and '*b*' are among such pigments which is present in green algae and cyanobacterias. Spectrophotometer is used to estimate these pigments and cell concentration in the culture medium. A wide range of wavelength (550–750 nm) is considered to identify algal cell concentration which is estimated at a specific time interval and algal growth curve formed (wave length vs time).

3.3.1.2 Biomass Yield

Besides algal cell concentration, growth rate can also be determined from its biomass growth which is estimated from the total biomass achieved from a specific culture time period. To obtain algal biomass, algal culture is washed repeatedly to remove the nutrients present in it and then moisture-free biomass is obtained through centrifugation followed by drying at approx. 108 °C. There are several kinetic equations available in literatures which were usually used to determine the biomass productivity and yield.

3.3.1.3 Kinetic Equations for Growth Study

Several kinetic equations are reported for algal growth estimation and among them specific growth rate in terms of algal density is quite popular. Besides, dry algal biomass has also been considered to estimate specific growth rate of algae (Xu and Boeing, 2014). 'μ' was determined from a linear fit in a semi-logarithmic plot of algal density against time.

$$..= \frac{\ln\left(W_t / W_0\right)}{t - t_0} \tag{3.1}$$

In equation (3.1), 'W_t' is algal biomass concentration (g L^{-1}) at the time '*t*', W_0 is initial biomass concentration (g L^{-1}) and '*t*' is the experimental time (d). Specific growth rate can also be determined from the slope of a semi-log plot of biomass concentration versus time.

3.3.2 Growth Medium and Algal Growth Characteristics

Algae are known to have different nutrient requirements not only by composition but also by concentration of the nutrients supplied which is basically composed of macro- and micro-nutrients, dissolved ions, trace metals and several vitamins (Rawat et al., 2013). Growth media for algae are grouped depending on fresh water or salt water species. There is no universal growth media recipe that works for all taxa, so researchers are forced to give great care on how growth media is composed, stored and used. In general, algal growth medium is composed of macro- and micro-nutrients. Macro-nutrients required by algae, diatoms and cyanobacteria include carbon, nitrogen, phosphorous, silicon and major ions including Na, K, Mg, Ca, Cl and SO_4 as a base media. Micro-nutrients are trace amounts of essential elements and these include iron, manganese, zinc, cobalt, copper, molybdenum and a small amount of metalloid selenium (He et al., 2013).

In the past few decades, tremendous efforts have been put into research of micro-algae cultivation in wastewaters for the removal of nitrogen, phosphorus and other elements (Cai et al., 2013). This process of algal bioremediation route towards biomass production makes algal cultivation process cost effective coupled with simultaneous wastewater treatment and biomass production for biofuel and other applications. To pursue algal bioremediation photoautotrophic/photoheterotrophic/mixotrophic cultivation system were followed. Recently, untreated and treated domestic secondary effluent, municipal wastewater, brewery wastewater, thin stillage, soy whey, carpet industry wastewater and other industrial wastewater were successfully used for microalgal-based bioremediation process.

3.3.2.1 Algal Growth in Conventional Medium

Conventional lab-grade synthetic mediums are widely studied at lab-scale or small scale experiments for screening and process optimization of various algal strains (Abou-Shanab et al., 2011). During the culture process nitrate and phosphate play a central role in microalgal cell physiology and growth where the optimum concentration enhances the growth rate of microalgal cultures (Bhola et al., 2011). In addition to nutrients, CO_2 enhances algal photosynthetic efficiency leading to increase of algal growth rate (Suali and Sarbatly, 2012). In a recent study, Basu et al. (2014) reported indigenous *Scenedesmus obliquus* SA-1 in BG11 medium and studied its growth charac-teristics. *S. obliquus* showed higher biomass concentration of 4.86 g L^{-1} which was achieved at 13.8 ± 1.5 % CO_2 supply. To study the nitrogen and phosphorous effect on algal growth they increased their concentration and achieved higher biomass concentration of 4.975 ± 0.003 g L^{-1} which justified that nitrogen and phosphorous play important role in uplifting algal growth rate.

Algae are known to have specific nutrient requirement for their growth for which their growth characteristics varies at different growth medium. Chaichalerm et al. (2012) grew six algal strain *Chlorococcum humicola*, *Didymocystis bicellularis*,

Monoraphidium contortum, *Oocystis parva*, *Sphaerocystis* sp., and *Scenedesmus acutus* in four different growth medium (3NBBM, BG-11, Kuhl and N-8) and the mode of algal growth effect was observed by varying nutrient compositions. Significant difference in their growth characteristic and lipid productivity with respect to its growth medium was observed for all the six strains. Among all, *C. humicola* had the highest biomass yield of 0.113 g $L^{-1}d^{-1}$ (in Kuhl medium), the highest lipid content of 45.94 % w/w (in BG-11 medium), and the highest lipid yield of 0.033 g $L^{-1}d^{-1}$ (in 3NBBM medium). The 3NBBM medium, which has the lowest nitrogen concentration among the four culture media, was considered the optimal culture medium for *C. humicola* for lipid production. This is due to the fact that nitrogen limitation does not inhibit algal growth and creates stress environment to cell which enhances synthesis of membrane and storage lipid (Cakmak et al., 2014). Some potential algal strains and their growth characteristics have been presented in Table 3.2.

Table 3.2 Growth characteristics of different algal strains grown in synthetic media at lab-scale

Algal strain	Specific growth rate (d^{-1})	Biomass productivity (g L^{-1})	Lipid (%w/w)	References
Nitzschia cf. pusilla YSR02	1.68±0.28	1.37±0.08	48±3.1	Abou-Shanab et al. (2011)
Chlorella ellipsoidea YSR03	1.42±0.02	1.48±0.04	32±5.9	
Micractinium pusillum YSW07	1.19±0.17	2.28±0.16	24±0.5	
Ourococcus multisporus YSW08	0.51±0.14	0.95±0.11	52±8.3	
Scenedesmus obliquus YSR04	1.32±0.05	1.98±0.04	21±1.1	
Scenedesmus obliquus YSR01	2.35±0.55	1.57±0.67	58±1.5	
Scenedesmus obliquus YSR05	1.06±0.03	1.75±0.34	27±1.9	
Scenedesmus obliquus YSW06	0.99±0.02	1.80±0.13	27±5.6	
Scenedesmus obliquus SA-1	–	4.86	33.04±0.46	Basu et al. (2014)
Nannochloropsis oculata	0.194–0.571	0.296–0.497 (g $L^{-1}d^{-1}$)	22.7–41.2	Xu and Boeing (2014)
Nannochloropsis sp. F&M-M24	–	0.18 (g $L^{-1}d^{-1}$)	30.9	Rodolfi et al. (2009)
Pavlova salina CS 49	–	0.16 (g $L^{-1}d^{-1}$)	30.9	
Chlorococcum humicola	–	0.113 (g $L^{-1}d^{-1}$)	45.94	Chaichalerm et al. (2012)

3.3.2.2 Algal Growth in Wastewater

Wastewater from fertilizer industry, brewery industry, untreated and treated domestic secondary effluent, municipal wastewater, etc. contain high amount of nitrogen, phosphorous and other elements which are major algal growth nutrients. Achieving high nutrient removal along with feasible biomass yield is rather difficult due to high nutrient load in wastewater which creates stress environment for algal growth. There are few algae which has inherent adaptability to sustain their growth in high nutrient load environment. Among them *Chlorella* sp. are more commonly studied in various types of wastewaters and achieved high nutrient removal in addition to feasible biomass yield (Cai et al., 2013). In a study by Li et al. (2011), isolated *Chlorella* sp. was grown in municipal wastewater where a higher growth rate of 0.677 d^{-1} was achieved without any lag phase. During the 14 d of cultivation period, *Chlorella* could achieve more than 80 % of nutrient removal (ammonia, total nitrogen, total phosphate) with 0.92 g L^{-1} of feasible biomass productivity. There is also report which claimed of achieving 9.8 g L^{-1} and 6.3 g L^{-1} biomass from thin stillage and soy while growing *Chlorella vulgaris* at mixotrophic conditions in a bioreactor. Besides high biomass yield *Chlorella vulgaris* could achieve high lipid yield of 43 % and 11 % (w/w) respectively (Mitra et al., 2012).

As mentioned earlier, nutritional requirement for algal growth varies from strain to strain for which different algal strains need to be screened. Sydney et al. (2011) conducted a screening experiment over 20 algal strains on secondary domestic wastewater treated effluents. Among 20 strains three strains—unknown LEM-IM 11, *Botryococcus braunii* and *Chlorella vulgaris*—have shown higher growth rate and high nutrient removal. During 2 L photobioreactor study unknown LEM-IM 11 has shown growth rate of 0.19 d^{-1} whereas 0.11 d^{-1} was observed in *Botryococcus braunii* and *C. vulgaris*. But greater nutrient removal was achieved by *Botryococcus braunii* with 79 % nitrate and 100 % phosphate in addition to higher biomass productivity of 0.68 g L^{-1} containing over 36.14 % lipid. There are also reports where two different wastewaters were mixed in different proportion to achieve higher algal growth and nutrient removal. In one such study, *Scenedesmus obliquus* CCAP 276/3A reported a maximum specific growth rate of 0.074 h^{-1}, volumetric biomass productivity of 4 mg L^{-1} h^{-1}, and net biomass generation 0.28 g L^{-1} when grown in a culture blend of 25 % (v/v) of urban wastewater from secondary treatment added to 5 % (v/v) olive-oil mill wastewater (Hodaifa et al., 2013). Table 3.3 represents various algal strains grown in different wastewater and their biomass and lipid productivity. This suggests algal strains represent variable growth characteristics for which screening process is highly essential to select potential algal strain to achieve high biomass yield with simultaneous nutrient removal.

Table 3.3 Biomass and lipid productivity of different algal strains studied in different waste water

Species	Source of wastewater	Biomass (g L^{-1})	Lipid content (%)	References
Chlorella sp.	Thin stillage	9.8±0.3	43	Mitra et al. (2012)
	Soy whey	6.3±0.1	11.1±1.1	
	Domestic secondary effluent	0.42	43	Yang et al. (2011)
	Brewery wastewater	2.28±0.09	220±0.02 (mg g^{-1})	Farooq et al. (2013)
	Municipal wastewater	1.75	–	Cho et al. (2013)
	Carpet industry wastewater	0.016±0.003 (g L^{-1} d^{-1})	17.00±2.89	Chinnasamy et al. (2010)
Scenedesmus obliquus CCAP 276/3A	Blend of 25 % urban wastewater and 5 % olive-oil mill wastewater	0.28 (g dm^{-3})	33.2	Hodaifa et al. (2013)
Dunaliella tertiolecta	Carpet mill untreated wastewater	28 (mg L^{-1} d^{-1})	15.20	Chinnasamy et al. (2010)
Botryococcus braunii	Treated domestic sewage	0.68	36.14	Sydney et al. (2011)
Chlamydomonas reinhardtii	Municipal wastewater	2 (g L^{-1} d^{-1})	25.25	Kong et al. (2010)
Chlamydomonas sp. TAI-2	Industrial wastewater	1.8	18	Wu et al. (2012)

3.3.3 Growth Study at Mass Cultivation Prospect

Algal biofuel are globally considered as 3rd and 4th generation biofuel due to its significant biomass productivity containing high cellular concentration of lipids, resources and economic sustainability and overall potential advantages over other sources of biofuels (Rawat et al., 2013). Productivity of algae, microalgae in general, can be twenty times than that of oil seed crops on per hectare basis and is thus a more viable alternative towards mass cultivation prospect (Chisti, 2007). Moreover, microalgae have faster growth rates than plants and are capable to grow in highly saline waters and utilize a large fraction of solar energy making them effective solar to chemical energy converters (Huber et al., 2006). Various algal strains were studied towards mass cultivation prospect using raceway pond and outdoor photobioreactors (Singh and Gu, 2010). But feasibility over economic viability has always been a greater challenge of algal cultivation at pilot/commercial scale (Acién et al., 2012). There are many vital steps which need to be critically analyzed at each stage such as, isolation of algal strains for mass cultivation prospect which can be either freshwater or marine algae, cultures selected from various culture repositories or indigenous wild types which might be best suited for large scale production (Suali and Sarbatly, 2012). Furthermore, the screening process should identify potential strains bearing faster growth rate, high biomass yield, and significant lipid productivity. The synergistic interactions that occur between naturally grown algae and other microorganisms cannot be ignored (Rawat et al., 2013). Besides isolation and identification of potential algal strains, reactor design and biomass harvest from mass cultivation system play vital role to make mass cultivation process cost effective (Acién et al., 2012).

3.3.3.1 Raceway Pond

Low cost, low maintenance, easy setup and large volume cultivation capacity make raceway pond more commercially popular among various researchers and industrialists (Suali and Sarbatly, 2012). This is one of the oldest mass cultivation systems which are widely studied for algal biomass production. Though it is commercially popular it has several drawbacks such as wide exposure to environment always contaminates algal culture and can't be recommended during rainy season and winter season with snowfall. Maintaining growth parameters such as temperature, pH and light intensity is also fairly possible. Besides several drawbacks, there are quite a few algal strains which have been successfully grown with a biomass productivity up to 14–50 g $m^{-2}d^{-1}$, which can be further enhanced with supply of external CO_2 (Putt et al., 2011). In a recent study by Ranga Rao et al. (2012), *Botryococcus braunii* has shown high growth rate leading to biomass concentration of 1.8 g L^{-1} in 18 days of cultivation in a raceway pond having culture capacity of 80 L. Similar biomass concentration (1.9 g L^{-1}) with specific growth rate of 0.419 d^{-1} was achieved in 18 days of cultivation in 2000 L raceway pond for *Botryococcus braunii* (Ashokkumar et al., 2014). This growth rate is comparatively four times higher than that observed by *Botryococcus braunii* in secondary domestic wastewater treatment effluents (Sydney et al., 2011). In another experiment, fast growing marine algae *Chlorella variabilis* was grown in 400 L raceway pond with a culture volume of 150 L where significant specific growth rate of 0.36 day^{-1} was observed. Cultivating the *C. variabilis* throughout the year, maximum biomass (1.76 ± 0.04 g L^{-1}) was achieved during summer season when high light intensity used to be observed causing higher photosynthesis and algal growth (De Bhowmick et al., 2014). Nurra et al. (2014) studied algal mass cultivation using *Nannochloropsis gaditana* in a 53 m^3 raceway pond on a biorefinery concept where significant growth rate with biomass productivity of 19.9 g $m^{-2}d^{-1}$ was achieved. Some of potential algal strain studied towards mass cultivation prospect using raceway pond and photobioreactor are depicted in Table 3.4.

3.3.3.2 Photobioreactor

Controlled growth environment, minimal contamination and high photosynthesis leading to higher biomass yield are promising features of a photobioreactor as an algal cultivation system. Among several photobioreactor designs, stirred tank and

Table 3.4 Growth characteristics of different algal strains at mass cultivation prospect

Cultivation system	Species	Biomass (g L^{-1})	Lipid content (%)	References
Raceway pond	*Botryococcus braunii*	2.31	–	Ranga Rao et al. (2012)
	Botryococcus braunii	1.9	48.40	Ashokkumar et al. (2014)
	Chlorella variabilis	1.76±0.04	–	De Bhowmick et al. (2014)
	Spirulina platensis	13.5 (g^{-2} d^{-1})	–	Jiménez et al. (2003)
	Anabaena sp. ATCC 33047	23.5 (g^{-2} d^{-1})	–	Moreno et al. (2003)
	Nannochloropsis gaditana	19.9 (g m^{-2} d^{-1})	22	Nurra et al. (2014)
Photobioreactor	*Nannochloropsis* sp. F&M-M24	0.30 (g L^{-1} d^{-1})	60	Rodolfi et al. (2009)
	Tetraselmis suecica	0.35±0.03 g L^{-1} d^{-1}	–	Michels et al. (2014)
	Chlorella zofingiensis	58.4 mg $L^{-1}d^{-1}$	54.5	Feng et al. (2011)
	Chlorella sp.	4.3 g $L^{-1}d^{-1}$	–	Doucha and Lívanský (2009)
	Botryococcus braunii	2.31	–	Ge et al. (2011)

vertical column photobioreactors are commonly used during lab-scale experiments, whereas high volume flat panel and tubular reactors are recommended during pilot-scale experiment (Wang et al., 2012). Due to controlled growth parameters most algal strains showed better growth rate and delivers higher biomass yield. *Botryococcus braunii* achieved significant growth rate when grown in photobioreactor and resulted 2.31 g L^{-1} of biomass on 25th day of cultivation with 20 % v/v CO_2 which was supplied in the form of flue gas (Ge et al., 2011). This is comparatively higher than the biomass obtained from raceway pond (Ranga Rao et al., 2012). Since algal growth rate is specific to strains and its potential towards mass cultivation prospect, research has been focused on cultivating potential high biomass productivity algal strains with a mass cultivation prospect.

Doucha and Lívanský (2009) cultivated fast growing *Chlorella* sp. in outdoor open thin-layer photobioreactor having culture volume of 2000 L where greater biomass productivity was achieved. During this maximum algal concentration of 4.3 g $L^{-1}d^{-1}$ was achieved whereas less growth rate was observed by *Chlorella zofingiensis* grown in 60 L flat plate photobioreactor. The highest specific growth rate and biomass productivity obtained was 0.994 d^{-1} and 58.4 mg $L^{-1}d^{-1}$ respectively (Feng et al., 2011). Besides fresh water algal strains, sea water and brackish water algal strains have also been studied on scale-up experiments using photobioreactors. High lipid producing *Nannochloropsis* sp. F&M-M24 has shown significant growth rate and average biomass productivity of 0.30 g $L^{-1}d^{-1}$ when grown in 110 L Green Wall Panel photobioreactors under nutrient sufficient and deficient conditions containing over 60 % w/w lipid (Rodolfi et al., 2009). Similar growth rate was observed from *Tetraselmis suecica* of 0.35 ± 0.03 g $L^{-1}d^{-1}$ with optimum biomass concentration of 0.7 g L^{-1} when grown in 40 L tubular photobioreactor (Michels et al., 2014). This suggests fresh water green algae have high potential towards mass cultivation prospect using photobioreactor.

3.4 Conclusion

Recent reports in the field of algal research suggest different algal species bear different growth characteristic which depends on their mode of cultivation in addition to medium of nutrition and CO_2 supply. Over several algal species, green algae have shown promising growth characteristics and have been widely studied. Potential green algae have shown sustainable growth in different wastewater with respect to high nutrient removal efficiency which makes algal cultivation process a cost effective process when coupled with simultaneous wastewater treatment. Fresh water algal strains have shown significant growth features towards mass cultivation prospect over saline water algal strains. However, nutritional stress can result in high lipid yield from algal biomass during mass cultivation process which has feasibility towards algal biorefinery concept.

References

Abou-Shanab, R.A.I., Hwang, J.-H., Cho, Y., Min, B. and Jeon, B.-H. (2011). Characterization of microalgal species isolated from fresh water bodies as a potential source for biodiesel production. *Appl. Energy*, 88, 3300–3306.

Acién, F.G., Fernández, J.M., Magán, J.J. and Molina, E. (2012). Production cost of a real microalgae production plant and strategies to reduce it. *Biotechnol. Adv.*, 30, 1344–1353.

Ashokkumar, V., Rengasamy, R., Deepalakshmi, S., Sivalingam, A. and Sivakumar, P. (2014). Mass cultivation of microalgae and extraction of total hydrocarbons: A kinetic and thermodynamic study. *Fuel*, 119, 308–312.

Basu, S., Roy, A.S., Mohanty, K. and Ghoshal, A.K. (2014). CO_2 biofixation and carbonic anhydrase activity in *Scenedesmus obliquus* SA1 cultivated in large scale open system. *Bioresour. Technol.*, 164, 323–330.

Bellou, S. and Aggelis, G. (2012). Biochemical activities in *Chlorella* sp. and *Nannochloropsis salina* during lipid and sugar synthesis in a lab-scale open pond simulating reactor. *J. Biotechnol.*, 164, 318–329.

Beneroso, D., Bermúdez, J.M., Arenillas and Menéndez, J. (2013). Microwave pyrolysis of microalgae for high syngas production. *Bioresour. Technol.*, 144, 240–246.

Bhola, V., Desikan, R., Santosh, S.K., Subburamu, K., Sanniyasi, E. and Bux, F. (2011). Effects of parameters affecting biomass yield and thermal behavior of *Chlorella vulgaris*. *J. Biosci. Bioeng.*, 111, 377–382.

Cai, T., Park, S.Y. and Li, Y. (2013). Nutrient recovery from wastewater streams by microalgae: Status and prospects. *Renew. Sustain. Energy Rev.*, 19, 360–369.

Cakmak, Z.E., Olmez, T.T., Cakmak, T., Menemen, Y. and Tekinay, T. (2014). Induction of triacylglycerol production in *Chlamydomonas reinhardtii*: Comparative analysis of different element regimes. *Bioresour. Technol.*, 155, 379–387.

Chaichalerm, S., Pokethitiyook, P., Yuan, W., Meetam, M., Sritong, K., Pugkaew, W., Kungvansaichol, K., Kruatrachue, M. and Damrongphol, P. (2012). Culture of microalgal strains isolated from natural habitats in Thailand in various enriched media. *Appl. Energy*, 89, 296–302.

Chinnasamy, S., Bhatnagar, A., Hunt, R.W. and Das, K.C. (2010). Microalgae cultivation in a wastewater dominated by carpet mill effluents for biofuel applications. *Bioresour. Technol.*, 101, 3097–3105.

Chisti, Y. (2007). Biodiesel from microalgae. *Biotechnol. Adv.*, 25, 294–306.

Cho, S., Lee, N., Park, S., Yu, J., Luong, T.T., Oh, Y.-K. and Lee, T. (2013). Microalgae cultivation for bioenergy production using wastewaters from a municipal WWTP as nutritional sources. *Bioresour. Technol.*, 131, 515–520.

Bhowmick, G., Subramanian, G., Mishra, S. and Sen, R. (2014). Raceway pond cultivation of a marine microalga of Indian origin for biomass and lipid production: A case study. *Algal Res.* DOI: 10.1016/j.algal.2014.07.005.

Doucha, J. and Lívanský, K. (2009). Outdoor open thin-layer microalgal photobioreactor: Potential productivity. *J. Appl. Phycol.*, 21, 111–117.

Farooq, W., Lee, Y.-C., Ryu, B.-G., Kim, B.-H., Kim, H.-S., Choi, Y.-E. and Yang, J.-W. (2013). Two-stage cultivation of two *Chlorella* sp. strains by simultaneous treatment of brewery wastewater and maximizing lipid productivity. *Bioresour. Technol.*, 132, 230–238.

Feng, P., Deng, Z., Hu, Z. and Fan, L. (2011). Lipid accumulation and growth of *Chlorella zofingiensis* in flat plate photobioreactors outdoors. *Bioresour. Technol.*, 102, 10577–10584.

Ge, Y., Liu, J. and Tian, G. (2011). Growth characteristics of *Botryococcus braunii* 765 under high CO_2 concentration in photobioreactor. *Bioresour. Technol.*, 102, 130–134.

He, P.J., Mao, B., Lü, F., Shao, L.M., Lee, D.J. and Chang, J.S. (2013). The combined effect of bacteria and *Chlorella vulgaris* on the treatment of municipal wastewaters. *Bioresour. Technol.*, 146, 562–568.

Hodaifa, G., Sánchez, S., Martínez, M.E. and Órpez, R. (2013). Biomass production of *Scenedesmus obliquus* from mixtures of urban and olive-oil mill wastewaters used as culture medium. *Appl. Energy*, 104, 345–352.

Huber, G.W., Iborra, S. and Corma, A. (2006). Synthesis of transportation fuels from biomass: Chemistry, catalysts, and engineering. *Chem Rev*, 106, 4044–4098.

Jiménez, C., Cossío, B.R. and Niell, F.X. (2003). Relationship between physicochemical variables and productivity in open ponds for the production of *Spirulina*: A predictive model of algal yield, *Aquaculture*, 221, 331–345.

Kirrolia, A., Bishnoi, N.R. and Singh, R. (2013). Microalgae as a boon for sustainable energy production and its future research and development aspects. *Renew. Sustain. Energy Rev.* , 20, 642–656.

Kong, Q-X., Li, L., Martinez, B., Chen, P. and Ruan, R. (2010). Culture of microalgae *Chlamydomonas reinhardtii* in wastewater for biomass feedstock production. *Appl Biochem Biotechnol.*, 160, 9–18.

Kumar, K., Dasgupta, C.N. and Das, D. (2014). Cell growth kinetics of *Chlorella sorokiniana* and nutritional values of its biomass. *Bioresour. Technol.* , 167, 358–366.

Li, Y., Chen, Y.-F., Chen, P., Min, M., Zhou, W., Martinez, B., Zhu, J. and Ruan, R. (2011). Characterization of a microalga *Chlorella* sp. well adapted to highly concentrated municipal wastewater for nutrient removal and biodiesel production. *Bioresour. Technol.*, 102, 5138–5144.

Michels, M.H.A., Slegers, P.M., Vermuë, M.H. and Wijffels, R.H.(2014). Effect of biomass concentration on the productivity of *Tetraselmis suecica* in a pilot-scale tubular photobioreactor using natural sunlight. *Algal Res.*, 4, 12–18.

Mitra, D., van Leeuwen, J. (Hans) and Lamsal, B. (2012). Heterotrophic/mixotrophic cultivation of oleaginous *Chlorella vulgaris* on industrial co products. *Algal Res.*, 1, 40–48.

Moreno, J., Vargas, M.A., Rodrı'guez, H., Rivas, J. and Guerrero, M.G. (2003). Outdoor cultivation of a nitrogen-fixing marine cyanobacterium, *Anabaena* sp. ATCC 33047. *Biomol. Eng.*, 20, 191–198.

Norton, T.A., Melkonian, M. and Anderson, R.A. (1996). Algal biodiversity. *Phycologia*, 35, 308–326.

Nurra, C., Torras, C., Clavero, E., Ríos, S., Rey, M., Lorente, E., Farriol, X. and Salvadó, J. (2014). Biorefinery concept in a microalgae pilot plant: Culturing, dynamic filtration and steam explosion fractionation. *Bioresour. Technol.*, 163, 136–142.

Putt, R., Singh, M., Chinnasamy, S. and Das, K.C. (2011). An efficient system for carbonation of high-rate algae pond water to enhance CO_2 mass transfer. *Bioresour Technol.*, 102(3), 3240–3245.

Rao, R., Ravishankar, G. and Sarada, R. (2012). Cultivation of green alga *Botryococcus braunii* in raceway, circular ponds under outdoor conditions and its growth, hydrocarbon production. *Bioresour. Technol.*, 123, 528–533.

Rawat, I., Ranjith Kumar, R., Mutanda, T. and Bux, F. (2013). Biodiesel from microalgae: A critical evaluation from laboratory to large scale production. *Appl. Energy*, 103, 444–467.

Rodolfi, L., Chini Zittelli, G., Bassi, N., Padovani, G., Biondi, N., Bonini, G. and Tredici, M.R. (2009). Microalgae for oil: Strain selection, induction of lipid synthesis and outdoor mass cultivation in a low-cost photobioreactor. *Biotechnol. Bioeng.*, 102, 100–112.

Roleda, M.Y., Slocombe, S.P., Leakey, R.J.G., Day, J.G., Bell, E.M. and Stanley, M.S. (2013). Effects of temperature and nutrient regimes on biomass and lipid production by six oleaginous microalgae in batch culture employing a two-phase cultivation strategy. *Bioresour. Technol.*, 129, 439–449.

Singh, J. and Gu, S. (2010). Commercialization potential of microalgae for biofuels production. *Renew. Sustain. Energy Rev.*, 14, 2596–2610.

Suali, E. and Sarbatly, R. (2012). Conversion of microalgae to biofuel. *Renew. Sustain. Energy Rev.*, 16, 4316–4342.

Sydney, E.B., da Silva, T.E., Tokarski, A., Novak, A.C., de Carvalho, J.C., Woiciecohwski, A.L., Larroche, C. and Soccol, C.R. (2011). Screening of microalgae with potential for biodiesel production and nutrient removal from treated domestic sewage. *Appl. Energy*, 88, 3291–3294.

Wang, B., Lan, C.Q. and Horsman, M. (2012). Closed photobioreactors for production of microalgal biomasses. *Biotechnol. Adv.*, 30, 904–912.

Wang, L., Li, Y., Chen, P., Min, M., Chen, Y., Zhu, J. and Ruan, R.R. (2010). Anaerobic digested dairy manure as a nutrient supplement for cultivation of oil-rich green microalgae *Chlorella* sp. *Bioresour. Technol.*, 101, 2623–2628.

Wijffels, R.H. and Barbosa, M.J. (2010). An outlook on microalgal biofuels. *Science*, 329, 796–799.

Wu, L.F., Chen, P.C., Huang, A.P. and Lee, C.M. (2012). The feasibility of biodiesel production by microalgae using industrial wastewater. *Bioresour. Technol.*, 113, 14–18.

Xu, Y. and Boeing, W.J. (2014). Modeling maximum lipid productivity of microalgae: Review and next step. *Renew. Sustain. Energy Rev.*, 32, 29–39.

Yang, J., Li, X., Hu, H., Zhang, X., Yu, Y. and Chen, Y. (2011). Growth and lipid accumulation properties of a freshwater microalga, *Chlorella ellipsoidea* YJ1, in domestic secondary effluents. *Appl. Energy*, 88, 3295–3299.

Chapter 4
Perspectives on Algal Engineering for Enhanced Biofuel Production

Namita Khanna

4.1 Introduction

Algae as photoautotrophs can trap the solar energy and convert it into usable form. Solar energy is the most abundant and ultimate energy source. The total amount of solar energy absorbed by the Earth's surface is 1.74×10^5 terawatts (TW) (Bhattacharya et al., 2005), which is a tremendous amount as compared to the world's energy consumption (~13 TW) (Walter et al., 2010). Therefore, conversion of solar energy to fuels may constitute the most sustainable way to solve the energy crisis.

The incessant use of petroleum over the last century has greatly polluted the environment. In the USA alone, 3780 million metric tonnes of CO_2 are emitted from stationary sources alone (DoE Report, USA). Thus, while debating the next generation of fuels, besides energy efficiency, carbon neutrality of the fuel must be taken into consideration. As solutions to handle the waste emissions are discussed, algae can provide a suitable solution. They have a unique metabolism of sequestering CO_2 that can be redirected into biofuel production through photosynthesis. Algal fuel production reactors can be stationed near waste emission sources that could feed algae its nutrition and in turn provide valuable biofuels.

With a 1.2 % growth in the world population every year, arable land for cultivation of food crop is another concern (Summary Report, FAO, UN, 2002). The use of energy crops such as sugarcane for production of ethanol is intensely debated. However, algae can be grown on deserted wasteland/marginal land saving the arable land for cultivation of food crops. Moreover, algae are better photosynthesizers that can harvest upto 10 % of the incident solar light as compared to the efficiency of the C3 and C4 plants that can harvest a maximum of only 4.5 to 6 % (Janssen et al.,

N. Khanna (✉)
Microbial Chemistry, Department of Chemistry, Ångström Laboratory, Uppsala University, Box 523, SE-75120 Uppsala, Sweden
e-mail: namitakhanna1@gmail.com

D. Das (ed.), *Algal Biorefinery: An Integrated Approach*,
DOI 10.1007/978-3-319-22813-6_4

2000). Cyanobacteria and red algae contain phycobilisomes which make them more robust in the efficient absorption of polychromatic visible and far red irradiation. Moreover, unlike corn ethanol that can require up to 15 gallons of fresh water to produce a single gallon of fuel (Wu et al., 2008), algal cultivation does not require fresh water. Contrarily, use of algae can help address issues of fresh water scarcity since they can thrive in salt water and under harsh conditions (Um and Kim, 2009).

Currently available alternatives that can be processed from algae include: biodiesel, high carbon fuels like isoprene, hydrogen and alcohols. The chapter will give an overall perspective of how pathways for each of these fuels have been genetically engineered or synthetically circuited *in vivo* to drive biofuel production from algae. Some of these fuels are already in various stages of development and commercialization. Particularly, significant progress has been made in applying synthetic biology strategies to optimize algal chassis for the development of these biofuels. Overall, this article attempts to incorporate the recent advances to optimize the algal chassis and the recent advances in the production of various biofuels.

4.2 Improving the Algal Chassis for Biofuel Production

Both cyanobacteria and green algae can hugely impact the biofuel industry. Proof of concept studies have been carried out by several research groups demonstrating their feasibility. However, for a scale-up of these processes and wide scale industrial applicability the present processes face several technical limitations such as low yield and inherent light saturation limitations. To resolve these technical issues, the algae chassis must be optimized for biofuel production. There are several key features that can be engineered while developing a suitable chassis. Autotrophic carbon reserves are the feedstock of all the biofuels that are produced from algae. However, the cell must intelligently decide how much it needs to partition between cell growth and other cellular processes. Therefore, one way towards improving the chassis would be to trick the cell to channel more energy towards the desired product as compared to cellular growth and maintenance. This can be achieved by:

(a) increasing the rate of carbon fixation; and
(b) biasing the metabolic flux towards biofuel production.

Suitable chassis engineering must incorporate the specific requirements of the targeted biofuel. Recent studies reveal that the production of certain biofuels is stressful for the cells. It is thus imperative to increase the tolerance of the algal species towards them. Integrated omics approach has laid a foundation to suggest the stress response that is induced when algae is used as chassis to develop biofuels (discussed more in Section 4.4).

4.2.1 *Increasing the Rate of Carbon Fixation*

The rate of carbon fixation can be improved by engineering factors that are involved in sequestering the carbon. These primarily include the key catalase ribulose-bisphosphate carboxylase-oxygenase (Rubisco) and other components of the carbon concentrating mechanism (CCM). It is well known that 50 % of the total photosynthesis on Earth is carried out by algae (Field et al., 1998). Yet, these organisms face several challenges in obtaining the CO_2 and converting them into fixed carbon via its photosynthetic enzymes. Firstly, Rubisco is not designed to scavenge low concentrations of CO_2. It is a sluggish enzyme with a low affinity (K_m) for CO_2. At atmospheric concentrations alone, Rubisco can function only at a quarter of its catalytic efficiency due to the low concentration of dissolved CO_2 and the relatively high concentration of O_2 which competes with CO_2 (Long et al., 2006).

Rubisco Was Not Always Inefficient

It is worth to note that in the Earth's evolutionary history, the ability to fix CO_2 was not always a problem. When Rubisco first evolved in cyanobacteria some 3.5 billion years ago, it did so in an environment where CO_2 levels were many folds higher than the present. Under those conditions CO_2 levels were saturating Rubisco which seemed to have been efficient in achieving relatively high rates of carboxylation. However, as oxygenic photosynthesis evolved and the levels of O_2 increased and that of CO_2 dropped, the catalytic efficiency of Rubisco faced a setback and reduction of CO_2 became a rate limiting step (Badger et al., 1998; Badger and Price, 2003). As an evolutionary response to the changed atmospheric conditions, two general strategies evolved: in eukaryotes (higher plants), there was an increase in the affinity of Rubisco towards CO_2 as compared to O_2 (Badger et al., 1998; Tcherkez et al., 2006) while in an alternate strategy in prokaryotes (cyanobacteria) the carbon concentration mechanism evolved to increase the supply of CO_2 to Rubisco although at a minor metabolic cost.

Secondly the diffusion of CO_2 in water is 10,000 times slower than the diffusion of CO_2 in air. Lastly, algae often experience significant fluctuations in inorganic carbon ($C_i = CO_2 + HCO_3^-$) availability due to change in the environment pH. At an acidic pH, the C_i is mainly in the form of CO_2, while at an alkaline pH, C_i is mostly in the form of HCO_3^- (Moroney and Ynalvez, 2007). Thus, to overcome the challenge of low CO_2 availability along with a slow enzyme algae have developed the CO_2 concentrating mechanisms (CCM) (Fig. 4.1).

It is a mechanism which enhances photosynthetic productivity in algal cells by augmenting the availability of inorganic carbon within the immediate surroundings

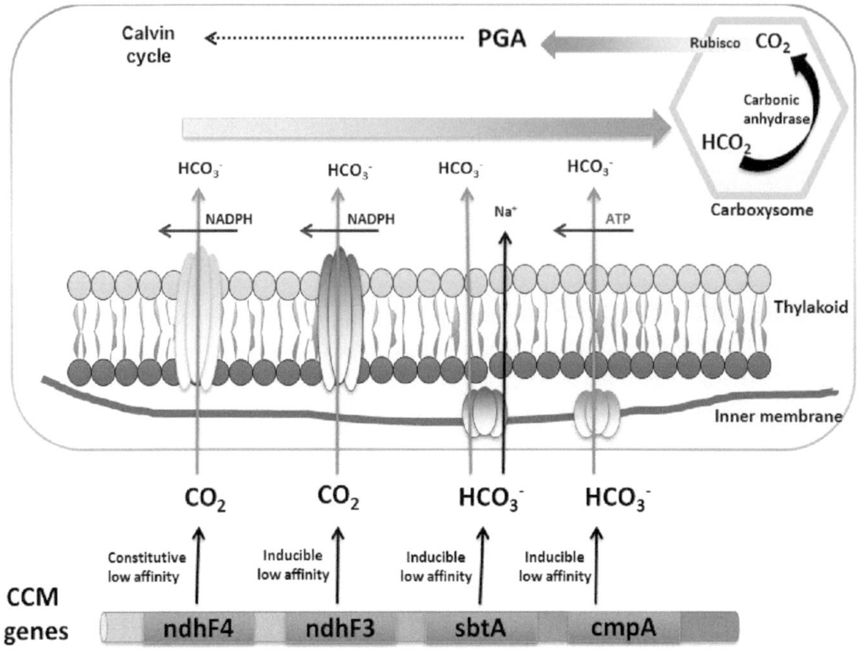

Fig. 4.1 Schematic of carbon concentrating mechanism in algae. NdhF3 and NdhF4, SbtA and CmpA have been putatively recognized as translocators located in the plasma membrane and the thylakoid membrane that pump HCO_3^- and CO_2 into the cytosol and thylakoid of a cyanobacterium. The internal gradient of HCO_3^- drive the inorganic carbon to the carboxysome. The carboxysomal carbonic anhydrase catalyzes the inter-conversion of HCO_3^- and CO_2 and, in so doing, increases the concentration of CO_2 around Rubisco, facilitating the carboxylation of ribulose 1,5-bisphosphate.

of the enzyme as compared to the atmospheric concentration. The components of CCM thus maybe the prime targets that could enhance the rate of carbon fixation.

Alternative pathways to Rubisco have been elucidated that fix approximately a quarter of the total fixed CO_2. This primarily includes the Phosphoenol pyruvate carboxylase (PepC) pathway. Besides the Calvin cycle, these pathways are also amenable to genetic engineering that could potentially increase the total amount/rate of CO_2 fixed. This is crucial to producing high yield biofuel. Below the recent updates on engineering of these pathways have been summarized.

4.2.1.1 Engineering the Rubisco Pathway

To date several studies have been conducted to increase the rate of photosynthetic efficacy by engineering Rubisco, the primary enzyme involved in the catalysis of CO_2 reduction. When Rubisco performs the carboxylation reaction, the first step of the photosynthetic carbon reduction, it uses CO_2 and ribulose-1,5-biphosphate

(RuBP) to produce two molecules of 3-phosphoglycerate (3PGA). One of these is recycled to regenerate ribulose 1,5-biphosphate whereas the other is diverted towards the carbon metabolism for biosynthesis of sugars, terpenoids and fatty acids. Rubisco are not the most efficient catalysts; in fact they are far from it. Owing to the low affinity of Rubisco towards CO_2, plants must devote up to one-half of their leaf-soluble protein to this enzyme to achieve acceptable rates of photosynthesis (Iwaki et al., 2006). To overcome this, the algal species evolved a carbon concentrating mechanism. In this design, Rubisco is localized in micro-cellular compartments called carboxysomes along with carbonic anhydrase. Carbonic anhydrase breakdown the accumulated cellular bicarbonate into CO_2 within the carboxysomes resulting in an elevated concentration of CO_2 around Rubisco. Other CCM components include the inorganic carbon transporters on plasma membrane as well as those on the thylakoids that uptake the CO_2 and bicarbonate. Since the CCM involves four modules: Rubisco, carbonic anhydrase, carboxysomes and Ci transporters, intuitively all the four features can be engineered to increase the rate of photosynthetic efficiency.

Rubisco, besides being a notoriously sluggish catalyst, cannot dis-criminate between its substrate CO_2 and O_2. Reaction with O_2 results in the formation of toxic phosphoglycolate that must be routed through the photorespiratory pathway which results in the loss of the previously fixed CO_2. Biotechnological advancements towards the improvement of Rubisco have proved challenging thus far. Attempts to increase the rate of catalysis has led to higher photorespiratory losses, while increasing selectivity towards the substrate CO_2 has led to partial loss in activity of the enzyme (Tcherkez et al., 2006). Earlier, Daniel et al. (1989) had accidentally obtained a mutant of *Anacystis nidulans* (now *Synechococcus*) that had high expressions and activity of Rubisco as compared to the wild type. Improved growth or photosynthetic rates could not be observed from these mutants. Experimental design revealed that though they used carbon dioxide saturating conditions, use of low intensity light appeared as a classic bottleneck in achieving high rate photosynthesis. In nature, variants of Rubisco are available, some intrinsically more efficient than their counterparts. Therefore, current strategies focus on replacing the sluggish native gene with a naturally more efficient counterpart. In one such study, Iwaki et al. (2006) expressed the Rubisco from *Allochromatium vinosum*, a purple sulphur bacterium in *Synechococcus elongates* sp. PCC 7942 which increased the CO_2 assimilation by almost 50 %. The study used two different strong promoters: psba from *Synechococcus elongatus* and psba2 from *Synechocystis* PCC 6803 (henceforth *Synechocystis*). However, the *A. vinosum* Rubisco was reported to be not fully incorporated into the carboxysomes.

Besides, heterologous expression, random mutagenesis and similar techniques including directed evolution and hybrid construction besides others have been in focus to increase the catalytic efficiency of Rubisco (Cirino and Frei, 2009). In the last decade, the technique of directed evolution has become a versatile tool for artificially mimicking the process of natural evolution (Bershtein and Tawfik, 2008). Directed evolution involves the creation of mutant library and then selecting the desirable phenotypes. This technique has been applied to improve both Rubisco and

Rubiscoactivase of various organisms (Mueller-Cajar and Whitney, 2008). Greene et al. (2007) considered the Rubisco from *Synechococcus* PCC7942 for directed evolution and examined it in *E. coli* hosts. They selected the strain with M262T mutation which conferred 26 % increase in carboxylation efficiency and increased Rubisco expression in *E. coli* by 5-folds. In another such study, trans-complementation of the *Rhodobacter capsulatus* (*R. capsulatus*) deletion mutant was carried out with a randomly mutagenized Rubisco from *Synechococcus elongatus* PCC 6301. The researchers screened for 5000 colonies. One of the selected engineered *R. capsulatous* strain carrying the randomly mutagenized single amino acid substitution showed higher growth rate compared to the strain carrying the wild type. Kinetic analysis of the engineered Rubisco showed a higher affinity to the substrate which was reasoned to provide growth advantages to the *R. capsulatus* strain (Smith and Tabita, 2003). It would be rather interesting to trans-complement the same mutagenized gene into its host *S. elongatus* to study if it would provide the same benefits (growth advantage) to the host strain.

So far it can be concluded that the extensive studies revolving around enhancing the catalytic efficiency of Rubisco have shown only modest success thus far. These results suggest that it may be possible that the Calvin cycle itself is not the major limiting factor for carbon fixation; instead it is possible that the downstream 'metabolic sinks' that utilize the product may be restricting carbon fixation to certain decree under specific environmental conditions (Stitt et al., 2010). It does appear that inspite of its sluggish catalytic action and non-specificity to substrate CO_2 it may be evolutionarily well optimized (Tcherkez et al., 2006). In view of this, it may perhaps be worth it to focus on the flux analysis and holistically look at the Calvin cycle in terms of multiple control points.

As mentioned earlier, carbon fixation in algae takes place in specialized compartments called 'carboxysomes' harbouring Rubisco and the carbonic anhydrase. Carbonic anhydrase deletion mutants with genetic disruption of the *icfA/ccaA* gene resulted in a high CO_2 requiring strain that fixed less than 5 % (v/v) CO_2 compared to the wild type (Fukuzawa et al., 1992). This indicates the crucial need of co-localizing carbonic anhydrase within the carboxysomes. In another interesting study, Lieman-Hurwitz et al. (2003) expressed the *ictB* gene involved in HCO_3^- from *Synechococcus* PCC 7942 in *Arabidopsis thaliana* and *Nicotianatabacum* to demonstrate faster photosynthetic rates in the engineered strain compared to the wild type under limiting but not saturating CO_2 concentrations. However, this did not alter the expression of Rubisco in the mutants. The authors suggest that the concentration of CO_2 in close proximity to Rubisco was nonetheless higher which could explain the increased photosynthetic rate. Similar expression studies in algae in a bid to improve the algal chassis may appear appealing. However, the expression of carbonic anhydrase is modulated by changes in the CcmM protein which is part of the structural shell of the beta-type carboxysomes. Transgeneic expression of the CcmM protein using a synthetic trc promoter led to higher expression of carbonic anhydrase in *S. elongatus* PCC 7942. The Rubisco activity still remained unchanged (Long et al., 2007). These results suggest that the amount of carbonic anhydrase is not limiting the activity of Rubisco*in vivo*. The study does suggest that carbonic anhydrase and CcmM58 are co-regulated and overexpressing carbonic anhydrase alone might lead to increased levels of carbonic anhydrase in the cytosol, which is likely to be detrimental to the cells (Price and Badger, 1989). Thus for effective increase of the carbonic anhydrase in carboxysome, co-expression with CcmM may be required.

To create a high partial pressure of CO_2 within the carboxysomes, the flow of the substrate is facilitated by the presence of several CO_2/HCO_3^- transporters in the wall of the carboxysomes. This ensures that the oxygeneic photosynthesis is not plagued by photorespiration. Several research groups across are attempting to increase the number of inorganic transporters to see if they can further modulate the efficiency of the carbon fixation process by increasing the availability of the substrate in the cell and the carboxysomes in particular. To achieve this, four different transport systems for dissolved carbon uptake have been identified in cyanobacteria. The first two HCO_3^- transporters are single-gene systems and includes SbtA which is an inducible, high-affinity Na^+ dependent HCO_3^- transporter (Shibata et al., 2002; Price et al., 2004) that apparently acts as an Na^+/HCO_3^- symporter with a relatively low flux rate. In addition, BCT1 is a high-affinity HCO_3^- transporter (uniporter), inducible under dissolved inorganic carbon limitation, belonging to the ATPase family, and, unlike SbtA, is encoded by four genes (*cmpABCD*) to complete a functional transporter complex. NdhF3 and NdhF4 are the other two transporters that have been recognized as importers for CO_2 uptake. The transcripts of their genes were up-regulated under CO_2 atmosphere (McGinn et al., 2003). These transporters are being further evaluated for their role and possibility of transgenic over expressions.

The recent advent of synthetic biology has explored a whole new set of possibilities. Bonacci et al. (2012) demonstrated transgene expression of functional carboxysomes assembly and packaging in *E. coli*. Characterization of purified synthetic carboxysomes showed that they were well formed and were capable of CO_2 fixation *in vitro*. This shows the extent of modularity of expressing non-native systems in desirable hosts. This work has demonstrated that protein- protein interactions using scaffolds can be used to 'rewire' signal networks and improve the metabolic pathways. This could prove beneficial to improve the rate and yield of cellular chemical reactions while lowering toxicity to the host.

4.2.1.2 Engineering Synthetic Carbon Fixation Pathways

Other than the main Calvin cycle there are several alternative pathways available that can increase the rate of carbon fixation within the cells. Recently, Bar-Even et al. (2010), inspired by the alternate carbon fixation cycles prevalent in nature, devised indigenous synthetic pathways. These pathways were based on the repertoire of enzyme information already available in the databases. The pathways were accredited based on their physicochemical criteria such as kinetics, energetics and topology. Based on different modelling algorithms, their study concluded that the malonyl-CoA-oxaloacetate-glyoxylate (MOG) pathway involving phosphoenolpyruvate carboxylase (PepC) was the most promising candidate for synthetic carbon fixation. In this pathway PepC is assisted by malic acid in the fixation of large amounts of CO_2 which is similar to that in C4 plants (Yang et al., 2002). However, instead of the final release of the CO_2, as occurs in the bundle sheath of the C4 plants, the metabolic bypasses of the pathway largely comprising eight or more enzymes would be used to convert malate to pyruvate without the loss of the fixed carbon (Bar-Even et al., 2010). This pathway appears as an attractive alternative because of the favourable properties of the carboxylation enzymes allowing much higher rates of carbon fixation as compared to the native Calvin cycle pathway.

To date there are limited reports about the effect of expression of these genes in prokaryotes and eukaryotes alike. In fact, prokaryotic PepC with the exception of that of *E. coli* were not well studied until recently when a deluge of information about them was available, mainly through the efforts of genome sequencing. In an attempt to study the effect of expression of *pepc* in plants, Chen et al. (2004) expressed the *pepc* from *Synechococcusvulcanus* in *Arabidopsis* using a strong constitutive promoter. In nature, most characterized PepC are found to be allosterically inhibited by the downstream product aspartate or malate. However, *Synechococcusvulcanus* harbours a unique enzyme almost insensitive to feedback inhibition at neutral pH. However, the *Arabidopsis* transgenic lines failed to respond along the expected lines of enhanced carbon metabolism. Instead, the T1 generation of plant transformants showed severe visible phenotypes such as leaf bleaching and were infertile when grown on soil. The growth inhibition of mutants was presumed to be primarily due to a decreased availability of phosphoenolpyruvate, one of the precursors for the shikimate pathway for the synthesis of aromatic amino acids and phenylpropanoids. However, it is still worth to attempt endogenous expression as well as transgeneic expression of these pathways in both cyanobacteria and green algae towards enhancing the rate of fixed carbon.

4.2.1.3 Improving the Light Harvesting Antennae

Another major problem of the algae when grown in reactors for the purpose of food or feed is the shading effect. For microalgae and photosynthetic bacteria, there exists a technical barrier caused by the 'cell-shading' phenomenon of these pigmented cells. The photosynthetic efficiency directly depends on the ability to harness the solar photons and store it in the form of fixed carbon. The algae have developed large harvesting complexes due to their evolution under low light regime. Presence of large antennas ensures maximum absorption of the incident light. However, at the same time, algae do not respond well to saturating light intensity. Under such conditions, most of the reaction centres are closed, and a mechanism that dissipates the excess absorbed light into heat is employed to digress part of the energy absorbed by the antennae, such that the excess energy does not reach the photosynthetic reaction centre. This prevents photo damage of the reaction centres. Considering the goal of maximizing fuel production from these organisms the incapability of harvesting saturating light intensity may limit the production efficiency from large scale mass culture. Towards this, Mussgnug et al. (2007) applied the technique of RNAi technology to down regulate the entire LHC gene family simultaneously to reduce energy losses by fluorescence and heat. The mutant Stm3LR3 had significantly reduced levels of LHCI and LHCII mRNAs and proteins while chlorophyll and pigment synthesis was functional. The mutant showed higher photosynthetic quantum yield and a reduced sensitivity to photoinhibition, resulting in an increased efficiency of cell cultivation under elevated light conditions. Collectively, these properties offer three advantages in terms of algal bioreactor efficiency under natural high-light levels: (i) reduced fluorescence and LHC-dependent heat losses and thus increased photosynthetic efficiencies under high-light conditions; (ii) improved light penetration properties; and (iii) potentially reduced risk of oxidative photodamage to PSII.

Polle et al. (2003) developed chlorophyll antenna truncation of the green alga *Chlamydomonas reinhardtii* (henceforth *C. reinhardtii*), which showed higher photosynthetic efficiency and biofuel production capacity compared with the wild-type. The photosynthetic apparatus (PS) of cyanobacteria in general is similar to that described for higher plants; however, the PS in cyanobacteria have several peculiarities. In addition to the chlorophyll protein complex, the cyanobacterial PS possess phycobilisomes with the pigment proteins phycocyanin and phycoerythrin primarily attached to the PSII dimers. The presence of these pigments extends the spectrum of visible light usable for photosynthesis (Rascher et al., 2003). However, cyanobacteria also suffer from similar light limitations and, hence, another approach to enhance their biofuel potential may be to develop truncated PS. Using chemical mutagenesis, Nakajima et al. developed a phycocyanin-deficient PD1 mutant of *Synechocystis* PCC 6714 (Nakajima et al., 2001). This mutant showed 1.5-times the photosynthetic productivity of the wild type under high light intensity. However, the potential of this mutant as a biofuel producer has not been reported.

4.2.2 Biasing the Metabolic Flux Towards Biofuel Production

The synthesis of glycogen is catalyzed by glucose 1-phosphate adenyl-transferase (GlgC), glycogen synthase and 1,4-alpha glucan branching enzyme. GlgC catalyzes the primary step of conversion of Glucose 1-phosphate into ADP glucose. Since the storage metabolite glycogen is the major metabolic source for spending fixed carbon and reducing power, decoupling of the native metabolism from glycogen synthesis could redirect the carbon to other biosynthetic processes which may favour increased cell growth and higher growth dependent product yield. This was illustrated in the study of Li et al. (2014) who demonstrated the development of a GlgC mutant that could redirect its carbon and reducing power towards the production of a non-native isobutanol. Under high light the GlgC mutant showed growth impairment; however, induction of the isobutanol pathway in the same cell rescued the growth defects. The study strongly suggests that growth of GlgC mutant alone under high light faced redox imbalance due to accumulation of reduced

NADPH. Induction of the isobutanol pathways facilitated the cell in finding an alternate pathway to release the redox stress.

Another aspect that can be utilized in implementing biotechnological applications including the production of biofuels is the utilization of these solar harnessing photosynthetic cells as sugar producing factories. Endogenous production can be enhanced by increasing the rate of sugar transport out of the cell, knocking out the competing pathways and stimulating production by altering cultivation conditions. Xu et al. (2013) showed that by deleting the *glgAI* and *glgAII* genes responsible for converting ADP-glucose to glycogen amylose, *Synechococcus* PCC 7002 accumulated 1.8 fold more sugar than the wild type and these cells spontaneously excreted soluble sugars into the medium at high levels without the need for additional transporters. These cells producing soluble sugar in the medium can then directly be utilized by bacteria (mesophiles/thermophiles) co-cultivated in reactors to produce biofuels and other valuable compounds. This will also bypass the need of pretreatment costs considering instances where the stored glycogen/starch of algae has been used to cultivate bacteria (Nayak et al., 2014; Roy et al., 2014).

Another excellent example to diverting the metabolic flux towards the desired product was provided by Matthias Rögner's research group in Germany. They have shown how the affinity of the electron carrier ferredoxin (Fd) can be modulated towards the hydrogen production pathway instead of the Calvin-Benson cycle. The reductants obtained by splitting of water enter into the Calvin cycle via Ferredoxin NADPH Reductase (FNR). This is a diverging point where the reductants can either enter into the Calvin cycle or be accepted by hydrogenase (in green algae) to produce the biofuel hydrogen. They have shown for the first time in the model organism, *Synechocystis* that the bias of the ferredoxin can be decreased towards FNR (Rögner, 2013). This in turn makes the reduced ferredoxin available to donate the electrons to a transgenetically available non-native [FeFe] hydrogenase integrated into the genetic circuit of the strain. This increases the possibility of higher hydrogen production as compared to biomass generation. This is one of the most critical points in increasing the photosynthetic hydrogen production in *Synechocystis* and is due to the much higher affinity of Fd for FNR than for the [Fe–Fe]-hydrogenase from *C. reinhardtii* (K_m differs by a factor of 10–40). This may also be due to the much higher concentrations of FNR in the cell as compared to the non-native hydrogenase. Changing bias of ferredoxin in favour of hydrogenase was possible by decreasing the Fd affinity to FNR. Preliminary studies have shown that K_d values could be shifted by a factor of more than 10 as determined with isolated FNR from WT and directed mutants (Rögner, 2013). This would be a great step in the direction of obtaining higher rate of hydrogen production from cyanobacteria.

4.3 Genetic and Metabolic Engineering of Algae to Obtain High Yield

4.3.1 *Alcohols*

Among the alcohols that have been reportedly produced from cyanobacteria using either the native system or the non-native system are 2–4 carbon alcohols including ethanol, 1, 2 propanediol, 2,3-Butanediol, 1-butanol and isobutanol (Fig. 4.2).

Besides, fatty alcohols and *n*-carbon alcohols derived from fatty acids have also been reported. The first four aliphatic alcohols (namely methanol, ethanol, propanol and butanol) are of interest as fuels because they can be synthesized both chemically or biologically and they have properties that allow them to be used as fuel in internal combustion engines. Blends of ethanol with other fossil fuels have increasingly gained momentum in the transport industry.

The major pathway for ethanol synthesis is catalyzed by two enzymes, pyruvate decarboxylase (Pdc) and alcohol dehydrogenase (Adh). Pdc catalyzes the non-

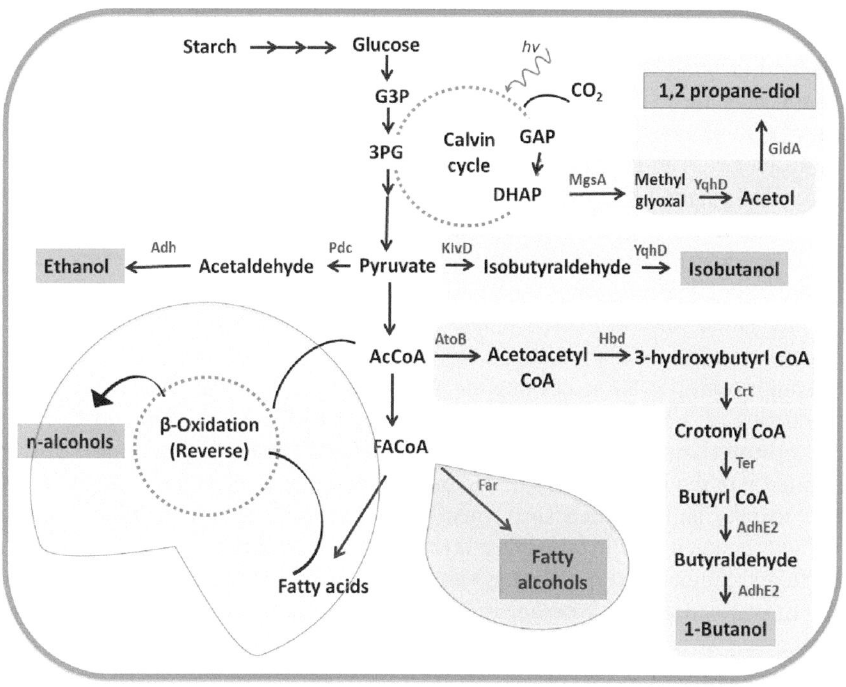

Fig. 4.2 Schematic of heterologous pathways that have been introduced into cyanobacteria for the production of various alcohols including 1,2 propanediol, isobutanol, 1-butanol, fatty alcohols, *n*-alcohols and ethanol. Abbreviations: G3P: Glucose-3-phosphate, 3PG: 3-phosphoglycerate, AcCoA: acetyl-CoA, FACoA: Fatty-acylCoA, GAP: Glyceraldehyde-3-phosphate, DHAP: Dihydroxyacetone-phosphate.

oxidative decarboxylation of pyruvate, which produces acetaldehyde and CO_2. Acetaldehyde is then converted to ethanol by Adh. This fermentation pathway plays a role in the regeneration of NAD^+ for glycolysis under anaerobic conditions in fungi, yeasts, and higher plants. Currently, the majority of alcohol is being produced by fermentation of starch or cellulose-derived sugars. Research groups have explored the possibility of utilizing the algal starch/glycogen as feedstock for bacterial ethanol production (Nguyen et al., 2009). However, recently researchers have come up with a more direct way of coupling photosynthetically fixed carbon to alcohol. Deng et al. demonstrated the production of ethanol from cyanobacterium *Synechococcus* sp. strain PCC 7942 as early as 1999. However, to date any significant improvements in yields have not been reported. The earlier report was a proof of concept design with high sensitivity to the oxygen due to splitting of water at PSII.

Recently, butanol has been explored for use as transport fuel. Butanol as fuel has advantage in the existing combustion engines since it has properties close to that of the combustion engines already in use. For a long time butanol production was reported by obligate anaerobes like *Clostridium*. Thus, the pathway is oxygen sensitive. The challenge then was to express the pathway in oxygenic autotrophs. This was successfully demonstrated by Lan and Liao (2012) where they reported the production of 1-butanol by a modified-CoA dependent pathway obtained from obligate and facultative bacteria into *Synechoccocus elongatus*. Five genes were integrated into the *Synechoccocus* genome including *hbd*, *crt*, *adhE2* from *C. acetobutylicum*, *ter* from *T. denticola*, and *atoB* from *E. coli*. The acetyl-CoA acetyltransferase (A to B) was selected to replace the acetoacetyl-CoA thiolase (Thl) from *Clostridium* due to its higher specific activity. The activities of these enzymes were determined by enzyme assays and it was found that the activities of ter and adh2 were lower than the other three enzymes in the pathway. This indicated that in cyanobacteria these enzymes were the rate limiting step. Addition of poly histidine tag increased the ter activity and thus the productivity of 1-butanol in cyanobacteria.

Another effort in the same direction was made by Oliver et al. (2013) towards the production of 2,3-butanediol production in *Synechococcus elongatus* PCC 7942. The operon comprising *alsS*, *alsD* and *adh* under the effect of plac promoter was inserted into the neutral site. They identified 2,3-butanediol as a biofuel with low host toxicity and designed an oxygen-insensitive, cofactor-matched biosynthetic pathway coupled with irreversible enzymatic steps to create a driving force toward the target. Similarly, very recently, Varman et al. (2013) demonstrated the production of isobutanol from *Synechocystis*. They improved the strain both by metabolic engineering as well as bioprocess optimization. They engineered a glucose tolerant *kivd* and *adhA* gene of the Ehrlich pathway. It is already well known that isobutanol is toxic to the cells and may also be degraded photochemically by hydroxyl radicals produced during the cultivation process. Therefore, a clever strategy to improve yields is to remove the product from the culture soon after its release into the medium. The study of Varman et al. demonstrated this with ease by simply placing a glass tube filled with oleyl alcohol placed within the cultivation flask. Due to the high volatility, the produced isobutanol in the headspace of the cultivation flask would get trapped in the oleyl alcohol and could easily be separated. This depicted a non-intrusive way of downstream processing.

In yet another proof of concept Li and Liao (2013) demonstrated the production of 1,2-propanediol from *Synechococcus elongatus* PCC 7942 by introduction of the genes encoding methylglyoxal synthase (*mgsA*), glycerol dehydrogenase (*gldA*) and aldehyde reductase (*yqhD*) . However, a comparable amount of the toxic intermediate acetol was also produced. To overcome this issue, the NADH-specific secondary alcohol dehydrogenases was expressed which minimized the accumulation of the incomplete reduction product acetol.

Another category of alcohol the 'fatty alcohols' have recently been produced from the photoautotrophic organisms. They are also considered as suitable fuel additives due to favourable fuel properties. Generally fatty alcohols can be produced by two pathways: the first pathway involves the reduction of the fatty acyl-CoA (acyl-ACP) to fatty alcohols. The alternate pathway involves two distinct enzymes, a fatty aldehyde forming acyl-CoA reductase which catalyzes a two-electron reduction of fatty acyl-CoA to a fatty aldehyde intermediate, and a fatty aldehyde reductase which catalyzes the reduction of fatty aldehyde to fatty alcohol (Metz et al., 2000). Cyanobacteria do not contain endogenous fatty alcohol producing pathway. Yao et al. (2014) investigated the production of these alcohols by transgenetically expressing the fatty acyl-CoA reductase (FAR) from higher animal models like mouse, and plant models like *Jojoba* and *Arabidopsis*. They reported $9.73 \pm 2.73\ \mu g\ OD_{730}^{-1}\ L^{-1}$ of fatty alcohol from *Synechocystis* strain expressing the *jojobaFar* gene.

Thus, several proof of concept studies have been reported in literature in the recent past, a boon made possible due to the advent of synthetic biology. Most precedents are recent and have occurred towards the end of the last decade. The issue now is how to move beyond the proof of concept study and identify the bottlenecks that are limiting further progress in this field. Expressing the pathway in optimal algal chassis would go a long way in realization of these fuels from phototrophic hosts.

Considering alcohol production, the general targets for improvements may include (1) tolerance of the developed strain towards the solvent produced. During fermentation alcohols are produced in the stationary phase. They are toxic due to their hydrophobic nature that increases the fluidity of the cell membranes. As a result the cell membrane permeability is compromised; transmembrane pH is affected resulting in energy shortage. Microbes in general react to the toxicity by producing heat shock proteins, and by activating efflux pumps for the solvents. This kind of toxicity leads to decreased productivity. Therefore, of foremost importance is to address the questions of how to make the cell more robust in terms of solvent production and prolonged cell viability. (2) Secondly, if these enzymes are linked to the photosynthetic chain the oxygen toxicity of these enzymes must be addressed due to the natural evolution of oxygen at PSII. (3) Competing pathways that require the same redox potential must be eliminated. In this case, native hydrogenase, lactate pathway may be deleted as they would dispose electrons/energy that could otherwise be channelled towards alcohol production. Also, elimination of carbon sinks can increase productivity. Impairment of the glycogen synthesis pathway by deletion of *glgC* the key enzyme that converts glucose 1-phosphate to ADP-glucose has been suggested. This is an interesting approach since production of alcohol comes from the catabolism of glucose through the glycolysis cycle in the dark. However, GlgC steals away the glucose 1-phosphate and reconverts it into ADP-glucose that leads to glycogen synthesis. A non-native pathway linked to such an engineered strain could already have a high driving force to release the redox imbalance built due to the accumulation of Glucose 1-phosphate stress. (4) Next, optimizing pathways to drive the flux towards product formation must be considered. An interesting example of this was shown by James Liao's group where they constructed a non-native Clostridial pathway for 1-Butanol production in *E. coli* (Shen et al., 2011).

To improve the production beyond the known titer limits they identified the metabolic bottleneck in the pathway reversible on both thermodynamic and enzymatic grounds. Therefore, they envisaged the presence of an artificial driving force to improve product yields. The *Clostridium* 1-butanol pathway requires both NADH and reduced ferredoxin as cofactors (Bcd-EtfAB complex). If the pathway could be dependent on only NADH, then its cellular accumulation could be used as the driving force towards production of 1-butanol. Concurrently, the elimination of all other pathways utilizing NADH including ethanol, lactate and succinate would be desirable. Such a strain would not be able to survive under anaerobic conditions due to redox imbalance till it produced butanol as the fermentative redox stress buster. Therefore they replaced the Bcd-EtfAB complex with Ter (trans-enoyl-coenzyme A reductase) from *Treponema denticola* (Lan and Liao, 2012). This NADH-dependent enzyme could potentially facilitate tighter coupling of 1-butanol production with the NADH driving force without flavoproteins or ferredoxin.

4.3.2 Hydrogen

Hydrogen as a fuel has been shown to be attractive primarily because on combustion it produces water as the only "waste emission". Photoautotrophic organisms are considered as attractive alternatives because they can generate their own feed stock reserves and require only water and sunlight to produce hydrogen. The enzyme catalyzing the deceivingly simple reaction $2H^+ + 2e^-\ H_2$ are called hydrogenase. They are classified depending on the presence of the transition metals in the active site. The need for catalytic transition metals is explained by the significant increase in the acidity of molecular hydrogen when bound to them (Collman, 1996). Accordingly there are three phytogenetically unrelated class of these enzymes that include [NiFe], [FeFe] and a third class in which hydrogen uptake is coupled with methanopterin reduction. Most known hydrogenases are oxygen labile, and cyanobacteria and green algae are the only known organisms so far that are capable of both oxygenic photosynthesis along with hydrogen production. While cyanobacteria contain the 'native hox operon' which is homologous to the [NiFe] hydrogenase for hydrogen production, the green algae *C. reinhardtii* hosts [FeFe] class of hydrogenase. Since they are the most promising hosts to produce this biofuel, a range of genetic engineering work including genetic knockouts, introduction of non-native genes and pathways and elimination of competing pathways utilizing the same reductant has been approached among others. However, to date these are only 'proof of concept' designs that are available and still research needs to go a long way before the system becomes commercially viable.

In algae, hydrogen production can occur through three pathways. Two of them, though essentially catalyzed by the same enzyme 'bidirectional hoxhydrogenase', differ in terms of source of reductants (Fig. 4.3), the third differs in terms of the enzyme catalyzing the reaction. In light, the source of reductants for the bidirectional hoxhydrogenase is obtained from the splitting of water at PSII that results in reduced NADPH via PSI, ferredoxin and ferredoxin:NADPHreductase (FNR). Notably, in green algae hosting [FeFe] hydrogenase the reduced Fd can directly act as an electron donor to it. This process is coupled to the oxygen evolving splitting of water at PSII and occurs only transiently when dark kept cells are exposed to light, the excess generated reductant is channelled out as molecular hydrogen. However, the rate of oxygen production is higher than the rate of consumption by respiration, hence the damage to the oxygen labile hydrogenase. In an alternate

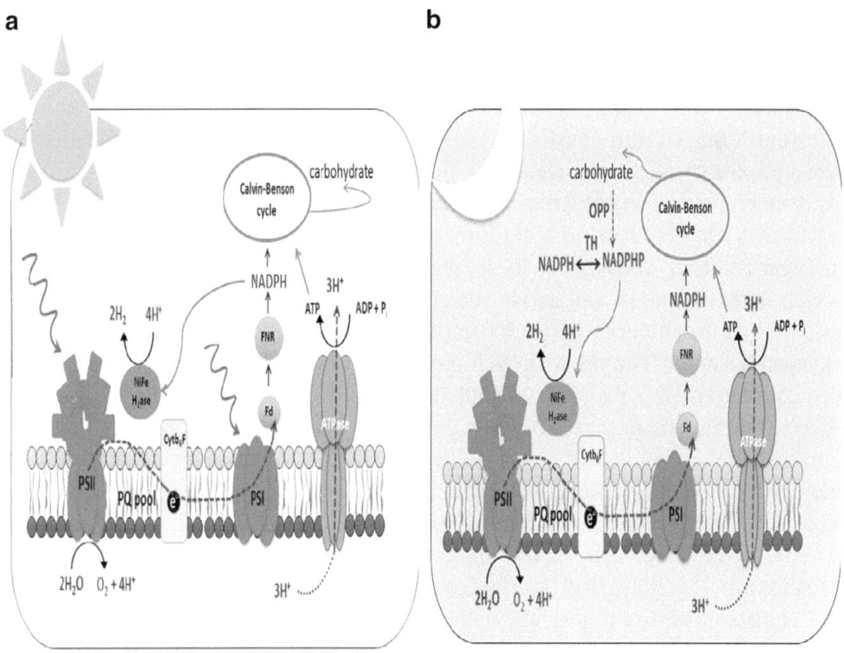

Fig. 4.3 Schematic of hydrogen production from cyanobacteria: (a) Under light the source of reductants comes from the splitting of water at PSII and (b) In dark the source of reductants is obtained from the catabolism of glycogen.

pathway, the hydrogen is produced in the dark where the source of the exogenous electrons is the stored carbohydrate. In the absence of light, catabolism of sugar occurs to feed the energy requirement of the cell and the reductant thus generated is employed by the hydrogenase, besides others. In the process of NADPH oxidation, the electrons are shuttled to the plastoquinone pool and subsequently used for hydrogen biosynthesis. Hydrogen production through this mode is prolonged.

The reversible bidirectional hoxhydrogenase is associated with the cytoplasmic membrane and likely functions as an electron acceptor from both NADH and H_2 (Boison et al., 1999). The reversible bidirectional hoxhydrogenase is a multimeric enzyme consisting of eight different subunits. Molecularly it is a [Ni–Fe]-hydrogenase of the $NAD(P)^+$ reducing type and consisting of a hydrogenase dimer coded by *hoxYH* gene. Maturation of reversible hydrogenases requires the action of several auxiliary proteins collectively termed as *hyp* (products of genes: *hypF*, *hypC*, *hypD*, *hypE*, *hypA* and *hypB*) (Casalot and Rousset, 2001). Unlike uptake hydrogenase, reversible hydrogenases are helpful in hydrogen production. While the uptake hydrogenase is present in all nitrogen-fixing strains screened so far, the bidirectional enzyme is distributed both among the nitrogen-fixing and the non-nitrogen-fixing cyanobacteria (although it is not a universal cyanobacterial enzyme).

The third pathway for hydrogen production is mediated by nitrogenase. Nitrogenase dependent production of molecular hydrogen occurs primarily in the heterocyst's nitrogen fixing cyanobacteria. The source of reductant for nitrogen fixation is obtained from the catabolism of sugar through the oxidative pentose phosphate (OPP) pathway in the vegetative cells. Hydrogen is produced as a by-product of nitrogen fixation and is an energy intensive process. Hydrogen production catalyzed by the nitrogenase occurs as a side reaction at a rate one-third to one-fourth that of nitrogen fixation, even in a 100 % nitrogen gas atmosphere. In close proximity of the nitrogenase is situated the uptake hydrogenase in the thylakoid membrane of the heterocyst where it transfers the electrons from the breakdown of hydrogen evolved from nitrogenase for the reduction of oxygen via the knallgas reaction. The enzyme itself consists of two subunits, a large subunit (hupL) which harbours the [NiFe] active site and a small subunit that is coded by (hupS) that acts as the molecular wire to transfer the electrons from the active site to an electron acceptor. The net hydrogen produced from such strains is lowered due to this counterproductive reaction. Higher hydrogen yields have been demonstrated by successful deletion of these enzymes. Previous studies have reported that hydrogen uptake-deficient mutants of *A. variabilis* AVM13 (ΔhupSL) (Happe et al., 2000), *N. punctiforme* NHM5 (ΔhupL) (Lindberg et al., 2002), *Anabaena* PCC 7120 AMC 414 (ΔxisC) (Carrasco et al., 2005), *Anabaena* PCC 7120 (ΔhupL, ΔhupL/ΔhoxH, ΔhupW) (Masukawa et al., 2002; Lindberg et al., 2012), *Nostoc* PCC 7422 (ΔhupL) (Yoshino et al., 2007) and *A. siamensis* TISTR 8012 (ΔhupS) (Khetkorn et al., 2012) have an ability to produce hydrogen at a significantly higher rate compared with the respective wild-types. Ekman et al. (2011) carried out whole scale proteomic analysis of one such deletion mutant developed by Lindberg et al. and found that the pathway for OPP was significantly up-regulated in the mutant suggesting that the deletion of uptake hydrogenase may increase production of the substrate via the OPP pathway.

It is to be noted that hydrogenase are extremely oxygen labile proteins. With this understanding it is imperative that their expression is under anaerobic conditions. Development of anaerobic conditions in pilot scale is another challenge both technically and economically. But how does nature protect its oxygen labile enzyme? It has developed the micro-anaerobic environment in the heterocyst of filamentous cyanobacteria. This protects its oxygen labile nitrogenase and uptake hydrogenase. For those organisms in which heterocysts are not present, the expression of the enzymes are metabolically or cicardially controlled. Moreover, the metabolism within the heterocyst consists of only PSI, without the presence of the oxygen evolving PSII. The nitrogenase mediated nitrogen fixation is a reductive process, and gets an electron supply from the vegetative cells via ferredoxin. Thus, an electron transport system suitable for [FeFe]-hydrogenases may already be present in the heterocyst. Getting inspiration from nature, the best way forward to realize a hydrogen economy would be to express the enzymes in nature's special anaerobic compartments. Gärtner et al. (2012) attempted to achieve it by expression of the *Shewenella oneidensis* MR-I dimerichydrogenase in *Anabaena* sp. strain PCC 7120 using the heterocyst *hetR* promoter for heterocyst specificity. However, they could demonstrate the hydrogen production only in *in vitro* studies. The study showed that though the enzyme was functional the source of reductants *in vivo* was limited to the developed system. This was clear, since activity could be clearly observed on addition of a promiscuous electron donor (methyl viologen).

Recently, Ducat et al. (2011) showed the expression of *Clostridium acetobutylicum* hydrogenase in unicellular, non-nitrogen fixing cyanobacteria, *Synechococcus elongates* sp PCC 7942 along with its accessory genes. They demonstrated that the hydrogenase are functional *in vitro* and *in vivo*. Under anoxic light conditions obtained by addition of DCMU (3-(3,4-dichlorophenyl)-1,1-dimethylurea), the mutants evolved 500-fold higher hydrogen as compared to the wild type. DCMU is a herbicide that inhibits photosynthesis by binding to the plastoquinone site of PSII thereby preventing the splitting of water and limiting the linear photosynthesis. However, the addition of DCMU has no effect on the activity of PSI.

Towards optimizing the cell for higher hydrogen production, several efforts have been made. Among these, consistent efforts have been made by the group of Matthias Rögner in Ruhr University, Bochum, Germany. The research group at Bochum is in the process of development of a suitable chassis for high rate hydrogen production using the model organism *Synechocystis* (Fig. 4.4).

The electrons released by splitting of water at PSII primarily govern the rate of hydrogen production. In *Synechocystis* the PSII to PSI ratio stands as 1:10. Moreover, the PSII reaction centre carries the phycobilisome (PBS) proteins that get over-saturated at high light intensities. For high rate hydrogen production this feature is a deterrent. In view of this, the German group was the first to show the development of olive mutant (mutant carrying deletion of phycocyanin-binding subunits) with reduced antennae size that was capable of 4-fold increase in linear electron transport. In a step further, 6-fold increase was obtained in the PAL mutant with

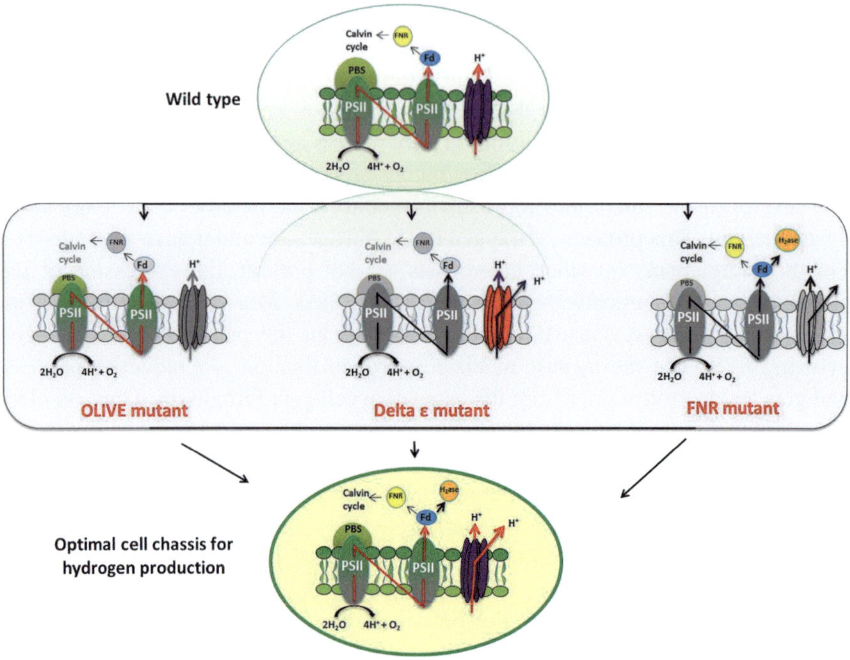

Fig. 4.4 Schematic for the development of an optimal chassis for hydrogen production. The optimized cell comprises (a) PAL mutant with completely deleted PBS (phycobilisome) genes, (b) Delta ε mutant with lower ATP synthesis rate as compared to the WT and (c) FNR mutant with decreased bias of the FerredoxinNADPH Reductase (FNR) towards ferredoxin (modified from Rögner, 2013).

completely deleted PBS genes (Bernat et al., 2009). Such truncated antennae mutants showed more tolerance to high light. The cellular metabolism of the phycobilisome mutant cells also saves considerable cellular energy currency. Considering the total protein pool, phycobilisomes constitute 63 % of all the soluble proteins in *Synechocystis* cells (Moal and Lagoutte, 2012).

In the next step towards the development of an optimal chassis they developed the delta ε mutant that had lower ATP synthesis rate compared to the wild type. This was developed with the hypothesis that the supply of electrons from water splitting is further limited by the tight coupling of luminal proton efflux with ATP synthesis. The partial uncoupling resulted in 2-fold higher linear electron transport (Imashimizu et al., 2011). Moreover, the impairment

did not affect growth or survival. This suggested the possibility of readjusting the cellular metabolism to the availability of NADPH and ATP required for carbon fixation and growth. It was postulated that the additional electrons available as a result of lower carbon fixation could be rechannelled towards biofuel production.

In yet another step towards chassis optimization, the group showed the decreased bias of FNR towards ferredoxin. Efforts are also ongoing to improve the green algae, *C. reinhardtii* chassis for hydrogen production. Towards this end, Melis

(2009) developed the antennae truncated mutants that showed higher rate of linear electron transport. Further, Reifschneider-Wegner et al. (2014) demonstrated that a codon-optimized native hydrogenase could be successfully expressed in the chloroplast. However, they observed a strong negative selection due to constitutive expression of the hydrogenase and which was resolved by using a vitamin induced gene expression system.

Thus efforts are ongoing towards the realization of a stable optimal hydrogen producing algae in an effort to replace a part of our current fossil fuel needs.

4.3.3 *Biodiesel*

Biodiesel is of significant importance because it has properties similar to that of petrol diesel and can be used with the current available infrastructure. Biodiesel is essentially fatty acids with long alkyl chains used by cells for energy storage and chemical production. It is a clean burning fuel made by the trans-esterification of the triacyl glycerides purified from photosynthetic plants or algae. Statistical analysis shows that in UK out of the 47 billion litres of transport fuel utilized, 53 % comprised petro-diesel (UK department of energy report, 2009). The world economy could greatly benefit if this could be achieved using sustainable sources. However, if this sustainability were to be achieved using plant oil such as rapeseed, it would require 17.5 Mha land, more than 50 % of the arable land area of UK (Scott et al., 2010). Algae may thus be the obvious alternative. Microalgae have the ability to accumulate oil in the form of triacylglycerides (TAGs) (Chisti, 2008). However, the lower yields limit their industrial application. An obvious choice would be to increase their oil content. Stress conditions are known to increase the oil content. Physiologically, this can be achieved by nutrient starvation such as nitrogen; however, this also decreases the biomass yields and hence lowers productivity. Therefore, indigenous studies around optimizing the lipid pathway and its recovery are required to increase productivity. The four obvious strategies would be (i) to increase the lipid accumulation, (ii) to decrease the lipid catabolism, (iii) to eliminate the competing pathways feeding on the same substrate and (iv) to optimize downstream processing.

Present efforts to obtain high yield biodiesel from photosynthetic microorganisms have been focused on the green algae, *C. reinhardtii* (Msanne et al., 2012). With the availability of the complete genomic information, many genes required for lipid and tri acycyl glycerol biosynthesis have been identified (Riekhof et al., 2005). Genome comparison and gene prediction analysis have shown that the pathways of fatty acid synthesis between plants and algae are largely conserved. The *in vivo* function of these fatty acids is to build membrane lipids and storage lipids. Acetyl CoA serves as the basic building block for fatty acid biosynthesis and in algae is derived mainly from glycolysis (Hu et al., 2008). The rate limiting step in this process is the formation of malonyl CoA from acetyl CoA catalyzed by acetyl CoA carboxylase. The malonyl group is then transferred to a small acyl carrier protein

(ACP) catalyzed by 3-keto acyly-ACP reductase and dehydration catalyzed by 3-hydroxylacyl-ACP dehydratase. Second round of reduction is catalyzed by enoyl-ACP reductase. These serial reactions result in the addition of two methylene carbons to the growing acyl chain. Fatty acids with a carbon chain length of 12–14 atoms are most suitable for the production of biodiesel. Most microalgae synthesize lipids with a chain length of 14–20 carbon atoms. The acyl ACP thiosterase enzyme is responsible for the length of the chain, and species specificity is typical for it.

To date most of the studies have focused on increasing the lipid biosynthesis by blocking competing pathways. Wang et al. (2009) reported that after 48 h of nitrogen starvation in the presence of acetate, the wild-type lipid body content has increased 15-fold. When starch biosynthesis was blocked in the *sta6* mutant, the lipid body content further increased to 30-fold, demonstrating that genetic manipulation can enhance lipid body production. However, on downside the cellular growth rate was much compromised. Thus to date strategies that have increased lipid production have resulted in decreased growth in the engineered strains. Maintaining high growth rates and high biomass accumulation is imperative for algal biofuel production on large economic scales, and engineering efforts that increase lipid content without decreasing growth or biomass can significantly reduce production cost and increase the economic viability of algal biofuels.

In *E. coli* the product of fatty acid biosynthesis is fatty-acyl-ACP (Acyl carrier protein), accumulation of which causes feedback inhibition of fatty acid biosynthesis. Lu et al. (2008) found that over expression of thioesterase 1 in *E. coli* bypass the feedback inhibition caused by product accumulation. Recently, researchers introduced a codon optimized acyl carrier protein thioesterase in *Synechocystis* with six different genetic mutations (Liu et al., 2011). These mutations resulted in weakened cell wall layers due to altered surface proteins and peptidoglycan layers allowing enhanced diffusion through phospholipid layers.

Downstream processing including (cell disruption, extraction and separation) is another area that is energy/cost intensive. In an indigenous study, lipase enzymes that degrade the membrane lipids and increase the cell wall permeability to release the oil were expressed using CO_2 inducible promoters. These promoters were previously tested by McGinn et al. (2003). They reported that aeration of illuminated cells with CO_2-free air for 30 min depleted the culture of CO_2 to near zero levels. Under these conditions the transcripts for three inducible inorganic carbon uptake systems *ndhF3*, *sbtA* and *cmpA* showed near maximal abundance after 15 min of CO_2 concentration reductions. Liu et al. (2011) used promoters of these genes to regulate the expression of lipase. For harvesting the culture was subjected to zero CO_2 environment for 30 min. This induced the lipolytic gene expression which permeabilized the cell wall for release of the molecules into the medium.

On the negative side, overexpression of lipid synthesis pathway genes affects microalgal proliferation. This can be overcome by use of an inducible promoter that can be activated once the microalgal cells have grown to a high density and have entered the stationary phase. There are several examples of inducible promoters in algae including copper-responsive elements in *C. reinhardtii* (Quinn and Merchant, 1995) and a nitrate-responsive promoter in diatoms (Poulsen and Kroger, 2005).

Inhibiting lipid catabolism may also cause problems with proliferation and biomass productivity since microalgae often rely on catabolic pathways to provide energy and precursors for cell division. Lately, it was shown that the targeted knockdown of lipase/phospholipase/acyltransferase increased lipid yields without affecting growth in the diatom *Thalassiosira pseudonana*. Further, antisense expressing knockdown strains 1A6 and 1B1 exhibited wild type like growth and increased lipid content under both continuous light and alternating light-dark conditions (Liu et al., 2011). Analyses of fatty acids, lipid classes, and membrane stability in the transgenic strains suggest a role for this enzyme in membrane lipid turnover and lipid homeostasis. These results demonstrate that targeted metabolic manipulations can be used to increase lipid accumulation in eukaryotic microalgae without compromising growth. Similar strategies in pathway engineering should be achievable also in cyanobacteria, where photosynthesis, rather than organic feedstock, would provide energy and carbon.

4.3.4 High Carbon Compounds

Alkanes with C4-C23 length possess higher energy density, hydrophobicity and compatibility with existing liquid fuel infrastructure and are the predominant constituents of gasoline, diesel and jet fuels (Peralta-Yahya et al., 2012). The capacity of cyanobacteria to produce alkanes was reported as early as the 1960's (Han et al., 1968). The pathway consists of an acyl–acyl carrier protein reductase and an aldehyde decarbonylase, which together converts intermediates of fatty acid metabolism to alkanes and alkenes (Fig. 4.5).

Heptadecane is the most common alkane found in algae. Heterologous expression of the alkane operon from cyanobacteria in *E. coli* led to the production and secretion of C13 to C17 mixtures of alkanes and alkenes (Schirmer et al., 2010).

Short and medium chain alkanes have the potential to be used directly as transportation fuel and have been reported to be secreted by a diverse group of organisms. Algae produce high amount of lipids. It is feasible to convert these lipids into desired alkanes via the formation of aldehydes. Aldehydes to alkanes is catalyzed by a decarboxylase enzyme (Li et al., 2012). Not all alkanes are produced naturally by algae. Ethylene synthesis via this route appears promising. It can be polymerized into gasoline, alcohol and diesel. It is currently produced exclusively from fossil fuels, and its production is the largest CO_2-emitting process in the chemical industry. Therefore, production of ethylene with simultaneous sequestration of CO_2 is no doubt a very attractive process. This feat was recently achieved by expression of the *efe* gene encoding an ethylene-forming enzyme from *Pseudomonas syringae* pv. In *Synechocystis* that led to continuous ethylene production. Interestingly, this gene was earlier used in other studies and was found to be unstable after 3 to 4 generations. Further investigations had revealed gene duplication of the sequence CTAT as the cause of gene inactivation. These sequences were found at three sites which were recognized as 'genetic hot spots'. Researchers introduced silent mutations at

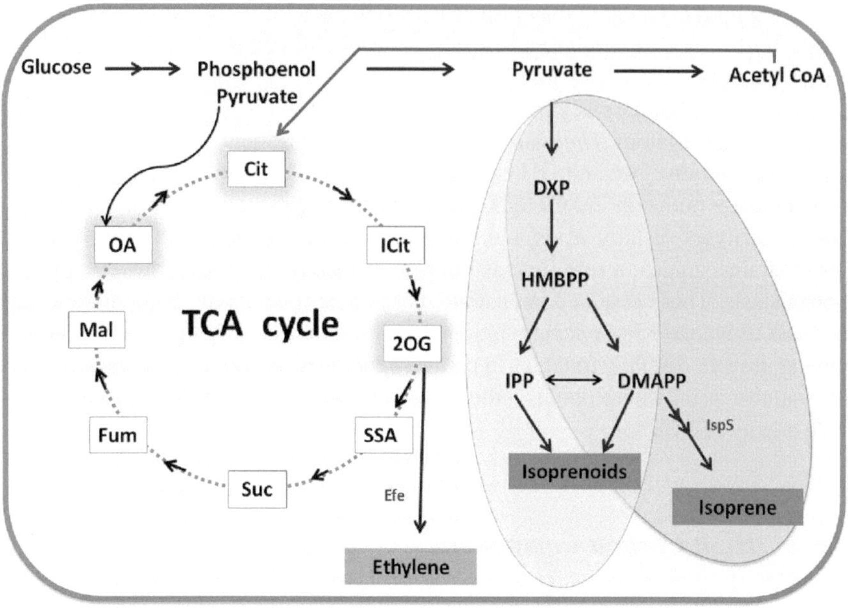

Fig. 4.5 Schematic of pathways in cyanobacteria for the production of various alkanes/alkenes. Abbreviations: Cit:Citrate, Icit:Isocitrate, 2OG: 2-Oxaloglycerate, SSA: Succinyl CoA, Suc: Succinate, Fum: Fumarate, Mal: Malate, OA: Oxaloacetate, DXP: 1-deoxyxylulose-5-phosphate, HMBPP: 1-hydroxy-2-methyl-2-butenyl-4-pyrophosphate, IPP: isopentenyl-pyrophosphate, DMAPP: dimethylallyl-pyrophosphate.

these sites and further codon optimized the sequence thereby resolving the stability issues and increasing yield. Up to 5.5 % of the fixed carbon was directed to ethylene synthesis, surpassing the published carbon-partition rate into the TCA cycle. Nitrogen and phosphorus enriched seawater can support both growth and ethylene production. Factors limiting ethylene production, including *efe* expression levels, light intensity and nutrient status, were identified and alleviated (Ungerer et al., 2012). Optimizing the expression of the alkane biosynthesis genes and enhancing the carbon flux through the fatty acid and alkane biosynthesis pathways can lead to the accumulation and/or secretion of notable amounts of alkanes. Further, it also becomes important to understand how to control the chain lengths of the produced alkane molecules.

Isoprenoids, e.g. the monoterpene pinene and the sesquiterpene farnesene, are considered precursors for future biodiesel or next-generation jet fuel. Cyanobacteria produce carotenoids and extending the carotenoid biosynthetic pathways by introduction of constructs for appropriate terpene synthases should allow the biosynthesis of selected mono and sesquiterpenes. Isoprenoids, also known as terpenoids, represent an incredibly diverse group of natural compounds, with more than 40,000 different molecules. In microalgae, isoprenoids are synthesized via the methylerythritol (MEP) pathway using glyceraldehydes-3-phosphate and pyruvate to generate the basic building blocks of isoprenoid biosynthesis, isopentyl diphosphate (IPP), and dimethylallyl diphosphate (DMAPP). Molecules that could potentially work as gasoline substitutes, including isopentenol, have been produced by *E. coli* using isoprenoid biosynthesis pathways. Two enzymes from *Bacillus subtilis* that utilize IPP and DMAPP for the biosynthesis of isopentenol were overexpressed in *E. coli*, resulting in production of 112 mg/litre isopentenol (107, 200).

Recently, Lindberg et al. (2010) showed the feasibility of producing isoprene hydrocarbons from *Synechocystis* by expression of non-native isoprene synthase gene in *Synechocystis*. Codon usage further optimized the *ispS* gene expression which was driven by the photosynthesis *psbA2* promoter to obtain light driven isoprene accumulation. Further, Bentley in a follow up work from the same group described the improvement of the photosynthetic partitioning between isoprene and biomass in *Synechocystis*. Expression of the non-native Mevalonic acid pathway genes in *Synechocystsis* endowing a non-native pathway for carbon flux amplification to isopentenyl-diphosphate (IPP) and dimethylallyl-diphosphate (DMAPP) precursors of isoprene. Heterologous expression of the isoprene synthase in combination with the MVA pathway enzymes resulted in photosynthetic isoprene yield improvement by approximately 2.5-fold, compared with that measured in cyanobacteria transformed with the isoprene synthase gene only. These results suggest that the MVA pathway introduces a bypass in the flux of endogenous cellular substrate in *Synechocystis* to IPP and DMAPP, overcoming flux limitations of the native MEP pathway.

To make these biofuels more promising, the low tolerance of algae to alkanes must be overcome. In-depth studies have shown that the oxidative stress of the organisms increases upon production of these hydrocarbons suggesting that they are possibly the major protection mechanism against the production of these hydrocarbons. Expression of transporters to facilitate the secretion of the product into the culture medium may help develop more robust cells for the synthesis of such compounds.

4.4 Application of 'Omics Technologies'—Genomics, Transcriptomics and Proteomics

Omic technology adopts a holistic view of the molecules that make up a cell. They are aimed primarily at the universal detection of genes (genomics), mRNA (transcriptomics), proteins (proteomics) and metabolites (metabolomics). The aim of these technologies is that the complex interaction between the organisms can be understood more thoroughly. The study of metagenomes is important with respect to the discovery of more robust pathways in different organisms. For instance, the [FeFe] hydrogenase is a promising candidate; however, the enzyme is difficult to work with because it involves the issues of oxygen sensitivity. Genetic engineering approaches to resolve the issues have produced limited success so far. A way around the issue would be to analyze the different metagenomes from complex ecosystems for different variants of this enzyme present in nature. In one such approach, Craig Venter's institute carried out large scale sequencing of millions of organisms from a myriad of locations to identify possible host candidates to improve hydrogen production.

Engineering algae into fuel factories may interfere with their overall cell homeostasis, which in turn might counteract sustainability and compromise product yields. Therefore, it is necessary to access the overall cellular response in terms of the production of the targeted molecule towards establishing more robust organisms. Cellular proteomics and transcriptomics find importance in holistic understanding of not only the wild type organisms, but also to study a comparative profile between the wild type and the engineered mutants. For example the deletion or over expression of a particular gene in a pathway may have several consequences. Pinto et al. (2012) compared the *Synechocystis* wild type proteome to that of the engineered strain where the hydrogen producing Hox operon had been deleted. The in-depth study showed that the deletion did not affect cell viability or any other pathways and that it could be used as another neutral site to express heterologous genes. Similar studies were carried out by Dienst et al. (2014) to study the cellular response towards the long-term production of ethanol from *Synechocystis*. Since the group observed specific physiological response to the long-term ethanol production such as a bleaching phenotype, slow growing biomass and down regulation of the light harvesting capacity, an omic study could indicate which stress factors were responsible. Future development of engineered strains could consider the effect of the same and engineer the associated responses. More such studies are encouraged so that the scientific community gets a holistic picture of the stress responses and other pathways that get affected by the introduction or deletion of non-native/native pathways. This knowledge will immensely contribute towards the development of more robust systems.

4.5 Conclusions

Algae are a phylogenetically diverse group of organisms and possess novel metabolic features that can be exploited for the production of sustainable biofuels. Several biotechnology companies have shown their interest in commercializing their products. Prominent among them are Algenol Biofuels that is currently commercializing algae-based ethanol produced by an engineered cyanobacteria which they claim can produce 10,000–12,000 gal/acre/year (Waltz, 2009). Exxon Mobil has developed expertise in free-fatty acid producing cyanobacteria and Joule unlimited has a patent for alkane producing cyanobacteria. These companies reflect the surging growth market for these biofuels as also their social acceptance. As new species are discovered and their genomes sequenced the possibility of implementation of a bio-based economy is highly probable. However, though many milestones have been accomplished there is still a long way to go. Towards this, advancement of fundamental knowledge is crucial. A deeper insight into our current setbacks and technical limitations is encouraged. Future accomplishments must endow a high driving force (high enzyme activities and irreversible reaction), stable enzyme (active under oxygenic photosynthesis), low toxicity of accumulated products and an abundant source of carbon substrate. Genetic engineering and synthetic biology together have a power to reform the world. Recent advances and a surging biofuel market show that it can happen. A century earlier, the discovery of fossilized algae brought around the industrial revolution; a century later they demonstrate the capacity to spin the wheel yet again.

References

Badger, M.R. and Price, G.D. (2003). CO_2 concentrating mechanisms in cyanobacteria: Molecular components, their diversity and evolution. *Journal of Experimental Botany*, 54, 609–622.

Badger, M.R., Andrews, T.J., Whitney, S.M., Ludwig, M., Yellowlees, D.C., Leggat, W. and Price, G.D. (1998). The diversity and coevolution of rubisco, plastids, pyrenoids, and chloroplast-based CO_2-concentrating mechanisms in algae. *Canadian Journal of Botany*, 76, 1052–1071.

Bar-Even, A., Noor, E., Lewis, N.E. and Milo, R. (2010). Design and analysis of synthetic carbon fixation pathways. *Proceedings of the National Academy of Sciences, USA*, 107, 8889–8894.

Bernat, G., Waschewski, N. and Rögner, M. (2009). Towards efficient hydrogen production: The impact of antenna size and external factors on electron transport dynamics in *Synechocystis* PCC 6803. *Photosynthesis Research*, 99, 205–216.

Bershtein, S. and Tawfik, D.S. (2008). Advances in laboratory evolution of enzymes. *Current Opinion in Chemical Biology*, 12,151–158.

Bhattacharya, S., Schiavone, M., Nayak, A. and Bhattacharya, S.K. (2005). Biotechnological storage and utilization of entrapped solar energy. *Applied Biochemistry Biotechnology*, 120, 159–167.

Boison, G., Bothe, H., Hansel, A. and Lindblad, P. (1999). Evidence against a common use of the diaphorase subunits by the bidirectional hydrogenase and by the respiratory complex I in cyanobacteria. *FEMS Microbiology Letters,* **174,** 159–165.

Bonacci, W., Teng, P.K., Afonso, B., Niederholtmeyer, H., Grob, P., Silver, P.A. and Savage, D.F. (2012). Modularity of a carbon-fixing protein organelle. *Proceedings of the National Academy of Sciences of the United States of America*, 109, 478–483.

Carrasco, C.D., Holliday, S.D., Hansel, A., Lindblad, P. and Golden, J.W. (2005). Heterocyst-specific excision of the *Anabaena* sp. strain PCC 7120 hupL element requires xisC. *Journal of Bacteriology*, 187, 6031–6038.

Casalot, L. and Rousset, M. (2001). Maturation of the [NiFe] hydrogenases. *Trends in Microbiology*, 9, 228–237.

Chen, L.M., Li, K.Z., Miwa, T. and Izui, K. (2004). Overexpression of a cyanobacterial phosphoenolpyruvate carboxylase with diminished sensitivity to feedback inhibition in *Arabidopsis* changes amino acid metabolism. *Planta*, 219, 440–449.

Chisti, Y. (2008). Biodiesel from microalgae beats bioethanol. *Trends in Biotechnology*, 26, 126–131.

Cirino, P.C. and Frei, C.S. (2009). Combinatorial enzyme engineering. *In:* Sheldon, J., Park and Cochran, J.R. (eds) Protein Engineering and Design. CRC Press, BC, pp. 131–150.

Collman, J.P. (1996). Coupling H_2 to electron transfer. *Nature Structural Biology*, 3, 213–217.

Daniel, H., Torres-Ruiz, J.A. and McFadden, B.A. (1989). Amplified expression of ribulosebisphosphate carboxylase/oxygenase in pBR 322-transformants of *Anacystis nidulans*. *Archives of Microbiology*, 151, 59–64.

Deng, M.D. and Coleman, J.R. (1999). Ethanol synthesis by genetic engineering in cyanobacteria. *Applied and Environmental Microbiology*, 65, 523–528.

Department of Energy, USA. Source: http://www.netl.doe.gov/research/coal/carbon-storage/natcarb-atlas/co2-stationary-sources

Dienst, D., Georg, J., Abts, T., Jakorew, L., Kuchmina, E., Borner, T. et al. (2014). Transcriptomic response to prolonged ethanol production in the cyanobacterium *Synechocystis* sp. PCC6803. *Biotechnology for Biofuels*, 7, 7–21.

Ducat, D.C., Sachdeva, G. and Silver, P.A. (2011). Rewiring hydrogenase-dependent redox circuits in cyanobacteria. *Proceedings of the National Academy of Sciences of the United States of America*, 108, 3941–3946.

Ekman, M., Ow, S.Y., Holmqvist, M. and Lindblad, P. (2011). Metabolic adaptations in a H_2 producing heterocyst-forming cyanobacterium: Potentials and implications for biological engineering. *Journal of Proteomic Research*, 10, 1772–1784.

Field, C.B., Behrenfeld, M.J., Randerson, J.T. and Falkowski, P. (1998). Primary production of the biosphere: Integrating terrestrial and oceanic components. *Science*, 281, 237–240.

Fukuzawa, H., Suzuki, E., Komukai, Y. and Miyachi, S. (1992). A gene homologous to chloroplast carbonic-anhydrase (*icfa*) is essential to photosynthetic carbon-dioxide fixation by *Synechococcus* PCC7942. *Proceedings of the National Academy of Sciences of the United States of America*, 89, 4437–4441.

Gärtner, K., Lechno-Yossef, S., Cornish, A.J., Wolk, C.P. and Hegg, E.L. (2012). Expression of *Shewanella oneidensis* MR-1 [FeFe]-hydrogenase genes in *Anabaena* sp. strain PCC 7120. *Applied and Environmental Microbiology*, 78, 8579–8586.

Greene, D.N., Whitney, S.M. and Matsumura, I. (2007). Artificially evolved *Synechococcus* PCC 6301 Rubisco variants exhibit improvements in folding and catalytic efficiency. *Biochemistry Journal*, 404, 517–524.

Han, J., McCarthy, E.D., Hoeven, W.V., Calvin, M. and Bradley, W.H. (1968). Organic geochemical studies, a preliminary report on the distribution of aliphatic hydrocarbons in algae, in bacteria, and in a recent lake sediment. *Proceedings of the National Academy of Sciences, USA*, **59, 29–33.**

Happe, T., Schütz, K. and Bohme, H. (2000). Transcriptional and mutational analysis of the uptake hydrogenase of the filamentous cyanobacterium *Anabaena variabilis* ATCC 29413. *Journal of Bacteriology*, 182, 1624–1631.

Hu, Q., Sommerfeld, M., Jarvis, E., Ghirardi, M., Posewitz, M., Seibert, M. and Darzins, A. (2008). Microalgal triacylglycerols as feedstocks for biofuel production: Perspectives and advances. *The Plant Journal*, 54, 621–639.

Imashimizu, M., Bernát, G., Isato, K., Broekmans, M., Konno, H., Sunamura, E.I., Rögner, M. and Hisabori, T. (2011). Regulation of F_oF_1-ATPase from *Synechocystis* sp. PCC 6803 by the γ and $\in$ subunits is significant for light/dark adaptation. *Journal of Biochemistry*, 286, 26595–26602.

Iwaki, T., Haranoh, K., Inoue, N., Kojima, K., Satoh, R., Nishino, T. and Wadano, A. (2006). Expression of foreign type I ribulose-1,5-bisphosphate carboxylase/oxygenase (EC 4.1.1.39) stimulates photosynthesis in cyanobacterium *Synechococcus* PCC 7942 cells. *Photosynthesis Research*, 88, 287–297.

Janssen, M., de Winter, M., Tramper, J., Mur, L. R., Snel, J. and Wijffels, R.H. (2000). Efficiency of light utilization of *Chlamydomonas reinhardtii* under medium-duration light/dark cycles. *Journal of Biotechnology*, 78, 123–137.

Khetkorn, W., Lindblad, P. and Inchroensakdi, A. (2012). Inactivation of uptake hydrogenase leads to enhanced and sustained hydrogen production with high nitrogenase activity under high light exposure in the cyanobacterium *Anabaena siamensis* TISTR 8012. *Journal of Biological Engineering*, 6, 19.

Lan, E.L. and Liao, J.C. (2012). ATP drives direct photosynthetic production of 1-butanol in cyanobacteria. *Proceedings of the National Academy of Sciences of the United States of America*, 109, 6018–6023.

Li, H. and Liao, J.C. (2013). Engineering a cyanobacterium as the catalyst for the photosynthetic conversion of CO_2 to 1,2-propanediol. *Microbial Cell Factories*, 12, 4.

Li, N., Chang, W.C., Warui, D.M., Booker, S.J., Krebs, C. and Bollinger, J.M. (2012). Evidence for Only Oxygenative Cleavage of Aldehydes to Alk(a/e)nes and Formate by Cyanobacterial Aldehyde Decarbonylases. *Biochemistry,* **51,** 7908–7916.

Li, X., Shen, C.R. and Liao, J.C. (2014). Isobutanol production as an alternative metabolic sink to rescue the growth deficiency of the glycogen mutant of *Synechococcus elongatus* PCC 7942. *Photosynthesis Research*, 120, 301–310.

Lieman-Hurwitz, J., Rachmilevitch, S., Mittler, R., Marcus, Y. and Kaplan, A. (2003). Enhanced photosynthesis and growth of transgenic plants that express ictB, a gene involved in HCO_3^- accumulation in cyanobacteria. *Plant Biotechnology Journal*, 1, 43–50.

Lindberg, P., Devine, E., Stensjö, K. and Lindblad, P. (2012). HupW protease specifically required for processing of the catalytic subunit of the uptake hydrogenase in the cyanobacterium *Nostoc* sp. strain PCC 7120. *Applied Environmental Microbiology*, 78, 273–276.

Lindberg, P., Park, S. and Melis, A. (2010). Engineering a platform for photosynthetic isoprene production in cyanobacteria, using *Synechocystis* as the model organism. *Metabolic Engineering*, 12, 70–79.

Lindberg, P., Schütz, K., Happe, T. and Lindblad, P. (2002). A hydrogen-producing, hydrogenase-free mutant strain of *Nostoc punctiforme* ATCC 29133. *International Journal Hydrogen Energy*, 27, 1291–1296.

Liu, X., Sheng, J. and Curtiss, R. (2011). Fatty acid production in genetically modified cyanobacteria. *Proceedings of the National Academy of Sciences of the United States of America*, 108, 6899–6904.

Long, B.M., Badger, M.R., Whitney, S.M. and Price, G.D. (2007). Analysis of carboxysomes from *Synechococcus* PCC 7942 reveals multiple Rubisco complexes with carboxysomal proteins CcmM and CcaA. *The Journal of Biological Chemistry*, 282, 29323–29335.

Long, S.P., Zhu, X.G., Naidu, S.L. and Ort, D.R. (2006). Can improvement in photosynthesis increase crop yields? *Plant Cell Environment*, 29, 315–330.

Lu, X., Vora, H. and Khosla, C. (2008). Overproduction of free fatty acids in *E. coli*: Implications for biodiesel production. Metabolic Engineering, 10, 333–339.

Masukawa, M., Mochimaru, M. and Sakurai, H. (2002). Disruption of the uptake hydrogenase gene, but not of the bidirectional hydrogenase gene, leads to enhanced photobiological hydrogen production by the nitrogen-fixing cyanobacterium *Anabaena* sp. PCC 7120. *Applied Microbiology Biotechnology*, 58, 618–624.

McGinn, P.J., Price, G.D., Maleszka, R. and Badger, M.R. (2003). Inorganic carbon limitation and light control the expression of transcripts related to the CO_2-concentrating mechanism in the cyanobacterium *Synechocystis* sp. strain PCC 6803. *Plant Physiology*, 132, 218–229.

Melis, A. (2009). Solar energy conversion efficiencies in photosynthesis: Minimizing the chlorophyll antennae to maximize efficiency. *Plant Science*, 177, 272–280.

Metz, J.G., Pollard, M.R., Anderson, L., Hayes, T.R. and Lassner, M.W. (2000). Purification of a *jojoba* embryo fatty acyl-coenzyme A reductase and expression of its cDNA in high erucic acid rapeseed. *Plant Physiology*, 122, 635–644.

Moal, G. and Lagoutte, B. (2012). Photo-induced electron transfer from Photosystem 1 to $NADP^+$: Characterization and tentative simulation of the *in vivo* environment. *Biochemistry Biophysics Acta*, 1817, 1635–1645.

Moroney, J. V. and Ynalvez, R.A. (2007). Proposed carbon dioxide concentrating mechanism in *Chlamydomonas reinhardtii. Eukaryotic Cell*, 6, 1251–1259.

Msanne, J., Xu, D., Konda, A.R., Casas-Mollano, J.A. and Awada, T. (2012). Metabolic and gene expression changes triggered by nitrogen deprivation in the photoautotrophically grown microalgae *Chlamydomonas reinhardtii* and *Coccomyxa* sp. C-169. *Phytochemistry*, 75, 50–59.

Mueller-Cajar, O. and Whitney, S.M. (2008). Directing the evolution of Rubisco and Rubiscoactivase: First impressions of a new tool for photosynthesis research. *Photosynthesis Research*, 98, 667–675.

Mussgnug, J.H., Thomas-Hall, S., Rupprecht, J., Foo, A., Klassen, V., McDowall, A., Schenk, P.M., Kruse, O. and Hankamer, B. (2007). Engineering photosynthetic light capture: Impacts on improved solar energy to biomass conversion. *Journal of Plant Biotechnology*, 5, 802–814.

Nakajima, Y., Fujiwara, S., Sawai, H., Imashimizu, M. and Tsuzuki, M. (2001). A phycocyanin-deficient mutant of *Synechocystis* PCC 6714 with a single-base substitution upstream of the cpc operon. *Plant Cell Physiology*, 42, 992–998.

Nayak, B.K., Roy, S. and Das, D. (2014). Biohydrogen production from algal biomass (Anabaena sp. PCC 7120) cultivated in airlift photobioreactor. *International Journal of Hydrogen Energy*, 39, 7553–7560.

Nguyen, M.T., Choi, S.P., Lee, J., Lee, J.H. and Sim, S.J. (2009). Hydrothermal acid pretreatment of *Chlamydomonas reinhardtii* biomass for ethanol production. *Journal of Microbiology and Biotechnology*, 19, 161–166.

Oliver, J.W.K., Machado, I.M.P., Yoneda, H. and Atsumi, S. (2013). Cyanobacterial conversion of carbon dioxide to 2,3-butanediol. *Proceedings of the National Academy of Sciences of the United States of America*, 110, 1249–1254.

Peralta-Yahya, P.P., Zhang, F.Z., Cardayre, S.B. and Keasling, J.D. (2012). Microbial engineering for the production of advanced biofuels. *Nature*, **488,** 320–328.

Pinto, F., van Elburg, K.A., Pacheco, C.C., Lopo, M., Noirel, J., Montagud, A. and Tamagnini, P. (2012). Construction of a chassis for hydrogen production: Physiological and molecular characterization of a *Synechocystis* sp. PCC 6803 mutant lacking a functional bidirectional hydrogenase. *Microbiology*, 158, 448–464.

Polle, J.E., Kanakagiri, S.D. and Melis, A. (2003). tla1, a DNA insertional transformant of the green alga *Chlamydomonas reinhardtii* with a truncated light-harvesting chlorophyll antenna size. *Planta*, 217, 49–59.

Poulsen, N. and Kroger, N. (2005). A new molecular tool for transgenic diatoms. *Journal of FEBS*, 272, 3413–3423.

Price, G.D. and Badger, M.R. (1989). Expression of human carbonic anhydrase in the cyanobacterium *Synechococcus* PCC 7942 creates a high-CO_2 requiring phenotype. *Plant Physiology*, 91, 505–513.

Price, G.D., Woodger, F.J., Badger, M.R., Howitt, S.M. and Tucker, L. (2004). Identification of a SulP-type bicarbonate transporter in marine cyanobacteria. *Proceedings of the National Academy of Sciences, USA*, 101, 18228–18233.

Quinn, J.M. and Merchant, S. (1995). Two copper-responsive elements associated with the *Chlamydomonas* Cyc6 gene function as targets for transcriptional activators. *Plant Cell*, 7, 623–638.

Rascher, U., Lakatos, M., Büdel, B. and Lüttge, U. (2003). Photosynthetic field capacity of cyanobacteria of a tropical inselberg of the Guiana Highlands. *European Journal of Phycology*, 38, 247–256.

Reifschneider-Wegner, K., Kanygin, A. and Redding, K.E. (2014). Expression of the [FeFe] hydrogenase in the chloroplast of *Chlamydomonas reinhardtii. International Journal of Hydrogen Energy*, 39, 3657–3665.

Riekhof, W.R., Sears, B.B. and Benning, C. (2005). Annotation of genes involved in glycerolipid biosynthesis in *Chlamydomonas reinhardtii*: Discovery of the betaine lipid synthase BTA1(Cr). *Eukaryotic Cell*, 4, 242–252.

Rögner, M. (2013). Metabolic engineering of cyanobacteria for the production of hydrogen from water. *Biochemical Society Transactions*, 4, 1254–1259.

Roy, S., Kumar, K., Ghosh, S. and Das, D. (2014). Thermophilic biohydrogen production using pretreated algal biomass as substrate. *Biomass and Bioenergy*, 61, 157–166.

Schirmer, A., Rude, M.A., Li, X.Z., Popova, E. and delCardayre, S.B. (2010). Microbial Biosynthesis of Alkanes. *Science,* **329,** 559–562.

Scott, S.A., Davey, M.P., Dennis, J.S., Horst, I., Howe, C.J., Lea-Smith, D.J. and Smith, A.J. (2010). Biodiesel from algae: Challenges and prospects. *Current Opinion in Biotechnology*, 21, 277–286.

Shen, C.R., Lan, E.I., Dekishima, Y., Baez, A., Cho, K.M. and Liao, J.C. (2011). Driving Forces Enable High-Titer Anaerobic 1-Butanol Synthesis in *Escherichia coli. Applied and Environmental Microbiology*, 77, 2905–2915.

Shibata, M., Katoh, H., Sonoda, M., Ohkawa, H., Shimoyama, M., Fukuzawa, H., Kaplan, A. and Ogawa, T. (2002). Genes essential to sodium-dependent bicarbonate transport in cyanobacteria—Function and phylogenetic analysis. *Journal of Biological Chemistry,* 277, 18658–18664.

Smith, S.A. and Tabita, F.R. (2003). Positive and negative selection of mutant forms of prokaryotic (cyanobacterial) ribulose-1,5-bisphosphate carboxylase/oxygenase. *Journal of Molecular Biology*, 331, 557–569.

Stitt, M., Sulpice, R. and Keurentjes, J. (2010). Metabolic networks: How to identify key components in the regulation of metabolism and growth. *Plant Physiology*, 152, 428–444.

Summary report: World agriculture: towards 2015/2013 (2002). Food and Agriculture Organization of the United Nation, Rome.

Tcherkez, G.G.B., Farquhar, G.D. and Andrews, T.J. (2006). Despite slow catalysis and confused substrate specificity, all ribulose bisphosphate carboxylases may be nearly perfectly optimized. *Proceedings of the National Academy of Sciences, USA*, 103, 7246–7251.

UK Department of Energy and Climate Change: Energy Trends March 2009, http://www.decc.gov.uk/en/content/cms/statistics/publications/publications.aspx.

Um, B.H. and Kim, Y.S. (2009). Review: A chance for Korea to advance algal-biodiesel technology. *Journal of Industrial and Engineering Chemistry*, 15, 1–7.

Ungerer, J., Tao, L., Davis, M., Ghirardi, M., Maness, P.C. and Yu, J. (2012). Sustained photosynthetic conversion of CO_2 to ethylene in recombinant cyanobacterium *Synechocystis* 6803. *Energy Environment Science*, 5, 8998–9006.

Varman, A.M., Xiao, Y., Pakrasi, H.B. and Tang, Y.J. (2013). Metabolic Engineering of *Synechocystis* sp. Strain PCC 6803 for Isobutanol Production. *Applied Environmental Microbiology*, 79, 908–914.

Walter, J.M., Greenfield, D. and Liphardt, J. (2010). Potential of light-harvesting proton pumps for bioenergy applications. *Current Opinion in Biotechnology*, 21, 265–270.

Waltz, E. (2009). Biotech's green gold? *Nature Biotechnology*, 27, 15–18.

Wang, Z.T., Ullrich, N., Joo, S., Waffenschmidt, S. and Goodenough, U. (2009). Algal lipid bodies: Stress induction, purification, and biochemical characterization in wild-type and starchless *Chlamydomonas reinhardtii. Eukaryotic Cell*, 8, 1856–1868.

Wu, M., Mintz, M., Wang, M. and Arora, S. (2008). Consumptive Water Use in the Production of Bioethanol and Petroleum Gasoline. Argonne National Laboratory's work report.

Xu, Y., Guerra, L.T., Li, Z., Ludwig, M., Dismukes, G.C. and Bryant, D.A. (2013). Altered carbohydrate metabolism in glycogen synthase mutants of *Synechococcus* sp. strain PCC 7002: Cell factories for soluble sugars. *Metabolic Engineering*, 16, 56–67.

Yang, C., Hua, Q. and Shimizu, K. (2002). Metabolic flux analysis in *Synechocystis* using isotope distribution from 13C-labeled glucose. *Metabolic.Engineering*, 4, 202–216.

Yao, L., Qi, F., Tan, X. and Lu, X. (2014). Improved production of fatty alcohols in cyanobacteria by metabolic engineering. *Biotechnology for Biofuels*, 7, 94.

Yoshino, F., Ikeda, H., Masukawa, H. and Sakurai, H. (2007). High photobiological hydrogen production activity of a *Nostoc* sp. PCC 7422 uptake hydrogenase-deficient mutant with high nitrogenase activity. *Marine Biotechnology*, 9, 101–112.

Chapter 5
Photobioreactors for Improved Algal Biomass Production: Analysis and Design Considerations

Ankush Karemore, Dineshkumar Ramalingam, Geetanjali Yadav, Ganeshan Subramanian, and Ramkrishna Sen

5.1 Introduction

Production of microalgal biomass has a long history due to its biotechnological potential for commercial applications such as high value nutraceutical products (PUFA, pigments, vitamins), human nutrition, animal nutrition, cosmetics, wastewater treatment, etc. (Spoalore et al., 2006). Most recently, microalgae received much attention as a potential source for biofuels to replace fossil fuel, and for CO_2 capture due to its high photosynthetic efficiency. Many of these commercial applications require a photobioreactor system in which monoculture of microalgal biomasses can be developed with high productivity for an extended period of time. A number of open ponds, outdoor and enclosed photobioreactor systems have been developed for growing phototrophic algae such as cyanobacteria and microalgae. Photobioreactor facilitate in maximization of solar energy capture and conversion, preferably using sunlight, atmospheric CO_2 and water, to chemical energy stored as organic carbon sources (such as carbohydrates and lipids) (Pulz and Scheinbenbogen, 1998).

Currently, cultivation technologies used at industrial level are open ponds, which are preferred for their low capital and operational costs. However, there are drawbacks like poor control of operational conditions, high contamination risk, poor mixing, low gas-liquid mass transfer rate, low volumetric productivity and biomass concentration, huge evaporation losses, poor temperature control and less light penetration (Pulz, 2001). Whereas, photobioreactor has several advantages when compared to open ponds like high gas-liquid mass transfer rate, better control of operational condition (such as pH, temperature, mixing and illumination), low contamination risk, high volumetric and areal productivity, low water loss and low

A. Karemore • D. Ramalingam • G. Yadav • G. Subramanian • R. Sen (✉)
Department of Biotechnology, Indian Institute of Technology Kharagpur,
Kharagpur, West Bengal, 721302, India
e-mail: rksen@yahoo.com

D. Das (ed.), *Algal Biorefinery: An Integrated Approach*,
DOI 10.1007/978-3-319-22813-6_5

harvesting cost. Photobioreactors have been developed at small and large scale since the 1950s and several design configurations have been proposed such as tubular, flat plate, bubble column and airlift. Extensive scientific and industrial efforts have been focused on the development of cost-effective and high-efficiency cultivation systems for the production of high cell density microalgal cultures (Wang et al., 2012).

In this chapter, we discuss the design principles of photobioreactors, important factors and the parameters influencing its performance. The issues such as gaseous exchange, mixing patterns, suitability of irradiance, nutrient supply, pH and temperature control, geometrical configuration and building material are considered important. Furthermore, recent developments in the functioning of photobioreactor systems have been discussed.

5.2 Photobioreactor Design Parameters

Open ponds and closed photobioreactors (PBR) are the cultivation systems used for growing algal biomass. The efficiency of the PBR performance is determined by photosynthetic rate, light utilization, biomass concentration, mixing, pH and temperature control, culture hydrodynamics and mass transfer rate. Efficient PBRs facilitate in (1) maximum light harvesting, distribution and utilization, (2) high productivities of biomass or product, (3) effective mixing and gas-liquid mass transfer, (4) allow precise control of operational parameters, (5) minimize the capital and operational cost, (6) ease of operation and scalability, (7) lower land area requirement and (8) low energy consumption during operation (Wang et al., 2012). Therefore, there is an urgent need for photobioreactor design and operation strategies which may enhance volumetric and areal productivities of algae.

5.2.1 *Design Considerations*

The most important parameter that affects PBR design is effective light penetration, which means a high surface-to-volume (S/V) ratio. This improves the photosynthetic efficiency, which in turn results in high product and biomass productivities (Wang et al., 2012). There are several designs of PBR developed to attain high S/V ratio (Fig. 5.1). Main designs types can be of cylindrical, flat plate or tubular type to achieve maximum light capture. Important design aspect of PBR systems are shown in Table 5.1. Most suitable choice of reactors are tubular and flat plate types, considering high area-to-volume ratio while maintaining reasonable working volume, mixing pattern and gas transfer. The volumetric productivity (V_p) of algal biomass or product (g L^{-1} day^{-1}) and areal productivity (A_p) (g m^{-2} day^{-1}) are important factors for determining the efficiency of PBR systems. Although solar energy is available in abundance, current light utilization efficiencies are restricted to 5–6 % (Jorquera et al., 2010).

The surface area-to-volume ratios of various reactor configurations have been elucidated in Fig. 5.1 and Table 5.2. Photobioreactors come in various shapes and sizes. Flat panel reactors (Fig. 5.1A) are cuboidal in shape. They have large lengths and heights, but narrow widths to allow greater light penetration. They are also categorized into vertical or inclined reactors (Molina Grima et al., 1999) based on their orientation to sunlight. Dome shape reactors (Fig. 5.1B) are hemispherical in shape and are not used very often. Tubular reactors (Fig. 5.1C) are cylindrical in shape, either horizontal or vertical columns (Molina Grima et al., 1999). Further, the vertical columns are divided into bubble column or air-lift reactors for better mixing and mass transfer characteristics (Sánchez Mirón et al., 2000). Pyramidal reactors (Fig. 5.1D) have also been reported in literature.

The important design aspect of any PBR is that the incident light should be perpendicular to the reactor surface. The tetrahedron has the highest illuminated surface area per unit volume ratio and the lowest is that of a spherical shape. Hence, the illuminated surface area to volume ratio decreases from pyramid > flat panel > tubular > dome shaped reactors. Further, the detailed discussion on PBR design considerations is described in section 5.3.

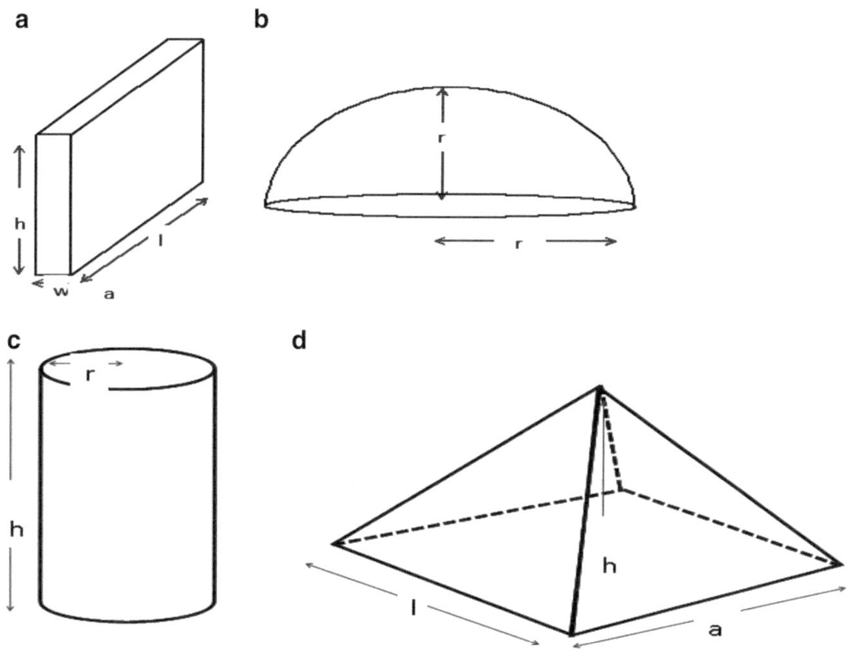

Fig. 5.1 Design configurations of (A) Flat panel, (B) dome, (C) tubular and (D) pyramidal shape reactors. H = height, l = length, a = side length, r = radius and w = width of reactor.

Table 5.1 Important design aspects of open and closed photobioreactors

PBR design aspect	Open systems	Closed systems
Area-to-volume ratio	Large	Small
Algal species	Restricted	Flexible
Productivities	Low	High
Cultivation period	Limited	Extended
Water loss due to evaporation	High	Low/prevented
Light harvesting efficiency	Poor/fair	Fair/excellent
Gas transfer	Poor	Fair/high
Temperature control	None	Excellent
Land area required	Large	Small
Scale up	Easy	Possible
Degree of control	Poor	Excellent
Capital investment	Small	High

Source: Carvalho et al., 2006

Table 5.2 Surface area to volume ratios of various reactor configurations

Reactor type	Illuminated surface area	Volume	Surface area/volume
Flat panel reactor	lh (single sided lighting)	lhw	1/w
	2lh (dual-sided lighting)		2/w
Dome reactor	$2\pi r^2$	$\frac{2}{3}\pi r^3$	3/r
Tubular reactor	πrh (single sided lighting)	$\frac{1}{3}\pi r^2 h$	3/r
	$2\pi rh$ (dual-sided lighting)		6/r
	$2\pi r_l h_l{}^*$ (annular lighting)	$\frac{1}{3}\pi r^2 h - \frac{1}{3}\pi r_l{}^2 h_l$	$(6r_l h_l)/(r^2 h - r_l{}^2 h_l)$
Pyramid reactor	$al + l\sqrt{\frac{a}{4}+h^2} + a\sqrt{\frac{l^2}{4}+h^2}$	$\frac{1}{3}alh$	$\frac{3}{h}+\frac{3}{ah}\sqrt{\frac{a}{4}+h^2}+\frac{3}{lh}\sqrt{\frac{l^2}{4}+h^2}$

*Annular lighting is when light is inserted inside the reactor. r_l and h_l are corresponding dimensions of lighting system.

5.2.2 *Factors Affecting Algal Production*

In addition to the reactor design parameters, other important factors that can affect algae culturing should be considered. Significant factors for high density algal production are nutrient supply, irradiance (of specific intensity, duration and distribution), CO_2 supply, O_2 removal, mixing and other control of environmental parameters like pH, temperature, salinity, etc.

Table 5.3 Interconversion of photon flux density units for visible light

Light source	μ mole photon m^{-2} s^{-1}	Lux	W m^{-2}
Sunlight	1	54	0.219
Metal halide lamps	1	71	0.218
Cool white fluorescent lights	1	74	0.218
Incandescent	1	50	0.2

Source: Thimijan and Heins, 1983

Light energy: Supply of light energy is essential for the phototrophic growth of algae. Light act as an energy source for algal cells and is measured in photon flux density (I). It can be expressed in either of these units: lux, W m^{-2} and μ mole photon m^{-2} s^{-1}. Photon flux density varies with the light source and light intensity, the relationships among the three different SI units are explained in Table 5.3. The total light energy available for algal growth is given by Eq. (5.1), where the photon flux density is incident on the illuminated surface area of photobioreactor (A) for a duration of T (h day^{-1}).

$$\text{Amount of energy supplied to algae} = I \times A \times T \tag{5.1}$$

Provision of irradiance is a key design challenge for efficient PBR. Algal photosynthetic efficiency directly gets influenced by the source of light, spectral quality and intensity of light. Light utilized by algae is defined by absorption spectrum in the range of 400 to 700 nm, called as photosynthetically active radiation (PAR), for the photosynthetic active pigments. An efficient light source is required to exploit commercial potential of algae. Different types of light source have been used such as incandescent lamps (tungsten or halogen lamps), discharge lamps (mercury, xenon, fluorescent lamps) and light-emitting diodes (LEDs). Levels of light intensities directly affect the growth and photosynthesis. High light intensity could stimulate photooxidation or photoinhibition, which partially damage the process of photosynthesis.

Nutrients enrichment: Nutrients supply is required to support primary growth of photosynthetic organisms. Media with varying composition and concentration are supplied to the cells in the form of CO_2, water and mineral salts in macro or micro quantities. The macronutrients considered important for normal growth include carbon, nitrogen, phosphorous, calcium, magnesium, hydrogen, sulphur, etc. The micronutrients needed in trace amounts include iron, boron, copper, cobalt, manganese, nickel, etc. The nutrient requirements can be determined from the elemental biomass composition or stoichiometry of growth. Grobbelaar (2004) gives the molecular formula of algae as $CH_{1.83}O_{0.48}N_{0.11}P_{0.01}$. Since carbon is supplied from carbon dioxide, and hydrogen and oxygen from water, the rate limiting nutrients are nitrogen and phosphorus. Molar ratio of N:P is crucial for the growth of algal cells and the optimal ratio generally supplied is 11:1 (N:P). Deficiency of critical nutrient decreases cell growth rates. Nitrogen deficiency has been reported to increase the lipid content, mainly accumulation of long chain fatty acids which improves biofuel production. On the other hand, oversupplying the medium with nutrients increases the operational costs. Different media have varying nutrient quantities that can significantly change the quantity of cell biomass produced during cultivation (Wang et al., 2011).

Carbon dioxide supply and oxygen removal: Carbon dioxide (CO_2) is the carbon source for photoautotrophic algae. In order for CO_2 to be available to algae, it should be first dissolved in water. CO_2 predominantly exists as CO_2 gas at pH <6, HCO_3^- between pH 6–10 and CO_3^{2-} at pH >10. In the presence of sunlight, the absorbed CO_2 is converted into glucose during the rate limiting step of photosynthesis using the enzyme Rubisco. Rubisco has both carboxylase and oxygenase activities, and has a higher affinity for O_2 than CO_2 (Giordano et al., 2005). A high photosynthesis requires Rubisco to use CO_2 for the Calvin cycle and not O_2 for photorespiration. Algae have therefore evolved carbon concentrating mechanisms using carbonic anhydrase within the cell to aid photosynthesis (Bhattacharya et al., 2004). Despite the evolution-selected mechanism, Giordano et al. (2005) states that photorespiration or high level of dissolved oxygen can inhibit biomass formation of around 50 %. Thus, in order to hasten the rate of photosynthesis, evolved O_2 is removed before it reaches inhibitory levels. Removal of excess O_2 is a mass transfer problem and is possible by proper mixing and aeration because the saturation concentration of dissolved O_2 is 10 ppm. However, photorespiration cannot be completely stopped because algal cultures may reach dissolved O_2 concentrations of 40 ppm (Pulz, 2001). Therefore, an efficient sparging and mixing pattern is required to increase dissolved CO_2 concentration in the reactor in order to increase CO_2 uptake by algal cells. However, increased CO_2 concentration reduces culture pH (Kumar et al., 2011) which has a detrimental effect on algal growth. Therefore, appropriate amount of CO_2 supply (1–5 %) is provided for growth of high-production cultures (Fulke et al., 2010). Alternatively, small amounts of pure CO_2 may be injected into high-cell-density microalgal cultures with biomass concentrations of 8–10 g dry weight (DW) L^{-1} (Cheng-Wu et al., 2001). Membrane-based technology for selective CO_2 supply and O_2 removal are also promising (Pulz, 2001).

Mixing: Adequate mixing of an algal culture is essential to keep the culture system in homogeneous environment, to enhance light capture efficiency, nutrient distribution, gas exchange and transfer. The effect of mixing has shown significant influence on the algal productivities and overall performance of the photobioreactor. When the environmental conditions are not growth limiting, mixing plays an important role to create turbulence flow pattern to get high biomass yield. Nature of mixing system employed directly influences the productivity of algal system and the cost of construction and operation (Suh and Lee, 2003).

Temperature control: Temperature and pH are important elements for growing algae as they strongly influence protein structure, nutrient solubility and uptake and algal growth rates. Growth rate increases with the increase in temperature/pH up to its optimal range and then decline drastically outside this range. Thus, the use of controller is required to maintain the optimal range of pH and temperature for algal growth. Algae are classified as thermophilic (grows at high temperatures) and mesophilic in terms of temperature tolerance of algae. Similarly, acidophilic, neutral and alkalinophilic algae exist in terms of pH tolerance. Salinity affects osmotic pressure in an algal cell and is dependent on the habitat of a species. Some species are halophiles, i.e. have high tolerance for salt.

5.2.3 Performance Evaluation Parameters of Photobioreactors

In order to obtain high volumetric productivities, various kinds of photobioreactors designs have been evaluated for cultivation of microalgae. Janssen et al. (2002) reviewed three types of photobioreactors mainly vertical column reactors (bubble column and airlift reactors), flat plate and tubular reactors. The areal and volumetric productivities of these engineered PBRs are of magnitude greater than raceway ponds. For production of high cell density cultures, it is important to make a choice of a suitable photobioreactor which gives better biomass productivity than others. The performance of various photobioreactors can be assessed by determining the values of critical parameters. Following parameters helps to determine the performance of a photobioreactor for culturing microalgae.

Volumetric productivity, P_X ($g\ L^{-1}\ d^{-1}$): can be defined as cell concentration per unit of reactor volume per unit time. It is calculated from:

$$P_x = \frac{\left(c_f - c_i\right)}{t} \tag{5.2}$$

c_f and c_i are the final and initial dry biomass concentrations ($g\ L^{-1}$) measured over a period of time, t (d) for batch test.

Areal productivity (A_x)($g\ m^{-2}\ day^{-1}$)*:* can be defined as productivity per unit of occupied-land area per unit of time. It can be reported as grams per square metre, tons per acre, tonnes per hectare gm^{-2}

Carbon dioxide fixation rate, F ($g\ L^{-1}\ d^{-1}$) can be calculated from:

$$F = a\ P_x \tag{5.3}$$

where a = carbon dioxide fixed by unit biomass (considered 50 % of carbon in the biomass) so,

$$a = 0.5 \times \frac{44}{12} = 1.833\ \text{g}\ \text{CO}_2 \left(\text{g}\ \text{dry}\ \text{cell}^{-1}\right)$$

Photosynthetic efficiency (PE)*:* can be defined as energy stored in biomass per unit of light energy supplied (Hu and Richmond, 1996).

$$\text{PE} = Y_{X/E} \frac{\Delta H_X}{N\ h\nu} \tag{5.4}$$

where ΔH_X = specific energy content of the biomass, N = Avogadro number and $h\nu$ = mean photon energy of the PAR according to the Planck law.

$Y_{X/E}$ is defined as the ratio between the biomass production rate and the irradiance absorbed by the culture and is expressed as microalgal dry mass per mole of photons:

$$Y_{X/E} = \frac{\mu \cdot V}{I^o A a_X^*} \tag{5.5}$$

where μ is the specific growth rate, V = culture volume, A = irradiated surface, and a_X^* = the specific biomass absorption coefficient (expressed as $m^2\ g^{-1}$).

The specific growth rate, specific absorption coefficient and photon flux density can be combined to obtain the *biomass yield on light energy* (*Y*prot, *E*), in grams of protein per mole of photons, in the range of photosynthetic active radiation (PAR 400 to 700 nm). This parameter is a good measure of the photosynthetic efficiency of phototrophic growth. Instead of protein, however, dry weight (dw) can also be used, as done by Pulz and Scheinbenbogen (1998), dw, *E*.

$$Y_{prot,E} = \frac{\mu}{a_{proe} PFD.3.6.10^{-3}} \tag{5.6}$$

Mixing time: This is a very important parameter in designing PBRs for effective CO_2 sequestration by microalgae. It is defined as the time taken to achieve a homogenous mixture after injection of tracer solution (Ugwu et al., 2008). While a good mixing improves gas-liquid mass transfer, reduces photo inhibition and increases biomass yield on light energy by reducing light/dark cycles and hence enhances the reactor performance (Hu and Richmond, 1996), poor mixing results in oxygen build up (inhibitory to microalgal cells), biofouling etc. It is calculated by estimating the time taken by a tracer dye to traverse the reactor.

Shear effects: With most microalgae, the growth rate increases with increasing aeration rate due to uniform distribution of light or CO_2. But after an optimum level of turbulence, the growth starts decreasing with any further increase in superficial gas velocity (Silva et al., 1987). This is believed to be caused by cell damage due to bubble rupture at the cell surface.

Gas hold-up: This is determined by measuring the aerated liquid height relative to the gas-free liquid level. It can be expressed analytically as the ratio of superficial gas velocity U_G to the mean terminal rise velocity U_T of the gas bubbles:

$$\varepsilon = \frac{U_G}{U_T} \tag{5.7}$$

This equation basically arises from the definition of gas hold up which is:

$$\varepsilon = \frac{A_G}{A} \tag{5.8}$$

where A_G and A are the actual or true cross-sectional area for gas flow and the total cross-section of the gas-liquid flow channel respectively. Also gas holdup measurements are necessarily related with specific power input; therefore it increases with power input and vice-versa. The following equation describes how performance of bubble column and air-lift reactor is dependent on gas holdup (Chisti and Moo-Young, 1989),

$$\text{Mass transfer } k_L a_L = \psi \frac{\varepsilon}{1-\varepsilon} \tag{5.9}$$

$$\text{Gas} - \text{liquid specific interfacial area}, a_L = \frac{6\varepsilon}{d_B(1-\varepsilon)} \tag{5.10}$$

where k_L = mass transfer coefficient (ms^{-1}), a_L = interfacial area per unit liquid volume (m^{-1}), ψ = constant (s^{-1}), ε = overall gas holdup and d_B = mean bubble diameter (m).

The hydrodynamics of flow in horizontal and vertical columns are quite different. Whereas the gas sparged airlift and bubble columns have greater gas holdups, the horizontal tubular reactors have smaller or are virtually free of gas or any bubbles. Therefore the flow in vertical reactors is more turbulent and chaotic. The difference in gas holdup and bubble size affect light penetration, mass transfer, mixing and shear stress and therefore are largely associated with the hydrodynamics of the algal cultivation system in engineered photobioreactors (Sánchez Mirón et al., 1999).

Volumetric mass transfer coefficient, k_La: This is the most critical parameter for assessing the performance of photobioreactors to attain optimum microalgal cell growth. It involves three phase mass transfer system: gas ($CO_{2(g)}$)-liquid (culture medium)-solid (microalgal cells). The volumetric mass transfer coefficient (k_La) is the product of mass transfer coefficient (k_L) and the interfacial area per unit volume of aerated reactor. Thus, it is affected mainly by superficial gas velocity, sparger type, agitation rate, temperature etc. (Ugwu et al., 2008; Kumar et al., 2011). Mass transfer from gas phase to liquid phase is given by following equation:

$$\frac{dC}{dt} = k_L a(C^* - C) \tag{5.11}$$

where $\frac{dC}{dt}$ = the mass transfer rate, k_L = the mass transfer coefficient, a = the interface area, C^* = the equilibrium gas concentration at the interface of the gas and liquid and C = the gas concentration in the liquid.

k_La is used to describe the overall volumetric mass transfer coefficient in photobioreactor. The k_La increases linearly with increase in superficial gas velocity up to a certain limit after which this trend starts to decline due to coalescence of bubbles

Table 5.4 Liquid circulation times (t_c) for different reactor configurations

Reactor type	Reactor dimensions	U_g (ms^{-1})	t_c (ms)
Flat panel	LP = 1.3 cm	–	87–130
	LP = 2.6 cm	–	173–260
Bubble column	T_v = 20 cm	0.05	960
Air-lift	A_d/A_r = 0.5	0.025	35600
	T_v = 20 cm		
	h_L = 500 cm	0.05	28600
	C_b = 11.8 cm		
	A_d/A_r = 0.5	0.025	43800
	T_v = 40 cm		
	h_L = 1000 cm	0.05	35000
	C_b = 47.2 cm		

LP = Light path, T_v = Vessel diameter, A_d = cross sectional area of downcomer, A_r = cross sectional area of riser, h_L = non-aerated liquid height, C_b = bottom clearance, U_g = superficial velocity.
Source: Janssen et al., 2002; Barbosa et al., 2003

which changes the interfacial area per unit volume of gas. Mass transfer is greater in bubble column and flat panel reactors which are pneumatically agitated than the horizontal tubular reactors which have plug flow kind of system and results in oxygen build up (Sánchez Mirón et al., 1999).

Light regime: In order to attain high cell density cultures, it is very important to efficiently use high light intensities in photobioreactors. A common characteristic of photobioreactors operated at high cell density is the existence of light gradient which results in formation of dark and light zones in photobioreactor. The cells circulate through these zones during growth and therefore receive light intermittently. The cycle time and ratio between light and dark periods in the cycle determines the photosynthetic efficiency (PE). Different PBRs have variable cycle time depending on the superficial gas velocities, U_G (ms^{-1}). Table 5.4 gives a comparison of liquid circulation times (t_c) for different reactor configurations.

Generally, reactors with short exposures to high light intensity and short optical paths attain high biomass productivities (Richmond, 2003). Therefore flat panels and bubble column reactors are promising photobioreactors having short optical path that allows fast liquid circulation and achieves highest biomass concentration and photosynthetic efficiencies. The comparison of various performance evaluation parameters for different PBRs is presented in Table 5.5.

5.2.4 *Different Photobioreactor Configurations for Microalgal Cultivation*

Microalgae are photosynthetic organisms and can be cultivated in open-culture systems like raceway ponds, lakes, rivers or under highly controlled closed culture systems called photobioreactors (PBRs). Both systems have their own advantages

Table 5.5 Comparison of the performances of basic algal photobioreactor designs

S. No.	Configu–rations	Species	Volume (L)	Px (g L^{-1} d^{-1})	Ax (g m^{-2} d^{-1})	X (g L^{-1})	μ (d^{-1})	References
1	Airlift tubular	*Porphyridium cruentum*	200	1.50		3.0		Camacho et al. (1999)
2	Vertical tubular	*Cyanobium* sp.	2	0.071		1.0	0.127	Henrard et al. (2011)
3	Vertical flat plate	Synechocystis *aquatilis SI-2*	192		30	1.2		Zhang et al. (2001)
4	Bubble column	*Chlorella* sp.	0.8	0.42		1.45	0.605	Chiu et al. (2008)
5	Airlift	Phaeod–actylum tricornutum	50	2.47		6.2		Sobczuk et al. (2000)

and disadvantages. Open systems are normally less expensive and have large production capacity but they suffer from a most important drawback i.e. contamination from other microalgae or bacteria. There is absolutely no control over weather conditions, light and evaporation losses are common. Also, raceway ponds consume much energy in order to provide uniform mixing by paddle wheels (Richmond, 2003). Closed photobioreactors have better process control over open systems and gives higher productivity as compared to open system. There is reduced contamination risk, uniform mixing, high area-to-volume ratio and better light utilization efficiency is achieved. In addition to this, temperature and light control can be achieved by artificially providing lights. Closed photobioreactors are generally made up of sophisticated materials that involve high investment and operation cost. A brief description of widely used photobioreactors configurations for microalgal cultivation is presented.

5.2.4.1 Tubular Photobioreactors (Vertical and Horizontal)

Tubular photobioreactors are most suited for outdoor algal cultivation since they provide large illumination surface area. They can be in the form of horizontal/serpentine, conical, inclined, and helical. The culture medium is circulated with the help of air bubbling. The main issue with these types of PBRs is in the buildup of oxygen inside since excess O_2 is inhibitory for microalgal growth (Molina Grima et al., 1999). Scale up becomes difficult due to poor mass transfer and adherence of cells on the walls of the tubes is a common problem in tubular photobioreactors.

5.2.4.2 Air-Lift Photobioreactors

Air-lift PBRs are characterized by high mass transfer, good mixing, low shear stress, less energy consumption, easy scale up, reduced photo inhibition and other better performance properties which makes it as better choice over other PBRs. ALRs are column reactors which consist of an inner draft tube and an external cylindrical tube. The gas is sparged at the bottom with the help of gas spargers (an aerating device) and liquid flows through riser and comes down as down riser. The liquid behaves in a circulatory motion and this ensures proper mixing of nutrients (Sánchez Mirón et al., 1999).

5.2.4.3 Bubble Column Photobioreactors

Bubble column reactors are simple column photobioreactors with no internal partition. The bubbles rise through spargers and disengage in due course of time with decreasing bubble size and finally fully collapsing in the liquid. These reactors are low cost as they lack complexity of instrument parts as well as provide relatively satisfactory heat and mass transfer (Kumar et al., 2011).

5.2.4.4 Flat-Panel Photobioreactors

These PBRs are characterized by having large illumination surface area-to-volume ratio, good biomass productivity, good light path, low oxygen buildup and open disengagement zones. Flat panels are considered good for outdoor cultures (Cheng-Wu et al., 2001) but they suffer from a major drawback that is high hydrodynamic stress and poor temperature control.

5.3 Photobioreactor Configurations: Recent Modification and Performance Comparison

Recently, the typical configurations of PBRs such as bubble column, airlift, horizontal-tubular and flat panel are suitably modified or designed to overcome the challenges of producing high-density microalgal cultures. The critical parameters considered while designing an efficient photobioreactor are irradiance, hydrodynamics, mass transfer rate and control of operational conditions (Olivieri et al., 2014).

5.3.1 Irradiance

Some recent modifications in supplying light to the PBRs for enhancing microalgal biomass production include the following: (i) decreasing the light path along the reactor; (ii) increasing the photosynthetic efficiency and (iii) increasing the biomass yield on light supplied.

Flat panel PBRs are considered as the best option for improving biomass productivity via short light-path based irradiance supply. The light path <2 cm is regarded as effective for obtaining high density microalgal cultures. Liu et al. (2013) introduced an attached microalgal cultivation technology, in order to facilitate the concept of narrow light-path and maximizing photosynthetic efficiency. They developed suitable supporting materials for growing *Scenedesmus obliquus* as biofilm and then arranged in an array fashion to improve the biomass yield on light energy supplied. This attached cultivated technology resulted in very high biomass productivity of 80 g m^{-2} d^{-1} with 17.3 % photosynthetic efficiency, which was manifold higher than that reported in open pond cultivation systems. A marine green alga *Chlorococcum littorale* cultivated in a short light-path (1 cm) flat PBR at 2000 μmol m^{-2} s^{-1}, resulted in a biomass productivity of 9.12 g L^{-1} d^{-1}, which was 6.4-fold higher than that obtained in 4-cm flat PBR (Hu et al., 1998). They also observed that the areal biomass productivity of short light-path reactor improved by 30 %, as compared to 4-cm reactor, irrespective of culture volume.

Jacobi et al. (2012) designed transparent-open pored sponges, which was installed in flat reactors for enhancing the light dilution and distribution in PBRs. This novel transparent-glass sponge based reactor improved the growth rate of *Chlamydomonas reinhardtii* by about 25 % and photosynthetic conversion efficiency by 20 %. The increased illumination surface area to volume ratio provided the maximum photosynthetic efficiency and growth rate. Benson et al. (2009) proposed an appropriate lighting system for hydraulically integrated serial turbidostat algal reactor (HISTAR) and calculated the photosynthetic efficiency, biomass productivity of *Selenastrum capricornutum* and lighting cost by implementing deterministic model simulations. They found that the average simulated biomass productivity was increased from 22.5 g m^{-3} d^{-1} to 112 g m^{-3} d^{-1} by optimizing the light regime and cost of HISTAR system. A novel PBR utilizing the concept of photo-acclimation for attaining high density *Synechocystis aquatilis* SI-2 was developed by Grobbelaar and Kurano (2003). They designed a novel multi-compartment reactor to separate alga, which are acclimated to different light intensities in a reactor. The areal biomass productivity of 67.7 g m^{-2} d^{-1} was obtained using this specific light regime based novel reactor.

In order to improve the biomass yield on light energy supplied, light-emitting diodes (LED) provide various advantages such as longevity, energy saving, negligible heat generation, and targeting appropriate wavelength spectrum for pigments. Fu et al. (2012) systematically studied the use of highly efficient 660 nm LEDs for improving the biomass productivity of *Chlorella vulgaris*, in the aspect of reducing lighting cost of PBRs. After optimizing the critical parameters such as superficial velocity, light distribution, CO_2 concentration and nutrient media, the maximum biomass productivity of 2.11 g L^{-1} d^{-1} with a light yield of 0.81 g biomass $Einstein^{-1}$ was obtained and this study may direct a new dimension in phototrophic microalgal cultivation. Cuaresma et al. (2009) conducted a study, which involved the use of short-light path (14 mm) and red LEDs, in order to enhance the biomass productivity and biomass yield on light energy. Under continuous illumination of 2100 μmol m^{-2} s^{-1}, the maximum biomass productivity of *Chlorella sorokiniana* was observed to be 7.7 g m^{-2} h^{-1} with a photosynthetic efficiency of 1 g mol $photons^{-1}$ at a dilution rate of 0.24 h^{-1}. Irradiating microalgal cultures by submerging the light source inside bubble column or airlift PBRs was regarded as first generation of internally irradiated PBRs and it offers the reduced light-path irradiance strategy. The use of optical fibres for transmitting light gives long-term performance of PBR and it is considered as second generation of internally irradiated PBRs (Olivieri et al., 2014). These recent improvements in irradiance supply to PBRs offer many advantages in obtaining high density microalgal biomass production. However, the scaling-up of these advanced PBRs seems difficult in terms of either cost-effective or long-term performance.

5.3.2 Hydrodynamics

Mixing and gas hold-up are regarded as important hydrodynamic parameters of PBRs and are strongly dependent on flow-regime of the reactor. These parameters describe the mixing pattern, nutrient utilization of algal cells and the fraction of reactor volume occupied by the gas. Hence, these hydrodynamic characteristics significantly affect the performance of any PBR. For instance, the phenomenon of flashing light effect for improved light utilization by microalgal cells was well characterized by using airlift photobioreactors (Barbosa et al., 2003; Kumar et al., 2011). However, the circulation time between the light and dark-zones of the reactor is in the order of seconds, which is considered as slightly high with respect to photosynthetic turn over time scale of 10–100 ms (Olivieri et al., 2014).

To overcome this challenge, a novel flat panel-airlift type PBR was proposed by Bergmann et al. (2013). This works on the principles of airlift reactor with distinctively incorporated baffles to facilitate very short flashing light effect (1 to 5 Hz) and also uses flat panel design to reduce the light decay. *Phaeodactylum tricornutum* was cultivated using this hybrid reactor and the maximum biomass concentration and productivity was observed to be 25 g L^{-1} and 1.2 g L^{-1} d^{-1}, respectively. Another interesting reactor, which utilized the concept of flashing light effect and mixing, was termed as Taylor vortex flow short light-path PBR (Kliphuis et al., 2010). Taylor vortex flow represents the mixing pattern induced by the rotation of the inner cylinder in the cultivation vessel, operated in the turbulent regime. The microalgae *Chlorella sorokiniana* was cultured in turbulently mixed short light-path PBR and the biomass yield on light energy was found to be of 0.8 g mol $photons^{-1}$. However, this yield was found to be comparable or slightly higher than the values reported for green microalgae in the literature. The above mentioned recent advancements in hydrodynamics strategy improved the performance of algal cultivation in PBRs in terms of effective light dilution and distribution.

5.3.3 Mass Transfer Rate

CO_2, one of the most important limiting nutrients, acts as a carbon source for the photosynthetic microorganisms like microalgae. Gas-liquid mass transfer in PBRs is essential for two important aspects such as CO_2 fixation and O_2 release by microalgae, as a result of photosynthesis. Olivieri et al. (2014) classified the gas-liquid mass transfer in PBRs into two cases: (i) dissolution of $CO_{2(g)}$ into the medium through separate gas exchange systems (e.g. horizontal tubular PBRs) and (ii) supplying $CO_{2(g)}$ via internal sparging systems as in the case of bubble column and airlift reactors.

The recent improvements in mass transfer rate involve the following strategies: reduce the loss of CO_2 from the reactor, increase the retention time of CO_2 along the reactor and the use of advanced control system to maintain sufficient concentration of $CO_{2(l)}$ in the medium to avoid carbon limitation and excess. A simple and effective way of reducing the loss of CO_2 was investigated by implementing reactor-in-series or multi-stage sequential reactor strategy. A three-stage serial PBR was examined by de Morais and Costa (2007) for enhancing the CO_2 fixation rate. At 6 % CO_2, the maximum daily CO_2 fixation rate of 53.29 % and 28.08 % was observed for *Spirulina* sp., and *Scenedesmus* sp., respectively, indicating the usefulness of this strategy. Cheng et al. (2013) investigated 9-stage and 14-stage sequential reactor strategy to improve the CO_2 utilization efficiency of *Chlorella* PY-ZU1. They found that the average CO_2 fixation efficiency was improved from 7.6 % to 70.48 % with the biomass concentration of 5.42 g L^{-1} and biomass productivity of 0.95 g L^{-1} d^{-1}. CO_2 sequestration from oil producing industrial flue gas was studied by Kumar et al. (2014). *Chlorella sorokiniana* grown in flue gas CO_2, which was passed through serially connected airlift reactors, significantly removed the CO_2 content from 15.65 % (v/v) to 4.1 % (v/v). These studies show the significant contribution towards effective utilization of CO_2 by microalgae.

In order to increase the retention time of $CO_{2(g)}$ along the PBR, various micro-porous hollow fibre membrane modules were proposed recently. These modules are implemented in two ways: (i) integrating with the reactor as a gas sparger or (ii) installed separately, then connected to the reactor via peristaltic pump to facilitate the maximum gas-liquid mass transfer. Fan et al. (2008) studied the enhancement of CO_2 fixation by using enclosed membrane PBR, wherein the micro-porous hollow fibre namely, polyvinylidene fluoride (PVDF) was fixed inside the airlift tube as sparger. The maximum rate of CO_2 fixation and O_2 evolution of *Chlorella vulgaris* was found to be 5.4-fold and 2-fold higher, respectively as compared to conventional reactors. Cheng et al. (2006) investigated a membrane PBR, in which the polypropylene hollow fibre membrane was integrated with the reactor externally. It was observed that the retention time of the gas bubbles in membrane PBR was increased from 2 s to 20 s, as compared with the ordinary PBR. The CO_2 fixation rate of *Chlorella vulgaris* was improved by 3.25-fold with the drastic reduction in dissolved oxygen level (by a factor of 30). This type of improving mass transfer rate strategy not only enhanced the rate of CO_2 fixation but also the oxygen evolution in

PBR, which is considered to be inhibiting the photosynthesis process. A hybrid type of membrane PBR namely, membrane-sparged helical tubular photobioreactor (MSTR) was developed by Fan et al. (2008). This reactor utilizes the helical tube design to facilitate good light regime and membrane sparger to increase the mass transfer rate of CO_2. This MSTR enhanced the CO_2 fixation rate of *Chlorella vulgaris* by 58 %. Hence, these contemporary improvements in mass transfer phenomenon of PBRs gave some valuable insights and seem promising for achieving high density microalgal cultures.

5.3.4 Modelling and Control

5.3.4.1 Photobioreactor Modelling

Photobioreactor modelling have been advanced since the pioneering contribution by Sheth et al. (1977). This paper and other reported literature have attempted to deal with the three basic issues in modelling of photobioreactors: (a) the light distribution field in the liquid culture phase, (b) the culture hydrodynamics in terms of irradiance field and (c) the photosynthetic kinetics as a function of irradiance (Olivieri et al., 2014).

Lambert-Beer law could not be used accurately in models as it does not describe light scattering well enough. However, Lambert-Beer law of light decay in opaque system is a good tool for approximate analysis. Irradiance field in photobioreactor is well described by the radiative transfer model. This model takes into consideration: the effect of light on growth of cells, transfer of light impinging on the wall, and light absorption by the photosynthetic pigments in algal cells. Radiative transfer based 1D two-flux model was developed by Cornet et al. (1992) emphasizing on two parameters: the absorption coefficient (Ea) and the scattering coefficient (Es). Further, the radiative transfer equation (RTE) was solved considering the light path and the biomass concentration as a function to estimate the irradiance decay (I/I_0) at a distance z from the wall of the PBR (Olivieri et al., 2014).

$$\frac{I(z)}{I_0} = 2\frac{(1+\alpha)e^{-\delta(z/LP-1)} + (1-\alpha)e^{\delta(z/LP-1)}}{(1+\alpha)^z \varepsilon^{\delta} - (1-\alpha)^z e^{-\delta}} \tag{5.12}$$

where $\alpha = \sqrt{Ea/(Ea+Es)}$ and $\delta = (Ea+Es).\alpha.X.LP$

A tubular photobioreactor was modelled by Acién Fernández et al. (1997) using Equation (5.12) considering change in angle of incidence of light on the tube surface due to diurnal change. Further, airlift photobioreactor equipped with internal and external irradiance was modelled by Li et al. with the two-flux model. The ideal productivity was calculated of a volumetrically irradiated PBR, in turn the radiative transfer model was applied by Cornet to optimize the internal irradiance in terms of space distribution.

The radiative transfer equation solution was improved with regard to Eq. (5.12) by considering (a) the light spectrum; (b) the refraction/reflection of light on the surface of gas-liquid bubble; and (c) the size of reduced antenna of the genetically modified microalgae (Berberoglu et al., 2007). Further, Berberoglu et al. (2007) selected the discrete ordinate method to solve the 1D model with respect to (a), (b) and (c) phenomena. And the inferences drawn were: (i) the scattering can be considered as isotropic for wild stains; (ii) the bubbles result in additional anisotropic scattering due to the large gas-liquid interfacial area; and (iii) the light scattering due to cells with reduced antennae size is anisotropic.

Recent advances proposed stochastic algorithms and the Monte Carlo method to solve the radiative transfer equation. Regardless of the authenticity obtained by these complex models, the calculation remains complicated to apply in the design and operation of PBRs. Mass transfer and hydrodynamic behaviour have been studied extensively in the photobioreactors. Mostly in multiphase systems like bubble column and airlifts, the hydrodynamic modelling has been carried out based on computational fluid dynamics (CFD). CFD simulation was analyzed on various aspects of the hydrodynamics of different configurations of PBRs. The focus of CFD simulation was to study the flow characteristics of culture medium and to track the microalgae in the PBR irradiance field. Various sophisticated diagnostic techniques, coupled with simulation was applied to collect data related to microalgae trajectories in various configuration of PBRs, such as particle image velocimetry (PIV), computer automated radioactive particle tracking (CARPT) 192–195 and laser Doppler anemometry (LDA). Despite the success of these models, it is not yet possible to match the irradiance history obtained by tracking with a photosynthetic model, due to the variation in time scale (from milliseconds to seconds to days) of photosynthesis. Therefore, there is a need for long acquisition times and high frequency to cope with the photosynthesis kinetics.

Models describing photosynthesis and microalgal growth could be classified into (i) relationship between growth rate, irradiance and substrate and product concentration; (ii) dynamic models based on photosynthetic units (PSU); (iii) physiologically detailed models and (iv) metabolic scale models based on genome sequence. There is a need for detailed chemical description of liquid phase for thorough microalgal cell modelling. Model must combine hydrodynamics and kinetics. So, the procedure to describe and solve intermediate case include eulerian-based partial differential equations, light dynamic imposition, flow field imposition, stochastic Lagrangian and energetic analysis.

Latest photobioreactor model development includes analyzing hydrodynamic and mass transfer phenomenon with CFD simulations, complex structured kinetics model of microalgal growth and photosynthesis, algorithm to solve the radiative transfer equation for the illuminated field. All these models development are in the direction to support and improve the design and scale-up of the photobioreactor

5.3.4.2 Photobioreactor Control

The role of control system in the photobioreactor is to ease the operation of reactor and suppress the environmental fluctuation. Outdoor cultivation tends to have fluctuating environmental condition such as temperature, irradiance and contamination. The control system is often integrated with the PBR models for better control of the desired conditions. Functioning of control system depends on the number as well as type of the controlled and manipulated variables. Important parameters for maintaining optimal values include pH, CO_2 concentration, substrate feeding for continuous and semi-continuous mode of operation, illumination intensity and frequency, etc. pH is an important factor to be maintained at optimal level, and generally neutral pH is the optimal condition for growth of various microalgal stains. An on-off controller approach can maintain the optimal pH and pH value can be controlled either by pumping in acid/base solution or by feeding CO_2 through gas stream, usually concentration of more than 0.1 %.

Physiological state of the microalgal biomass is measured by photosynthetic efficiency. The term 'physiostat' was developed by Marxen et al.(2005), where the photosynthetic efficiency of PSII (Φ_P) was measured on line by using pulsed amplitude modulated (PAM) fluorescence technique. UV radiation was set by a feedback PI control loop to keep Φ_P constant. Physiostat was employed successfully to optimize TFA and EPA production by *N. salina* (Hoffmann et al., 2010).

Light acts as an important source of energy responsible for photosynthesis of microalgae. In outdoor condition control of light intensity is difficult, whereas in indoor cultures the irradiance can be regulated to optimize the photosynthesis and, in turn, the growth rate of microalgae. This modulated light system to control photosynthesis is known as 'luminostats', where light is considered like a substrate. One of the reports used luminostat in the airlift reactor equipped with eight internal fluorescent lamps, and to keep the average irradiance constant, the lamp numbers were increased gradually with higher biomass density. The radiative transfer model was used to assess the irradiance field. Further, control parameters like specific light uptake rate, specific growth rate and photon flux density transmitted through PBR was included in the control system. Melnicki et al. (2013) used stirred tank PBR and the two combined control system, turbidostat and luminostat mode, in which the dilution rate and the light intensity was adjusted to keep biomass concentration and irradiance transmittance constant.

The typical 'turbidostat' is a type of culture operating mode, either continuous and semi continuous, to maintain the biomass concentration constant. In continuous culture the dilution rate is adjusted whereas in semi-continuous culture fresh medium is replaced with the culture periodically. These control strategies was developed to increase the volumetric and area biomass productivity. Different approaches were selected like SI/SO mode: a single input and a single output for observed and manipulated variable respectively. Further MI/MO control strategy was reported by Ifrim et al. (2013), a multiple input and multiple output system measuring and controlling pH and biomass concentration. In this multi variable non-linear feedback control system, pH and biomass concentration were controlled by adjusting both the CO_2 flow and the dilution rate. Basically control system helps controlling of growth kinetics of microalgae. Typically, batch and continuous culture have been used for growth characterization.

5.4 Conclusion

In last few decades, advancement in the bioengineering of photobioreactor for efficient microalgal mass production has been made. Of the various types of microalgal culture systems proposed, closed culture system show promise for diverse application in large scale like pharmaceuticals, food and feed, nutraceutical, bioenergy, etc. Even after the profound progress in the photobioreactor technology, there are challenges such as huge capital-cost and operating-cost. There are difficulties with the scale-up mainly due to the change in reactor dimensions, like tube diameter, plate thickness, large land area required, etc. All these changes majorly affect the volume of light and dark zones, photosynthetic efficiency, gas exchange rate and hydrodynamics pattern. However, recent developments have improved irradiance distribution, hydrodynamics and mass transfer rate. New models systems and control strategies gives better description and control of the process. All these extensive efforts show great promise for scale-up, low-cost and efficiency of the culture systems. The future of microalgal biotechnology looks promising.

References

Acién Fernández, F.G., García Camacho, F., Sánchez Pérez, J.A., Fernández Sevilla, J.M. and Molina Grima, E. (1997). A model for light distribution and average solar irradiance inside outdoor tubular photobioreactors for the microalgal mass culture. *Biotechnology and Bioengineering*, 55, 701–714.

Barbosa, M.J., Janssen, M., Ham, N., Tramper, J. and Wijffels, R.H. (2003). Microalgae cultivation in air-lift reactors: Modeling biomass yield and growth rate as a function of mixing frequency. *Biotechnology and Bioengineering*, 82(2), 170–179.

Benson, B.C., Gutierrez-Wing, M.T. and Rusch, K.A. (2009). Optimization of the lighting system for a Hydraulically Integrated Serial Turbidostat Algal Reactor (HISTAR): Economic implications. *Aquacultural Engineering*, 40, 45–53.

Berberoglu, H., Yin, J. and Pilon, L. (2007). Simulating light transfer in a bubble sparged photobioreactor for simultaneous hydrogen fuel production and CO_2 mitigation. *International Journal of Hydrogen Energy*, 32, 2273–2285.

Bergmann, P., Ripplinger, P., Beyer, L. and Trosch, W. (2013). Disposable flat panel airlift photobioreactors. *Chemie Ingenieur Technik*, 85, 202–205.

Bhattacharya, S., Nayak, A., Schiavone, M. and Bhattacharya, S.K. (2004). Solubilization and Concentration of Carbon Dioxide: Novel Spray Reactors with Immobilized Carbonic Anhydrase. *Biotechnology and Bioengineering*, 86(1), 37–46.

Camacho, R.F., Acien Fernandez, F.G., Sanchez Perez, J.A., Garcia Camacho, F. and Molina Grima, E. (1999). Prediction of dissolved oxygen and carbon dioxide concentration profiles in tubular photobioreactors for microalgal culture. *Biotechnology and Bioengineering*, 62(1), 71–86.

Carvalho, A.P., Meireles, L.A. and Malcata, F.X. (2006). Microalgal reactors: A review of enclosed system designs and performances. *Biotechnology Progress*, 22(6), 1490–1506.

Cheng, J., Huang, Y., Feng, J., Sun, J., Zhou, J. and Cen, K. (2013). Improving CO_2 fixation efficiency by optimizing *Chlorella* PY-ZU1 culture conditions in sequential bioreactors. *Bioresource Technology*, 144, 321–327.

Cheng, L., Zhang, L., Chen, H. and Gao, C. (2006). Carbon dioxide removal from air by microalgae cultured in a membrane-photobioreactor. *Separation and Purification Technology*, 50, 324–329.

Cheng-Wu, Z., Zmora, O., Kopel, R. and Richmond, A. (2001). An industrial-size flat glass reactor for mass production of *Nannochloropsis* sp. (Eustigmatophyceae). *Aquaculture*, 195, 35–49.

Chisti, Y. and Moo-Young, M. (1989). On the calculation of shear rate and apparent viscosity in airlift and bubble column bioreactors. *Biotechnology and Bioengineering*, 34, 1391–1392.

Chiu, S., Kao, C., Chen, C., Kuan, T., Ong, S. and Lin, C. (2008). Reduction of CO_2 by a high-density culture of chlorella sp. in a semicontinuous photobioreactor. *Bioresource Technology*, 99, 3389–3396.

Cornet, J.F., Dussap, C.G. and Dubertret, G. (1992). A structured model for simulation of cultures of the cyanobacterium Spirulina platensis in photobioreactors. II. Identification of kinetic parameter under light and mineral limitations. *Biotechnology and Bioengineering*, 40, 826–834.

Cuaresma, M., Janssen, M., Vılchez, C. and Wijffels, R.H. (2009). Productivity of *chlorella sorokiniana* in a short light-path (slp) panel photobioreactor under high irradiance. *Biotechnology and Bioengineering*, 104, 352–359.

de Morais, M.G. and Costa, J.A.V. (2007). Biofixation of carbon dioxide by *Spirulina* sp. and *Scenedesmus obliquus* cultivated in a three-stage serial tubular photobioreactor. *Journal of Biotechnology*, 129, 439–445.

Fan, L.H., Zhang, Y.T., Zhang, L. and Chen, H.L. (2008). Evaluation of a membrane sparged helical tubular photobioreactor for carbon dioxide biofixation by *Chlorella vulgaris. Journal of Membrane Science*, 325, 336–345.

Fu, W., Gudmundsson, O., Feist, A.M., Herjolfsson, G., Brynjolfsson, S. and Palsson, B.Ø. (2012). Maximizing biomass productivity and cell density of *Chlorella vulgaris* by using light-emitting diode-based photobioreactor. *Journal of Biotechnology*, 161, 242–249.

Fulke, A.B., Mudliar, S.N., Yadav, R., Shekh, A., Srinivasan, N., Ramanan, R., Krishnamurthi, K., Devi, S.S. and Chakrabarti, T. (2010). Bio-mitigation of CO_2, calcite formation and simultaneous biodiesel precursors production using *Chlorella* sp. *Bioresource Technology*, 101, 8473–8476.

Giordano, M., Beardall, J. and Raven, J.A. (2005). Mechanisms in algae: Mechanisms, environmental modulation, and evolution. *Annual Review of Plant Biology*, 56, 99–131.

Grobbelaar, J.U. (2004). Algal nutrition. *In:* Richmond, A. (editor). Handbook of microalgal culture: Biotechnology and applied phycology. Blackwell, 97–115.

Grobbelaar, J.U. and Kurano, N. (2003). Use of photoacclimation in the design of a novel photobioreactor to achieve high yields in algal mass cultivation. *Journal of Applied Phycology*, 15, 121–126.

Henrard, A.A., de Morais, M.G. and Costa, J.A. (2011). Vertical tubular photobioreactor for semicontinuous culture of *Cyanobium* sp. *Bioresource Technology*, 102(7), 4897–4900.

Hoffmann, M., Marxen, K., Schulz, R. and Vanselow, K.H. (2010). TFA and EPA productivities of *Nannochloropsis salina* influenced by temperature and nitrate stimuli in turbidostatic controlled experiments. *Marine Drugs*, 8, 2526–2545.

Hu, Q., Kurano, N., Kawachi, M., Iwasaki, I. and Miyachi, S. (1998). Ultrahigh-cell density culture of a marine green alga *Chlorococcum littorale* in a flat-plate photobioreactor. *Applied Microbiology and Biotechnology*, 49, 655–662.

Hu, Q. and Richmond, A. (1996). Productivity and photosynthetic efficiency of *Spirulina platensis* as affected by light intensity, algal density and rate of mixing in a flat plate photobioreactor. *Journal of Applied Phycology*, 8, 139–145.

Ifrim, G.A., Titica, M., Barbu, M., Boillereaux, L., Cogne, G., Caraman, S. and Legrand, J. (2013). Multivariable feedback linearizing control of *Chlamydomonas reinhardtii* photoautotrophic growth process in a torus photobioreactor. *Chemical Engineering Journal*, 218, 191–203.

Jacobi, A., Bucharsky, E.C., Schell, K.G., Habisreuther, P., Oberacker, R., Hoffmann, M.J., Zarzalis, N. and Posten, C. (2012). The application of transparent glass sponge for improvement of light distribution in photobioreactors. *Journal of Bioprocessing and Biotechniques*, 2, 1–8.

Janssen, M., Tramper, J., Mur, L.R. and Wijffels, R.H. (2002). Enclosed Outdoor Photobioreactors: Light Regime, Photosynthetic Efficiency, Scale-Up, and Future Prospects. *Biotechnology and Bioengineering*, 81(2), 193–210.

Jorquera, O., Kiperstok, A., Sales, E., Embiruçu, M. and Ghirardi, M. (2010). Comparative energy life-cycle analyses of microalgal biomass production in open ponds and photobioreactors. *Bioresource Technology*, 101, 1406–1413.

Kliphuis, A.M.J., de Winter, L., Vejrazka, C., Martens, D.E., Janssen, M. and Wijffels, R.H. (2010). Photosynthetic efficiency of *Chlorella sorokiniana* in a turbulently mixed short lightpath photobioreactor. *Biotechnology Progress*, 26, 687–696.

Kumar, K., Banerjee, D. and Das, D. (2014). Carbon dioxide sequestration from industrial flue gas by *Chlorella sorokiniana. Bioresource Technology*, 152, 225–233.

Kumar, K., Dasgupta, C.N., Nayak, B., Lindblad, P. and Das, D. (2011). Development of suitable photobioreactors for CO_2 sequestration addressing global warming using green algae and cyanobacteria. *Bioresource Technology*, 102, 4945–4953.

Liu, T., Wang, J., Hub, Q., Cheng, P., Ji, B., Liu, J., Chen, Y., Zhang, W., Chen, X., Chen, L., Gao, L., Ji, C. and Wang, H. (2013). Attached cultivation technology of microalgae for efficient biomass feedstock production. *Bioresource Technology*, 127, 216–222.

Marxen, K., Vanselow, K.H., Lippemeier, S., Hintze, R., Ruser, A. and Hansen, U.P. (2005). A photobioreactor system for computer controlled cultivation of microalgae. *Journal of Applied Phycology*, 17, 535–549.

Melnicki, M.R., Pinchuk, G.E., Hill, E.A., Kucek, L.A., Stolyar, S.M., Fredrickson, J.K., Konopka, A.E. and Beliaev, A.S. (2013). Feedback-controlled LED photobioreactor for photophysiological studies of cyanobacteria. *Bioresource Technology*, 134, 127–133.

Molina, G.E., Acién Fernández, F.G., García Camacho, F. and Chisti, Y. (1999). Photobioreactors: Light regime, mass transfer, and scaleup. *Journal of Biotechnology*, 70, 231–247.

Olivieri, G., Salatinoa, P., and Marzocchellaa, A. (2014). Advances in photobioreactors for intensive microalgal production: Configurations, operating strategies and applications. *Journal of Chemical Technology and Biotechnology*, 89, 178–195.

Pulz, O. (2001). Photobioreactors: Production systems for phototrophic microorganisms. *Applied Microbiology and Biotechnology*, 57, 287–293.

Pulz, O. and Scheinbenbogen, K. (1998). Photobioreactors: Design and performance with respect to light energy input. *Advances in Biochemical Engineering/Biotechnology*, 59, 123–152.

Richmond, A. (2003). Growth characteristics of ultra-high cell density microalgal cultures. *Biotechnology and Bioprocess Engineering*, 8, 349–353.

Sánchez Mirón, A., Contreras Gómez, A., García Camacho, F., Molina Grima, E. and Chisti, Y. (1999). Comparative evaluation of compact photobioreactors for large-scale monoculture of microalgae. *Journal of Biotechnology*, 70, 249–270.

Sánchez Mirón, A., Garcia Camacho, F., Contreras Gomez, A., Grima, E. M., & Chisti, Y. (2000). Bubble-column and airlift photobioreactors for algal culture. *AIChE Journal*, *46*(9), 1872–1887.

Sheth, M., Ramkrishna, D. and Fredrickson, A.G. (1977). Stochastic models of algal photosynthesis in turbulent channel flow. *AIChE Journal*, 23, 794–806.

Silva, H.J., Cortinas, T. and Ertola, H.J. (1987). Effect of hydrodynamic stress on *Dunaliella* growth. *Journal of Chemical Technology and Biotechnology*, 40, 41–49.

Sobczuk, T.M., Camacho, F.G., Rubio, F.C., Fernandez, F.G.A. and Grima, E.M. (2000). Carbon dioxide uptake efficiency by outdoor microalgal cultures in tubular airlift photobioreactors. *Biotechnology and Bioengineering*, 67(4), 465–475.

Spoalore, P., Joannis-Cassan, C., Duran, E. and Isambert, A. (2006). Commercial applications of microalgae. *Journal of Bioscience and Bioengineering*, 101, 87–96.

Suh, I.S. and Lee, C.G. (2003). Photobioreactor engineering: Design and performance. *Biotechnology and Bioprocess Engineering*, 8(6), 313–321.

Thimijan, R.W. and Heins, R.D. (1983). Photometric, radiometric and quantum light units of measure: A review of procedures for interconversion. *Hort Science*, 18, 818–822.

Ugwu, C.U., Tatara, M., Fukui, H., Makiuchi, T., Goto, M. and Sode, K. (2008). Production for mass cultivation of algae. *Bioresource Technology*, 99, 4021–4028.

Vasumathi, K.K., Premalatha, M., and Subramanian, P. (2012). Parameters influencing the design of photobioreactor for the growth of microalgae. *Renewable and Sustainable Energy Reviews*, 16(7), 5443–5450.

Wang, B., Lan, C.Q. and Horsman, M. (2012). Closed photobioreactors for production of microalgal biomasses. *Biotechnology Advances*, 30, 904–912.

Wang, S.T., Pan, Y.Y., Liu, C.C., Chuang, L.T. and Chen, C.N.N. (2011). Characterization of a green microalga UTEX 2219–4: Effects of photosynthesis and osmotic stress on oil body formation. *Botanical Studies*, 52, 305–312.

Zhang, K., Miyachi, S. and Kurano Zhang, N. (2001). Evaluation of a vertical flat-plate photobioreactor for outdoor biomass production and carbon dioxide bio-fixation: Effects of reactor dimensions, irradiation and cell concentration on the biomass productivity and irradiation utilization efficiency. *Applied Microbiology and Biotechnology*, 55, 428–433.

Chapter 6
Scale-up Problems for the Large Scale Production of Algae

Teresa Lopes da Silva and Alberto Reis

6.1 Introduction

Microalgae are a natural source of high-value compounds for pharmaceutical and food industry and can also be a source of biofuels. Usually naturally occurring algae live completely submerged in aquatic environments, as low density cellular suspension. To have significant social, environmental and economic impact on human society, algae must be grown in large scale systems, in order to significantly increase the production volume. The objective of a process scale-up is to enlarge the production quantities with similar or higher productivity and product quality.

In fact, scale-up is a crucial topic in process development. Expanding microalgal cultivations from a lab-scale unit to a commercial one is a challenge due to the difficulty in assessing the factors affecting the scale-up process during the cultivation. As a result, most large-scale cultivations give a lower yield than is expected in the laboratory (Hsu and Wu, 2002). In addition, large scale algal growth benefits the environment since algae is a valuable carbon capture source and feeds on atmospheric nitrogen oxides, another prominent harmful greenhouse gas (Emeka et al., 2012).

There is a gap between the theoretical biological potential of microalgae and the biomass productivities obtained outdoors, especially in mass culture in photobioreactors (PBR). The present chapter will describe the key factors explaining microalgal performance far above expectations and ways to overcome it in the near future. The basic technical (material durability, others) and biological problems (mixing, oxygen buildup, temperature control, biofouling, contamination) that should be solved for successful full-scale operation after detailed study and research at an

T.L. da Silva • A. Reis (✉)
Laboratório Nacional de Energia e Geologia, I.P., Estrada do Paço do Lumiar, 22, 1649-038 Lisbon, Portugal
e-mail: alberto.reis@lneg.pt

D. Das (ed.), *Algal Biorefinery: An Integrated Approach*,
DOI 10.1007/978-3-319-22813-6_6

adequately lab and scaled pilot level will be covered and critically discussed along the text.

6.2 Factors Involved in Scale-up Problems

6.2.1 Light/Light Modulation

A survey of relevant publications in the last decades clearly showed that if large biomass yields were claimed at the laboratory level, high outputs over a long period were generally not achieved in outside full scale culture units. As suggested by Richmond (1986) this was likely caused by reduced growth rate related to light limitation. More important than the total light impinging on the culture surface was the quantity of light and regime (and energy) per cell. Soeder (1980) used for the very first time the concept areal density accounting the self-shading effect. High areal biomass densities imply low productivities as consequence of significant light limitation and a high contribution of respiration and endogenous biomass consumption. Low areal biomass densities imply damaging effects of radiation in excess (photo-inhibition and photo-oxidation) and lower productivities. This means that best productivities require a strict control of areal biomass densities over a narrow optimum interval which should be determined for each species (strain), location and season of the year.

Dense cultures are exposed to overcome the phenomenon of light limitation and to obtain the best light regime, high levels of mixing are necessary to reach a turbulent flow of the culture. If insufficient for photosynthesis, slow growth occurs. If too high, photo-inhibition (reversible) and photo-oxidative (irreversible) damage could take place with detrimental effect over the cultures. Critical key parameters are: reactor surface:volume (S/V) ratio, geometry, orientation, and inclination of reactor, material and thickness of reactor walls, culture depth and density and mixing.

6.2.2 Temperature

Along with light intensity, temperature is one of the most difficult parameters to control and optimize in large-scale outdoor culture systems. Fluctuations in temperature, both daily and seasonally, can lead to significant decreases in productivity. The optimal growth temperature for microalgae is species (strain) specific but seldom is attained outdoors during long periods of time. The problem of overheating especially in closed PBR will be addressed and several case studies will be presented.

A combination of high dissolved oxygen (DO), high temperature and high irradiance of a low density culture may result in photo-oxidative damage to algal cells.

Blocking of Photosystem II at high temperatures in the presence of CO_2 sensitizes the algae to photo-oxidative death. Photo-oxidative death and bleaching of photosynthetic pigments are separable phenomena. Photo-oxidative conditions were demonstrated in Israeli fish ponds using *Anacystis nidulans* as the test organism during dense summer blooms, when dissolved CO_2 is low, and in winter, when water temperatures generally drop below 15 °C. This finding suggested that photo-oxidative death may be responsible for the sudden decomposition of blue-green blooms in summer, and may be a factor in the absence of blue-green blooms in winter.

Elevated temperatures have a significant effect on microalgae, causing photo-inhibition before other cell functions are impaired (Harrison and Platt, 1986; Davison, 1991). Extreme temperatures limit electron transport and carbon fixation by reducing the ability of the algae to use light. This results in excess light energy and causes photo-inhibition by damaging the Photosystem II (PSII) apparatus (Levasseur et al., 1990; Anning et al., 2001). PSII is the most thermo-sensitive component of photosynthesis (Falkowski and Raven, 1997). Temperature influences algae photosynthesis by changing the photosynthetic rate, or by inducing phenotypic or genotypic changes among algae species (Davison, 1991).

Cooling is crucial during the hot season of the year and during the period of the day when sun is high. Several methods have been proposed, such as blocking a part of radiation impinging the reactor, covering it with an opaque material (this technical option affects light and temperature simultaneously), spraying the reactor surface with cool water, spreading cool water through a cascade acting as a heat exchanger or submerging the reactor in a pool containing cool water acting as a heat dissipator.

Low prevailing temperatures in winter may require additional culture heating. Near-by industries, boilers and furnaces can provide the needed energy at low cost. Renewable energies, such as geothermal, solar and photovoltaic should not be underestimated.

6.2.3 Mixing

It is crucial to avoid poor mass transfer, biomass settling and anaerobic zones ("dead zones"). But higher mass transfer (many times together with bubbles bursting) causes shear (hydrodynamic) stress and some cell damage with known detrimental effects due to cell fragility and high energy use, increasing production costs. Mixing times from lab/bench bioreactors to large-scale production facilities increase dramatically affecting the overall bioreactor performance. Spatial heterogeneities in the concentration of nutrients are cause of cellular stress and affects overall productivities. Kolmogorov (1949) and Hinze (1955) developed a theory for bubble or drop breakup in turbulent flows. They suggested that a bubble breaks as a result of interactions with turbulent eddies that are of approximately the same size as the bubble. Barbosa et al. (2004) concluded that bubble formation was the main cause for cell death in gas-sparged photobioreactors and that gas entrance velocity could be used as a measure for estimating cell damage in these reactors. It was proved that neither bubble bursting nor bubble rising contributed to cell death. Cell damage was strain-dependent and each strain had a critical gas entrance velocity above which cell death occurred.

These findings contribute to optimal design, operation, and scale-up of photobioreactors in terms of minimizing shear-related cell death for any given microalgae strain. Based on these results, the strategy to follow, whenever high superficial gas velocities need to be used, is to determine the critical gas velocity at the sparger for the strain of interest and to keep the gas velocity at the sparger lower than the critical value by increasing the number of nozzles and/or increasing the nozzle diameter. The standardization of the critical gas entrance velocity for any given strain also implies knowledge of the death mechanisms at the sparger. Study of the effect of cell concentration and viscosity on cell death rate of microalgae cells is also of great interest for high-density cultures, where a high mixing rate is a requisite to achieve high light efficiencies.

The gap between the theoretical biological potential of microalgae and the biomass productivity obtained with algal culture in tubular photobioreactors is due to a reduced growth rate related to hydrodynamic stress of pumping (Gudin and Chaumont, 1991). High levels of mixing are necessary to reach a turbulent flow of the culture, in order to optimize the light regime. The optimal conditions of pumping to produce this significant liquid mixing may produce some cell damage. Factors affecting this hydrodynamic stress (geometry of the bioreactor involved, type of pump utilized, morphology of algal cells, physiological conditions of microalgae, etc.) have been discussed by several authors (Gudin and Chaumont, 1991).

Results obtained outdoors suggested the need to develop an explanation besides the “light per cell” and “areal density” concepts but taking into account the cell fragility (Gudin and Chaumont, 1991). To overcome light limitation and in order to achieve the best light regime together with a good mass transfer, high levels of mixing, inducing a turbulent flow, are necessary. In most photobioreactors this is achieved by means of pumps. Paradoxically, optimal conditions of pumping for adequate mixing (and optimal light utilization) could produce partial or total cell damage by hydrodynamic stress. Pumping is a major concern. Too much shear will damage more delicate algae species, especially dinoflagellates. The type of pump involved induces different damaging effects. According to Gudin and Chaumont (1991), volumetric pumps (characterized by rotor revolution speed proportional to the culture flow rate) were less detrimental than centrifugal pumps (with constant and high rotation speed and long retention times of the culture suspension inside the stator). The same authors ordered different pumps according to their increasing microalgal growth rate based on experimental results: centrifuge pump, eccentric pump, trilobs pump and Archimedes screw pump.

Centrifugal pumps can affect the growth of microorganisms because they generate high shear stresses, which are not supported by cell walls, resulting in high cell mortality which was quantified by Ramírez-Duque et al. (2012). Only diaphragm and peristaltic pumps are recommended, as they are less damaging to cells than centrifugal pumps (Jaouen et al., 1999). Aiming at an increase in cell concentration, usually a decrease in the thickness of the culture is employed and mixing issues become more important, highlighting the concept of "cell fragility" as a primary growth-limiting factor.

Significantly less is known on the effects of hydrodynamic forces encountered by dinoflagellates in bioprocesses (Gallardo-Rodríguez et al., 2011). Hydrodynamic shear forces can inhibit growth, alter the cell cycle, initiate production of reactive oxygen species (ROS) in the cells, cause peroxidation of membrane lipids, affect calcium mobilization, alter the fluidity of the cell membrane, and lead to apoptosis. Some of these effects are interdependent. *Protoceratium reticulatum* and many other dinoflagellate algae have proved difficult to mass culture because they are inhibited by low-level turbulence (Gallardo-Rodríguez et al., 2011). The only exceptions appear to be the nontoxic species of the genus *Crypthecodinium*. For example, *Crypthecodinium cohnii* can withstand an energy dissipation rate of 5.8–107 W/m^3 without lysis. In comparison with this, an energy dissipation rate of a little above 7 W/m^3 is sufficient to damage *P. reticulatum* (Gallardo-Rodríguez et al., 2011).

Work is needed on methods for mitigating the damaging effects of turbulence. Certain shear-protective chemicals can be added to the culture of sensitive cells to dramatically improve their tolerance to shear stresses (Gallardo-Rodríguez et al., 2011). This allows culture of the otherwise sensitive algae in relatively turbulent conditions in a photobioreactor. *P. reticulatum* and other non-dinoflagellate microalgae have also been at least partly protected against turbulence-associated damage by the use of the shear protectant Pluronic F68 (PF68), a surface-active polymer.

Finally, low flow rates will bleach algae from too much light and dissolved oxygen will build up, both of which greatly inhibit cell reproduction.

6.2.4 Nutrient Provision

Growth inhibition due to toxicity in media composition, CO_2 provision and O_2 removal issues (mass transfer, sparging and degassing mechanisms, gas concentration and flow rate, headspace, gas holdup volume) are key factors in any bioreactor for microalgal mass production (Griffiths, 2013) especially in closed photobioreactors. The choice of medium is fundamental for the success of microalgal cultivation at pilot scale phase (Rawat et al., 2013). The choice of media can be made based on the findings and recommendations of other researchers in literature if the nutritional requirements of the target microalgal strain are known. However, if the requirements of the selected microalgal strain are not known, it is recommended to do a trial and error approach using the different modified media that can promote the growth of a wide range of microalgae. A wide range of artificial media have been formulated and the recipes are available in literature. However, most of these media are best suited for strain maintenance at lab scale and are prohibitively expensive for mass culture. Preferably, nutrients should be obtained from a cheap source e.g. urea as a source of nitrogen or wastewater as a complete medium, especially for low value and high volume products such as biofuels.

The cost of medium for microalgal biomass production in a 100 ton per annum was estimated to be approximately \$3000/ton. The cost of conventional medium for microalgae production is not feasible for low value products such as oil; thus other nutrient sources must be considered.

The use of artificial media at large scale is not economically viable, therefore, it is desirable to formulate cheap and readily available alternative media such as the use of domestic wastewater streams for sustainable microalgal biomass accumulation. Wastewater constitutes important macronutrients that are vital for the growth of a wide variety of microalgae. Important macronutrients in wastewater are nitrates, phosphates, ammonium, urea and essential trace elements such as vitamins (biotin and thiamine) and certain trace metals. The unique nutritional composition of municipal domestic wastewater makes it a valuable medium for the growth of microalgae with the added advantage of phycoremediation of the wastewaters to prevent eutrophication.

The pH and the dissolved CO_2 concentration in the wastewater are ideal for the growth of several microalgae. The use of post-chlorinated wastewater is more suitable at large scale cultivation of microalgae since the microbial load in the wastewater is greatly reduced, consequently minimising the risks of bacterial contamination. The problem of eutrophication is of global concern and is mainly exacerbated by a wide range of anthropogenic activities such as release of copious amounts of partially treated wastewater into water bodies. In order to ameliorate this problem, it is desirable to aggressively use the wastewater for microalgal growth. As the wastewater provides nutrients for microalgal growth, the main pollutants in the wastewater are concurrently removed by the microalgae enabling safe disposal into the receiving water bodies in the environment. The use of wastewater as medium for microalgal growth has not been well documented especially at large scale commercial cultivation of microalgae. Wastewater is cheap and readily available and is an excellent medium whose feasibility as substrate at large scale microalgal cultivation requires serious assessment. Coupling of wastewater treatment with the production of microalgae for biofuels has the potential to significantly improve the economics of biomass production. Secondary and tertiary wastewaters contain nitrates and phosphates in sufficient levels to support microalgal growth with little supplementation. Wastewater utilization can reduce the nitrogen by 94 % and eliminate the need to the addition of elements such as potassium, magnesium, and sulphur. CO_2-rich wastewater promotes the growth rates of microalgae as it balances the ratio of carbon:nitrogen:phosphorus. This further decreases harvesting cost due to higher biomass concentrations and overall costs by increased lipid production. This reduces the cost of treatment that would normally be incurred for nutrient removal by conventional methods. The use of wastewater reduces the need for enormous amounts of freshwater in microalgae cultivation, improving the economic viability whilst being an environmentally friendly means to renewable microalgal biomass production. Growth of microalgae on wastewater provides a means removal of organic contaminants, heavy metals and pathogens, thus saving on the costs of chemical remediation. The cost of conventional removal of nitrogen and phosphorus is reported to be \$4.4/kg N and \$3.05/kg P removed. It was shown that a 70–110 ton/(ha annum) facility using wastewater can result in a saving of \$48,400–\$74,800/(ha annum) for nitrogen removal and \$4575–\$7625/(ha annum) for phosphorus removal. The combination of saving from wastewater treatment and reduction of microalgae production costs is thus a win–win strategy when used for the production of energy or liquid fuels.

Despite the favourable outlook for the use of wastewater mediated biomass production, the real potential must be explored practically at large scale. A potential problem associated with wastewater utilisation is the viral and bacterial contamination that may or may not negatively affect the production process. The composition of wastewater varies and may impact on growth rates. This factor cannot be controlled and close monitoring and adjustment of nutrient levels may be required. Utilisation of wastewater will further necessitate frequent cleaning of the culturing system.

Many studies have noted that microalgae flocculate at high pH, a process that is often referred to as 'auto-flocculation', regardless of the type of culture media and salinity. This phenomenon occurs quite often under CO_2 limitation. Auto-flocculation is induced at pH 10–10.5 in many marine species and at pH 10.5–11 in many freshwater species. The flocs formation could be detrimental, hindering mass transfer, enhancing heterogeneities and lowering biomass productivities.

6.2.5 Contamination/Infestation/Predation

A longtime prevailing problem outdoors especially under open raceways ponds but can also occur quite often in closed systems as sanitation does not solve all potential problems. Some microalgae can grow easily under open growth conditions such as *Chlorella* sp. with minimal contamination from non-target microalgae due to its high biomass volumetric output rates together with the release of biocidal substances. Moreover, microalgae such as *Chlorella* sp. are known to grow under diverse conditions as a mixed population with *Scenedesmus* sp. in open reactors for wastewater treatment. The choice of media is also important if the whole exercise is to succeed such as the use of artificial media against supplemented municipal domestic wastewater. Under open growth conditions a wide variety of microalgae have been shown to depict seasonal variations; hence a thorough study of microalgal population dynamics is crucial. Economic feasibility of the whole exercise from upstream to downstream processing is important to establish if it is feasible and profitable.

In addition, careful manipulation of pH can be done in order to prevent contamination by grazers such as zooplankton, especially when wastewater is used in the culture broth. Moreover, a detergent and phenol procedure may be employed as a prevention method for bacterial contamination.

Spirulina, *Chlorella* and *Dunaliella* are amongst the most important commercially produced microalgae and cyanobacteria. Open system cultivation of these strains is possible as they grow in highly selective media and/or environment and therefore can be cultivated with relatively limited contamination by other microorganisms. N-fixing cyanobacteria culture in N-deprived culture media offers also a possibility of using open reactors outdoors with low risk of contamination by other microalgae and cyanobacteria.

Special concern must be devoted to the heterotrophic culture of microalgae in fermenters. Bacteria, yeasts and fungi can spoil one biomass production cycle in few hours as they undergo by far higher biomass specific growth rates and productivities.

6.2.6 Biomass Film Formation (Fouling) and Loss of Reactor Wall Transparency

The single most important factor in limiting algae productivity is light. Therefore, the photostage in closed photobioreactors primarily consists of transparent vessels columns or tubes, made of specialty plastics or glass. An algae bearing fluid is slowly pumped through the system, presenting all the algae to the sun absorption zone. The algae grow inside the closed system, reducing the possibility of infestation. A controlled amount of CO_2 is injected into the photo-bioreactor to maximize photosynthesis. The formation of a film is undesirable as difficult mass and light transfer, decreasing the available radiation impinging the reactor surface as well as creating zones where gas and nutrient exchange is less efficient, with biomass productivity loss. The detrimental effect of material deterioration in PBR walls affects the decrease in transparency, affecting also the available radiation entering the reactor.

Fouling at the inner surfaces of the bioreactors blocks sunlight, restricting photosynthesis—a crucial factor for algae productivity. Several technical solutions have been proposed in order to overcome this drawback. The injection of solid particles into culture flowing together with the culture broth has been used by many groups worldwide as a way to prevent or, at least, minimize biomass fouling. Ultrasonic cleaning—applied from the outside—removes this algae film efficiently. Hielscher ultrasonic devices (http://www.hielscher.com/algae_reactor_cleaning_01.htm) have been used in a novel method to clean the inner surface of glass reactor tubes effectively. An ultrasonic sonotrode is pressed against the outer surface. This couples ultrasonic vibrations to the glass, which lead to a vibration of the glass itself. This removes the fouling from the inner surface. The loose algae are flushed away and the sunlight can reach into the reactor again. In order to clean multiple long tubes, a single ultrasonic system can be moved along each tube and from tube to tube.

Unwanted adhesion of microalgae on surfaces is a ubiquitous problem across many bioprocesses. Ren et al. (2014) explored the strategy of developing a silver nanoparticle (AgNP) coating for antifouling applications in marine and freshwater environments. *In situ* growth of AgNPs was achieved by a polydopamine (PDA)-based method. A range of most used industrial materials, including glass, polystyrene, stainless steel, paint surface, and even cobblestone, were employed, on which AgNP coatings were built and characterized. Ren et al. (2014) described the fouling-resistant behaviour of these AgNP-modified surfaces against two typical fouling organisms: a marine microalga *Dunaliella tertiolecta* and a freshwater green alga community. The PDA-mediated AgNP deposition strategy was demonstrated applicable for all the above materials. The resulting AgNP coatings showed a significant surface inhibitory effect against the adhesion of microalgae by above 85 % in both seawater and freshwater environments. They observed that contact killing was the predominant antifouling mechanism of AgNP-modified surfaces, and the viability of the microalgae cells in bulk media was not affected. In addition, silver loss from PDA-mediated AgNPs was relatively slow; it could allow the coating to persist for long-term usage in FBR. This study showed the potential of preparing environmentally friendly surfaces that can effectively manage biofouling through the direct deposition of AgNP coatings as a suitable material for PBR.

The German engineering and consulting company GICON (www.gicon.de), in collaboration with the Anhalt University of Applied Sciences, has successfully developed a scalable photo bioreactor based on silicone materials in combination with a new pulsation principle, in order to keep bio-fouling at a minimum, resulting in production with very limited contamination in a closed system. Biomass productivities of about 1 g L^{-1} d^{-1} (dry weight) and concentrations of 10 g L^{-1} (dry weight) have been achieved based on an input of approximately 50 W $m–^{3}$ of electric energy.

Materials for photobioreactors must resist to weathering (durability) besides other ideal properties such as high transparency, high mechanical strength, lack of toxicity, chemical stability, antifouling surface, resistance to scratching, and low cost, bearing in mind that materials required for building the photostage represent a major cost component of the photobioreactor (Tredici, 1999). A preliminary investigation of the durability of six different plastic tubing materials exposed to natural weathering in central Italy has been shown by Tredici (1999). The conclusions were summarized as follows: polyethylene and polypropylene tubes were inexpensive, but both lost transparency very quickly and presented serious limitations as regards biofouling and mechanical strength. Polycarbonate tubes were seriously weakened by alkaline solutions and became brittle and broke easily after 2–3 years of exposure outdoors. They lost transparency mainly in the 300- to 500-nm range. Tubes of polymethyl methacrylate or Teflon had been shown to possess high resistance to weathering and good overall properties. PVC tubes lost transparency significantly after hydration. Some of the flexible plastic materials tested have shown limited mechanical strength that has led to trivial problems (such as leakage due to punctures inflicted by cats and wild animals) that could become rather serious in large-scale systems. Glass might be an excellent material for photobioreactors, due to its high transparency, chemical stability, and durability. Glass tubular systems, however, would require mostly field assembly and numerous connections, and this would greatly increase installation costs; furthermore, fragility still constitutes a serious limitation to the use of glass tubes in large-scale installations.

6.2.7 Oxygen Build-up

High biomass densities (especially in closed thin photobioreactors), under high CO_2 concentrations, and high irradiances, favour oxygen build-up that imposes new challenges for scaling-up outdoors. According to Molina et al. (2001), under outdoor conditions, for given levels of oxygen supersaturation, the productivity decline is greater outdoors than indoors, suggesting that under intense outdoor illumination photo-oxidation contributes to loss of productivity in comparison with productivity loss due to oxygen inhibition alone. It is not uncommon to record dissolved oxygen values at the outlet of the solar photostage above 400 % of air saturation. Any practice conducting to oxygen stripping will be beneficial to the culture. Degassing tanks or release valves carefully inserted along the culture path have been proposed by several authors. As a rule of thumb for tubular closed photobioreactors, at least at any 50 m of tube length a release valve should be placed in order to keep dissolved oxygen in the comfort zone.

6.3 Scale-up Problems in Heterotrophic Cultures

The approach of large-scale production of microalgae in fermenters is typical for any fermentative process and has been widely studied by several authors in the last 40 years, especially for bacteria and yeasts.

Fermentation can be performed in batch, fed-batch or continuous mode. Continuous culture is not common in the pharmaceutical industry because the probability of mutation and contamination is higher and that the definition of a batch size is not evident.

Batch processes are simple and robust but the only way to reach a high cell density is the fed-batch mode, which is more complex but allows the metabolism of the strain to be controlled (Thiry and Cingolani, 2002). For fed-batch processes, exponential feeding allows the cells to be grown at a constant growth rate. However, the dissolved oxygen is often a limiting factor if a high growth rate is reached.

The factors affected by scale are the number of generations, the mutation probability, medium sterilization, the quality of temperature and pH regulations, agitation, aeration and pressure. The best way to prepare the scaling-up of a process is to first scale-down to the pilot scale of the conditions of culture that will be used at the final scale of production. Then, when the scale is increased, the broth will become more and more heterogeneous. In large fermenters, oxygen can be depleted in some areas of the reactor.

Since it is impossible to keep all physical or chemical parameters constant during the scale-up course, it is essential to know the critical factor that affect the implementation of successful scale-up (Hu et al., 2013). In the aerobic fermentation process, oxygen supply is usually a limiting factor owing to the low solubility of oxygen in the fermentation broth. The traditional method for scaling-up a microbial system is usually based on empirical criteria such as constant power input per unit volume, a constant mass transfer coefficient, constant mixing time and constant impeller tip velocity, in the case of conventional bioreactors used for heterotrophic microorganisms (Shuler and Kargi, 1992). It is assumed that physiological and physical properties of the cells and the broth are the same in geometrically similar conditions and fully baffled bioreactors (Chopra, 2004). However, this assumption often fails as cells are exposed to various changes in the environment where they grow, due to gradients present in bioreactors, as a result of insufficient mixing which changes the local concentrations of nutrients and pH.

The scale-up of fed-batch fermentation process for the DHA (docosahexaenoic acid)-rich microbial lipids production by *Schizochytrium* sp. from shake flasks to bench-scale (10 L and 50 L) and pilot-scale (1500 L and 7000 L) stirred fermenters was studied by Hu et al. (2013). The fermentation process was successfully scaled-up from experimental-scale to bench-scale and pilot-scale at matched oxygen volumetric mass transfer coefficient K_La value (88.9 h^{-1}) estimated by the K_La model developed in this study and the K_La empirical equations. The fermentation performances of the bench-scale and pilot-scale (DHA content of 13.96, 14.44 and 14.16 g L^{-1} in the 10 L, 50 L and 1500 L fermenters respectively) processes were similar to that of the experimental-scale process (DHA content of 13.84 g L^{-1}) and the fermentation results obtained in the 7000 L stirred fermenters (DHA content of 19.72 g L^{-1}) were much higher. These results indicated that K_La was applicable for the scale-up of the fed-batch fermentation process of DHA production by *Schizochytrium* sp. and this scale-up strategy enabled a maximum of 150-fold quantitative scale-up of the fed-batch fermentation process.

An integrated approach of biodiesel production from heterotrophic *Chlorella protothecoides* focused on scaling up fermentation in bioreactors was reported by Li et al. (2007). Through substrate feeding and fermentation process controls, the cell density of *C. protothecoides* achieved 15.5 g L^{-1} in 5 L, 12.8 g L^{-1} in 750 L, and 14.2 g L^{-1} in 11,000 L bioreactors, respectively. Resulted from heterotrophic metabolism, the lipid content reached 46.1 %, 48.7 % and 44.3 % of cell dry weight in samples from 5 L, 750 L and 11,000 L bioreactors, respectively. Transesterification of the microalgal oil was catalyzed by immobilized lipase from *Candida* sp. 99–125 with lipase (12,000 Ug^{-1}, based on lipid quantity) and 3:1 molar ratio of methanol to oil batch-fed at three times, 98.15 % of the oil was converted to monoalkyl esters of fatty acids in 12 h. The expanded biodiesel production rates were 7.02 g L^{-1}, 6.12 g L^{-1} and 6.24 g L^{-1} in 5 L, 750 L and 11,000 L bioreactors, respectively. The properties of biodiesel from *Chlorella* were comparable to conventional diesel fuel and comply with the US Standard for Biodiesel (ASTM 6751). These results suggest that it is feasible to expand heterotrophic *Chlorella* fermentation for biodiesel production at the industry level. An attempt to scale-up a biodiesel production plant from a heterotrophic culture of microalgae, together with an evaluation of feasibility was published by Tabernero et al. (2012). The plant utilized the microalgae *Chlorella protothecoides* to obtain biomass, being the subsequent oil extraction done with supercritical carbon dioxide. Accordingly with the authors, with a surface of only 7500 m^2 it would be possible to manufacture 10,000 ton of biodiesel per year. Based on previous studies, it was possible to determine the mass and energy balances and to design the equipment of the main process. A manufacture cost bigger than the soy-based biodiesel was obtained and with a high investment, but without economic and social problems in connection with the raw material.

A non-conservative study revealed the infeasibility of the production plant unless the residues were sold (two different and real prices were chosen), providing with an investment recovery in both cases. On the other hand, the estimations from a conservative study showed the non-viability of the process even if the residues are sold. This is mainly due to the large quantity of bioreactors required by the plant and the oil extraction yield. An alternative process with a complete extraction was also considered, which became viable in a conservative study only if the residues are sold at the highest price. All these results highlighted the future potential of a plant with these characteristics in the current energetic context.

6.4 Scale-up Problems in Autotrophic and Mixotrophic Cultures

Autotrophic microalgae are photosynthetic organisms that undergo the conversion of light into chemical energy as a form of a wide range of organic compounds through its photosynthetic machinery. The cultivation of microalgae brings environmental advantages, bearing in mind the capability of nutrient recycling in wastewaters together with the fixation of greenhouse gases such as CO_2. Mixotrophic microalgae undergo the dual capability to perform autotrophy and heterotrophy cooperatively, consuming inorganic carbon (CO_2) and/or organic carbon.

These microorganisms have been widely recognized as having huge potential as feedstock for food and feed industries, as "nutraceutical" agents (carotenoids, antioxidants, polyunsaturated fatty acids, single-cell proteins (SCP), phycobiliproteins, polysaccharides, vitamins, phytosterols, minerals), for the cosmetic industry, bioplastics, agriculture biofertilizers and recently as an energetic vector towards the production of a wide range of biofuels. Microalgae exhibit clear advantages when compared with higher plants, such as higher photosynthetic efficiency, higher areal biomass productivities, higher CO_2 biofixation rates (many polluting focus such as cement and thermoelectric plants can be used), higher O_2 production rates, non-competition for agricultural areas (marginal lands such as deserts, rocky areas and salt pans can be used), non-competition for drinking waters (saltwater, brackish water and wastewaters can be used), harvesting routines can be carried out daily with a better equipment and resources management trimming storage costs.

Several constraints should be overcome in order to achieve a cost-effective microalgal biofuel production, such as high energy inputs and still prohibitive production costs (currently around 5000 €/ton, far above the desired threshold target of 700 €/ton). But the main bottleneck is, by far, scale up problems for the large scale production of algae, reducing the number of successfully mass-culture microalgae cases to few such as pigments, fine highly polyunsaturated fatty acids and microalgal biomass as food supplement (*Chlorella*, *Spirulina*).

Regardless of the type of photobioreactor (tubular, thin film, bag, tank, etc.) there are limiting factors that are relevant. Certain requirements of closed photobioreactors, including the need for cooling, the need for strict control of oxygen accumulation and biofouling, and the need for frequent replacement of the material of the photostage, make these systems more expensive to build and operate than ponds. The technical difficulty in sterilizing the photoreception unit of any closed photobioreactor has hindered their application for algae culture for specific end-products such as high value pharmaceutical products. As alternative, several Clean-In-Place (CIP) routines have been developed in order to achieve sanitation but no full control of the culture against undesired alga competitors, infesters and predators is guaranteed. This is even more challenging for mixotrophic cultures. The use of organic carbon in transparent photobioreactors imposes more risky conditions in terms of contaminations. As plastic and glass cannot be easily sterilized for large volumes, and sanitation cannot be 100 % secure for keeping apart bacteria, yeast, virus and fungi, these organisms can grow quickly as compared to microalga.

Photobioreactors have different designs and configurations as compared to conventional bioreactors. For large-scale monoculture of microalgae, photobioreactors have conventionally been designed as devices with large surface–to–volume ratios (S/V). These reactors occupy vast land areas, they are expensive to build, difficult to maintain, and only somewhat scalable. For these reasons, large-scale tubular photobioreactors can usefully satisfy only medium level production demands (Mirón et al., 1999). Indeed although high S/V ratio photobioreactors work well in the laboratory, they may become extremely inefficient when scaled up. In high S/V ratio photobioreactors all the volumetric activities that depend on the input of light energy per unit volume (such as O_2 evolution, CO_2 absorption, nutrient depletion, metabolite excretion, and heat production) change at a high rate, and this could have long-term negative effects on the stability of the process. Therefore, it is not possible to fix an optimal S/V ratio for large-scale photobioreactors since this parameter depends on the organism to be grown, on the type of mixing and on other specific conditions (Tredici, 1999).

6.5 Scale-up Case Studies

The scale-up cultivation of *C. ellipsoidea* was studied in big bioreactors from indoor to outdoor conditions (Wang et al., 2014). The microalgae could adapt to the outdoor conditions, although with a lower biomass production. The overall cost of the biomass produced by the 200 L outdoor cultivation (58.70 US$/kg-dry weight) was estimated to be more than seven times lower than that of the 20 L indoor cultivation (431.39 US$/kg-dry weight) (Wang et al., 2014).

Oncel and Sabankay (2012) were focused on a scale-up procedure considering two vital parameters (light energy and mixing) for microalgae cultivation, taking *Chlamydomonas reinhardtii* as the model microorganism. Applying two stage hydrogen production protocol to 1 L flat type and 2.5 L tank type photobioreactors hydrogen production was investigated with constant light energy and mixing time. The conditions that provided the shortest transfer time to anaerobic culture (light energy of 2.96 kJ s^{-1} m^{-3} and mixing time of 1 min) and highest hydrogen production rate (light energy of 1.22 kJ s^{-1} m^{-3} and mixing time of 2.5 min) were applied to 5 L photobioreactor. The final hydrogen production for 5 L system after 192 h was measured as 195 ± 10 mL that was comparable with the other systems was a good validation for the scale-up procedure.

Passell et al. (2013) attempted to bridge the gap between laboratory-scale and larger scale biodiesel production by using cultivation and harvesting data from a commercial algae producer with 1000 m^2 production area (the base case), and compared that with a hypothetical scaled up facility of 101,000 m^2 (the future case). Environmental impacts were quantified as NER (energy in/energy out), among others. Results for the base case and the future case showed NER of 33.4 and 1.37, respectively, both values away from energetic profitability. A critical feature in this work was the low algal productivity (3 g/m^2 day) reported by the commercial producer, relative to the much higher productivities (20–30 g/m^2 day) reported by other sources, revealing a productivity loss when going from small to bigger reactors outdoors. Notable results included a sensitivity analysis showing that algae with an oil yield of 0.75 kg oil/kg dry biomass in the future case can bring the NER down to 0.64, more comparable with petroleum diesel and soy biodiesel which is unlikely to occur in the near future as high lipid yields are obtained sacrificing biomass productivities.

The consequences of large-scale production of biodiesel from microalgae for eutrophication in four large European seas were analyzed by Blaas and Kroeze (2014). To this end, scenarios for the year 2050 were analyzed; assuming that in the 27 countries of the European Union fossil diesel will be replaced by biodiesel from algae. Estimates were made for the required fertilizer inputs to algae parks, and how this may increase concentrations of nitrogen and phosphorus in coastal waters, potentially leading to eutrophication. The Global NEWS (Nutrient Export from Water Sheds) model had been used to estimate the transport of nitrogen and phosphorus to the European coastal waters. The results indicated that the amount of nitrogen and phosphorus in the coastal waters may increase considerably in the future as a result of large-scale production of algae for the production of biodiesel, even in scenarios assuming effective wastewater treatment and recycling of wastewater in algae production. To ensure sustainable production of biodiesel from microalgae, it is important to develop cultivation systems with low nutrient losses to the environment. Pertinent questions, however, need to be answered on the commercial viability of large scale production of biodiesel from microalgae (Rawat et al., 2013). Vital steps need to be critically analysed at each stage.

Significant displacement of petroleum fuels will require that algae feedstock production reach large volumes that will put demands on key resources (Pate et al., 2011). This scenario-based analysis provides a high-level assessment of land, water, CO_2 and nutrient (nitrogen and phosphorus) demands resulting from algae biofuel feedstock production reaching target levels of 10 billion gallons per year (BGY), 20 BGY, 50 BGY and 100 BGY for four different geographical regions of the United States. Different algae productivities are assumed for each scenario region, where relative productivities are nominally based on annual average solar insolation. The projected resource demands are compared with data that provide an indication of the resource level potentially available in each of the scenario regions. The results suggest that significant resource supply challenges can be expected to emerge as regional algae biofuel production capacity approaches levels of about 10 BGY. The details depend on the geographic region, the target feedstock production volume, and the level of algae productivity that can be achieved. The implications are that the supply of CO_2, nutrients, and water, in particular, can be expected to severely limit the extent to which US production of algae biofuel can be sustainably expanded unless approaches are developed to mitigate these resource constraints in parallel to emergence of a viable algae technology. Land requirements appear to be the least restrictive, particularly in the Western half of the country where larger quantities of potentially suitable classes of land exist. Within the limited scope and assumptions of this analysis, sustainable photosynthetic microalgae biofuel feedstock production in the US in excess of about 10 BGY will likely be a challenge due to other water, CO_2 and nutrient resource limitations.

Developing algae production approaches that can effectively use non-fresh water resources and minimize both water and nutrient requirements will help reduce resource constraints. Providing adequate CO_2 resources for enhanced algae production appears the biggest challenge, and could emerge as a constraint at oil production levels below 10 BGY.

Taylor et al. (2013) presented a techno-economic analysis of carbon-negative algal biodiesel production routes that used available technologies. The production process included the following stages: carbon-neutral renewable electricity generation for powering the plant, algal growth in photobioreactors, algae dewatering and lipid extraction, and biofuel conversion and refining. As carbon dioxide is consumed in the algal growth process, side products are not burned (with CO_2 release), and the energy supplied to the entire production process was obtained from concentrated solar power, the whole system was assumed carbon footprint negative. Under assumptions related to economics of scale, the techno-economic model was extended to account for varying industrial scales of production. Verified data from a selection of commercially available technologies were used as inputs for the model, and the economic viability of the various production routes was assessed. From the various routes investigated, one scheme involving combined gasification and Fischer–Tropsch of algal solids to produce biodiesel along with conversion of algal lipids into biodiesel through transesterification was found to be promising. Assuming a typical economic scaling factor of 0.8, an algal biodiesel process with an annual production rate of 100 Mt/year was identified to achieve a biodiesel price comparable to the current conventional diesel price (approximately £1.39/litre at the pump, or $114/barrel of crude) with a discounted break-even time of six years.

The high algal biodiesel fuel price could also be explained as the pilot plant scale technologies are not designed to be economically feasible. Scale-up of the original pilot plant might be another option to reduce the biodiesel production cost. However, there they considered only scale-out of existing technology. Three larger biodiesel production capacities were presented to illustrate the effect of scale-out on reducing the biodiesel price. The resulting prices with respect to different industrial scales in the process were shown, where the discounted break-even point was set as 15 years and was set to be 0.9 (Taylor et al., 2013).

6.6 Harvesting and Downstream Processing in Scale-up

Harvesting at large scale poses major challenges (Rawat et al., 2013). Harvesting of 25–33 % of the reactor volume may be required daily for viable production of biodiesel. Greater microalgal biomass production effectively decreases the cost of harvesting/dewatering steps and the high cost of biomass recovery. The most rapid and effective universally accepted method for total biomass separation is by continuous centrifugation. However, for large scale biomass harvesting in the biofuels arena, this is generally not practiced due to process being energy intensive and not typically economically feasible. Gravity sedimentation is preferred due to low cost. However, the efficacy of gravity sedimentation is strongly influenced by the density and radius of algal cells.

Flocculation is used to enhance the settling characteristics by increasing particle density of culture that may be unsuccessfully separated due to low particle density. The process may be enhanced by the use of lamella separators and sedimentation tanks. Ancient technique of filtration is another commonly used method. Upgrading the two basic techniques of sedimentation and filtration may be used in combination for dewatering with flocculation. Vacuum filtration is effective for the recovery of larger algae (greater than 70 μm) when operated under required pressure combination with a filter aid. This has led to more compact solid biomass harvesting. Microalgae of size greater than 30 μm have to be found to be effectively harvested by filtration. The potential use of membrane microfiltration or ultra-filtration is reliable for cells of smaller size. This may be impacted by replacement of low-cost membrane due to rapid fouling membrane and pumping of the biomass without back flushing. Microalgal biomass harvesting is one of the major steps in upstream processing. The end results of many methods tend to be highly energy intensive and more complex. Drying or dewatering of biomass is generally required as a pretreatment prior to final product extraction or use in various conversion techniques. Moisture in the biomass will negatively interfere with the downstream processing and greatly influence the cost of product recovery. Furthermore, the biomass can spoil in a matter of hours post harvest should it not be rapidly processed which is an extra challenge outdoors. Drying may be achieved by spray drying, drum drying, freeze-drying, solar drying, as well as various forms of oven drying. The extraction of products can be facilitated by means of mechanical cell disruption methods such as cell homogenizers, bead mills, ultrasounds, autoclave, and spray drying or non-mechanical methods such as freezing, utilization of organic solvents, osmotic shock, acid and base as well as enzyme reactions. Many of these methods are still not feasible at large scale due to high energy input requirements.

6.7 Scale-up Guidelines

The importance of accurate piloting has been suggested by Tredici (1999) as the establishment of a commercial activity based on the cultivation of phototrophic microorganisms requires realizing that a photobioreactor design that is efficient at the laboratory level is not necessarily so at the level of industrial production. First of all, it is imperative to do a full optimisation study at lab scale before considering scaling up (Rawat et al., 2013). The one-factor-at-a-time approach for optimisation experiments is frequently used for optimization studies despite being associated with drawbacks such as being time consuming and labour intensive. To date some workers have used the response surface methodology for optimization studies and this is reported to be fast and a large set of experiments can be done simultaneously. However, this approach is not amenable at large scale since a number of experimental trials must be run which is nearly impossible at large scale. The optimization of physico-chemical parameters is technically difficult under open growth conditions since it is impossible to control factors such as temperature, light intensity, quantity, spectral quality and photoperiod which are prone to environmental variations and fluctuations.

The move from laboratory to large scale microalgal cultivation requires careful planning (Rawat et al., 2013). It is imperative to do extensive pre-pilot demonstration trials and formulate a suitable trajectory for possible data extrapolation for large scale experimental designs. Rawat et al. (2013) presented a review with an empirical and critical analysis on the potential of translating research findings from laboratory scale trials to full scale application.

The seed culture propagation is an important exercise which can determine the success or failure of the scaling up process. The stepwise propagation of the seed culture is the methodology of choice that has been demonstrated by other researchers and the seed inoculum should be 20–25 % of the final culture volume. The viability, robustness and vigour of the chosen microalgal strain must be routinely checked and analysed at all stages of the scaling up process. The cell density is important and it should be above 1 E7 cells per ml to avoid an unwanted long lag phase after inoculation.

A few simple guidelines should be followed during piloting:

1. Experiments must last long enough to reveal the "aging" problems, including contamination, biofouling, and material deterioration, that become evident only after several months of continuous operation.
2. The units experimented with should be similar in size and type to the cultivation units envisaged for the full-scale plant.
3. The productivity of the system must be correctly evaluated and extrapolated.
4. Experiments must be carried out under the climatic and operating conditions that will prevail in the final commercial process.
5. Unless true axenic conditions are adopted, the possibility of contamination must not be underestimated.
6. Appropriate solutions to the main problems commonly encountered in photobioreactors (overheating, biofouling, oxygen accumulation, etc.) must be provided before beginning operation of the commercial plant.
7. The data obtained during the scaling-up stage must be subjected to rigorous analysis and used to reevaluate the economics of the process.
8. The commercial units to be built for further operation should not be very large. A typical mistake is to build very large to keep operating costs low, depriving the systems of necessary flexibility and negated the possibility of making improvements by trial and error. Under these conditions, any problem that might arise (e.g. contamination or toxic oxygen tensions) affects the whole culture and affects all of production.

Today the efforts of many researchers, from both academia and industry, are directed toward development of new photobioreactor designs. In the majority of cases, such designs are anything but new and remain an academic exercise with little or no practical value. What we really need is a serious evaluation of the performance of photobioreactors at a relatively large scale. Through such analysis, and together with likely further improvements in engineering design, the development of new materials with suitable characteristics, and the improvement of cultivation techniques derived from a better knowledge of the physiology of phototrophs in mass culture, we can reasonably expect to achieve a degree of reliability in large-scale cultivation of phototrophic microorganisms such as to bridge the gap between laboratory experimentation and commercial application. Most of scale-up problems should be avoided if the previous guidelines are strictly followed.

6.8 Tools for Studying the Stress of Large Scale Microalgal Cultivations: The Use of Flow Cytometry

Stress conditions such as starvation, inefficient aeration/gas exchange, or changes in carbon dioxide tension and pH often occur with scale-up of microalgal cultivations, from laboratory to large scale production, inducing cell heterogeneities that reduce product quantity and quality. These stress conditions can damage or kill the cells. A large number of dead or dormant cells (which are partially or completely metabolically inactive) present during any part of the microalgal cultivation development can affect the production, impair the product quality and/or raise costs. Precise control and optimization of microalgal cultivations are, therefore, crucial.

The most used large scale systems for autotrophic microalgae cultivations are open ponds, tubular, flat and column photobioreactors (Emeka et al., 2012). The cheapest systems, the open ponds, show poor mixing, and light and carbon dioxide reduced utilization. These problems are strongly intensified as the scale system increases. As a result, the cells are exposed to nutrient and carbon dioxide gradients, and to a non-homogeneous distribution pattern of light within the culture, as a consequence of the self-shading effect among cells. Such adverse conditions will reduce the process performance, as cells are not cultivated at the optimal growth conditions.

Microalgal cells growing in tubular photobioreactors (PBR) are exposed to gradients of pH, carbon dioxide and light limitation along the tubes (Emeka et al., 2012). To overcome the light limitation, high levels of mixing are necessary to reach a turbulent flow of the culture. In fact optimal conditions of light exposure require rapid liquid velocity which generates cell damage, and thus limits the biomass productivity of the culture system (Gudin and Chaumont, 1991). Wall growth (biofilm) is also common in these systems, generating stagnant regions, which are increased as the cultivation scale increases. In these regions, soluble nutrients, oxygen and carbon dioxide cannot diffuse through it. Therefore, most of the cells located in the inner part of the biofilm are dead, and thus do not participate in the biotransformation, reducing the process yield.

Temperature and pH control are difficult to achieve in flat PBR (Emeka et al., 2012) resulting in temperature and pH gradients which disturb the cells grown in these systems. Wall growth is also common. Due to the reactor configuration, hydrodynamic stress is also present in these systems, which affects the microalgae. According to Gudin and Chaumont (1991) hydrodynamic stress depends on the geometry of the bioreactor involved, type of pump utilized, morphology of algal cells, physiological conditions of microalgae, etc. Cells grown in column PBR are usually affected by both hydrodynamic and shear stress. The latter is considered a key problem of microalgae cultivated in PBR and has been intensively studied. Shearing action in sparged photobioreactors, such as bubble columns and flat panels, is necessary for mixing, heat elimination, mass and light transfer, and its importance increases with scale-up. However, excessive shear stress can lead to impaired cell growth, cell damage, and eventually cell death.

Heterotrophic microalgae are cultivated in conventional closed bioreactors. In large-scale heterotrophic bioreactors, stress conditions such as aeration, starvation, changes in oxygen tension and pH and glucose concentration gradients also occurred. In fed-batch processes, the most used in industry, the limiting substrate concentration fluctuates due to the difficulties in mixing the feed solution fast enough, as the feed solution is usually applied in as concentrated form as possible, to minimize the dilution effects. As a result, when the cells circulate around the reactor, they will experience microenvironments with different substrate concentrations, which will affect their metabolic activity (Enfors et al., 2001).

Despite the high stress level experienced by the microalgae cells when growing in large scale systems, there are only a few studies evaluating the cell stress response in large scale microalgal cultivations. Flow cytometry offers a powerful and effective methodology for on line monitoring cellular status and growth in microbial cultivation systems, by rapidly characterizing complex cultures at the single-cell level. This technique allows evaluating, near real time, the physiological cell status when grown in such systems. With this information, it is possible to change the process control strategy (by changing the operational conditions such as the speed/aeration rate, feed strategy, etc.) during the time course of the cultivation, so that the proportion of stressed/dead cells may be reduced. This is not possible when using traditional analytical methods for growth monitoring such as dry cell weight or colony forming unit, as the results are available at considerable time period after the sample harvesting. On the other hand, despite the optical density being a fast method for growth evaluation, it gives limited information, as it does not give knowledge on the cell status or metabolism.

Camacho et al. (2011) reported the scale-up of a dinoflagellate *Phaeodactylum tricornutum* culture, from a 2 L stirred-tank PBR to a 15 L photobioreactor. They concluded that, notwithstanding the *P. tricornutum* extreme shear sensitivity, a good potential for mass scale culture of this microalga is expectable in suitably designed PBR. The authors used flow cytometry to measure the relative mean cell size and the reactive oxygen species (ROS) concentration during the microalga growth, using the reactive fluorescent dye 20,70-dichlorodihydrofluorecein diacetate to evaluate *P. tricornutum* stress response. This dye is nonfluorescent until the acetate groups are removed by intracellular esterases and oxidation occurs within the cell. The measured fluorescence is then proportional to the amount of ROS in the cell. One mechanism of cell damage that has been suggested is that hydrodynamic shear forces somehow trigger a metabolic cascade that leads to an increase in the intracellular concentration of reactive oxygen species which ultimately damage the cellular organelles. On the other hand, increase in *P. reticulatem* cell size on exposure to inhibitory shear forces has been documented (Rodríguez et al., 2009).

Barbosa et al. (2003) studied the effect of superficial gas velocities and gas entrance velocities on continuous and batch cultures of three *Dunaliella* strains developed in gas-sparged photobioreactor column at different scales (lab and pilot), evaluating the cell viability by staining the cells with fluorescein diacetate (FDA) using flow cytometry. This non-fluorescent fluorochrome is taken up by the cell and converted inside the cell to a fluorescent substance by non-specific esterases, therefore, giving information on cell enzymatic activity. The authors concluded that the cell damage was shown to be strain-dependent and the cell wall provided a crucial protection against hydrodynamic shear. The authors also demonstrated that the main parameter causing cell death and damage was the gas entrance velocity at the sparger.

But flow cytometry offers other applications to large-scale microalgal cultivations, beyond those here reported (Hyka et al., 2012). DNA content, cell cycle, membrane potential and integrity, mitochondrial membrane potential and microalgae products concentration (lipids, carotenoids, cellulose, protein) can be detected, near real time, during the time course of large-scale cultivations by this technique (Gouveia et al., 2009; Lopes da Silva et al., 2009a, b). This allows getting detailed information on the microalgal culture performance, and saving time and reagents usually needed, when traditional techniques are used to detect microalgal products that last a few days (extraction using organic solvents, gravimetry). Therefore, flow cytometry becomes the ideal technique for large-scale microalgal cultivations monitoring.

This technique also allows early detection of contaminants in microalgae cultivations, throughout cell size and complexity discrimination (Hyka et al., 2012). Microbial contamination remains a major economic challenge for large scale microalgal cultivations, particularly those conducted under non-aseptic conditions as open ponds (in closed systems like photobioreactors these risk factors are more controlled). Therefore, it is critical that microbial contamination be detected as early as possible, and controlled in the initial processing steps of microalgal cultivations. Failure to do so can lead to an eventual loss of production time until the system can be purged of contaminants and reinoculated with the microalgae strains, to restart the cultivation. This issue is especially critical in large-scale microalgal cultivations.

6.9 Challenges at Large Scale

Commercial microalgal cultivation at large scale has its associated problems (Rawat et al., 2013). Grobbelaar (2009) outlined six challenges they experienced at their large scale Spirulina cultivation facility in Musina, South Africa. The challenge of deviations from design and raceway specifications is a result of construction companies with little experience in raceway pond construction as well as biologists having little or no engineering skills. This causes significant delays to the inception of the project as well as successful completion according to the project time frames. Furthermore, time constraints and requirements due to insufficient pre-construction planning may prolong optimisation time thereby seriously delaying the whole exercise. According to Grobbelaar (2009), commissioning, start-up and basic optimisation require time and a three-year period is the minimum. The question of scale and seed culture preparation is a real challenge for a large scale microalgal handling facility of up to 300,000 L. The equipment for handling large volumes of broth must be available. Harvesting, seed culture preparation and inoculation facilities must be well planned in advance of the project inception. Lack of skills in microalgal cultivation operations is a major stumbling block for successful operations. Therefore, all personnel involved in the project must be fully trained and provided with all the essential skills on handling microalgal growth, parameter monitoring and the use of equipment such as pH meters, spectrophotometers, and probes to avoid system failures.

Site specific problems can also be a major challenge such as unexpected power and water outages, seepages, contamination, water evaporation, staff absenteeism and possible red tape from the investors. Last but not the least, the challenge of product quality and consistency must be addressed for successful operation and execution of the project. Daily culture sampling and analyses is required to quality and check product consistency as this is an ongoing challenge (Grobbelaar, 2009). A lot of work has been done to avoid system crashes and it is critical to keep an eye on all possible sources of system failure. A lot of money is injected in microalgal cultivation and any slight misfortune may put all these resources to waste if there is no careful planning, sound scientific approach, balanced chemical and mechanical engineering principles and good time management.

6.10 Conclusions

According to Rawat et al. (2013), much data has been generated at laboratory scale as an academic exercise but not much has been published in way of technology transfer to large scale. In recent years, although significantly higher photosynthetic efficiencies and a higher degree of system reliability have been achieved due to better understanding of the growth requirements of phototrophic microorganisms in mass culture only few photobioreactors have yet proved to be commercially successful, even for very-high value products. There exist several valid basic concepts and myriad variations. The principle obstacle to commercial application of this biotechnology seems instead to lie at the level of scaling-up. Transferring those processes developed at the laboratory scale to the industrial scale is never a simple proposition, and it is particularly difficult in the case of photobioreactors because of the peculiar problems of these systems. One of the major causes of the failures of photobioreactor technology at commercial scale is a lack of adequate piloting. The old rule of thumb used in the chemical industry and still valid in fermentation technology, that each individual increase of scale should not surpass a factor of 10, has been ignored in practice. At the small scale, the photobioreactor performance is usually evaluated in terms of productivity, whereas issues such as contamination, material degradation, and the reliability and sustainability of the production process, which more than the productivity determine the failure or success of the industrial-scale activity, are ignored. But even when productivities are compared, many times, piloting expectations are based on promising experimental results published at small scale that never are attained when going bigger. For instance, published biomass output rates are overestimated quite often, as correspond to short periods of time and many times collected at the best season of the year (summer), when high solar radiation and temperatures are less limiting. On the other hand, biomass output rates are calculated from dry weight data instead of ash-free dry weight data, overestimating productivities as non-organic matter (including salts and ashes) account for the calculation, even when careful biomass pellet washing with an acidic buffer is carried out.

Scale-up problems should be solved in order to extend the range of cost-effective products from microalgae. Massive investment in R & D is needed in the near future, mainly in the application of new long lasting materials for PBR as well as the use of innovative and sustainable antifouling coatings. The integration of unit operations, mass and energetic balances in a whole multiproduct and multiservice process in the frame of a biorefinery is also mandatory.

Acknowledgements FCT supported SIMBIOALGA project (PTDC/AAC-AMB/112954/2009) FCOMP-01-0124-FEDER-013935 (also supported by FEDER funding through COMPETE-Programa Operacional Factores de Competitividade).

References

Abeliovich, A. and Shilo, M. (1972). Photooxidative death in blue-green algae. *J. Bacteriol.* 111(3), 682–689.

Anning, T., Harris, G., & Geider, R. (2001). Thermal acclimation in the marine diatom Chaetoceros calcitrans (Bacillariophyceae). *European Journal of Phycology*, *36*(3), 233–241.

Barbosa, M., Albrecht, M. and Wijffels, R. (2003). Hydrodynamic stress and lethal events in sparged microalgae culture. *Biotechnol Bioeng*, 25, 112–120.

Barbosa, M.J., Hadiyanto and Wijffels, R.H. (2004). Overcoming Shear Stress of Microalgae Cultures in Sparged Photobioreactors. *Biotechnology and Bioengineering*, 85(1), 78–85.

Blaas, H. and Kroeze, C. (2014). Possible future effects of large-scale algae cultivation for biofuels on coastal eutrophication in Europe. *Science of the Total Environment*, 496, 45–53.

Camacho, F.G., Rodríguez, J.J., Mirón, A.S., Belarbi, E.H., Chisti, Y. and Grima, E.M. (2011). Photobioreactor scale-up for a shear-sensitive dinoflagellate microalga. *Process Biochem.*, 46, 936–944.

Chopra, M. (2004). Scale-up in biochemical processes. *PharmaChem, Biotechnology*, January-February 2004, 48–53.

Davison, I. R. (1991). Environmental effects on algal photosynthesis: temperature. *Journal of phycology*, *27*(1), 2–8.

Emeka, U., Ndukwe, G.I., Mustapha, K.G. and Ayo, R.I. (2012). Constrains to large scale algal biomass production and utilization. *J. Algal Biomass Utln.*, 3, 14–21.

Enfors, S.-O., Jahic, M., Rozkov, A., Xu, B., Hecker, M., Jürgen, B., Kruger, E., Scheweder, T., Hamer, G., O'Beirne, D., Noisommit-Rizzi, N., Reuss, M., Boone, L., Hewitt, C., McFarlane, C., Nienow, A., Kovacs, T., Trägard, C., Fuchs, L., Revstedt, J., Friberg, P.C., Hjertager, B., Blomsten, G., Skogman, H., Hjort, S., Hoeks, F., Lin, H.-Y., Neubauer, P. van der Lans R., Luyben, K., Vrabel, P. and Manelius, A. (2001). Physiological responses to mixing in large scale bioreactors. *J. Biotechnol.*, 85, 175–185.

Falkowski PG and Raven J (1997) Aquatic Photosynthesis. Blackwell Science, Oxford

Gallardo-Rodríguez, J.J., García-Camacho, F., Sánchez-Mirón, A., López-Rosales, L., Chisti, Y. and Molina-Grima, E. (2011). Shear-Induced Changes in Membrane Fluidity During Culture of a Fragile Dinoflagellate Microalga. *Biotechnol. Prog.*, 28(2), 467–473.

Gouveia, L., Marques, A., Lopes da Silva, T. and Reis, A. (2009). *Neochloris oleabundans* UTEX # 1185: A suitable renewable lipid source for biofuel production. *J. Ind. Microbiol. Biotechnol.*, 36, 821–826.

Griffiths, M.J. (2013). Microalgal Cultivation Reactor Systems. *In:* Biotechnological Applications of Microalgae. Chapter 5, Faizal Bux (ed). CRC Press, pp. 51–76.

Grobbelaar, J.U. (2009). From laboratory to commercial production: A case study of a Spirulina (Arthrospira) facility in Musina, South Africa. *J. Appl. Phycol.*, 21, 523–527.

Gudin, C. and Chaumont, D. (1991). Cell Fragility - The Key Problem of Microalgae Mass Production in Closed Photobioreactors. *Bioresour. Technol.*, 38, 145–151.

Harrison, W. G., & Platt, T. (1986). Photosynthesis-irradiance relationships in polar and temperate phytoplankton populations. *Polar biology*, *5*(3), 153–164.

Hinze, J. O. (1955). Fundamentals of the hydrodynamic mechanism of splitting in dispersion processes. *AIChE Journal*, *1*(3), 289–295.
Hsu, Y. and Wu, W. (2002). A novel approach for scaling-up a fermentation system. *Biochem. Eng. J.*, 11, 123–130.
Hyka, P., Lickova, S., Tribyl, P., Melzoch, K. and Kpvar, K. (2012). Flow cytometry for the development of biotechnological processes with microalgae. *Biotechnol Adv.*, 31, 2–16.
Jaouen, P., Vandanjon, L. and Quéméneur, F. (1999). The shear stress of microalgal cell suspensions (Tetraselmis suecica) in tangential flow filtration systems: The role of pumps. *Bioresour. Technol.*, 68, 149–154.
Kolmogorov, A. N. (1949). On the disintegration of drops in a turbulent flow. In *Dokl. Akad. Nauk SSSR* (Vol. 66, No. 825–828, p. 30).
Kossen, N.W.F. and Oosterhuis, N.M.G. (1992). Modelling and Scaling-up of Bioreactors. *In:* Rehm, H.-J. and Reed, G. (eds), Biotechnology. 1st ed., Vol. 2, pp 571–605.
Li, X., Xu, H. and Wu, Q. (2007). Large-Scale Biodiesel Production from Microalga Chlorella protothecoides Through Heterotrophic Cultivation in Bioreactors. *Biotechnology and Bioengineering*, 98(4), 764–771.
Lopes da Silva, T., Santos, C. and Reis, A. (2009a). Multi-parameter flow cytometry as a tool to monitor heterotrophic microalgal batch fermentations for oil production towards biodiesel. *Biotechnol. Bioproc. Eng.*, 14, 330–337.
Lopes da Silva, T., Reis, A., Medeiros, R., Oliveira, C. and Gouveia, L. (2009b). Oil production towards biofuel from autotrophic microalgae semi-continuous cultivations by flow cytometry. *Appl. Biochem. Biotechnol.*, 159, 568–578.
Mirón, A.S., Gómez, A.C., Camacho, F.G., Grima, E.M. and Chisti, Y. (1999). Comparative evaluation of compact photobioreactors for large-scale monoculture of microalgae. *J Biotechnol.*, 70, 249–270.
Molina Grima, E. (1999). Photobioreactors: Light regime, mass transfer, and scale-up. *Journal of Biotechnology*, 70, 231–247.
Molina, E., Fernández, J. Acién, F.G. and Chisti, Y. (2001). Tubular photobioreactor design for algal cultures. *Journal of Biotechnology*, 92, 113–131.
Oncel, S. and Sabankay, M. (2012). Microalgal biohydrogen production considering light energy and mixing time as the two key features for scale-up. *Bioresource Technology*, 121, 228–234.
Passell, H., Dhaliwal, H., Reno, M., Wu, B., Ben Amotz, A., Ivry, E., Gay, M., Czartoski, T., Laurin, L. and Ayer, N. (2013). Algae biodiesel life cycle assessment using current commercial data. *Journal of Environmental Management*, 129, 103–111.
Pate, R., Klise, G. and Wub, B. (2011). Resource demand implications for US algae biofuels production scale-up. *Applied Energy*, 88, 3377–3388.
Qu, L., Ren, L.-J. and Huang, H. (2013). Scale-up of docosahexaenoic acid production in fed-batch fermentation by Schizochytrium sp. based on volumetric oxygen-transfer coefficient. *Biochemical Engineering Journal*, 77, 82–87.
Ramírez-Duque, J.L., Marín-Quintero, D.A. and García-Pulido, C.H. (2012). Evaluation of Microalgal Mortality in a Centrifugal Pump of a Tubular Photobioreactor. *Ing. Univ. SciELO. (Colombia)*, 16(2), 333–347.
Rawat, I., Ranjith Kumar, R., Mutanda, T.F. and Bux, F. (2013). Biodiesel from microalgae: A critical evaluation from laboratory to large scale production. *Applied Energy*, 103, 444–467.
Ren, J., Han, P., Wei, H. and Jia, L. (2014). Fouling-resistant behavior of silver nanoparticle-modified surfaces against the bioadhesion of microalgae. *ACS Appl Mater Interfaces*, 6(6), 3829–3838.
Richmond, A. (1986). Outdoor mass cultures of microalgae. *In:* A. Richmond (ed.), Handbook of Microalgal Mass Culture. CRC Press. Boca Raton, pp. 285–330.
Rodríguez, J.J., Mirón, A.S.A., Camacho, F., Garcia, M. and Belarbi, E. (2009). Causes of Shear Sensitivity of the Toxic Dinoflagellate *Protoceratium reticulatum. Biotechnol. Prog.*, 25, 792–800.

Salleh, S., McMinn, A., Mohammad, M., Yasin, Z. and Tan, S.H.A. (2010). Effects of Temperature on the Photosynthetic Parameters of Antarctic Benthic Microalgal Community. *ASM Sci. J.*, 4(1), 81–88.

Shuler, F.L. and Kargi, F. (1992). Bioprocess Engineering: Basic concepts. 2nd ed, Prentice-Hall, Englewood Cliffs, NJ.

Soeder, C.J. (1980). Massive cultivation of microalgae: Results and prospects. *Hydrobiologia*, 72, 197–209.

Tabernero, A., Martín del Valle, E.M. and Galán, M.A. (2012). Evaluating the industrial potential of biodiesel from a microalgae heterotrophic culture: Scale-up and economics. *Biochemical Engineering Journal*, 63, 104–115.

Taylor, B., Xiao, N., Sikorski, Yong, M., Harris, T., Helme, T., Smallbone, A., Bhave, A. and Kraft, M. (2013). Techno-economic assessment of carbon-negative algal biodiesel for transport solutions. *Applied Energy*, 106, 262–274.

Thiry, M. and Cingolani, D. (2002). Optimizing scale-up fermentation processes. *Trends in Biotechnology*, 20(3), 33–35.

Tredici, M.R. (1999). Bioreactors, Photo. *In:* Flickinger, M.C. and Drew, S.W. (eds), Encyclopedia of Bioprocess Technology: Fermentation, biocatalysis, and bioseparation. ISBN 0-471-13822-3. John Wiley & Sons, Inc. pp. 364–419.

Wang, S-K., Hua, Y-R., Wang, F., Stiles, A.R. and Liu, C.-Z. (2014). Scale-up cultivation of Chlorella ellipsoidea from indoor to outdoor in bubble column bioreactors. *Bioresource Technology*, 156, 117–122.

Chapter 7
Large Scale Algal Biomass (*Spirulina*) Production in India

D. Selvendran

7.1 Introduction

large-scale farms around the world, using a wide range of algae production systems like *Spirulina* growing in natural lakes, commercial farms using outdoor raceway pond systems, photo bioreactors, integrated commercial production farms, village farms in the developing world with appropriate technology, micro farms, family and community size production systems. *Spirulina*, the spiral shaped microscopic aquatic plant, is the richest and the most complete nutrition discovered so far. These tiny green spiral coils harvest the energy of the sun, growing a treasure of bioavailable nutrients. *Spirulina* has the natural tendency to grow in alkaline water bodies, where growth of other algae and micro organisms is minimal. *Spirulina* cultures are almost monocultures and much less susceptible to contamination as compared to many other algal cultures.

In India, research and development on *Spirulina* production started by 1977. Even though, pilot capabilities were achieved in 1980, and further developmental works were delayed due to investor perception of *Spirulina* as a futuristic product. Commercial production was started in 1944 and at present there are many major producers of *Spirulina* in India. *Spirulina* is marketed in India mainly as formulated products like tablets and capsules by several pharmaceutical companies.

Millions of people worldwide eat *Spirulina* cultivated in scientifically designed algae farms. Current world production of *Spirulina* for human consumption is more than one thousand metric tonnes annually. The United States leads world production followed by Thailand, India and China (Henrikson, 1997). There are no apparent production data available about the quantity of *Spirulina* produced from India and the rest of the world; however, it was estimated that the world production was about

D. Selvendran (✉)
Antenna Nutritech Foundation, Ganesh complex, 120 Feet road
[Surveyor Colony – Mattuthavani], Madurai 625 007, TN, India
e-mail: d_selvendran@antennanutritech.org

D. Das (ed.), *Algal Biorefinery: An Integrated Approach*,
DOI 10.1007/978-3-319-22813-6_7

3000 metric tonnes per annum in 2004 and now it may be close to about 7000 metric tonnes and from India it may be close to 1000 tonnes per annum in 2014 across large, medium and small *Spirulina* farms.

7.2 Large, Medium and Small Scale *Spirulina* Production

7.2.1 Industries Producing in Large Scale

Spirulina has been produced by industries on a large scale in raceway ponds adopting the mechanised agitation, filtering and drying systems with relatively higher capital investments. Their minimum capacity is about 10 tonnes per year and the biggest producer in India has the present capacity of about 300 metric tonnes a year (Fig. 7.1).

The best examples are the leading Murugappa group's Parry Nutraceuticals, Pudukkota; Dabur group's Sannat Products, Tamil Nadu; and NB Laboratories, Nagpur, Maharashtra. There are about ten such industries producing about 600 tonnes of *Spirulina* per annum The most parts of the large scale *Spirulina* production is being exported directly by the producers or indirectly by the traders as *Spirulina* powder or the value added products like capsules, tablets or flakes to Europe, North America and Australia.

7.2.2 Small and Medium Scale Spirulina *Production*

In medium and small scale, *Spirulina* production is carried out by the small entrepreneurs. The NGOs (Nongovernmental Organisations) and social institutions are using the manpower like SPRTC model. In this model the producers adopt different types of pond construction like the concrete tanks (square or rectangular), tanks made of tarpaulins and even multiple small tanks. This type of tank is mostly

Fig. 7.1 Large scale *Spirulina* production by Parry Neutraceuticals and SPRTC, Madurai.

manually operated; thus the daily harvest, pond agitation and drying (sun drying in most farms) are carried out by the farm workers.

Though this model is very attractive and less profitable, this type of *Spirulina* production is not capital intensive and offers the employment to the self help groups [SHGs] and encourages them to come forward to start their own *Spirulina* production units at remote villages. They are also now facilitated by the banks and NBFIs (non banking financial institutions) through micro-credit projects. There are many such running models established in the north and south Tamil Nadu and the women of the SHGs are able to increase their income considerably. The few examples are SPRTC-Madurai, CHDP-Nagarkoil, Aurospirul-Pondycherry, SURYA-Thindivanam, and WED-PMS-Madurai (http://www.antenna.ch/en/research/malnutrition/spirulina-Production).

7.2.3 *Advantages of* Spirulina *Local Production*

The local production of *Spirulina* gives communities access to a local and sustainable source. The cultivation of *Spirulina* has its humanitarian aspects because it brings with it a growing awareness by stakeholders of the causes of malnutrition and of its own role in enhancing the nutritional state of a country. By deciding to grow their own *Spirulina*, developing countries can gradually integrate it into eating patterns. This route will allow it to become not only an excellent weapon in the fight against malnutrition but also a real tool for development. The local communities are highly motivated to be involved in a commercial way. This can be done by establishing an efficient network for distribution and communication as well as a targeted education strategy with the local community on the nutritional qualities of *Spirulina*. There are considerable advantages of the small and medium scale *Spirulina* production over large scale production farms like: *Spirulina* is produced locally with the available facilities, the technology is simple, and not capital intensive (even with about USD 10,000 or INR 6,00,000 capital, the small *Spirulina* farm can produce about 1000 kg *Spirulina*).

7.2.4 *Cost Estimation to Establish a Rural* Spirulina *Farm with 400 m² of Production Area and Outreach*

7.2.4.1 A Model Rural *Spirulina* Farm and Outreach

Production size: About 400 m^2
Spirulina produced (Annual): 1000 kg a year (dry)
Employment possibility: 10–12 women (Production and promotion)
Possible reach of children: About 6000 children [Assuming 1 gm/child for 100 days]
Possible reach of adults: About 2000 adults [Assuming 3 gm per adult for 100 days]

This can also be used to reach about 12,000 children for 100 days with one gram *spirulina* a day.

Through this supplementation the malnutrition among children and women will substantially be reduced.

Note: There was a clinical experiment among children of 1–6 years of age with just one gram a day *Spirulina* supplementation for six weeks which had proven effects as in the study abstract as below (Table 7.1).

A clinical study abstract: *Spirulina* – A nutrition booster (Thinakar Vel et al., 1999)

STUDY SUPPORTED BY ANTENNA TECHNOLOGY, GENEVE AND ANTENNA TRUST, MADURAI.

SPIRULINA : A NUTRITION BOOSTER

Study by Mr.Thinakarvel, Prof. Dr. N.Edwin department of paediatrics Madurai Medical College and Government Rajaji Hospital, Madurai.

(Study paper Presented at the 7th World Congress on Clinical nutrition)

A GENERAL SUMMARY

- To study the nutritional benefits of Spirulina in a group of malnourished pre-school children (1-6yrs.) a total of 60 children were selected from ICDS block and Rajaji Government Hospital Madurai, of which 30 children were taken as test group and 30 were taken as control group
- The Study Period was for Six Weeks while the test group children were given Spirulina at a Dose of 1 gram per day. The control group children were given placebos.

The results are as follows:

PARAMETARS	TEST GROUP FED WITH SPIRULINA.	CONTROL GROUP NOT FED WITH SPIRULINA
	(% OF CHILDREN SHOWING IMPROVEMENT)	
Weight	63.33%	43.00%
Haemo globin	93.33%	23.33%
Serum Proteins	93.33%	16.66%
Serum total Iron	63.33%	40.00%
Serum Ferritin level	93.33%	16.66%
Serum retinol level	90.00%	26.6 %

The study duration of 6 weeks, assumes significance as the beneficial effects over a longer period will be greatly increased.

Apart from proteins Spirulina also gives Iron and Retinol. Thus Spirulina provides us with a cheap and widely available method of preventing nutritional deficiency amongst children in developing countries and helps them attain full growth and development potential.

Table 7.1 Expenditures involved in *Spirulina* cultivation

S. No	As on June 2014	INR	US$*
1	*Spirulina* cultivation training and seed culture cost	30,000	500
2	*Spirulina* cultivation tank construction and material cost	293,000	4883
3	Cost of production related accessories	32,120	535
4	Technical equipments cost	10,000	167
5	Cost of nutrients required for starting the culture	18,900	315
6	Cost for land preparation (leveling, fencing, water source, etc.)	90,000	1500
7	Technical support for one full year (12 months × INR 10000)	120,000	2000
	Total cost	594,020	9900

*For calculation 1 $ = 60 INR

7.3 Cultivation of *Spirulina*

In open *Spirulina* cultivation, the environmental suitability and the requirements are similar for the small, medium and large scale *Spirulina* farms even though the process differs. In order to cultivate *Spirulina*, one must provide all the necessary elements, in a situation which will permit the plant to absorb and utilize these elements. The situation involves a range of suitable temperature and illumination to provide the energy needed for photosynthesis, water and nutrients (Vonshak, 1997).

Normally *Spirulina* producing farms choose tropical region so that the required temperature range of 30–37 °C in day time and 25–30 °C in night time is obtained. Also rich source of water, 10 h of illumination (minimum), low humidity, low wind have to be ensured for good production. Regarding nutrient source requirement, one must provide organic/inorganic sources of carbon, nitrogen and minerals.

7.3.1 *Tank Construction*

Building *Spirulina* cultivation tank is very important since many parameters are to be considered in tank construction to ensure the efficient *Spirulina* culturing. Any water-tight, open container can be used to grow *Spirulina*, provided it will resist corrosion and be non-toxic. Its shape is immaterial, although sharp angles should be avoided to facilitate agitation and cleaning. *Spirulina* can be grown in tarpaulin (transferable) or permanent cement tanks. Most current commercial farms over the past 30 years have been designed with shallow raceway ponds circulated by paddle wheels. Ponds vary in size up to 5000 m^2 (about 1.25 acres) or larger, and water depth is typically 15 to 25 cm. They require more capital investment than lake farms, and operate at higher efficiency and quality control.

7.3.2 *Culture Medium*

Spirulina thrives in alkaline, brackish water. The culture medium provides all essentials to grow *Spirulina* in a suitable environment. It is composed of sodium carbonate, a source of nitrogen, phosphorus, iron and trace metals. The composition of growing medium is shown in Table 7.2.

In addition, the solution contains traces of all micronutrients necessary to support plant life. Such solution is obtained by dissolving various combinations of chemicals. Only food grade chemicals are used. The choice of chemicals and concentrations that are used depends on cost and the compatibility of the chemicals with the raw water. Cost of nutrients accounts for about 15–25% of the total production cost.

Table 7.2 Composition of the medium for the growth of *Spirulina*

Carbonate	2800 mg L^{-1}
Bicarbonate	720 mg L^{-1}
Nitrate	614 mg L^{-1}
Phosphate	80 mg L^{-1}
Sulphate	350 mg L^{-1}
Chloride	3030 mg L^{-1}
Sodium	4380 mg L^{-1}
Potassium	642 mg L^{-1}
Magnesium	10 mg L^{-1}
Calcium	10 mg L^{-1}
Iron	0.8 mg L^{-1}
Total dissolved solids	12,847 mg L^{-1}
Density @ 20 °C	1010 g L^{-1}
Alkalinity	0.105 N (moles strong base L^{-1})
pH @ 20 °C	10.4

7.3.3 Strain Selection

The major determinants in the selection of strains for commercial production are growth rate, biochemical composition and resistance to mechanical and physiological stress. A wide variety of species and strains of *Spirulina* have been screened by several people in various countries. Continuous mass production of a particular strain depends on its suitability and stability under prevailing conditions of the farm environment.

7.3.4 Scaling-up of the Process

The culture is scaled up from mother culture strains. The scale-up follows a roughly 5:1 dilution ratio through successive volumes up to the required culture in the production ponds. In the second mode of operation, culture expansion to the entire volume of a production pond can be done from seedling at the concentration of 1 g L^{-1}. The scaling up process is the stage in the process where contamination by other algae and bacteria poses the greatest problem because of the initial dilute nature of the *Spirulina* inoculum. There is a direct relationship between the density of *Spirulina* in the culture and the density of contaminants (Vonshak, 1997). Contamination by green algae is high when the initial density of the inoculum is low. Conversely, the amount of contamination decreases as the *Spirulina* culture builds up in density. It is speculated that extra cellular products of *Spirulina* may have some allelopathic properties. Threshold concentration of these substances is probably reached at high cell densities. Light limitation of green algae by the positively buoyant trichomes of *Spirulina* may also account for the observed phenomenon. Through careful manipulation of the nutrient concentration and useful natural predators, it has been possible to maintain a mono algal culture even during the initial period of inoculation.

7.4 Maintenance of Open Air Cultures

Proper culture maintenance calls for a routine monitoring of various physical, chemical and biological parameters as follows.

7.4.1 Effect of Light

Outdoor algal cultures are exposed to two rhythms of the dark and light regime. The first is relatively fast. It is induced by the mixing in the pond which results in a turbulent flow of the culture, dictating the frequency of the light cycle. In this cycle algae cells are shifted between full solar radiation when located at the upper culture surface and complete darkness when reaching the bottom of the culture, usually at a depth of 12–15 cm. The time scale of such a cycle is measured in fractions of a second. The other, relatively slower regime is the change in solar irradiance during the day from sun rise to sun set. These two light cycles impose a unique physiological regime on the adaptation or acclimatization of outdoor algal cells to light. When growing algae at a depth of 12–15 cm in open tanks, self-shading governs the light availability to the single cell in the culture (Vonshak, 1997). Unless one uses a much diluted culture which allows penetration of light throughout the water column, a certain part of the culture will always fail to receive enough light to saturate photosynthesis. Thus almost by definition this kind of culture will be light limited. Increasing the cell concentration of culture, this increases self-shading and results in decrease of the growth rate of *Spirulina*. An increase in productivity and highest production at a higher cell concentration are the results of increased turbulence flow by agitators.

7.4.2 pH Control

The pH of the medium is one of the most important factors in culturing *Spirulina*. Maintaining the pH over 9.0 is mandatory in *Spirulina* cultures in order to avoid contamination by other algae. pH adjustment is made by increasing carbon dioxide level by addition of carbonate salts in to the culture. The main areas of loss of CO_2 are exchange with the atmosphere, precipitation as $CaCO_3$ and loss to the part of the medium that is not recycled back to the ponds. It is rare that the pH of the medium falls below about 9.0 (once the medium is much diluted). However, increase in pH is quite common due to human error like improper/lack in addition of nutrients. When this occurs, it is accompanied by precipitation of $CaCO_3$, which is sometimes followed by flocculation and sedimentation of algae (Vonshak, 1997).

7.5 Harvesting, Drying and Packaging

The best time for harvesting is early morning for various reasons:

- The % proteins in the *Spirulina* is highest in the morning.
- The cool temperature makes the work easier.
- More sunshine hours will be available to dry the product.

The harvesting process involves two stages of filtration. In the first, filtration to obtain a "biomass" containing about 10 % w/v dry matter and 50 % w/v residual culture medium and removal of the residual culture medium to obtain the "fresh *Spirulina* biomass", ready to be consumed or dried and practically no residual culture medium.

Filtration is simply accomplished by passing the culture through a fine weave cloth, using gravity as the driving force. Synthetic fibre cloth with a mesh size of about 30 to 50 microns is the preferred filtering medium (Venkatraman, 1983). Supporting the filtration cloth by a fine net will accelerate somewhat the filtration and protect the cloth against rupturing. The filter can be installed above the pond to directly recycle the filtrate. The culture to be harvested should be passed through a sieve (mesh size about 200 μm) to remove any foreign matter such as insects, larvae, leaves and lumps of polysaccharide or mud.

The filtration is accelerated by gently moving or scraping the filter. When most of the water has filtered through, the biomass will often agglomerate into a "ball" under the motion, leaving the cloth clean (this desirable condition happens mostly when the biomass is richer in spiral forms and the culture medium is clean). The slurry obtained (8–10% w/v of dry weight) after filtration is washed with potable filtered water. This step is used for washing excess salts from the biomass. It amounts to 20–30% w/v of dry weight.

The final dewatering is accomplished by pressing the biomass enclosed in a piece of filtration cloth plus a strong cotton cloth, either by hand or in any kind of press. The simplest is to apply pressure (0.15 kg cm^{-2} is enough) by putting a heavy stone on the bag containing the biomass. The "juice" that is expelled comes out first colourless, later it turns green and the operation must then be discontinued otherwise too much product will be lost. For the usual thickness of cake (about one inch after pressing), the pressing time is about 15 min.

Pressed biomass contains twice as much dry matter as unpressed biomass, which reduces the drying time. The efficiency of harvest depends on the Trichome size and the mesh size of the filters used at each stage. The smaller the mesh size, the higher is the efficiency (Fox, 1996). However, flow rates are invariably lower at the higher efficiencies associated with smaller mesh filters. Increasing the force of the water

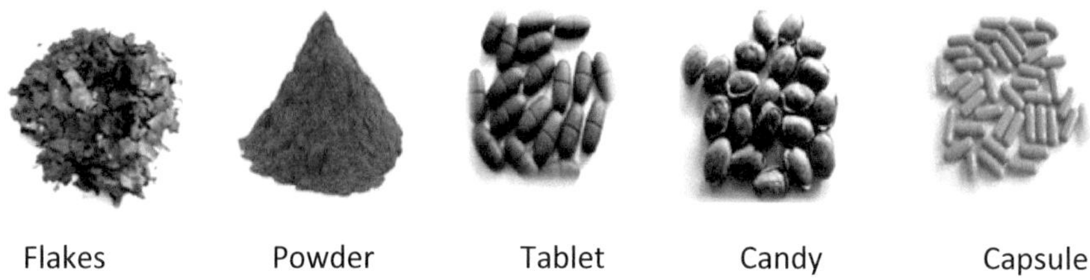

Fig. 7.2 *Spirulina* products: Various value added Spirulina forms are promoted in the market.

flow often results in breakage of cells and hence loss in efficiency and the return water has undesirable consequences in the pond culture.

The washed biomass is further concentrated by simple pressing system before being dried. Proper and quick drying is an essential feature of high-quality *Spirulina* production. Various types of drying systems are used in the industry for drying *Spirulina*. For economic reasons, the dryer of choice in large-scale *Spirulina* production facilities is the spray dryer. Freeze drying would give better overall product quality, but the cost is rather prohibitive. In spray drying, *Spirulina* droplets are sprayed in to the drying chamber just long enough to flash to the bottom. This quick spray drying process guarantees preservation of heat sensitive nutrients, pigments and enzymes (Venkataraman and Becker, 1985). Sun drying is one of the most economical systems for small projects. In this, the concentrated biomass is spread thinner in the food grade polythene sheet and is kept under hot sunlight for about 4 to 5 h. After being dried the flakes are collected and stored in the clean, moistureless, opaque containers. The grinding process is done by simple mixer grinder. Improper drying often results in high moisture content in the product. Moisture content in excess of 8 % w/w will result in the growth of moulds and bacteria in the product. Oven drying often results in the loss of some essential components like vitamins and pigments. Since the dried powder has a high sorption characteristic, the product is immediately packaged in a dry environment. The end product should have a maximum ash content of 7 %. For good preservation and storage for long period, moisture should not exceed 3–4% w/w (Fig. 7.2).

7.6 Life Cycle Analysis (LCA) of *Arthrospira plantensis*

SPRTC has carried out the life cycle assessment of the algal strain *Arthrospira platensis* grown in medium scale level at SPRTC (in the rectangular ponds and the circular ponds) and two of SPRTC's net-work *Spirulina* farms (in rectangular ponds) located at Madurai District, Tamil Nadu, India. According to the LCA-requirements, the parameters of the *Spirulina* ponds are listed as a first step and allotted the specific ponds from the regular cultivation tanks of SPRTC for regular monitoring. It is also decided that besides SPRTC's Madurai ponds, two more *Spirulina* farms in the outskirts of Madurai are included as stated above. The observation was started in all mentioned pond parameters on daily basis from November '10 in 72 m^2 (about 800 ft^2) culture area. From December '10, experimental area of culture increased to 136 m^2 (about 1500 ft^2) with the addition of new circular tank in SPRTC, Anthaneri.

Two *Spirulina* cultivation farms of SPRTC (each of 180 m^2 of production area) in March '11 at Chellampatti and Solavanthan were considered to analyse the differences for the same species at different locations and environment factors (about 35 kms from SPRTC, Anthaneri). Also in May 2012 farms at Chellampatti and Solavanthan expanded to each of 360 m^2 of production area.

Following are the important points explained in details of the technical progress:

- Maintenance of open air culture of around 150 m^2 (around 1600 ft^2) of production area in every production unit.
- Maintenance of indoor culture-mother culture in small level
    - Maintaining mother culture by daily addition of 20 % nutrient medium
    - Maintaining mother culture by daily addition of 10 % nutrient medium
- Regular monitoring and recording of pond maintenance parameters and production details
    - Inoculation details
    - Daily production details (wet weight, dry weight, water wash weight, after pressing weight, dry weight)
    - Calculating actual % of yield from wet to dry weight. Production from every square metre of area
    - Amount of nutrients added (inoculation, after harvest, troubleshooting)
    - Agitation (mixing of culture)/rotation
    - Microscopic observations (shape, size, colour, coil of *Spirulina* and presence of protozoa, rotifer, other algae)
    - Average day time temperature of pond cultures
    - Avg. viscosity of culture medium
    - Avg. pH of culture
    - Avg. culture depth
    - Dust removal/cleaning of pond
- Comparing productivity of regular tanks with that of natural resource utilized circular tanks (SPRTC has a new circular tank besides its rectangular tanks; hence it is also decided by the technical team to include it to read the differences in LCA of the same species).
- Harvest, nutrient addition, agitation, dust collection done manually in rectangular tanks
- Rotation and dust collection done automatically by wind power – Circular tanks
- Comparing productivity and other related parameters of culture maintained in various production units.

Continuous observation of open air *Spirulina* culture for the past sixteen months delivered in-depth knowledge on growth parameters of *Spirulina* such as to find out the more suitable pH, light, temperature, viscosity, average agitation per day, water quality to yield the maximum dried *Spirulina*.

The observations (Table 7.3) and the reasons are thoroughly discussed in order to have a clear understanding about the rectangular ponds of SPRTC.

- Daily production (dry)/m^2 shows that productivity of *Spirulina* was lower in monsoon and winter season (for Madurai region: October–December) and higher in summer (Figs. 7.3 and 7.4).
- Reason: *Spirulina* receives optimum range of culture temperature of 30 °C–35 °C; thus the yield is optimum at summer. Normally even in mon-

Table 7.3 Average readings/calculation monitored from the rectangular *Spirulina* ponds related yield and growth parameters of the SPRTC's Farm-1 from November 2010 to September 2012

Months	Average day production/ m^2 (g)	Avg. % of yield (%)	Average pH	Average agitation/d	Avg. day time temperature of the pond culture (°C)	Average viscosity of the culture (cp)
Nov-10	2	5.2	9.9	22	27.0	1.032
Dec-10	3.3	6.2	9.4	23	29.0	1.026
Jan-11	4.1	7.3	9.5	22	33.0	1.010
Feb-11	5.1	7	9.4	26	33.0	1.011
Mar-11	6.1	6.8	9.8	24	33.1	1.012
Apr-11	8	7.7	10.1	25	33.6	1.015
May-11	9.7	8.4	10.1	25	33.8	1.019
Jun-11	9.2	8.3	10.2	24	33.2	1.020
Jul-11	9.3	8.4	10.2	23	33.1	1.021
Aug-11	8.9	8.2	10.1	26	33.2	1.022
Sep-11	8.5	8.3	9.9	27	33.1	1.019
Oct-11	6.9	8.2	9.9	28	33.1	1.019
Nov-11	5.3	7.9	9.9	24	29.0	1.019
Dec-11	5.3	6.9	10	22	30.0	1.018
Jan-12	5.9	6.2	10	23	31.8	1.018
Feb-12	7.2	8.4	9.9	26	32.9	1.019
Mar-12	7.4	8.2	9.9	22	32.9	1.026
Apr-12	8.6	8.2	10	23	33.7	1.028
May-12	9.6	8.6	10.1	20	33.7	1.032
Jun-12	9.4	8.5	10.1	20	33.3	1.034
Jul-12	9.3	8.6	10.2	16	32.6	1.035
Aug-12	8.2	8.2	10.1	18	32.6	1.030
Sep-12	8.2	8.1	10.1	17	32.5	1.028
Oct-12	5.9	7.8	10.2	15	30.4	1.030
Nov-12	5.7	7.9	10.2	17	29.8	1.031

soon and winter we get optimum temperature (in Madurai region) but we get low light intensity which affects the productivity.

- Though productivity differs according to the factors like temperature, water quality, nutrient availability, light intensity and contaminants growth, it also

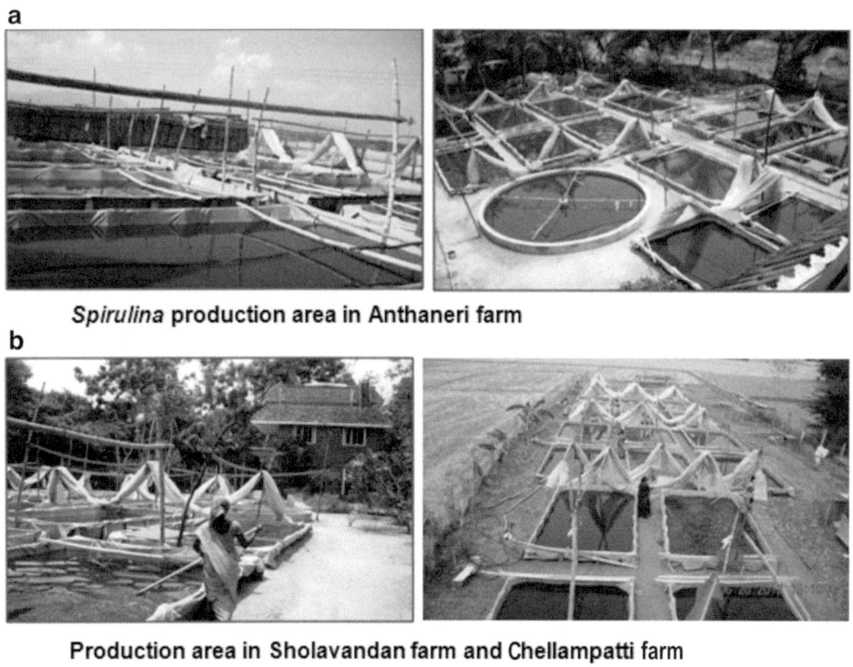

Fig. 7.3 Three different locations where LCA was carried out by SPRTC.

Fig. 7.4 Circular concrete tank of 3 m radius.

depends on other important factors like labour availability. Though it was cloudy and rainy days in monsoon like previous years, double productivity was achieved by adapting proper dual power drying system from the month September 2011.

- Also productivity in 2012 monsoon is little higher than 2011, since in 2012 monsoon was very less (7 rainy days).
- % of yield (% of dry weight from wet): It has the range 5.2–8.3 % but it adds nutrients by means of considering it as 9 %.
    - Reason: Actual is lower than the considered one since we certainly lose weight while washing and pressing processes after harvest and wastages while drying it.
    - While using solar drier the wastage ratio is very less.
- pH of the *Spirulina* culture (9.4–10.2) maintained in required normal range (9.5–10.5)
    - Reason: Ratio of addition of nutrients and utilization by *Spirulina* is quiet balanced. It ensures using good quality fertilizers (commercial grade).
- Temperature of the culture: Manual mixing of culture done frequently (22 to 28 times/d)
    - Reason: Frequent mixing of culture really helps all *Spirulina* in pond to get even sunlight, temperature and fertilizers. In spite of seasonal variations agitation helps *Spirulina* to sustain.
- Viscosity – Quiet in normal range except in monsoon seasons
    - Reason: Less productivity, less harvest, contaminants (like insect larvae, rotifer, protozoa)
        - It is necessary to take regular steps to maintain the viscosity in the optimum range by doing partial refreshment process of growing medium once in 6–8 months period.
        - It had come across some troubles like rain water mix up, windy days, invasion of other algae (*Chlorella*), insects and protozoa. It took some troubleshooting methods like refreshing the tanks with fresh medium/mother culture/healthy live mass/fertilizer addition (Figs. 7.5, 7.6 and 7.7).

The observations (Table 7.4) and the reasons are thoroughly discussed in order to have the clear understanding about the circular ponds of SPRTC.

- *Spirulina* yield/m^2 in circular tank is higher than normal rectangular tanks.
- Reason: Gentle and more frequent mixing of culture and continuous dust removal (automatic) increase the production. Temperature, viscosity and pH of culture are maintained in normal range. The average yield was 8 % but after addition of nutrients it was 9 %.
- Viscosity – Quiet in normal range (1.010–1.029 cp)

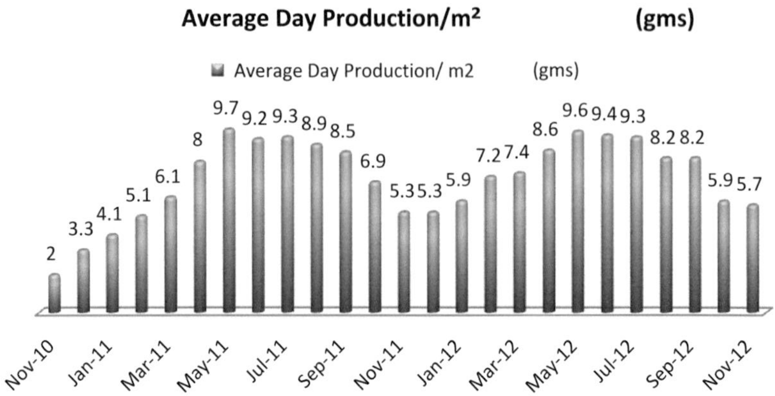

Fig. 7.5 The variation of algal biomass productivity in different months.

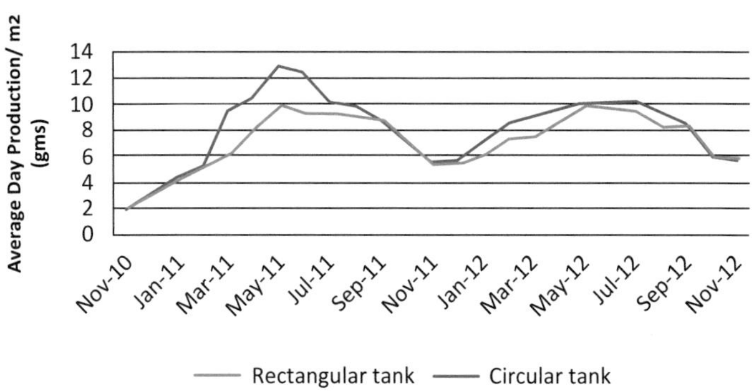

Fig. 7.6 Comparison of production between rectangular and circular tanks.

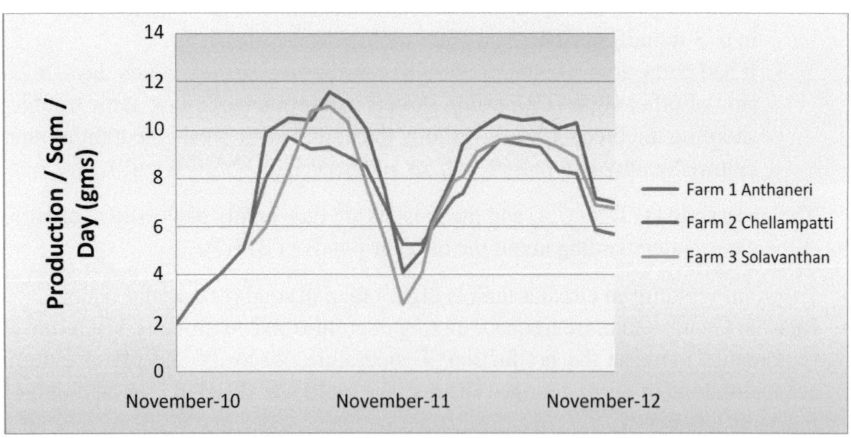

Fig. 7.7 Comparison of the yield of dry Spirulina from three different farms.

Table 7.4 Productivity details and other related parameters of Farm-1 *Spirulina* (grown in circular tank)

Month	Average day production/ m² (g)	Avg. percentage of yield (%)	Average pH	Agitation /d	Avg. day time temperature of the pond culture (°C)	Average viscosity of the culture (cp)
10-Nov						
10-Dec	3.2	2.3	9.5	37	29	1.01
11-Jan	4.4	7.7	9.5	41	33	1.01
11-Feb	5.1	8.6	9.6	51	33	1.01
11-Mar	9.5	7.8	9.8	60	33	1.011
11-Apr	10.5	8.3	10.1	101	33.8	1.016
11-May	12.9	8.5	10.1	102	33.6	1.016
11-Jun	12.4	8.4	10.2	102	33.5	1.016
11-Jul	10.2	8.3	10.1	118	33.2	1.022
11-Aug	9.8	8.3	10.1	132	32.6	1.021
11-Sep	8.7	8.4	10.3	146	34.5	1.021
11-Oct	6.9	8.2	10.3	146	34.4	1.021
11-Nov	5.6	7.9	10.2	128	29.8	1.02
11-Dec	5.7	7.5	10.1	120	29.7	1.02
12-Jan	7.2	8.5	10.2	138	31.5	1.021
12-Feb	8.4	8.6	10.3	141	33.1	1.022
12-Mar	8.9	8.3	10.1	126	33.1	1.024
12-Apr	9.7	8.4	10.1	119	33.6	1.027
12-May	10.1	8.2	10	118	33.8	1.024
12-Jun	10.3	8.2	10.1	117	33.4	1.028
12-Jul	10.2	8.1	10.2	64	32.7	1.029
12-Aug	9.1	8.2	9.8	74	31.4	1.028
12-Sep	8.4	8	9.7	75	31.2	1.028
12-Oct	7.7	7.8	10	80	30	1.027
12-Nov	7	7.6	10	82	29.7	1.028

- The culture maintained in the regular tanks requires refreshment of the medium once in 6–8 months time period. But for the culture maintained in the circular tank, it does not require medium refreshment (partial/complete).
- pH of the *Spirulina* culture (9.5–10.3) maintained is required in normal range (9.5–10.5)
- Reason: Ratio of addition of nutrients and utilization by *Spirulina* is quiet balanced. It also ensures using good quality fertilizers (commercial grade) and mixing.
- Manual mixing of culture done frequently (avg. 103 times/d).
  - Reason: Frequent rotation of culture really helps all *Spirulina* in pond to get even sunlight, temperature and fertilizers. We could manage rotation per day in circular tank much higher than normal agitation per day in normal rectangular tanks. So, it helps a lot to maintain the culture's viscosity for many more days in normal range.
- Yield at Farm 2, Chellampatti is higher than other two farms though the temperature, viscosity, and nutrient addition and agitation are maintained same as other farms.
  - Reason: Water quality and sunlight availability are better than other farms.
- Annual average of yield at Farm 3 is quiet low as same as Farm 1, though the water quality and other environmental factors match with Farm 2.
  - Reason: Lack of labour availability – regular maintenance (Table 7.5).

7.7 Summary and Conclusion

As SPRTC begun monitoring its proposed LCA of the algal strain, *Arthrospira platensis* since November 2010, this was initially carried out with the rectangular ponds of SPRTC, then the circular pond parameters were also included and compared during this research. Later from March two new farms located in the outskirts of Madurai have been included and all the parameters were monitored at both locations. The data recorded are being compared with SPRTC's Madurai unit and two other units which give a clear understanding about the factors influencing the growth of the algal strains. The important points and findings from November 2010 to November 2012 are thoroughly listed in the tables and discussed with the explanations at all the relevant places. SPRTC is happy to learn from the records of the past twenty five months that the research carried out is more helpful to optimize the production, reduce the expenditures on feeding some nutrients as the environment factors are the main regulators of the growth of the algal strains in most times.

Table 7.5 Yield comparison of dry *Spirulina* from three different *Spirulina* farms

	Production, g m^{-2} d^{-1}		
Month	Farm 1 Anthaneri	Farm 2 Chellampatti	Farm 3 Solavanthan
November-10	2		
December-10	3.3		
January-11	4.1		
February-11	5.1		
March-11	6.1	6.2	5.3
April-11	8	9.9	6
May-11	9.7	10.5	8.3
June-11	9.2	10.4	10.8
July-11	9.3	11.6	11.0
August-11	8.9	11.1	10.3
September-11	8.5	10.0	7.8
October-11	6.9	7.6	5.5
November-11	5.3	4.1	2.8
December-11	5.3	5.0	4.1
January-12	5.9	6.3	6.1
February-12	7.2	8.2	7.9
March-12	7.4	9.4	8.1
April-12	8.6	10	9.1
May-12	9.6	10.6	9.6
June-12	9.4	10.4	9.7
July-12	9.3	10.5	9.6
August-12	8.3	10.1	9.3
September-12	8.2	10.3	8.9
October-12	5.9	7.2	6.9
November-12	5.7	7	6.8

References

Fox, Ripley D. (1996). *Spirulina*: Production and Potential. EDISUD, France.

Henrikson, Robert (1997). Earth Food *Spirulina*. Ronore Enterprise Inc, Hawaii, USA (www.spirulinasource.com/PDF.cfm/EarthFoodSpirulina.pdf).

Thinakar Vel, M., Von der Weid, D. and Edwin, N. (1999). Spirulina, a nutrition booster. 7th World Congress on Clinical Nutrition, New Delhi, India.

Venkatraman, L.V. (1983). A monograph on *Spirulina plantensis* – Biotechnology and Application. Department of Science and Technology, CFTRI Press, Mysore, India.

Venkataraman, L.V. and Becker, E.W. (1985). Biotechnology and Utilization of Algae – The Indian Approach. Dept of Science & Technology, New Delhi, India.

Vonshak, Avigad (1997). *Spirulina plantesis* (*Arthispira*): Physiology, Cell Biology and Biotechnology. Taylor & Francis Ltd, London, UK.

Chapter 8
Improvement of Harvesting Technology for Algal Biomass Production

Supratim Ghosh and Debabrata Das

8.1 Introduction

Global demands for biomass utilization as food, feed, biofuels and chemical production have been increased to a great extent. For a sustainable future it is necessary to minimise the environmental impact of our activities keeping in mind the socio-economic parameters along with operational efficiency. Our continuous dependence on fossil fuels is unsustainable because of its dwindling world reserves and global warming due to its use. Recent research has focussed on the development of renewable and potentially carbon neutral biofuels. First generation biofuels derived from terrestrial crops has impacted the environment in a big way by hastening deforestation and environmental pollution. The food vs. fuel debate has also come into force. Replacing them with second generation biofuels which is derived from lignocellulosic feedstock has addressed majority of the problems. But a concern over land usage and competition still remains. Third generation biofuels derived from microalgae seem to be the solution to the demand for alternative energy sources which is devoid of the major drawbacks associated with first and second generation of biofuels.

Microalgae are photosynthetic microorganisms with simple growing requirements (light, sugars, CO_2, N, P, and K) that can produce lipids, proteins and carbohydrates in large amounts over a short period of time. These products can be processed into both biofuels and valuable co-products. Research on microalgal cultivation has focussed on high-value products. High rate algal biomass production for low-value applications can be realized only when the technologies are cost-efficient and environment friendly. This led to the scientists focussing on the increase of algal biomass production by utilising different methods. The research concentrated on improvement of photobioreactors for biomass production (Morweiser et al., 2010),

S. Ghosh (✉) • D. Das
Bioprocess Engineering Laboratory, Department of Biotechnology, Indian Institute of Technology Kharagpur, Kharagpur 721302, India
e-mail: supratim4737@gmail.com

D. Das (ed.), *Algal Biorefinery: An Integrated Approach*,
DOI 10.1007/978-3-319-22813-6_8

selection of suitable strains for different products (Larkum et al., 2012) and genetic engineering of metabolic pathways (Georgianna and Mayfield, 2012). To reduce the cost of microalgal biomass production, research on downstream processing of biomass is essential (Greenwell et al., 2010).

Today, microalgal production is rapidly moving from laboratory to pilot scale and commercial scale demo installations (Georgianna and Mayfield, 2012), prompting the need for cost and energy efficient downstream processing technologies. A major hurdle to cost effective production of algal biomass is the harvesting of microalgae. The microalgal cells are very small in size. So, it is required to separate a very small amount of biomass from a large volume of culture medium. A high biomass concentration leads to mutual shading of the microalgal cells and thus a reduction in productivity; therefore, biomass concentrations in microalgal cultures are usually low: 0.5 g L^{-1} in open pond reactors and 5 g L^{-1} in photobioreactors. This indicates that a large volume of water has to be removed to harvest the biomass.

8.2 Surface Characteristics of Microalgae

Algae are unique eukaryotic microorganisms, which convert sunlight, water and CO_2 to biomass resource with the process of photosynthesis. The factors which determine the stability of algal cells in the medium are surface charge, size and density. These factors influence the separation characteristics of the algal cells from the aqueous medium. The interactions between the algal cells as well as interactions with the surrounding medium govern the stability of the cells (Tenney et al., 1969). The settling rate of biomass is governed by the size and cell density which plays an important role in separation through sedimentation and centrifugation.

Suspended particles usually carry a positive or negative surface charge in water. They attract ions with an opposite charge from the solution to maintain their electrical neutrality. This is known as the counter ion effect. Together they form an electrical double layer consisting of the particle surface charge and the counter ions. A dense layer of ions is formed by the counter ions called the Stern layer. It is inaccessible to other counter ions. A balance is created further away from the particle surface between electrostatic attraction and thermal diffusion. This leads to the formation of a diffuse layer further away from the particle surface which results in an exponential decrease of the potential difference between the particle surface and the bulk solution with distance from the particle surface. An electrical repulsion between the particles is a result of the cloud of counter ions surrounding the particle. The ζ potential is the potential difference between the bulk fluid and the layer of counter ions that remains associated with the charged particle when the particle is moving through the solution (the slipping plane). The ζ potential can be estimated from the mobility of the charged particles in an electric field; therefore, it is a useful indicator of the degree of repulsion between charged particles in a suspension. A high ζ potential (>25 mV, positive or negative) leads to strong electrical repulsion between particles and the suspension is said to be stable. When the ζ potential is close to zero, particles can approach each other to a point where they will be

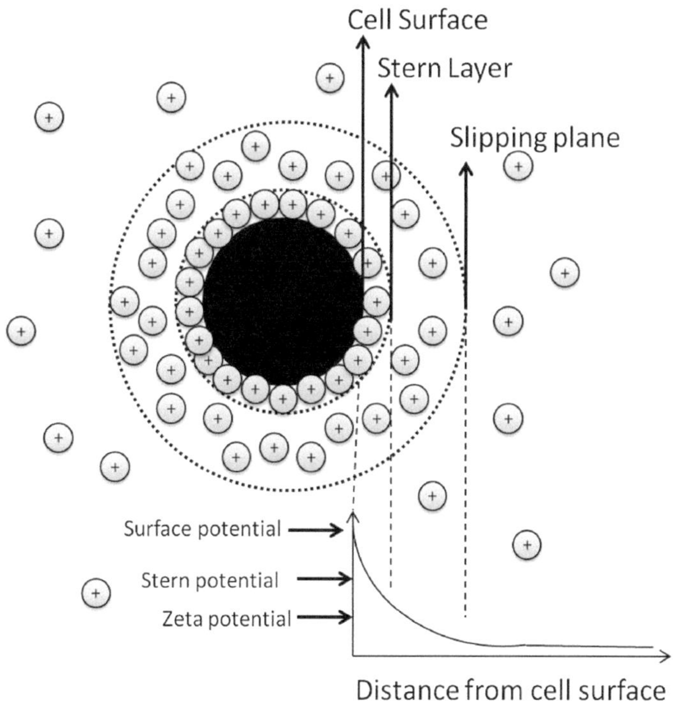

Fig. 8.1 Negatively charged microalgal cells surrounded by an electric double layer of charged ions (modified from Vandamme et al., Trends in Biotechnology, 2013).

attracted by Van der Waals forces. This leads to aggregation of particles. A similar observation is for microalgal cells where they are stabilized in media by the surface charge of the cells. The cell surface predominantly has carboxylic (-COOH) and amine (-NH_2) groups which leads to a surface charge. The carboxylic groups dissociate and are negatively charged above pH 4–5, whereas the amine groups are uncharged at this pH. This results in a net negative surface charge above pH 4–5 (Fig. 8.1).

8.3 Methods of Algae Harvesting

The further processing of algal biomass to value added products requires the algal culture to be devoid of water. This is an important step in the life cycle assessment (LCA) of the process as harvesting of biomass incurs the highest of the total production costs (20 %–30 %) (Rawat et al., 2011). An efficient harvesting method is required for large scale extraction of products from microalgae (Amaro et al., 2011; Uduman et al., 2010b). Selection of the appropriate harvesting method is of great importance to the economics of biofuels production. The different factors which determine the appropriate method of harvesting depends on the microalgal characteristics, i.e., the density and size of algal cells as well as the product derived from the algal biomass (Amaro et al., 2011). The ideal harvesting method should not be dependent on the algal species, should be energy efficient and environment friendly and may also release the desired product from the algal biomass (Chen et al., 2011). In order to emerge as a sustainable process, microalgae have to be primarily dewatered and concentrated from dilute cultures. This is one of the major challenges which one has to overcome (Uduman et al., 2010a). Secondly the algal cells are highly stable in the medium owing to their negative charge and excess algogenic organic materials they produce during their metabolism (Amaro et al., 2011). Many harvesting strategies, like centrifugation, sedimentation, flocculation, flotation and micro-filtration, can be used to harvest microalgae (Amaro et al., 2011), electrophoresis (Amaro et al., 2011; Uduman et al., 2010a) and any combination of these (Rawat et al., 2011).

Microalgae harvesting can generally be divided into a two-step process (Fig. 8.2), including:

- Bulk harvesting: The purpose of this is to separate microalgal biomass from the bulk suspension. By this method, the total solid matter can reach 2–7 % w/v using flocculation, flotation, or gravity sedimentation (Brennan and Owende, 2010).
- Thickening: The purpose of this harvesting is to concentrate the slurry with filtration and centrifugation. This step needs more energy than bulk harvesting (Brennan and Owende, 2010).

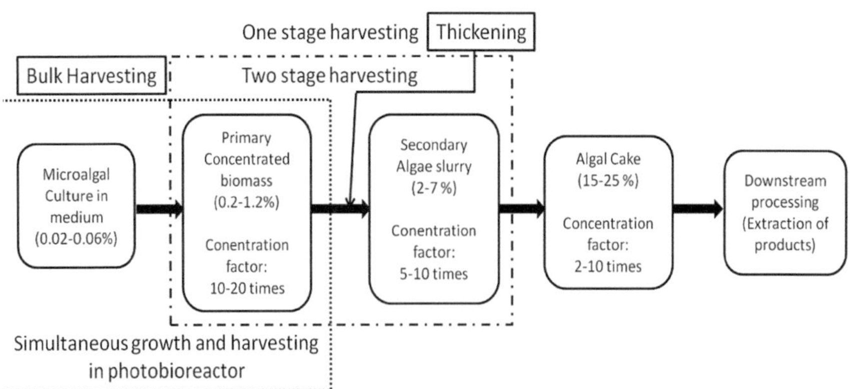

Fig. 8.2 Schematic representation of the essential steps of harvesting of microalgal biomass.

8.3.1 Screening

Two primary screening devices used for microalgal harvesting are microstrainers and vibrating screen filters because of their mechanical simplicity and availability in large unit sizes. A microstrainer is an array of rotating filters with very fine mesh screens which are frequently backwashed. They have several advantages, such as simplicity in function and construction, easy operation, low investment, negligible abrasion as a result of absence of quickly moving parts, being energy intensive and having high ion ratios. The concentration of microalgae plays an important part in determining the filtration efficiency of the device. A high microalgal concentration leads to clogging of the filtration screens whereas a low microalgal concentration may lead to inefficient separation (Wilde et al., 1991). Studies on microstraining by Molina Grima et al. (2003) confirmed this result and concluded that it would be necessary to flocculate the cells before microstraining. A highly efficient method for microalgal harvesting may be tangential flow filtration where 70–89 % recovery of microalgal biomass is possible (Petrusevski et al., 1995). In addition, tangential flow filtration retains the structure, properties and motility of the collected microalgae. Although the successful laboratory studies for concentrating microalgae, used in downstream fractionation (Rossignol et al., 1999; Rossi et al., 2004), a definitive study on large-scale algal harvesting is yet to be studied. Lazarova et al. (2006) showed that the cost of microfiltering river water can be as low as 0.2 kW h^{-1} m^{-3} of water processed. Decreasing the process volume by at least a factor of 100 significantly lowers the costs of disruption and fractionation stages downstream. Several variables associated with the choice of membranes and type of organisms could increase this cost, and there is a considerable scope for optimization of this process. As a guide to potential improvement, the costs of desalination by reverse osmosis, where a far higher pressure process is used, have fallen dramatically (85 %) over the past decade to give a total production cost of about $1 m^{-3} and with desalination energy costs being as low as 3 kW h^{-1} m^{-3}. This is largely down to a better membrane technology; greater membrane longevity, increased scale of operation and better system management and such advances might also be expected in membrane separation processes for harvesting of microalgae (Greenwell et al., 2010).

8.3.1.1 Microstraining

The rotary drum of microstrainers is covered by a straining fabric, stainless steel or polyester. Concentrations of algae through microstrainers was very low which required further dewatering. Continuous backwashing was required in order to increase the concentration of the harvested microalgal biomass from high rate clarification ponds. Small microalgae could not pass through the microstrainers leading to very dilute slurry which required further concentration (Koopman et al., 1978; Shelef et al., 1980). Successful studies were conducted on *Microactinium* and *Scenedesmus* species with the smallest mesh size of 23 μm. Greater success in reducing suspended solids of a stabilization lagoon effluent from about 80 to 20 mg l^{-1} or less by rotary microstrainers mounted with 1 μm screen has been achieved (Wettman and Cravens, 1980). Thickening of *Coelastrum proboscideum* to about 1.5 % suspended solids by microstrainers was reported when operating at a cost of about DM 0.02 m^{-3} and an energy consumption of 0.2 kW h^{-1} m^{-3} (Mohn, 1980). Problems encountered with microstrainers included low harvesting efficiency and difficulty in handling particle fluctuations. These problems may be overcome in part by varying the drum rotation speed (Reynolds et al., 1975). Another problem associated with microstraining was the build up of bacterial and algae biofilm slime on the fabric or mesh. Periodic fabric or mesh cleaning may help inhibit this biomass growth.

8.3.1.2 Vibrating Screen

Earlier studies by Mohn (1980) reported the harvesting of *Coelastrum* algae by vibrating screen. Higher algae solids concentration of 7–8 % has been harvested under batch operation in comparison with lower algal solids contents of 5–6 % when operated in continuous mode. In a commercial production, vibrating screens were used for harvesting *Spirulina* as a food source which are multicellular and filamentous blue-green microalgae belonging to two separate genera *Spirulina* and *Arthrospira* (Habib et al., 2008). The vibrating screen filtration used for harvesting achieved very high algae removal efficiency of up to 95 % for harvesting up to 20 m^3 h^{-1} from which algal slurry of 8–10 % solid contents was produced. Compared to inclining screens counterpart with a filtration area of 2–4 m^2 per unit, the vibrating screens required only one-third of the area.

8.3.2 *Coagulation–Flocculation*

Coagulation–flocculation is a process in which algal cells aggregate together to form larger clumps. These larger clumps can be easily filtered or rapidly settle down to be harvested easily. Chemicals used as algal coagulants could be broadly grouped into two categories – inorganic and long-chain organic coagulants. Flocculation can be achieved in several ways and a wide range of approaches for flocculating microalgae have been explored in recent years. These approaches range from traditional flocculation methods that are widely used in other fields of industry (e.g., chemical flocculation) to novel ideas based on the biology of microalgae (e.g. bioflocculation) and the use of emerging technologies (e.g. use of magnetic nanoparticles).

8.3.2.1 Chemical Flocculation

Water treatment and mining industries widely use chemical flocculants like ferric chloride and alum. Although metal salts are being applied for harvesting microalgae (e.g., *Dunaliella*; Ben-Amotz and Avron, 1990), their use results in high concentrations of metals in the harvested biomass. These metals remain in the biomass residue after extraction of lipids or carotenoids (Rwehumbiza et al., 2012). Further use of the protein residue of algal biomass as animal feed is restricted as the accumulated metal ions may interfere in the metabolism of the animals. The valorization of the protein fraction as animal feed is said to be important for making microalgal biofuels economically viable (Wijffels et al., 2010). Despite this shortcoming, metal coagulants provide a good model system to study the interaction between flocculants and microalgal cells because their properties are well understood (Wyatt et al., 2012; Zhang et al., 2012). Other commonly used chemical flocculants in other industries are synthetic polyacrylamide polymers. These may however contain traces of toxic acrylamide and thus also contaminate the microalgal biomass (Bratby, 2008). Flocculants based on natural biopolymers are therefore a safer alternative. As microalgal cell surfaces are negatively charged, the biopolymers used for flocculation should be positively charged, which is rare in nature. A well-known positively charged biopolymer is chitosan, which is derived from chitin, a waste product from shellfish production. Chitosan is a very efficient flocculant but it works only at low pH, but pH in microalgal cultures is relatively high (Chang and Lee, 2012). An alternative to chitosan is cationic starch, which is prepared from starch by addition of quaternary ammonium groups. The charge of those quaternary ammonium groups is independent of pH and therefore cationic starch works over a broader pH range than chitosan (Vandamme et al., 2010). Other examples of biopolymers that can be used to flocculate microalgae are poly-γ glutamic acid (an extracellular polymer produced by *Bacillus subtilis*) (Zheng et al., 2012) or polymers present in flour from *Moringa oleifera* seeds (Teixeira et al., 2012). Coiling of polymer flocculants at high ionic strengths is a general problem faced which makes it ineffective (Uduman et al., 2010a). Therefore, they are less suitable for harvesting microalgae cultivated in seawater.

8.3.2.2 Autoflocculation

An increase in pH above 9 leads to the spontaneous flocculation of microalgae (Spilling et al., 2011). The property of CO_2 sequestration by microalgae leads to the decrease in pH of the medium during photosynthesis. This causes the spontaneous aggregation of the cells which leads to flocculation. Autoflocculation is associated with the formation of calcium or magnesium precipitates. Depending on the conditions, these precipitates carry positive surface charges and can induce flocculation through charge neutralization and/or sweeping flocculation. When calcium ions are in excess in the medium, they interact with the negatively charged surface of the cells resulting in positively charged aggregates of calcium phosphate (Christenson and Sims, 2011; Schlesinger et al., 2012). This type of flocculation requires a high concentration of phosphates in the medium. So as to make the process sustainable, microalgae growing in wastewater containing high amounts of phosphate may be harvested by this method (Lundquist et al., 2010). Magnesium hydroxide or brucite also precipitates at high pH. These precipitates are positively charged up to pH 12 and can therefore also interact with the microalgal cell surface to cause flocculation (Vandamme et al., 2012; Wu et al., 2012). Most waters contain sufficiently high background concentrations of magnesium for this process to occur. Calcium carbonate or calcite also precipitates at high pH but whether it can induce flocculation of microalgae remains to be demonstrated. Flocculation at high pH is caused by formation of inorganic precipitates and not by pH as such; therefore, the harvested biomass contains high concentrations of minerals (Show et al., 2013). Although these have a low toxicity, it is nevertheless preferable to remove them from the biomass.

8.3.2.3 Physical Flocculation Methods

The above mentioned methods highly contaminate the algal biomass by introducing different chemical compounds. This can be minimized by using only physical forces to harvest the biomass. For instance, flocculation of microalgae can be accomplished by applying a field of standing ultrasound waves. On a laboratory scale ultrasonic methods may be feasible but large scale applications are still to be tested (Bosma et al., 2003). In electro coagulation flocculation, flocculation is induced through electrolytic release of metal ions from a sacrificial anode (Vandamme et al., 2011). The efficiency of this method might be improved by changing the polarity of the electrodes (Kim et al., 2012). It is similar to flocculation by metal salts but the contamination of algal biomass by metals is to a minimum in case of electroflocculation. OriginOil claims to have developed a solution for this problem. Its method uses only electromagnetic pulses to neutralize the surface charge of microalgal cells and induce flocculation (Gouveia, 2011). Recently, several studies have explored the use of magnetic nanoparticles to harvest microalgae. Magnetite (Fe_2O_3) nanoparticles may adsorb directly on the microalgal cells, upon which the cells can be separated from the medium by applying a magnetic field. This method thus combines flocculation and separation in a single process step (Cerff et al., 2012). Magnetite nanoparticles seem to adsorb more easily on some microalgal species than on others (Xu et al., 2011). Adsorption can be improved by coating the nanoparticles with cationic polymers (Lim et al., 2012; Liu et al., 2009). An advantage of using magnetite nanoparticles for harvesting microalgae is that the nanoparticles can be recovered after harvesting and subsequently reused (Cerff et al., 2012).

8.3.2.4 Bioflocculation

Spontaneous flocculation of microalgae is a common phenomenon observed in microalgal blooms in lakes, rivers or ponds. This spontaneous flocculation is assumed to be caused by extracellular polymer substances in the medium and is called bioflocculation. Bioflocculation is often successfully used for harvesting microalgae in facilities where micro-algae are used in wastewater treatment (Craggs et al., 2012). Further research is required to understand the underlying mechanism associated with bioflocculation. This may lead to a sustainable and cost effective process of harvesting of algal biomass. Some microalgal species flocculate more readily than others and such naturally bioflocculating microalgae can be mixed with other species to induce flocculation (Schenk et al., 2008; Taylor et al., 2012). There are indications that bioflocculation may be initiated by info chemicals (Eldridge et al., 2012). Recently, an info chemical isolated from a senescent and flocculating culture of a *Skeletonema* species was found to be capable of inducing flocculation in a culture of another species of microalgae (Salim et al., 2012). Bacteria or fungi can also induce bioflocculation of microalgae. Some fungi, for instance, have positively charged hyphae that can interact with the negatively charged microalgal cell surface and cause flocculation (Zhou et al., 2012; Zhang and Hu, 2012). Specific consortia of bacteria can also induce flocculation of microalgae (Gutzeit et al., 2005; Lee et al., 2008). Co-flocculation using bacteria or fungi require an organic carbon source in the medium. This can be provided by using wastewater for growth of microalgae. The presence of an organic carbon source in wastewater allows both the organisms to thrive together. This results in a culture of mixed algal–bacterial flocs that can be easily harvested (Van Den Hende et al., 2011; Su et al., 2011). The use of bacteria or fungi as a flocculating agent avoids chemical contamination of the biomass but results in microbiological contamination, which may also interfere with food or feed applications of the microalgal biomass. But if the biomass is being used for biofuels, specific species of fungi high in intracellular contents may be grown with microalgae which do not need separation later on.

8.3.3 Filtration

Filtration is a method in which the algal culture is run through filters which hold them and let the water pass through them. This is a continuous process which results into a thick paste of algae. Microfiltration, dead end filtration, vacuum filtration, pressure filtration, ultra filtration, and tangential flow filtration (TFF) are a few different filtration forms (Harun et al., 2010). The application of filtration is generally limited to the laboratory scale. Large scale applications often lead to membrane clogging, formation of compressible filter cakes and high maintenance costs limiting its acceptability. Separation of small microalgae through filtration is costly which limits its usage to filamentous or large microalgae. Generally tangential flow filtration is used for separation of microalgae. Another advantage of TFF is that it maintains the structure, properties and motility of the filtered microalgae. But the costs incurred during pumping and replacement of filtration membranes limits its utility in the large scale. Application of pressure or vacuum is used for recovering large microalgae but a higher concentration of biomass is required for this. A higher power consumption for these operations is required (in the range of 0.3–2 kW h^{-1} m^{-3}) which is almost similar to the power consumed during centrifugation (Molina Grima et al., 2003). Larger algae can be effectively recovered by vacuum filtration in combination with filter aid, while micro-filtration or ultra filtrations are effective in recovering smaller algae. Another filtration method called tangential flow filtration is a high rate method. About 70 %–89 % algae was recovered using this method (Rawat et al., 2011). Considering the output and initial feedstock concentration, according to recent studies, TFF and pressure filtration are energy efficient harvesting methods. Issues like back mixing make simple filtration methods, for example dead end filtration, inadequate for dewatering microalgae culture. However, when used along with centrifugation, give better separation. Filtration methods, in spite of being an attractive dewatering option, have extensive running costs and hidden pre-concentration requirements (Harun et al., 2010).

8.3.4 Gravity Sedimentation

Common applications of sedimentation include separation of microalgae in water and wastewater treatments. Density and radius of algae cells and the induced sedimentation velocity influence the settling characteristic of suspended solids (Brennan and Owende, 2010). Sedimentation is a very simple process but the rate of sedimentation is very slow (0.1–2.6 m h^{-1}) (Choi et al., 2006). High temperature environments such as temperate climates may lead to the degradation of algal biomass. Enhanced microalgal harvesting by sedimentation can be achieved through lamella separators and sedimentation tanks (Uduman et al., 2010a). The success of solids removal by gravity settling depends highly on the density of microalgal particles. Studies by Edzwald (1993) showed that low-density microalgal particles are unsuccessfully separated by settling. Flocculation is frequently used to increase the efficiency of gravity sedimentation.

8.3.4.1 Clarification in Simple Sedimentation Tank or Pond

Reports on algal sedimentation in ponds were always accompanied by flocculation methods. Operation involving fill-and-draw cycle for secondary pond gave rise to significant removal of algae from facultative oxidation pond effluent (Benemann et al., 1980). Similar secondary ponds were used for algae settling from high rate oxidation pond effluent (Adan and Lee, 1980; Benemann et al., 1980). Well clarified effluent and algae slurry up to 3 % solids contents were achieved attributable to algae autoflocculation which enhanced the settling. Coagulant dosing in a settling tube to promote algae sedimentation was investigated by Mohn (1980). The batched operation achieved an algal concentration of 1.5 % solids content. Separation of algae in sedimentation ponds is a simple and inexpensive process. However the rate of sedimentation was influenced by the use of flocculants. So an in-depth study of the flocculating nature of microalgae is required in order to use sedimentation tanks or ponds for harvesting of microalgae. Different flocculants can be screened for this purpose in order to determine the best one for the process.

8.3.4.2 Lamella Type Sedimentation Tank

Modifications to the simple sedimentation tanks were applied in which flat inclined plates were introduced in the tanks. The slopes of the plates were designed such that the down gliding of the settled algal particles were collected in a sump from which they were removed by pumping (Mohn, 1980; Shelef et al., 1984). Algae were concentrated to 1.6 % solids content, and coagulant dosing was suggested if suspension of tiny algae such as *Scenedesmus* was fed into the system (Mohn, 1980). Operational reliability of this method was fair and further thickening of algae slurry was required.

8.3.4.3 Flocculation–Sedimentation

A flocculation followed by gravity sedimentation process for algae separation has been studied (Golueke and Oswald 1965). Treating high rate oxidation pond effluent, the process achieved up to 85 % of the algal biomass using alum as a coagulant. This was found to be an effective process where algae slurry with 1.5 % w/v of solid content was achieved. A comparison of the flocculation–sedimentation process with flocculation-flotation method indicated that the latter exhibited very clear optimal operating conditions for algae separation (Friedman et al., 1977; Moraine et al., 1980).

8.3.5 Floatation

Some algae may not have a settling velocity which may not be significant for gravitational separation. In turn they float on the water surface and are hard to settle down. This problem may be solved by use of floatation techniques for harvesting. Flotation was simply gravity thickening upside down. Instead of waiting for the particles to settle to the bottom, liquid–solid separation was brought about by introducing air bubbles at the bottom of a flotation tank. The combined buoyancy of the particulate matter and the bubbles encouraged them to rise to the surface. Once the particles had floated to the surface, a layer of thickened slurry will be formed which could be collected by a skimming operation (Chen et al., 1998). A coagulant in an optimal dose was required for efficient harvesting of algal biomass by the floatation method (Bare et al., 1975). Different coagulants had been used in flotation systems. Chemicals such as aluminium and ferric salts, and polymers were used to facilitate the flotation with the overall objective to increase allowable solids loadings, percentage of floated solids, and clarity of the effluent. The time factor which was limiting in the case of sedimentation was overcome using the floatation technique. Flotation systems also offered higher solids concentrations and lower initial equipment cost. There are three basic variations of the flotation thickening systems: dissolved-air flotation, electro-flotation (or called electrolytic flotation), and dispersed-air flotation.

8.3.5.1 Dissolved-Air Flotation (DAF)

In the dissolved-air flotation (DAF) system, a liquid stream saturated with pressurized air is added to the DAF unit where it is mixed with the incoming feed. As the pressure returns to atmosphere, the dissolved air comes out of solution forming fine bubbles bringing fine particles with them, which rise to the surface where they are removed by a skimmer. Use of DAF process for algae separation in conjunction with chemical flocculation has been reported (McGarry and Durrani, 1970; Bare et al., 1975; Sandbank, 1979). The parameters responsible for determining the quality of the clarified effluent were recycling rate, air tank pressure, hydraulic retention time and particle floating rate (Bare et al., 1975; Sandbank, 1979). The concentration of the slurry depended on the skimmer speed and its overboard above water surface (Moraine et al., 1980). DAF was used for cleaning algae pond effluents with a high efficiency of thickening (up to 6 %). DAF along with floatation could further increase the concentration of the harvested algae (Bare et al., 1975; Friedman et al., 1977; Moraine et al., 1980). Optimization of parameters for DAF yielded higher separation efficiency of the biomass.

8.3.5.2 Electro-Flotation

During electrolysis the water splits into hydrogen. Hydrogen in the form of gas bubbles attaches to algal particles. This helps the algae particles to float on the water where they can be removed by skimming the surface. Further discussion of research on electro-flotation will be presented in Section 8.3.7.

8.3.5.3 Dispersed Air Flotation

Introduction of non-pressurized air into the floatation tank could provide us with an alternative to DAF. DAF could be modified by combining agitation with air injection to form foam. The algal cells could then be separated by froth floatation or foam floatation. Different factors such as aeration rate, pH of the algae suspension and temperature of operation were significant in determining the separation efficiency by foam floatation (Phoochinda and White, 2003; Phoochinda et al., 2004). *Scenedesmus quadricauda* was used to study dispersed air floatation. Surfactants such as cetyl trimethyl ammonium bromide (CTAB) and sodium dodecylsulfate (SDS) were also added to increase the separation efficiency. A very high algal removal was observed using CTAB (90 %) as compared to SDS (16 %). Algal removal efficiency could be increased to 80 % by adjusting the pH of the algal suspension. Reports by Golueke and Oswald (1965) suggested that very few reagents (2 out of 18) gave high removal efficiency. Another study suggested that dispersed air floatation was governed by the controlled pH of the medium (Levin et al., 1962). Critical pH level was recorded at 4.0, being attributed to the changes in the algae surface characteristics.

8.3.5.4 Ozone Flotation

Few reports studied the effect of ozone floatation for algae recovery (Betzer et al., 1980; Benoufella et al., 1994; Jin et al., 2006; Cheng et al., 2010, 2011). Ozone gas modified the cell surface of the algae which promoted the floatation of the cells and release of some agents which aided in floatation. A pilot plant study was conducted using ozone floatation on *Microcystis* (Benoufella et al., 1994). Different aspects of ozone floatation such as oxidising properties of ozone and floatation properties of the cells were studied. They indicated that ozone was responsible for inactivation of the cells. Ozone floatation in association with coagulation flocculation was found to be an efficient process for removal of cyanobacteria. Ferric chloride was found out to be the most potent coagulant. Preozonation also influenced the enhancement of the coagulation efficiency. *Scenedesmus obliquus* FSP-3, a species with excellent potential for CO_2 capture and lipid production, was harvested using dispersed ozone flotation. Ozone produces effective solid–liquid separation through flotation while air does not (Cheng et al., 2011). The ozone dose required for harvesting of algae was similar to that used for purification of water. During ozonation, the algae removal rate, surface charge, and hydrophobicity of algal cells and fluorescence characteristics and proteins and polysaccharides contents of algogenic organic matter (AOM) were determined. The proteins released from the AOMs altered the hydrophobicity of the bubble surfaces leading to formation of a froth layer which helped in harvesting of the microalgae. Humic substances in the suspension scavenge ozone adversely affected the ozone flotation efficiency of algal cells.

8.3.6 Centrifugation

Centrifugation can be described by Stokes' law, which predicts that its velocity is proportional to the difference in density between the cells and medium on the one hand and on the square of the radius of the cells (Stokes radius) on the other hand. Although for bacteria gravitational force-based methods are not easy to apply, for yeast and microalgae with diameters >5 μm and relatively thick cell walls they are feasible. High operating costs incurred during centrifugation negate the reliability and efficiency of the method. Laboratory centrifugation tests were conducted on pond effluent at 500–1,000 g and showed that about 80–90 % microalgae can be recovered within 2–5 min (Molina Grima et al., 2003). Centrifuges are analogous to sedimentation tanks except that the suspended particles were accelerated in their separation from the suspension by a centrifugal force that was higher than the gravity force. Several centrifugation devices were examined for application in algae separation (Mohn and Soeder, 1978, 1980; Moraine et al., 1980; Shelef et al., 1980, 1984). Some of them were very efficient as one-step separation process while others were found either inefficient or required thickened feed slurry. Centrifuges operated in batch mode were less attractive, as their operation had to be stopped for the solids to be removed. Knuckey et al. (2006) also states that the exposure of microalgal cells to high gravitational and shear forces can damage cell structure. According to Molina Grima et al. (2003), centrifugation is a preferred method, especially for producing extended shelf-life concentrates for aquaculture; however, they agree that this method is time-consuming and costly. Energy costs of about 1 kW h^{-1} m^{-3} have been quoted for centrifugation.

8.3.6.1 Solid-Bowl Decanter Centrifuge

The solid-bowl decanter centrifuge is a horizontal conical bowl which contains a screw conveyor that rotates in the same direction. Feed slurry enters at the centre and is centrifuged against the bowl wall. Settled solids were moved by the screw conveyor to one end of the bowl before discharge, while separated water formed a concentric inner layer which flowed over an adjustable dam plate and was discharged out of the centrifuge. The helical screw conveyor that pushed the deposited slurry operated at a higher rotational speed than the bowl. A solid bowl screw centrifuge was used to separate various types of algae (Mohn, 1980). A 22 % solids concentration was obtained in the separated algae when the feed suspension contained 2 % solids. Although the reliability of this centrifuge seemed to be excellent, the energy consumption was far too high. An attempt to concentrate an algal feed of 5.5 % solids derived from a flotation process by a co-current solid-bowl decanter centrifuge was not successful (Shelef et al., 1980). Subsequently, algae slurry concentration was improved to 21 % w/v TSS by reducing the screw conveyor speed to 5 rpm (Shelef et al., 1984). The solid-bowl decanter centrifuge was recommended for use concurrently with polyelectrolyte coagulant to increase the efficiency (Shelef et al., 1984).

8.3.6.2 Nozzle-Type Centrifuge

Continuous discharge of solids as a slurry was possible with the nozzle-type disc centrifuge. The shape of the bowl was modified so that the slurry space had a conical section which provided sufficient storage volume and afforded a good flow profile for the ejected cake (Shelef et al., 1984). The bowl walls sloped toward a peripheral zone containing evenly spaced nozzles. The number and size of the nozzles were optimized to avoid cake build up and to obtain reasonable concentrate of algal biomass. The application of nozzle-type disc centrifuge for algae harvesting, suggested by Golueke and Oswald (1965), investigated the influence of nozzle diameter on flow rate, algae removal efficiency and resultant slurry concentration. By comparing with other algae harvesting methods, it was concluded that the nozzle-type centrifuge seemed to be promising albeit it was less attractive because of power requirements and capitalization costs. In other studies, the centrifuge appeared to be more effective to harvest *Scenedesmus* than *Coelastrum* (Mohn and Soeder, 1978, 1980). By returning the centrifuge underflow to the feed, the solids content of the algae suspension (0.1 %) could be concentrated by a factor of

15–150 %. The reliability of this device could be ensured as long as the clogging of the nozzles was avoided.

8.3.6.3 Solid-Ejecting Disc Centrifuge

Solid-ejecting disc centrifuge provided intermittent solids ejection by regulating its valve-controlled peripheral ports by timer or an automatic triggering device. The advantage of this centrifuge for algae harvesting was its ability to produce algal cake in a single step without dosing with chemicals (Mohn and Soeder, 1978, 1980; Shelef et al., 1984). This centrifuge concentrated various types of microalgae effectively, achieving algal cake of 12–25 % solids (Mohn, 1980; Moraine et al., 1980). The extent of the algae suspension separation increased with increasing residence time (decreasing feed rate), and the ejected cake concentration was affected by the intervals between successive desludging (Shelef et al., 1984). Solid-ejecting disc centrifuges were found to be very reliable as the only setback reported was that solids finer than algae may be retained in the overflow which reduced the separation efficiency (Moraine et al., 1980). High capital and energy costs rendered this separation method unappealing.

8.3.7 Electrophoresis, Electroflotation and Electroflocculation Techniques

The electrolytic method is another potential approach to separate algae without the need to add any chemicals. In this method, an electric field drives charged algae to move out of the solution (Mollah et al., 2004). Water electrolysis generates hydrogen that adheres to the microalgal flocs and carries them to the surface. Electrocoagulation mechanisms involve three consecutive stages:

- Generation of coagulants by electrolytic oxidation of the sacrificial electrode
- Destabilization of particulate suspension and breaking of emulsion
- Aggregation of the destabilized phases to form flocs.

Electrical approaches to algae thickening included exploiting electrophoresis, electro-flocculation and electro-flotation. In a water solution, however, both electrophoresis and electroflocculation could occur under the same set of circumstances. If a vessel of algae in its growth medium was exposed to an electric field by placing metallic electrodes on two sides of the vessel and energizing them with a dc voltage, algae concentrations would occur at both electrodes (electrophoresis) and at the bottom of the tray (electro-flocculation). Research focussed on assessment of the factors influencing electrophoresis and electro-flocculation of algae in its growth medium was conducted (Pearsall et al., 2011) and the results showed that electrophoresis does occur but was complicated by the effects of the fluid motion. It appeared that the coupling of the algal cell and the fluid could be sufficiently strong such that fluid motion effects could influence or dominate behaviour. Electroflocculation appeared to be a robust process (Azarian et al., 2007). It does, however, inherently leave electrically induced trace metal flocculants in the dewatered algae. In electro-flotation or electrolytic flotation, fine hydrogen gas bubbles formed during the electrolysis which would cause the algal particles to float to the surface where they would be removed. Efficient bench scale electro-flotation system for algae flocculation by using the magnesium hydroxide formed in the electrolysis to enable precipitation and consequently flocculation was reported (Contreras et al., 1981). Laboratory and field scale electro-flotation units for algae removal from wastewater oxidation pond effluent was also reported (Sandbank et al., 1974; Kumar et al., 1981). A 2 m^2 pilot scale unit was tested for clarification of high-rate oxidation pond effluent (Shelef et al., 1984). For satisfactory algae separation, electro-flotation was followed or operated concurrently with alum flocculation (Sandbank et al., 1974).

Azarian et al. (2007) investigated the removal of microalgae from industrial wastewater using continuous flow electro-coagulation. Different from electrolytic coagulation, electrolytic flocculation does not require the use of sacrificial electrodes. Electrolytic flocculation works based on the movement of microalgae to the anode to neutralize the carried charge and then form aggregates. Poelman et al. (1997) showed that the efficiency of algal removal is 80–95 % when electrolytic flocculation is applied. There are several benefits to use electrochemical methods, including environmental compatibility, versatility, energy efficiency, safety, selectivity, and cost effectiveness (Mollah et al., 2004). An investigation into the removal of microalgae electrolytically in batch and continuous reactors by flotation was conducted by Alfafara et al. (2002). The results for a batch system showed that by increasing the electrical power input, the rate of chlorophyll removal increased and the electrolysis time decreased. Gao et al. (2010a, b) studied the algae removal by electro-coagulation–flotation (ECF) technology and indicated that aluminum was an excellent electrode material for algae removal when compared with iron. The optimal parameters determined were current density = 1 mA cm^{-2}, pH = 4–7, water temperature = 18–36 °C, algae density = 0.55×10^9–1.55×10^9 cells L^{-1}. Under the optimal conditions, 100 % of algae removal was achieved with the energy consumption as low as 0.4 kW m^{-3}. The ECF performed well in acid and neutral conditions. At low initial pH of 4–7, the cell density of algae was effectively removed in the ECF, mainly through the charge neutralization mechanism; while the algae removal worsened when the pH increased (7–10), and the main mechanism shifted to sweeping flocculation and enmeshment.

Furthermore, initial cell density and water temperature could also influence the algae removal. Overall, the results indicated that the ECF technology was effective for algae removal, from both the technical and economical points of view (Gao et al., 2010a, b). Recently, OriginOil company is employing several next-generation technologies to greatly enhance algae cultivation and oil extraction (OriginOil, 2010), by going on to control the harvesting and oil extraction cycles in a high-speed, round-the-clock, streamlined industrial production of algae oil. In the process, mature algae culture is injected through the OriginOil device, where Quantum Fracturing™, pulsed electromagnetic fields and pH modification (using CO_2) combine to break the cell walls, thereby releasing the oil within the cells. The processed culture now travels into a settling tank, or gravity clarifier, to fully separate into oil, water and biomass. Algae oil increases to the top for skimming and refining, while the remaining biomass settles to the bottom for further processing as fuel and other valuable products (OriginOil, 2010).

8.3.8 *Ultrasonic Methods*

Application of ultrasound technique to harvest microalgae has been reported in a laboratory scale study (Bosma et al., 2003). Algae separation process based on acoustically induced aggregation followed by enhanced sedimentation was carried out. Efficiencies higher than 90 % were recorded at high biomass concentrations and flow rates between 4 and 6 l d^{-1}. As much as 92 % of the algae biomass could be harvested with a concentration factor of 11. Attempts to harvest at higher efficiency were unfruitful due to small size and low particle density of the microalgae. Feed flow rate, biomass concentration and ratio between harvest and feed flows had a significant effect on the concentration factor. Use of ultrasound to improve the removal by coagulation of *M. Aeruginosa*—a common species of toxic algae—was investigated (Zhang et al., 2009). The results showed that sonication significantly enhanced the reduction of algae cells, solution UV 254, and chlorophyll-*a* without increasing the concentration of aqueous microcystins. The main mechanism involved the destruction during ultrasonic irradiation of gas vacuoles inside algae cells that acted as 'nuclei' for acoustic cavitation and collapse during the "bubble crush" period, resulting in the settlement of cyanobacteria. The investigation revealed that coagulation efficiency depended strongly on the coagulant dose and sonication conditions. With a coagulant dose of 0.5 mg L^{-1} and ultrasonic irradiation for 5 s, algae removal efficiency increased from 35 % to 67 %. Optimal sonication time was determined at 5 s, and further sonication would only marginally enhance the coagulation efficiency. The most effective sonication intensity was found to be at 47.2 W cm^{-2}, and the highest removal of the algae was recorded at 93.5 %. The authors recommended that this method could be successfully applied to natural water containing multiple species of algae. Ultrasonic harvesting on lab- or pilot-scale experiments has shown the merits that in addition to small footprint, the process could be operated continuously without evoking hydrodynamic shear stress on algal cells thus maintaining integrity of the algae (Bosma et al., 2003). The authors pointed out, however, that for industrial scale harvesting, centrifuges could better be used over the ultrasound aggregation sedimentation process because of lower energy requirement, better algae separation efficiencies and higher concentration factors.

8.4 Comparative Studies Amongst the Algae Separation Processes

Harvesting of microalgal biomass is one of the bottlenecks for biofuel production from microalgae (Li et al., 2011). It can be inferred from the above different harvesting methods that each of them have their own advantages and disadvantages and it also shows that efficiency of one method can be increased if integrated with another method, for example, integrating sedimentation with flocculation. Another such efficient method, which integrated electro-flocculation with dispersed-air flotation, was used for harvesting *Botryococcus braunii* (Li et al., 2011). According to another author, flocculation in combination with flotation or sedimentation followed by centrifugation or filtration is the most energy and cost efficient choice (Salim et al., 2011). Thus, integration of different methods is an efficient technology for harvesting microalgae. While undertaking research on harvesting, oil extraction, and refining processes for biofuel production from microalgae, nature and type of microalgal strain should be considered. Shape of algal cells, cell wall structure and oil composition vary from one algal strain to another; even two different cultures of the same strain are not similar in nature (Li et al., 2011). Although there are several biomass harvesting methods, Richmond (2004) suggested one main criterion for selecting a proper harvesting procedure is the desired product quality. In one hand for low value products, gravity sedimentation may be used, possibly enhanced by flocculation. Sedimentation tanks or settling ponds are also possible, e.g. to recover biomass from sewage-based processes. On the other hand for high-value products, to recover high-quality algae, such as for food, feed and nutraceuticals, it is often recommended to use continuously operating centrifuges that can process large volumes of biomass (Table 8.1).

8.5 Future Scope of Studies

Many research efforts are currently directed towards genetic modification of microalgae. Most recently published studies and granted patents in this field are aimed at increasing biomass productivity or increasing production of specific metabolites, most often lipids (Larkum et al., 2012; Georgianna and Mayfield, 2012). However, genetic modification may also be a promising way to harvest microalgae (Christenson and Sims, 2011; Georgianna and Mayfield, 2012). Here, achievements in genetic modification of yeast may be used as an example. In yeast, genetically modified strains have been developed that express flocculin proteins in their cell walls, causing the cells to aggregate (Govender et al., 2008). The expression of these proteins can be induced by an environmental trigger or during a specific growth stage. Sapphire Energy has described a method for flocculating microalgae in which ligand–receptor pairs can be expressed in different strains that are mixed to induce flocculation, or that are expressed sequentially in the same strain (Mendez et al., 2010). Genetic modification or selection may also be aimed at facilitating flocculation by other methods. For instance, a cell wall-deficient mutant of *Chlamydomonas*

Table 8.1 Comparative analysis between different methods of harvesting

Method	Advantages	Disadvantages
Screening	• Methods of screening are very simple	• Clogging of screens due to microalgal slime
	• Easy to operate	• High operational costs
Coagulation-flocculation	• Simple and easy to establish • High efficiency	• Chemical flocculants cause accumulation of harmful compounds in biomass
	• Sustainable and low in energy input	• Not suitable for high value added products
Filtration	• High recovery efficiency of biomass	• Fouling of membranes
	• Can be scaled up	• Energy intensive process
	• Easy for bulk harvesting	
Gravity sedimentation	• Cost effective process	• Slow rate of separation
	• Easy to operate	• Not suitable for very small microalgae
	• Bulk harvesting	
Floatation	• Simple and effective process	• Low harvesting efficiency
	• Can be made efficient in combination with other harvesting methods	• Some methods may damage the cells
Centrifugation	• Highly efficient process	• Energy intensive process
	• Up to 98 % recovery of biomass	• High maintenance costs
	• Can be scaled up	
Electrophoresis	• High efficiency	• Energy intensive process
	• Combines different methods	• Non-sustainable
		• Scale up not possible
Ultrasonication	• Simultaneous recovery of product possible	• Energy intensive process
	• Easy to operate	• May lead to damage of cells
	• Can be scaled up	

has been found to flocculate much more easily under alkaline conditions than the wild type strain (Scholz et al., 2011). This indicates that minor genetic modifications may greatly facilitate flocculation. Future studies should concentrate on the genetic modifications of microalgae. The flocculating substances produced by different organisms could be used for cost efficient harvesting of microalgae.

8.6 Conclusion

Microalgae are considered as efficient feedstock for production of third generation biofuels. The varied uses of microalgal biomass make it an important commodity of the future. Research on decreasing the cost of algal biomass production and its subsequent processing is on the rise. In order to come to a sustainable and cost efficient solution the problem of harvesting needs to be addressed. The cost of downstream processing of algal biomass alone accounts for most of the price required for the production. Efficient harvesting methods need to be developed in order to achieve the goal. The different techniques used for harvesting of microalgal biomass provide a roadmap for efficient separation of biomass. Some methods are too energy intensive to be used in a large scale. Use of bioflocculants for efficient separation of microalgal biomass is on the rise. Different flocculants such as guar gum, cationic starch and extracts have been used for efficient harvesting of microalgae. The biomass obtained after bioflocculation is devoid of any chemicals and can be used for further processing of the biomass.

New efforts concentrating on nanotechnology are also coming into focus. Magnetic nanoparticles are being used for effective harvesting of microalgal biomass. However, the large scale application of these methods still needs to be studied in depth. Genetic engineering of algae to produce external flocculants may be the path to the future. The algal cells can be modified as such they can express some surface agents which may help them to coagulate. This may probably present a cost effective and sustainable solution to the problem of dewatering of microalgal biomass. Fundamental research into infochemicals that induce flocculation in microalgae is urgently needed, because this may lead to a highly controllable method for inducing flocculation that avoids contamination. The same holds true for approaches to induce flocculation through genetic modification. Future studies should not only look at the efficiency of harvesting under specific conditions, but should also investigate how harvesting is influenced by properties of the microalgal cells or by culture conditions, particularly interference by organic matter in the culture medium. Cost is an important factor to consider when evaluating new flocculation methods for microalgae. Cost evaluation should not only take the cost of flocculation step itself into account, but also the influence on the entire production process.

References

Adan, B. and Lee, E.W. (1980). High rate algae growth pond under tropical conditions. Presented at a workshop on waste treatment and nutrient recovery. Singapore, 27–29 February, 1980.

Alfafara, C.G., Nakano, K., Nomura, N., Igarashi, T. and Matsumura, M. (2002). Operating and scale-up factors for the electrolytic removal of algae from eutrophied lake water. *Journal of Chemical Technology and Biotechnology*, 77, 871–876.

Amaro, H.M., Guedes, A.C. and Malcata, F.X. (2011). Advances and perspectives in using microalgae to produce biodiesel. *Applied Energy*, 88(10), 3402–3410.

Azarian, G.H., Mesdaghinia, A.R., Vaezi, F., Nabizadeh, R. and Nematollahi, D. (2007). Algae removal by electro-coagulation process, application for treatment of the effluent from an industrial wastewater treatment plant. *Iranian Journal of Public Health*, 36, 57–64.

Bare, W.F.R., Jones, N.B. and Middlebrook, E.J. (1975). Algae removal using dissolved air flotation. *Journal of Water Pollution Control Federation*, 47, 153–169.

Ben-Amotz, A. and Avron, M. (1990). The biotechnology of cultivating the halotolerant alga *Dunaliella. Trends in Biotechnology*, 8, 121–125.

Benemann, J.R., Kopman, B.L., Weismsman, D.E., Eisenverg, D.E. and Goebel, R.P. (1980). Development of microalgae harvesting and high rate ponds technologies in California. *In:* Shelef, B., Solder, C.J. (Eds), Algae Biomass. Elsevier, Amsterdam.

Benoufella, F., Laplanche, A., Boisdon, V. and Bourbigot, M.M. (1994). Elimination of microcystis cyanobacteria (blue-green algae) by an ozoflotation process—A pilot-plant study. *Water Science & Technology*, 30, 245–257.

Betzer, N., Argaman, Y. and Kott, Y. (1980). Effluent treatment and algae recovery by ozone-induced flotation. *Water Research*, 14, 1003–1009.

Bosma, R., van Spronsen, W.A., Tramper, J. and Wijffels, R.H. (2003). Ultrasound, a new separation technique to harvest microalgae. *Journal of Applied Phycology*, 15, 143–153.

Bratby, J. (2008). Coagulation and Flocculation in Water and Wastewater Treatment. IWA Publishing.

Brennan, L. and Owende, P. (2010). Biofuels from microalgae—A review of technologies for production, processing, and extractions of biofuels and co-products. *Renewable and Sustainable Energy Reviews*, 14, 557–577.

Cerff, M., Morweiser, M., Dillschneider, R., Michel, A., Menzel, K. and Posten, P. (2012). Harvesting fresh water and marine algae by magnetic separation: Screening of separation parameters and high gradient magnetic filtration. *Bioresource Technology*, 118, 289–295.

Chang, Y.R. and Lee, D.J. (2012). Coagulation–membrane filtration of *Chlorella vulgaris* at different growth phases. *Drying Technology*, 30, 1317–1322.

Chen, C. Y., Yeh, K. L., Aisyah, R., Lee, D. J., & Chang, J. S. (2011). Cultivation, photobioreactor design and harvesting of microalgae for biodiesel production: a critical review. *Bioresource technology*, *102*(1), 71–81.

Chen, M.Y., Liu, J.C. and Ju, Y.H. (1998). Flotation removal of algae from water. *Colloids and Surfaces B*, 12, 49–55.

Cheng, Y.L., Juang, Y.C., Liao, G.Y., Ho, S.H., Yeh, K.L., Chen, C.Y., Chang, J.S., Liu, J.C. and Lee, D.J. (2010). Dispersed ozone flotation of *Chlorella vulgaris*. *Bioresource Technology*, 101, 9092–9096.

Cheng, Y.L., Juang, Y.C., Liao, G.Y., Tsai, P.W., Ho, S.H., Yeh, K.L., Chen, C.Y., Chang, J.S., Liu, J.C., Chen, W.M. and Lee, D.J. (2011). Harvesting of *Scenedesmus obliquus* FSP-3 using dispersed ozone flotation. *Bioresource Technnology*, 102, 82–87.

Choi, S.K., Lee, J.Y., Kwon, D.Y. and Cho, K.J. (2006). Settling characteristics of problem algae in the water treatment process. *Water Science & Technology*, 53, 113–119.

Christenson, L. and Sims, R. (2011). Production and harvesting of microalgae for wastewater treatment, biofuels, and bioproducts. *Biotechnology Advances*, 29, 686–702.

Contreras, S., Pieber, M., del Rio, A., Soto, M.A., Toha, J. and Veloz, A. (1981). A highly efficient electrolytic method for microalgae flocculation from aqueous cultures. *Biotechnology and Bioengineering*, 23, 1165–1168.

Craggs, R., Sutherland, D. and Campbell, H. (2012). Hectare-scale demonstration of high rate algal ponds for enhanced wastewater treatment and biofuel production. *Journal of Applied Phycology*, 24, 329–337.

Edzwald, J.K. (1993). Algae, bubbles, coagulants, and dissolved air flotation. *Water Science & Technology*, 27, 67–81.

Eldridge, R.J., Hill, D.R.A. and Gladman, B.R. (2012). A comparative study of the coagulation behaviour of marine microalgae. *Journal of Applied Phycology*, 24, 1667–1679.

Friedman, A.A., Peaks, D.A. and Nichols, R.L. (1977). Algae separation from oxidation pond effluents. *Journal of Water Pollution Control Federation*, 49, 111–119.

Gao, S., Du, M., Tian, J., Yang, J., Ma, F. and Nan, J. (2010b). Effects of chloride ions on electro-coagulation-flotation process with aluminum electrodes for algae removal. *Journal of Hazardous Materials*, 182, 827–834.

Gao, S., Yang, J., Tian, J., Ma, F., Tu, G. and Du, M. (2010a). Electro-coagulation-flotation process for algae removal. *J Hazard Mater*, 177, 336–343.

Georgianna, D.R. and Mayfield, S.P. (2012). Exploiting diversity and synthetic biology for the production of algal biofuels. *Nature*, 488, 329–335.

Gouveia, L. (2011). Microalgae as Feedstock for Biofuels. Springer. (http://dz.doi.org/10.1063/1.3294480)

Govender, P., Domingo, J.L., Bester, M.C., Pretorius, I.S. and Bauer, F.F. (2008). Controlled expression of the dominant flocculation genes *FLO1*, *FLO5*, and *FLO11* in *Saccharomyces cerevisiae. Applied Environmental Microbiology*, 74, 6041–6052.

Greenwell, H.C., Laurens, L.M.L. , Shields, R.J., Lovitt, R.W. and Flynn K.J. (2010). Placing microalgae on the biofuels priority list: A review of the technological challenges. *Journal of the Royal Society Interface*, 7, 703–726.

Gutzeit, G., Lorch, D., Weber, A., Engels, M. and Neis, U. (2005). Bioflocculent algal-bacterial biomass improves low-cost wastewater treatment. *Water Science & Technology*, 52, 9–18.

Habib, M.A., Parvin, M., Huntington, T.C. and Hasan, M.R. (2008). A review on culture, production and use of *Spirulina* as food for humans and needs for domestic animals and fish. FAO Fisheries and Aquaculture Circular No. 1034, Food and Agriculture Organization of the United Nations, FAO Fisheries and Aquaculture Department Rome, Italy.

Harun, R., Singh, M., Forde, G.M. and Danquah, M.K. (2010). Bioprocess engineering of microalgae to produce a variety of consumer products. *Renewable and Sustainable Energy Reviews*, 14, 1037–1047.

Jin, P.K., Wang, X.C. and Hu, G. (2006). A dispersed-ozone flotation (DOF) separator for tertiary wastewater treatment. *Water Science & Technology*, 53, 151–157.

Kim, J., Ryu, B.G., Kim, B.K., Han, J.I. and Yang, J.W. (2012). Continuous microalgae recovery using electrolysis with polarity exchange. *Bioresource Technology*, 124, 164–170.

Knuckey, R.M., Brown, M.R., Robert, R. and Frampton, D.M.F. (2006). Production of microalgal concentrates by flocculation and their assessment as aquaculture feeds. *Aquaculture Engineering*, 35, 300–313.

Koopman, B.L., Thomson, R., Yackzan, R., Benemann, J.R. and Oswald, W.J. (1978). Investigation of the pond isolation process for microalgae separation from woodlands waste pond effluents. Final Report, U.C. Berkeley.

Kumar, H.D., Yadava, P.K. and Gaur, J.P. (1981). Electrical flocculation of the unicellular green algae *Chlorella vulgaris. Aquacultural Botany*, 11, 187–195.

Larkum, A.W.D, Ross, I.L., Kruse, O. and Hankamer, B. (2012). Selection, breeding and engineering of microalgae for bioenergy and biofuel production. *Trends in Biotechnology*, 30, 198–205.

Lazarova, V., Phillippe, R., Sturny, V. and Arcangell, J.P. (2006). Evaluation of economic viability and benefits of urban water reuse and its contribution to sustainable development. *Water Practical Technology*, 1, 1–11.

Lee, *A.K., Lewis,* D.M. and Ashman, P.J. (2008). Microbial flocculation, a potentially low-cost harvesting technique for marine microalgae for the production of biodiesel. *Journal of Applied Phycology*, 21, 559–567.

Levin, G.V., Clendenning, J.R., Gibor, A. and Bogar, F.D. (1962). Harvesting of algae by froth flotation. *Applied Microbiology*, 10, 169–175.

Li, Y.G., Xu, L., Huang, Y.M., Wang, F., Guo, C. and Liu, C.Z. (2011). Microalgal biodiesel in China: Opportunities and challenges. *Applied Energy*, 88(10), 3432–3437.

Lim, J.K., Chieh, D.C.J., Jalak, S.A., Toh, P.Y., Yasin, N.H.M., Ng, B.W. and Ahmad, A.L. (2012). Rapid magnetophoretic separation of microalgae. *Small*, 8, 1683–1692.

Liu, *D.*, Li, F. and Zhang, B. (2009). Removal of algal blooms in freshwater using magnetic polymer. *Water Science and Technology*, 59, 1085–1092.

Lundquist, T.J., Woertz, I.C., Quinn, N.W.T. and Benemann, J.R. (2010). A Realistic Technology and Engineering Assessment of Algae Biofuel Production. Energy Biosciences Institute, University of California, Berkeley.

McGarry, M.G. and Durrani, S.M.A. (1970). Flotation as a method of harvesting algae from ponds. Research program report No. 5. Asian Institute of Technology, Bangkok.

Mendez, M., Behnke, C., Poon, Y. and Lee, P. (2010). Induction of flocculation in photosynthetic organisms. WO 2009158658.

Mohn, F.H. (1980). Experiences and strategies in the recovery of biomass from mass cultures of microalgae. *In:* Shelef, B., Solder, C.J. (Eds), Algae Biomass. Elsevier, Amsterdam, pp. 547–571.

Mohn, F.H. and Soeder, C.J. (1978). Improved technologies for harvesting and processing of microalgae and their impact on production costs. *Archives in Hydrobiology Bech Ergebn Lemnology*, 11, 228–253.

Molina Grima, E., Belarbi, E.H., Acién Fernández, F.G., Robles Medina, A. and Chisti, Y. (2003). Recovery of microalgal biomass and metabolites: process options and economics. *Biotechnology Advances*, 20, 491–515.

Mollah, M.Y.A., Morkovsky, P., Gomes, J.A.G., Kesmez, M., Parga, J. and Cocke, D.L. (2004). Fundamentals, present and future perspectives of electrocoagulation. *Journal of Hazardous Materials*, 114, 199–210.

Moraine, R., Shelef, G., Sandbank, E., Bar Moshe, Z. and Schwarbard, L. (1980). Recovery of sewage borne algae: Flocculation and centrifugation techniques. *In:* Shelef, G., Solder, C.J. (Eds), Algae Biomass. Elsevier, North Holland.

Morweiser, M., Kruse, O., Hankamer, B. and Posten, C. (2010). Developments and perspectives of photobioreactors for biofuel production. *Applied Microbiology and Biotechnology*, 87, 1291–1301.

OriginOil (2010). http://www.originoil.com November 2010.

Pearsall, R.V., Connelly, R.L., Fountain, M.E., Hearn, C.S., Werst, M.D., Hebner, R.E. and Kelley, E.F. (2011). Electrically dewatering microalgae. *IEEE Transactions on Dielectrics and Electrical Insulation*, 18, 1578–1583.

Petrusevski, B., Bolier, G., van Breemen, A.N. and Alaerts, G.J. (1995). Tangential flow filtration: A method to concentrate freshwater algae. *Water Research*, 29, 1419–1424.

Phoochinda, W. and White, D.A. (2003). Removal of algae using froth flotation. *Environmental Technology*, 24, 87–96.

Phoochinda, W., White, D.A. and Briscoe, B.J. (2004). An algal removal using a combination of flocculation and flotation process. *Environmental Technology*, 25, 1385–1395.

Poelman, E., DePauw, N. and Jeurissen, B. (1997). Potential of electrolytic flocculation for recovery of micro-algae. *Resources Conservation and Recycling*, 19, 1–10.

Rawat, I., Kumar, R.R., Mutanda, T. and Bux, F. (2011). Dual role of microalgae: Phycoremediation of domestic wastewater and biomass production for sustainable biofuels production. *Applied Energy*, 88, 3411–3424.

Reynolds, J.H., Middlebrooks, E.J., Porcella, D.B. and Grenney, W.J. (1975). Effects of temperature on oil refinery waste toxicity. *Journal of Water Pollution Control Federation*, 46, 2674–2693.

Rossi, N., Jaouen, O., Legentilhomme, P. and Petit, I. (2004). Harvesting of cyanobacterium *Arthospira platensis* using organic filtration membranes. *Food and Bioproducts Processing*, 82, 244–250.

Rossignol, N., Vandanjon, L., Jaouen, O. and Quemeneur, F. (1999). Membrane technology for the continuous separation microalgae/culture medium: Compared performances of cross flow microfiltration and ultrafiltration. *Aquaculture Engineering*, 20, 191–208.

Rwehumbiza, V.M., Harrison, R. and Thomsen, L. (2012). Alum-induced flocculation of preconcentrated *Nannochloropsis salina*: Residual aluminium in the biomass, FAMEs and its effects on microalgae growth upon media recycling. *Chemical Engineering Journal*, 200–202, 168–175.

Salim, S., Bosma, R., Vermue, M.H. and Wijffels, R.H. (2011). Harvesting of microalgae by bio-flocculation. *Journal of Applied Phycology*, 23, 849–855.

Salim, S., Vermuë, M.H. and Wijffels, R.H. (2012). Ratio between autoflocculating and target microalgae affects the energy-efficient harvesting by bio-flocculation. *Bioresource Technology*, 118, 49–55.

Sandbank, E. (1979). Harvesting of microalgae from wastewater stabilization pond effluents and their utilization as a fish feed. D.Sc. thesis presented to the senate of the Technion – Israel Institute of Technology.

Sandbank, E., Shelef, G. and Wachs, A.M. (1974). Improved electroflotation for the removal of suspended solids from algae pond effluents. *Water Research*, 8, 587–592.

Schenk, P.M., Thomas-Hall, S.R., Stephens, E., Marx, U.C., Mussgnug, J.H., Posten, C., Kruse, O. and Hankamer, B. (2008). Second generation biofuels: High-efficiency microalgae for biodiesel production. *Bioenergy Research*, 1, 20–43.

Schlesinger, A., Eisenstadt, D., Bar-Gil, A., Carmely, H., Einbinder, S. and Gressel, J. (2012). Inexpensive non-toxic flocculation of microalgae contradicts theories; overcoming a major hurdle to bulk algal production. *Biotechnology Advances*, 30, 1023–1030.

Scholz, M., Hoshino, T., Johnson, D., Riley, M.R. and Cuello, J.L. (2011). Flocculation of wall-deficient cells of *Chlamydomonas reinhardtii* mutant cw15 by calcium and methanol. *Biomass and Bioenergy*, 35, 4835–4840.

Shelef, G., Azov, Y., Moreine, R. and Oron, G. (1980). Algae mass production as an integral part of a wastewater treatment and reclamation system. *In:* Shelef, B., Solder, C.J. (Eds.), Algae Biomass. Elsevier, North Holland.

Shelef, G., Sukenik, A. and Green, M. (1984). Microalgae harvesting and processing: A literature review. Report prepared for the US Department of Energy, Technion Research and Development Foundation Ltd., Haifa, Israel.

Show, K.Y., Lee, D.J. and Chang, J.S. (2013). Algal biomass dehydration. *Bioresource Technology*, 135, 720–729.

Spilling, K., Seppälä, J. and Tamminen, T. (2011). Inducing autoflocculation in the diatom *Phaeodactylum tricornutum* through CO_2 regulation. *Journal of Applied Phycology*, 23, 959–966.

Su, Y., Mennerich, A. and Urban, B. (2011). Municipal wastewater treatment and biomass accumulation with a wastewater-born and settleable algal-bacterial culture. *Water Research*, 45, 3351–3358.

Taylor, *R.L.*, Rand, J.D. and Caldwell, G.S. (2012). Treatment with algae extracts promotes flocculation, and enhances growth and neutral lipid content in *Nannochloropsis oculata* – A candidate for biofuel production. *Mar. Biotechnol.*, 6, 774–781.

Teixeira, C.M.L.L., Kirsten, F.V. and Teixeira, P.C.N. (2012). Evaluation of *Moringa oleifera* seed flour as a flocculating agent for potential biodiesel producer microalgae. *Journal of Applied Phycology*, 24, 557–563.

Tenney, M.W., Echelberger, W.F., Schuessler, R.G. and Pavpni, J.L. (1969). Algal flocculation with synthetic organic polyelectrolytes. *Applied Bacteriology*, 18, 965–971.

Uduman, N., Qi, Y., Danquah, M. K., Forde, G. M., & Hoadley, A. (2010a). Dewatering of microalgal cultures: a major bottleneck to algae-based fuels. *Journal of renewable and sustainable energy*, 2(1), 012701.

Uduman, N., Qi, Y., Danquah, M.K. and Hoadley, A.F. (2010b). Marine microalgae flocculation and focussed beam reflectance measurement. *Chemical Engineering Journal*, 162, 935–940.

Van Den Hende, S., Vervaeren, H., Saveyn, H., Maes, G. and Boon, N. (2011). Microalgal bacterial floc properties are improved by a balanced inorganic/organic carbon ratio. *Biotechnology and Bioengineering*, 108, 549–558.

Vandamme, D., Foubert, I., & Muylaert, K. (2013). Flocculation as a low-cost method for harvesting microalgae for bulk biomass production. *Trends in biotechnology*, *31*(4), 233–239.

Vandamme, D., Foubert, I., Fraeye, I., Meesschaert, B. and Muylaert, K. (2012). Flocculation of *Chlorella vulgaris* induced by high pH: Role of magnesium and calcium and practical implications. *Bioresource Technology*, 105, 114–119.

Vandamme, D., Foubert, I., Meesschaert, B. and Muylaert, K. (2010). Flocculation of microalgae using cationic starch. *Journal of Applied Phycology*, 22, 525–530.

Vandamme, D., Pontes, S.C.V., Goiris, K., Foubert, I., Jan Pinoy, L.J. and Muylaert, K. (2011). Evaluation of electro-coagulation-flocculation for harvesting marine and freshwater microalgae. *Biotechnology and Bioengineering*, 108, 2320–2329.

Wettman, J.W. and Cravens, J.B. (1980). Cost effective lagoon upgrading with microscreens. Proceedings of the 3rd Annual Pollution Control Association, Oklahoma, June 5, 1980.

Wijffels, R.H., Barbosa, M.J. and Eppink, M.H.M. (2010). Microalgae for the production of bulk chemicals and biofuels. *Biofuels, Bioproducts and Biorefining*, 4, 287–295.

Wilde, E.W., Benemann, J.R., Weissman, J.C. and Tillett, D.M. (1991). Cultivation of algae and nutrient removal in a waste heat utilization process. *Journal of Applied Phycology*, 3, 159–167.

Wu, Z., Zhu, Y., Huang, W., Zhang, C., Li, T., Zhang, Y. and Li, A. (2012). Evaluation of flocculation induced by pH increase for harvesting microalgae and reuse of flocculated medium. *Bioresource Technology*, 110, 496–502.

Wyatt, N.B., Gloe, L.M., Brady, P.V., Hewson, J.C., Grillet, A.M., Hankins, M.G. and Pohl, P.I. (2012). Critical conditions for ferric chloride-induced flocculation of freshwater algae. *Biotechnology and Bioengineering*, **109,** 493–501.

Xu, L., Guo, C., Wang, F., Zheng, S. and Liu, C.Z. (2011). A simple and rapid harvesting method for microalgae by in situ magnetic separation. *Bioresource Technology*, 102, 10047–10051.

Zhang, J. and Hu, B. (2012). A novel method to harvest microalgae via co-culture of filamentous fungi to form cell pellets. *Bioresource Technology*, 114, 529–535.

Zhang, X. et al. (2012). Influence of growth phase on harvesting of *Chlorella zofingiensis* by dissolved air flotation. *Bioresource Technology*, 37, 166–176.

Zhang, G., Zhang, P., & Fan, M. (2009). Ultrasound-enhanced coagulation for Microcystis aeruginosa removal. *Ultrasonics sonochemistry*, *16*(3), 334–338.

Zheng, H., Gao, Z., Yin, J., Tang, X., Ji, X. and Huang, H. (2012). Harvesting of microalgae by flocculation with poly (γ-glutamic acid). *Bioresource Technology*, 112, 212–220.

Zhou, W., Cheng, Y., Li, Y., Wan, Y., Liu, Y., Lin, X. and Ruan, R. (2012). Novel fungal pelletization-assisted technology for algae harvesting and wastewater treatment. *Applied Biochemistry and Biotechnology*, 167, 214–228.

Chapter 9
Prospect of Marine Algae for Production of Industrially Important Chemicals

Ramasamy Anbuchezhian, Valliappan Karuppiah, and Zhiyong Li

9.1 Introduction

Marine algae are simple, non-flowering plants with diverse group, consisting of unicellular (3–10 μm) to multicellular forms (kelps up to 70 metres). In the marine environment, they play an important role as a source of food and forming habitats. The origin of alga date back to several million years and becoming the oldest member in the kingdom of plant. As a largest primary producer in the marine environment, they support life in the marine ecosystem through the production of oxygen, by building food webs, and by providing habitats for different organisms. More than 90 % of global photosynthesis is contributed by algae and in the marine ecosystem algae are one of the leading primary producers. They consist of diverse group of large macro and micro algae. Marine microalgae include blue green algae and phytoplanktons. Marine macroalgae (seaweeds) are taxonomically diverse group of plants from which the terrestrial plants diverged over five hundreds millions years ago. As a plant, seaweeds have chlorophyll but do not have real stems, roots, leaves, and vascular tissue. They form an important food source for fishes, invertebrates and also providing breeding ground for several marine organisms. Microalgae are distributed throughout the ocean. Macroalgae mainly grow in the littoral zone, where they are constantly exposed to diverse fluctuating environmental conditions like oxidative stress, temperature and salinity which make them have great adaptations. So, they synthesize numerous valuable antioxidants including carotenoids, tocopherol, ascorbic acid, chlorophyll derivatives, phlorotannins, polyphenols and mycosporine-like amino acids. Marine algae are also a significant source of minerals, vitamins, proteins, fibres and poly unsaturated fatty acids. The beneficial effects

R. Anbuchezhian • V. Karuppiah • Z. Li (✉)
Marine Biotechnology Laboratory, State Key Laboratory of Microbial Metabolism, School of Life Sciences and Biotechnology, Shanghai 200240, People's Republic of China
e-mail: zyli@sjtu.edu.c

D. Das (ed.), *Algal Biorefinery: An Integrated Approach*,
DOI 10.1007/978-3-319-22813-6_9

upon the consumption of marine algae have also been demonstrated in different investigations.

Historically, the people living in coastal area used algae for long time as food, minerals, medicine, insulation, fertilizer and fodder. In Asia, seaweeds are the common ingredients in cuisine, but in western the direct consumption is less common. The food products, including nutraceuticals and functional foods are the largest market. Marine algae used in the human consumption as well as industries are mainly obtained from natural environment and when demand exceeds the availability they are produced through aquaculture. They produce plethora of industrially important chemicals. Now, marine algae have turned into a multibillion dollar industry. The most important industrial chemicals extracted from marine algae include agar, carrageenan, alginates, ulvan, furcellaran, hypnean, funoran, iridophycan, phyllophoran, laminaran, fucoidan and mannitol. Marine algae produce diverse fascinating metabolites with novel structures and biological activities. The isolated marine algal natural products include carotenoids, polyketides, polyphenols, peptides, halogenated compounds, fatty acids, steroids, lectins, etc. These different compounds exhibit diverse biological activities against several disease causing agents. They also become an excellent source of bioenergy as well as biofertilizers. The biofertilizers derived from marine algae form great biofertilizers and improve the soil quality and yields considerably. algae are raw feedstock for different biofuels and bioplastics production. In this chapter, we review the useful applications of marine algae in the production of industrially important chemicals.

9.2 Classification of Marine Algae

Marine algae are the photosynthetic organisms living in the marine environments. The photosynthetic mechanism of marine algae is like that of plants in terrestrial environment. The aquatic plants are more effective in the conversion of solar energy into biomass. They receive nutrients directly from the surrounding water through their tissues.

9.2.1 Microalgae

Marine microalgae include diatoms, dinoflagellates, green algae and blue-green algae. They occupy the bottom of the food chain. Even though, the estimated diversity is about 200 to 800 thousand species, only a fraction have been described (35,000). They are one of the important sources of food for various organisms in the marine environment. Marine environment represents the majority of algae classes.

9.2.2 Macroalgae

Macroalgae (seaweeds) can be grouped into three different classes based on their pigmentats as *Chlorophyceae, Phaeophyceae* and *Rhodophyceae*. The phycoerythrin and phycocyanin are responsible for the red colour in Rhodophyta, chlorophyll *a* and *b* present in the higher plants are responsible for the green colour in Chlorophyta, fucoxanthin and xanthophyll carotenoids are responsible for the brown colour in Phaeophyta. In general, brown seaweeds (30 cm to 70 metres) are larger than green and red seaweeds (few cm to about one metre). The Rhodophyceae sometimes appears as purple or brownish red in colour. In marine environment seaweeds are a potential renewable resource and they consist over 6000 species.

9.3 Industrial Chemicals

9.3.1 Scope of the Marine Algae Chemical Industry

Marine algae used in the industries are exploited either from natural environment or from cultivated crops. The rapid increase of industrial applications of algae and their products has escalated the aquaculture seaweed production. More than thirty five countries across the world (cold to tropical waters) are involved in commercial algal exploitations. The overall value of the seaweeds have been estimated at US$10 billion per year through the products of direct or indirect human uses (Bixler and Porse, 2011; FAO, 2013). The total production of macroalgae rises every year about 5.7 % and in 2011, from the natural and farming more than eighteen million tons were produced. Of these, over 96 % was produced through aquaculture and Asian countries contributed over 99 % of the total biomass (FAO, 2014). The global production of seaweeds by cultivation constitute around 98 % represented by five major genera including *Saccharina, Undaria, Porphyra, Eucheuma/Kappaphycus,* and *Gracilaria*. Marine algae based chemical industry turned into a multibillion industry because of their wide variety of applications (Fig. 9.1).

9.3.2 Phycocolloids (Hydrocolloids)

Phycocolloids are seaweed colloids made up of simple sugars. These gelling and viscous materials have been extracted from marine algae, especially red and brown algae, for centuries and usually used in food preparations. These materials can form colloidal nature in water. These include red algae derivatives agar and carrageenan and brown algae derivative alginates (Table 9.1). These polysaccharides are important as they exhibit high molecular weights, high viscosity and excellent gelling, stabilizing and emulsifying properties. All are water soluble and are extracted with

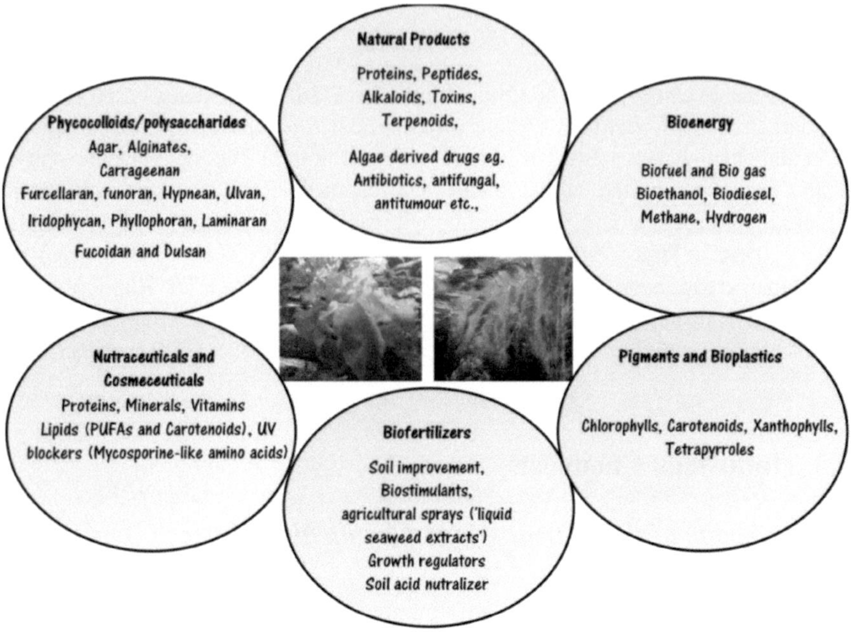

Fig. 9.1 The applications of marine algae and their chemicals.

Table 9.1 Marine algae and their hydrocolloids

Class	Hydrocolloids
Phaeophyceae	
Ascophyllum	
Durvillea	
Ecklonia	
Eisenia	Alginates
Laminaria	
Lessonia	
Macrocystis	
Nereocystis	
Sargassum	
Rhodophyceae	
Ahnfeltia	
Chondrus	
Eucheuma	
Furcellaria	Carrageenans
Gigartina	
Gymnogongrus	
Hypnea	
Iridaea	
Gelidium	
Pterocladia	Agar

hot water or alkaline solution. The four major functions in food applications are thickening, gelling, emulsification and stabilization. The seaweed polysaccharides are economically significant and have diverse applications.

9.3.2.1 Agar

Agar or agar-agar is a gelling hydrocolloid derived from marine algae and is important in the structural features of cell walls in Rhodophyta. Agar producing algae are generally known as agarophytes. The use of agar in culture medium of microbes was the first commercial application other than consumption as food. Seaweeds like *Gelidiales* and *Gracilariales* are important source of agar. Agar is the mixture of agarose and agaropectin (Fig. 9.2). The gelling properties of agar depend on the different factors like molecular weight, type, etc. Seaweeds belonging to the order *Gracilariales* are the largest source of agar. They have broad applications, because of their abundance and chemical characteristics (Abbott, 1996). *Gracilariales* are distributed along the temperate and tropical waters with high efficiency in aquaculture production than *Gelidiales* and at first, it was used to overcome the supply deficiency of *Gelidiales.* The gelling properties of the *Gracilaria* agar is lower than that of the *Gelidium* because of their high sulphate content. In order to increase the gelling ability of this agar it is necessary to remove sulphate groups by alkaline hydrolysis. Mass industrial scale production must have

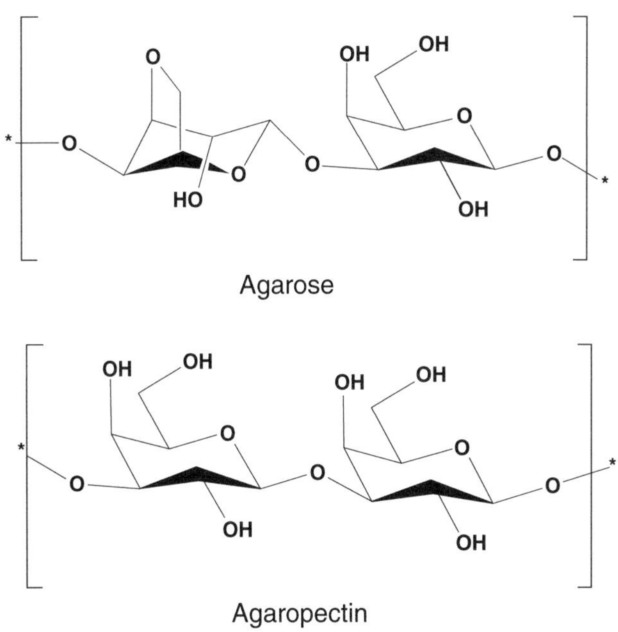

Fig. 9.2 Structure of agarose and agaropectin.

high attention in the alkali pre-treatment to exclude unwanted contamination through alkaline residues and sodium agaropectinates. In the extraction of the agar from *Gracilaria,* the environmental condition especially temperature and time affect the chemistry of agar. A considerable reduction in the production and the gelling ability are recorded associated to the elevated temperature or long extraction time, which also depends on the species of seaweeds used.

The excellent thickening and gelling properties of this colloid make them industrially important components. It is soluble only in boiling water, but not in cold water. When it is boiled the liquid transforms as a gel having stabilizing and thickening abilities. Among the hydrocolloids it is outstanding because of their gelling strength. The significant properties of agar include great gelling properties at low concentration, great transparency and viscosity in solution etc. Quality of agars are determined based on their gel strength and the international market referred the high quality agar with gelling property greater than 700 g cm^2 in a 1.5 % solution (Armisen, 1995). So, the agar with less than this value will be used in traditional applications and also used in some industrial applications like food, cryoprotectants, fat replacer, etc. Generally, in a solution the concentration of agar is directly proportional to its gelling strength. The loss of water and changes in their physical properties are associated with the lowering of agar. To produce gel with reduced loss of water, low gelling agar is important.

Agar is used in biotechnological (electrophoresis, chromatography etc.), medical (as laxative, anticoagulant etc.) and dental (impression materials) applications. It is also used in plant tissue culture, biochemistry, molecular biology and drug industry. Gelling properties of agar at moderate temperatures make it much useful in bakery products, confectionery making, and in puddings, creams, pastries, desserts, salad dressings and jellied products. Agar is used in different food items like milk products, confectionery, sweets, beverages, bakery, meat products, canned meats and fish, and as a clarifying agent in wines and beers. It is also used in the preparation of microbial culture media, tissue culture of plants, therapeutic agents to treat some diseases, medicine carrier, suspension agent of barium sulphate in radiology, stabilizer in cholesterol solutions, moulding, chromatography, electrophoresis, and in several types of emulsions as a suspension agent.

9.3.2.2 Alginate

Alginates (algin) are linear polyuronic acid hydrocolloids. It is a glycuronan of significant commercially important polysaccharide derived from brown seaweeds. The cell wall and in the matrix of the brown algae contain alginates; it gives both mechanical strength and flexibility to the algae. Even though, some bacteria contain polysaccharides similar to the alginates, brown algae are the important source of alginates. About 40 % of the dry weight of brown seaweeds contain alginates. Alginates consist of (1→4) linked β-D-mannuronic acid (M) and α-L-guluronic acid (G) residues (Fig. 9.3) of differing composition and structure. The sequence of monomers widely change with different species.

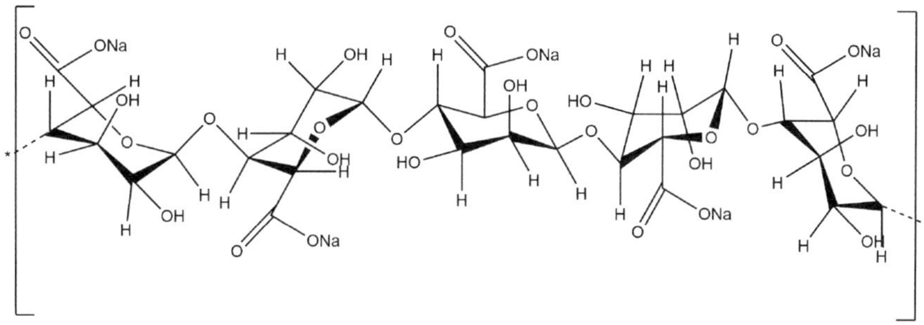

Fig. 9.3 Bonding of alginate repeating units mannuronate (M) and guluronate (G).

The major seaweeds used in the commercial productions include *Ascophyllum, Laminaria, Lessonia, Ecklonia, Macrocystis* and *Durvillea*, where they reach up to 40 % of the dry weight (Draget et al., 2002; Rinaudo, 2007). In water, they form viscous gel; they are also used as a stabilizer, binder, emulsifier etc. Its molecular weight ranges between 500 and 1000 kDa. The solubility of alginates depends on the different factors like pH, concentration, presence of solution ions and divalent ions. High content of alginates were observed in the seaweeds distributed in turbulent waters when compared to the normal environments. The chemical structure of the alginates varies between brown algae of genus. As the properties of alginate differ between seaweeds, the selection of candidate species for alginates extraction depends on both availability as well as the potential of the alginate production by the species. It is believed that the alginate act as a structure-forming component in brown seaweeds. The gelling strength of alginate depends upon its quality. Brown seaweeds contain alginic acid as calcium salt of alginic acid. The important properties of alginic acids are their high viscosity in solution, gel forming properties in divalent or multivalent cations, biocompatible and adhesive nature. Their capability of water retaining, gelling, viscosifying, and stabilizing properties determine their industrial applications. However, alginates do not have any nutritional value; they are often used as additives to modify and stabilize its texture. They are widely used as stabilizers, viscosifiers, and gelling agents in diverse products such as food, beverages, pharmaceuticals, cosmetics and various biotechnology industries. Moreover, based on their biochemical and biophysical properties several new types of food application are also emerging continuously. The other applications of alginate include pharmaceutical preparations such as Gaviscon, Bisodol, Asilone and in impression-making material in dentistry and prosthetics.

9.3.2.3 Carrageenan

The phycocolloid carrageenan is a linear sulphated polysaccharide derived from red algae of the genus *Gigartina, Hypnea, Eucheuma, Chondrus* and *Iridaea*. The content largely depends on the type of the seaweeds. Carrageenan can be classified in to three types (Fig. 9.4) based on their structure and chemical properties viz., kappa-carrageenan has one sulphate group per disaccharide (rigid and brittle gel, thermo-reversible, high gel strength, showing syneresis), iota-carrageenan has two sulphates per disaccharide (elastic gel, thermo-reversible, no syneresis, thixotropic) and lambda-carrageenan has three sulphates per disaccharide (cold soluble, non-gelling, high viscosity). They are composed of a linear galactose backbone with a varying degree of sulfatation (between 15 % and 40 %). The different carrageenan types vary in composition and conformation, resulting in a wide range of rheological and functional properties.

The carrageenan in industrial seaweeds content varies from 30 % to 60 % of dry weights. It totally depends upon the species of seaweeds and the environmental conditions, such as luminosity, nutrients, water temperature and oxygenation. Some species of seaweeds produce mixed type carrageenan such as kappa/iota, kappa/lambda or iota/lambda. The kappa-type carrageenan are extracted from seaweeds *Hypnea musciformis*, *Gigartina stellata*, *Eucheuma cottonii*, *C. crispus* and *Iridaea*. Seaweed species *G. teedi* and *E. spinosum* produce iota-type carrageenan. The species that produce lambda type carrageenan generally belong to the *Gigartina* class.

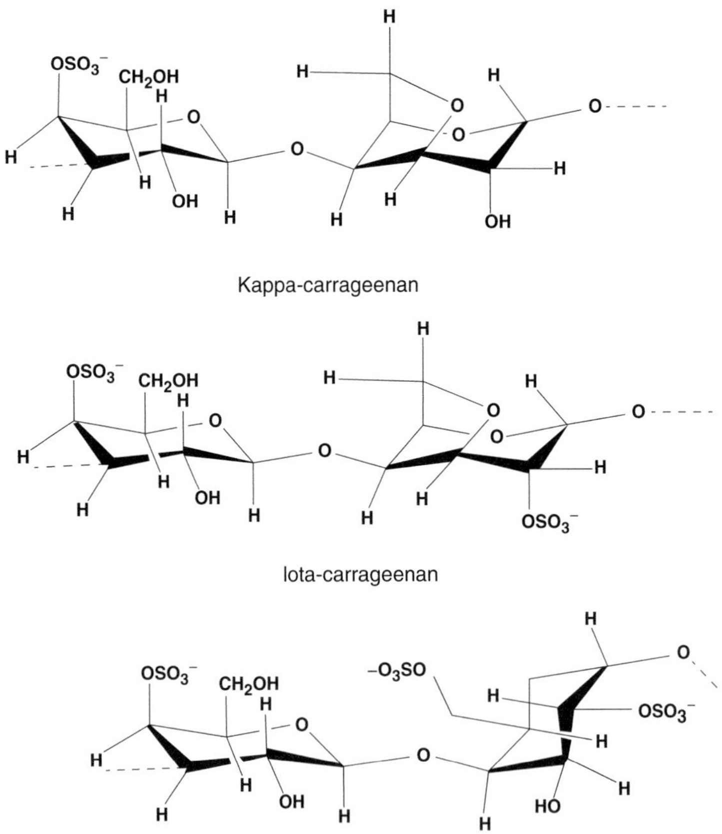

Fig. 9.4 Different types of carrageenans.

Carrageenans are used in various food items as gelling, thickening, and stabilizing agents. It is also used as a gelling agent to increase viscosity in different foods, clarifier to remove haze-causing proteins in beer and processed meats, stabilizer to prevent constituents separating in toothpaste, ingredient in the encapsulated gel of fruit gushers, thickener to cause foam to become sticky in fire fighting foam, thickener in shampoo, in air freshener gels, in the ancient art of paper and fabric marbling, in shoe polish to increase viscosity, in biotechnology to immobilize cells/enzymes, in pharmaceuticals as an inactive excipient in pills/tablets, plant milks and soy milks used as a thickening agent, in diet sodas as texture and suspend flavours enhancer, pet foods and cosmetics.

9.3.3 Other Seaweed Hydrocolloids/Polysaccharides

9.3.3.1 Dulsan

The marine red alga *Rhodymenia palmate* also known as the dulse is the source of Dulsan. Dulsan does not form gels and it is also water soluble and acid hydrolysis gives principally D-xylose, but also galactose and glucose with (1–3) and (1–4) linkages. It is well known as food. It also has been an important source of fibre throughout the centuries.

9.3.3.2 Fucoidan

Fucoidan is one of the well studied fucans from brown algae, which was first isolated by Kylin in 1913. These sulphated polysaccharides are composed of a fucose backbone (Fig. 9.5). The total content of fucoidan may vary between 20 and 30 % of algae dry weight, and it depends on the type of seaweed. There are two different forms of fucoidan such as F-fucoidan, which is >95 % composed of sulphated esters of fucose, and U-fucoidan, which is approximately 20 % glucuronic acid. The sulphated polysaccharide in fucoidan consists mainly of L-fucose units; at the same time it can also contain minor amounts of sugars such as galactose, mannose, xylose, or uronic acid and sometimes proteins. This is varying between different species. It has an α-(1,3)-backbone of repeating disaccharide units of α-(1,3)- and α-(1,4)-linked fucose residues. *Fucus vesiculosus* derived commercially available fucoidan is a heterogeneous mixture of more than 15 different fucans with varied properties of industrial monosaccharide moieties. Fucoidan occurs in all brown algae in different ratio and it is in the intercellular tissues and is considered to be a substance used by the weeds to protect themselves against the effects of drying out when exposed. It is extremely viscous after extraction, even in very low concentration, but is highly susceptible to aging, diluted acids and bases. This instability and its quality inconsistency have prevented its commercial significance of fucoidan. It is used as an ingredient in some dietary supplement products. Fucoidan has been consumed for a

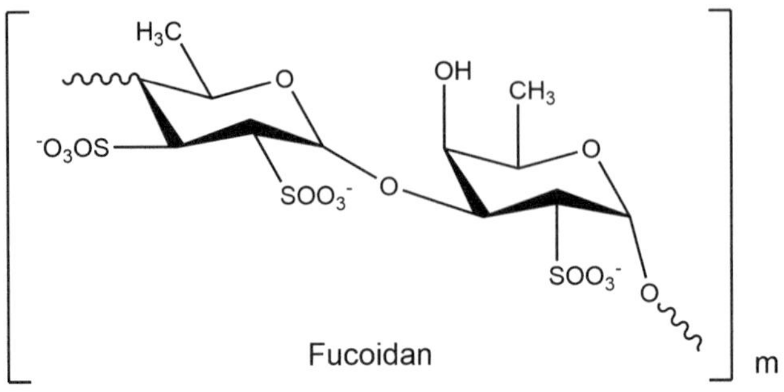

Fig. 9.5 Structure of fucoidan.

long time in Japan, China, and Korea as part of whole seaweed, and it is used as nutraceuticals in Australia and the United States. The fucoidan materials are also used in cosmetics because their ability to absorb directly by the human skin with the following different effects includes whitening, preserving moisture, removing freckles. It has also significant medical applications and the higher sulphation increases its therapeutic potential. It also acts as modulator of coagulation, as an alternative to the anticoagulant heparin. Fucoidan exhibits many biological activities which include anti-inflammatory, anticell proliferation, antiadhesion, antiviral, anticoagulant, antitumour and antiviral activities.

9.3.3.3 Funoran

Funoran is a sulphated galactans extracted from the red seaweed, *Gloiopeitis* species and distributed along the coasts of Japan, China and Pacific coasts of North America. Funoran isolated from *G. furcata* consists of several components which can be fractionated in terms of the solubility of their quaternary ammonium salts in aqueous potassium chloride. Apart from this main structure, an unfractionated funoran from a closely related species, *G. cervicornis*, has been proposed to possess a precursor moiety that is converted into the agarose sulphate under alkaline conditions. Funoran is marketed in the form of dried condition and not extracted. The mixing of whole seaweed in hot or lukewarm water gives a clear, viscous colloidal solution of excellent adhesive and sizing properties. It is used as an adhesive in several Japanese industries such as pottery and textiles. The major uses of funoran are as a constituent in hair waving and hair dyeing preparations. It is also a paper and textile sizing agent, and has widespread uses as a household adhesive.

9.3.3.4 Furcellaran

Furcellaran also known as Danish agar is an anionic sulphated polysaccharide extracted from the red alga *Furcellaria lumbricalis* and *F. fastigiata*. These species are distributed in the cold waters around Northern Europe and Asia. It is currently considered to be a copolymer of β- and κ-carrageenan and usually represented structurally as a repeating unit of alternating 3-linked β-D-galactopyranose and 4-linked α-D-galactopyranose residues, with part of the latter existing as a 3,6-anhydro derivate. The hydroxyl groups in polysaccharide chain may be substituted (sulphated, methylated, etc.) and other monomer residues such as xylose and glucose may be found. It is a polyelectrolyte that carries sulphate groups and is negatively charged over a wide range of pH. The charge density is usually one sulphate per three or four monomer units. The formation of 3,6-anhydro-α-D-galactose units from α-d-galactose 6-sulphate residues by alkaline treatment is an important and well known reaction of the carrageenans. After alkali treatment of the algae, the polysaccharide is isolated by using hot water. Furcellaran contains significant sulphate (12 to 16 %). It is composed of D-galactose (46 to 53 %), 3,6-anhydro-D-galactose (30 to 35 %), and sulphated portions of both of these sugars (16 to 20 %). It is used commercially to enhance gelation behaviour and, in the laboratory, for the quantitative determination of α-galactose 6-sulphate residues. Furcellaran is also used as a natural colloid, gelling agent, viscosity control agent primarily in food products, pharmaceuticals, in products for diabetics, proprietaries for reducing excess body weight, toothpastes, as carrier for food preservatives, bactericides and in bacteriological culture media.

9.3.3.5 Hypnean

Hypnean is derived from *Hypnea musciformis* and is often considered to be a type of carrageenan. Species of these seaweeds are fairly abundant in tropical waters, notably southern Florida, the Caribbean, the Gulf of Mexico, the northeast coast of Brazil, Australasia, parts of the South China Sea and South Africa and Gulf of Oman. The gel strength is far exceeding that of other phycocolloids. In addition, properties of hypnean are susceptible to a high degree of control and chemical modification. The high commercial importance of this phycocolloid is because of their ability to form, often in combination with other components, gels with great firmness and elasticity. Hypnean is mainly used for their gelling properties for food applications and they are known as a vermifuge in Indonesia, Greece and Turkey.

9.3.3.6 Iridophycan

Iridophycan is extracted from the seaweed species belonging to the genus *Iridea*. These algae are most prevalent in the waters off central California, but also are distributed in the coasts of South Africa, Japan, Chile and the Falkland Islands. It is extracted using boiling water and filtered to isolation. The sulphated galactan gives only D-galactose after hydrolysis. Iridophycan is used as a cold-mix stabilizer, for its adhesive properties, and it prevents the blood coagulation. There are some similarities in properties of funoran and carrageenan. In U.S.A., it has found particularly widespread use as a stabilizer in chocolates and some beverage mixes. It is also used as an excellent multipurpose adhesive and as a paper and textile size. The aqueous extract has pharmaceutical value as an agent to prevent blood coagulation.

9.3.3.7 Laminaran

Laminaran (laminarin) is isolated from species of *Laminaria*, *Ascophyllum* and *Fucus*. The species of *Laminaria* are distributed in the coasts of Norway, Scotland, Ireland, France, Spain and eastern Canada. Laminaran (Fig. 9.6) is a small glucan of 5 kDa with a degree of polymerization (DP) between 20 and 25. Among the two types of Laminaran based on the solubility, the soluble form is completely soluble in cold water and insoluble form only soluble in hot water. It consists of approximately upto 35 % of the dried weight of the algae. The solubility of laminaran increases when the degree of branching increases. There are two types of laminaran described, one with chains that are terminated by D-mannitol residues (M-series) and another type with chains terminated by D-glucose residues (G-series). In *Laminaria* and *Fucus* species, 40–75 % of the reducing end groups are linked to one of the primary hydroxyl groups of D-mannitol. Laminaran does not form gel or any viscous solution and its potential applications are in the medical and pharmaceutical uses. The content of Laminaran varies between species and seasons. The (β1-3,1-6)-linked D-glucans have the ability to enhance and stimulate the immune system of humans. The laminaran from *Alaria praelonga* or *L. coriacea* promotes adhesion of human skin fibroblast cells, and increase proliferation of human osteoblast cells. It was projected as elicitors of phytoalexins and flavonoids in alfalfa cotyledons. It is also used as a safe surgical dusting powder, and may have value as a tumour-inhibiting agent and, in the form of a sulphate ester, as an anti-coagulant. Laminaran

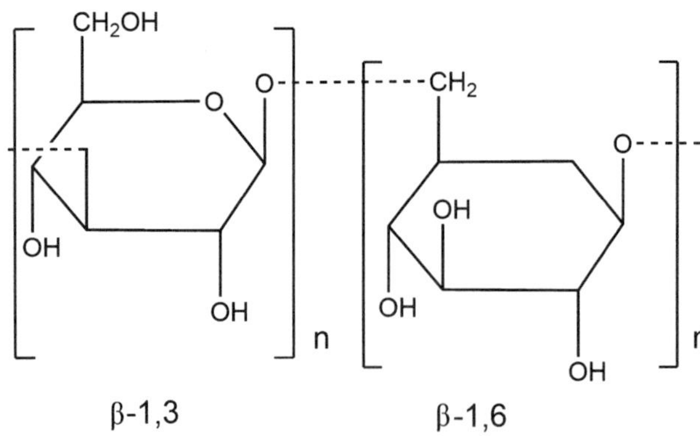

Fig. 9.6 Structure of laminaran.

has shown various biological activities which include anti-apoptotic and antitumoral activities.

9.3.3.8 Phyllophoran

Phyllophoran is extracted from red seaweeds belonging to the genus *Phyllophora*. These species are distributed in coasts of temperate and cold seas, including the eastern and western Atlantic, Baja California and Mexico, but abundant in the Black Sea. The properties of Phyllophoran are intermediate between those of agar and carrageenan.

9.3.3.9 Ulvan

Ulvales (Chlorophyta) are very common seaweeds distributed worldwide. Ulvales (*Ulva* and *Enteromorpha* sp.) are the source for complex sulphated polysaccharide ulvans, which are present in the cell walls, and represent a potential source of new functional biopolymer due to their peculiar composition and structure. The name ulvan is derived from the original terms ulvin and ulvacin introduced by Kylin. They represent 8–29 % of the algae dry weight. The major repeating disaccharide units reported are ulvanobiouronic acid 3-sulphate types containing either glucuronic or iduronic acid (Fig. 9.7). Moreover, minor repeat units have been reported that contain sulphated xylose replacing the uronic acid or glucuronic acid as a branch on *O*-2 of the rhamnose-3-sulphate (Lahaye and Ray, 1996). It contained 23 to 35 % sulphate ester groups, 10 to 15 % uronic acid, and 3.8 to 4.5 % protein. The variation in sugar composition can have methodological, taxonomic, and/or ecophysiological origins. It is an economical value gelling polysaccharide which forms a weak gel at 1.6 % (w/v) in deionized water. Ulvan inhibit the activity of cellulase, which indicates a protective role towards cell wall amorphous α-cellulose, protecting it from marine bacterial attack. It is also used as a nutritional dietary fibre. Ulvan exhibits different pharmacological properties like antiviral, anti-oxidant,

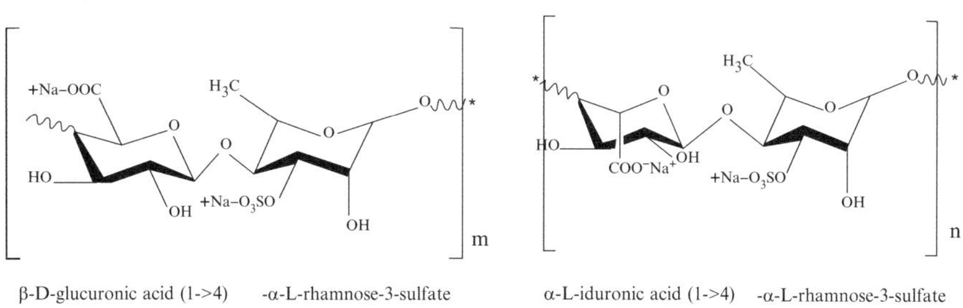

Fig. 9.7 Structure of ulvan.

anti-coagulant, anti-hyperlipidemic, immunostimulating and anti-proliferative towards cancer cells.

9.3.4 *Bioactive Natural Products*

Nature is continuously a broad arsenal of structurally diverse and pharmacologically active compounds that are used as highly effective drugs or lead structures for the development of novel synthetically derived drugs to combat a multitude of diseases. Natural products have been the most successful source of potential drug leads. It has estimated that over 50 % of the current drugs in use are either harvested from natural sources or from synthetic compounds based on natural products. Approximately 60 % of all drugs now in clinical trials are either natural products or their derivatives. The oceans, covering more than 70 % of the earth's surface, represent an enormous resource for the discovery of novel chemotherapeutics. The biodiversity in marine environment is remarkably high and their unique organisms also are exclusively living in sea. Marine algae represent a relatively untapped resource for discovering novel natural products. In order to survive in the highly competitive environments, marine algae have developed unique defense strategies that result in a tremendous diversity of compounds from different metabolic pathways.

The emergence of drug resistant pathogens and new diseases necessitated finding out novel drugs. The discovery of this chemical diversity to support finding of novel drugs is important. Marine algae offered a wide variety of novel natural products with different biological activities. They live in an unique environment, where they compete for space and nutrients and have the great adaptation strategy for living in different physico-chemical conditions. So, they have evolved with a wonderful defence strategy, and produce novel secondary metabolites (natural products). Generally, the production of bioactive substances differ in the same species of seaweed which depends on factors like environmental conditions, seasons, geographic locations, the stage of life cycle, and part of the seaweed. The synthesis of secondary metabolites mostly for chemical defence against the biotic pressure by predators, consumers, epibionts and the abiotic pressure from the surrounding environment (e.g. desiccation, nutrient availability, UV) can influence their synthesis.

The role of these secondary metabolites in marine algae is defense against herbivores, fouling organisms and pathogens. They also play a role in reproduction, protection from UV radiation and as allelopathic agents. Eventhough, the secondary metabolites are varied genetically and environmentally, transplant experiments showed that environmental conditions are able to alter the concentrations of secondary metabolites, but the types of compounds are genetically fixed. The production of secondary metabolites in seaweeds is different in their life history and maximum concentration of defense chemicals were found in new Halimeda species. Even though, marine algae offered huge number of chemicals which still remained largely unexploited. The bioactive compounds isolated from marine algae include proteins, peptides, terpenoids, phlorotanins, fucoidans, sterols and glycolipids, and the crude extracts or the compounds from algae exhibits a wide range of pharmacological properties which include anticancer, antibacterial, antifungal, antiviral, anti-inflammatory, antcoagulant, anti-oxidant, hypoglycaemic, hypolipidemic, anti-melanogenic, anti-bone loss, hepatoprotective and neuroprotective activities. Marine algae are a potential renewable resource in the marine environment. They are an excellent source of bioactive compounds, carotenoids, dietary fibre, protein, essential fatty acids, vitamins and minerals.

9.3.5 Nutraceuticals

Seaweeds are also commonly known as sea vegetables. Seaweeds are extensively used as food by coastal peoples, particularly in East Asia such as China and Vietnam, but also in some parts of Europe including Scotland. An archaeological investigation in Chile recoded that seaweeds have been used by humans for about 14,000 years. The use of seaweeds as food for their richness in protein, vitamin, trace minerals, and dietary fibre content has gained importance in many countries for exploitation of this natural renewable resource. Some food products directly made from seaweeds have a long tradition, and it was consumed during the fourth century in Japan and the sixth century in China (Tseng, 1987; Mc Hugh, 2003). The Japanese has the world's largest seaweed consumption per capita, with 10–15 % of their diet consisting of algae, and is also associated with a significantly lower rate of cancer, thyroid diseases, heart disease and dementia than western cultures. Marine algae are still considered an underexploited plant resource despite being used in diets and traditional remedies for centuries. They are also referred as treasure house of novel healthy food ingredients and biologically active compounds, due to their phenomenal biodiversity. In addition, they are a rich source of dietary fibre (DF), with a content ranging from 33 to 50 g/100 g dry weight placing them as an important candidate in the development of new functional foods.

The studies on nutritional potential of marine algae showed that they contain great nutritional potential and it can, therefore, be used as alternative dietary sources. The nutritional composition can vary with season, environment, and geographical regions. Moisture content of fresh seaweed is high and can account up to 94 % of the algal biomass. That of ash content is up to 55 % dry weight, macrominerals (Na, K, Ca, Mg) and trace elements (Fe, Zn, Mn, Cu, etc.), and it is much higher in algae than in terrestrial plants. The minerals are attributed to various ions associated with charged polysaccharides. The bioavailability of minerals depends on the type of linkage between the polysaccharide and the mineral, and also on polysaccharide digestibility. The halogenated compounds such as chloride are naturally produced by red and brown seaweeds. Generally the protein content is high and it also depends on the type of seaweeds used. The green seaweeds' protein content ranges from 8 to 26 %, in brown seaweeds it ranges from 5 to 26 % and in red seaweeds from 7 to 47 % algal dry weight.

Most of the seaweed contain all essential amino acids and they are also rich source of the aspartic and glutamic acid. Amino acid score of a seaweed protein ranges from 60 to 100 and this is higher than that of the proteins in cereal and vegetables. The protein content in marine algae is comparable to those found in high-protein vegetables such as soybeans. In the cultured seaweed *Saccharina latissima* the protein content was higher than the wild seaweeds. The fat content is generally lower than 2 % of dry weight and there are also some exceptions. But, this fat is high-quality oil, comprising mainly essential and polyunsaturated fatty acids (including omega-3 and omega-6). Marine algae are an excellent source of dietary fibre with a high ratio of soluble to total dietary fibre fraction. The dietary fibre in seaweeds is mainly formed from nondigestible polysaccharides which are resistant to human digestive enzymes. Majority of dietary fibre from edible seaweeds comprises soluble anionic polysaccharides that are scarcely degraded or not fermented at all by the human colonic microbiota.

9.3.6 Cosmeceuticals

The seaweed extracts are often found on the ingredients of cosmetic packages, particularly in face, hand and body creams or lotions. This usually refers to the use of phycocolloids alginate or carrageenan in the product. It is useful in various ways, including relief of rheumatic pain or the removal of cellulite. The marine algal compounds such as phlorotannins, sulphated polysaccharides and tyrosinase inhibitors are useful in cosmeceutical applications. The bioactive substances isolated from marine algae have diverse functional roles as a secondary metabolite, and these properties can be utilised in the development of cosmeceuticals. The fucoxanthin isolated from *Laminaria japonica* has reported to suppress tyrosinase activity in UVB-irradiated guinea pig and melanogenesis in UVB-irradiated mice. Fucoxanthin treated orally, significantly suppressed skin mRNA expression related to melanogenesis, suggesting that fucoxanthin negatively regulated melanogenesis factor at transcriptional level. Brown algal secondary metabolites phloroglucinol exhibits tyrosinase inhibitory activity because of their ability to chelate copper in this enzyme. The brown algae polyphenols, phlorotannins act as potential cancer chemopreventive agents against photocarcinogenesis and other adverse effects of UVB exposure. The majority of the phlorotannins and sulphated polysaccharides reported were from the members of the species *Ecklonia* and *Eisenia*.

9.3.7 Bioenergy

The ever rising need of energy is linked with rapid increases in industrialization and population. The depletion of fossil fuel has led to a search for alternate bioenergy sources. At present, throughout the world, energy crisis and environmental issues are the most important concern. Approximately 81 % of the energy supply is from fossil fuels, followed by 16 % renewable energy and 2.8 % nuclear energy. The utilization of fossil fuel as energy has disadvantages and affect the environment (Demirbas, 2010). Another issue is their uneven distribution in the world where 63 % of the global petroleum fuel resources are located in the Middle East with equity, environmental, economic and geopolitical implications. So, the biomass-based energy can serve as an alternative energy source to meet the present and future demand, including transportation fuel, although presently only about 0.6 % of transportation fuel is supplied as biofuels. Major sources of fossil fuels are petroleum, natural gas, coal, hydro and nuclear. Energy from the biological sources is one of the great alternatives to overcome the negative environmental effects like greenhouse gas emissions and substitute of fossil fuel. In addition, biodiesel is non-toxic and biodegradable with low pollutant. Generally, crops such as soya bean, jatropha, castor, coconut, animal fats and fish oils have been used to produce biodiesel but the price and supply of those sources are unstable. Marine algae are identified as one promising source of biofuels and they would be harvested and turned into a carbon neutral fuel source. So, the use of algae can be a suitable alternative as a potential source of the future renewable bioenergy.

In the past several years, as global petroleum supplies have diminished and serious environmental problems have arisen from greenhouse gas emissions, renewable energy sources have received much interest as a solution to the large reliance on fossil fuels and nuclear power. The renewable green energy includes wind, solar, tidal, biomass, and geothermal energies. In these sources, algae, in particular, are now considered one of the most important sources of biomass, because they can provide a significant contribution to liquid fuels. Marine algae are the source for the production of biofuels such as biodiesel, ethanol, and hydrogen gas (Table 9.2). The use of marine biosources as an alternative feed stock for bioethanol and biogas production could also reduce environmental problems in the sea because some sea pollutants could be utilized as bio-ethanol biomass. The algae contain carbon, hydrogen, oxygen, nitrogen and phosphorus. Typical ratios of these elements are reported as 106:263:110:16:1. They contain both carbohydrates and lipids. Carbohydrates can be converted to bioethanol through fermentation. Algal lipids can be converted to biodiesel through transesterification. The algae as biofuel feedstock provide economically and environmentally beneficial effects such as: they do not compete with food and water resources; they grow significantly faster than land crops used for biodiesel and they can treat wastewaters; they are carbon-neutral and can capture carbon dioxide; their low-temperature fuel properties and energy density of algae fuel make it suitable as jet fuel; they ensure a continuous supply; they can also provide valuable by-products like protein-rich feed for farm animals, organic fertilizer, and feedstock for producing biogas.

The application of microalgae as a source of fuel is not new, but it is now being taken seriously because of the increasing price of petroleum fuel, more significantly, the emerging concern about global warming that is associated with burning fossil fuels. Nowadays, the potential value of microalgal biodiesel is widely recognized. Bio-refinery of microalgae can be used to reduce the cost of making microalgal

Table 9.2 Biochemical composition of some marine algal biofuel feedstock (% w/w)

Algae	Carbohydrate	Protein	Lipid
Macroalgae			
Caulerpa lentillifera	44–46	11–12	1–2
Ulva lactuca	70	7.06	1.64
Eucheuma cottonii	35–36	10–12	1–2
Gracilaria cervicornis	63	19.7	0.427
Hypnea japonica	57.4	19	1.42
Sargassum vulgare	61	13.6	0.491
Laminaria hyperborea	50–52	8.9	<1
Ascophyllum nodosum	45–55	4.8–9.8	1–5
Microalgae			
Botryococcus brauni	2	40	33
Prymnesium parvum	25–33	28–45	22–38
Isochrysis sp.	15–5	29.5	23.4
Chlorella vulgaris	12–17	51–58	14–22
Porphyridium cruentum	40–57	28–39	9–14

biodiesel and microalgal-based carbon sequestration technologies cover the cost of carbon capture and sequestration. Generally biodiesel is similar to petroleum-derived diesel in its major characteristics such as cetane number, energy content, viscosity and phase changes and can be blended in any proportion with fossil. It is estimated that the use of biodiesel can decrease approximately by 90 % air toxicity and by 95 % cancers resulting from fossil diesel use. The marine microalgae appear to be one of the important sources to capture solar energy as they are sunlight-driven cell factories that convert carbon dioxide to potential biofuel. Some species contain much higher percentage of oil than conventional oil crops. They can multiply their biomass in few days, whereas higher plants take many months or years. Moreover, their chemical composition can be manipulated by altering the growth environment of the algal species.

The industrial CO_2 emission can be used as a source of carbon for algal growth. They can be cultivated in seawater or brackish water, raceway ponds on non-arable land and do not compete for resources with conventional agriculture. Their biomass can be harvested during all seasons. The seaweeds lack lignin but contain high amount of carbohydrates which makes them potentially suitable feedstock for the production of bioethanol. The seaweed phycocolloids can be converted into bioethanol, but direct use of these phycocolloids for bioethanol production will not be economically feasible. The industrial wastes of seaweeds derived from remaining pulp after extraction of the high value polysaccharides, still contain high amount of carbohydrates which may be used as a source of raw material for ethanol production. As the pulp contains high amount of carbohydrates and other organic materials, these can be converted into bioethanol through saccharification and fermentation. Bioethanol production from the seaweeds should be integrated with the higher value components. *Saccharomyces cerevisiae* and *Zymomonas mobilis* are the two most important microorganisms used for bioethanol production, but they have a very narrow substrate range. The *Pichia angophorae* is a more appropriate organism for ethanol production from seaweed. It can utilise both substrates such as mannitol as well as laminaran, simultaneously.

The principle selection of crops for bioenergy production is high growth rate, lower cost and biomass properties (moisture, ash, alkali and sugar contents). The green alga *U. lactuca* has been evaluated as a potential feedstock for energy production in USA since 1978 because of its high growth rate and high sugar contents. *U. lactuca* has been used to produce butanol (4 g L^{-1}) in the fermentation broth. Marine algae utilization as a substrate for biogas production is not well practised probably due to the availability of other important economic avenues. The seaweeds can be converted to biofuels by various processes including thermal treatment and fermentation, but the direct route to obtain energy from macroalgae is via its anaerobic digestion (AD) to biogas (~60 % methane). The biochemical conversion pathway is an anaerobic digestion of biomass by methanogenic bacteria, producing a mixture of gases which consist approximately two-thirds CH_4, one-third CO_2, water vapours and some impurities. Hydrogen is a future energy carrier and the rise in fossil fuel prices and their environmental destruction has drawn attention to renewable energy including hydrogen produced from renewable sources. Hydrogen is a clean and renewable energy source that does not produce CO_2 as a by-product, when used in fuel cells for electricity generation. It is evidenced from the recent research that the red algae *Gelidium amansii* and the brown algae *Laminaria japonica* are both potential biomass sources for biohydrogen production through anaerobic fermentation.

9.3.8 Pigments

There are different classes of pigments in marine algae generally occurring in bound and non-bound forms in the cells. Chlorophylls are greenish pigments which contain a porphyrin ring around which are electrons free to migrate. Because the electrons transfer freely, the porphyrin ring has the potential to gain or lose electrons easily, and thus the potential to provide energized electrons to other molecules. This is the fundamental process by which chlorophyll captures the energy of sunlight. On the other hand, carotenoids are usually red, orange, or yellow pigments, do not dissolve in water and must be attached to membranes within the cell. Carotenoids cannot transfer sunlight energy directly to the photosynthetic pathway, but pass their absorbed energy from one chlorophyll molecule to another. For this reason, they are called accessory pigments. The carotenoids are natural pigments derived from five-carbon isoprene units that are polymerized enzymatically to form regular highly conjugated 40-carbon structures (with up to 15 conjugated double bonds). Carotenoids are essential constituents of the photosynthetic organisms, and they are as accessory pigments for light-harvesting processes during photosynthesis, as structural stabilizers for protein assembly in photosystems, and as inhibitors of both photo and free radical oxidation provoked by excess light exposure. In human nutritions some carotenoids offer provitamin A activity and they directly provide photoprotection against UV light photooxidation in the skin. The ketocarotenoid astaxanthin play a key role in the prevention of several human pathological processes, such as skin UV-mediated photooxidation, inflammation, prostate and mammary carcinogenesis, ulcers due to *Helicobacter pylori* infection and age-related diseases. The results of several investigations have confirmed that these compounds can play important roles in prevention (and even treatment) of human diseases and health conditions, e.g., cancer, cardiovascular problems, atherosclerosis, rheumatoid arthritis, muscular dystrophy, cataracts and some neurological disorders. Carotenoids are effective antioxidants, they can quench singlet oxygen ($^1O_2$), suppress lipidperoxidation and prevent oxidative damage. The consumption of carotenoid-rich vegetables and fruits could protect humans against cardiovascular disease, certain cancers and other degenerative diseases as evidenced from several investigations (Hu et al., 2008). The average concentration of carotenoids in most of the algae is only 0.1–2 %, but *Dunaliella* when grown under the right conditions of high salinity and light intensity will produce up to 14 % beta-carotene.

9.3.9 Biofertilizer

Marine algae are used as manure in farming practice by coastal people. As the wet seaweeds are heavy in most part they are used only in farms near to the coastal areas. In some part of the world, seaweeds are mixed with sands and applied in the farms. In Brittany, France, farmers regularly collect algae from over a few hundred kilometres and use in inlands up to a kilometre. Marine algal species *Ascophyllum*, *Ecklonia* and *Fucus* are commonly utilized as fertilizer. They contain comparable amount of nitrogen and potassium like animal manure and organic fertilizers, but the phosphorus content is low. The carbohydrates content in the brown seaweeds are high. So, they are potent candidates in improving the soil quality in various ways like increasing the aeration, retention of moisture etc. “Afrikelp” is commercially available dried seaweed (*E. maxima*) collected from Africa and Namibia.

In the controlling of top soil loss, the brown seaweed *Ascophyllum* is used. They form strong gels when added with calcium, because of the alginate content. Application of seaweed fertilizers showed many favourable results such as increasing health and yield. They also increase the biochemical potential of plants. The seaweed extract contains micro-nutrients, auxins, cytokinins, betaines and other growth promoting constituents. These hormones play an important role in enhancement of cell size and cell division, and together they complement each other as cytokinins are effective in shoot generation and auxins in root development and micro-nutrients improve soil health.

The application of seaweed extracts to the soil, stimulated the plant growth (Blunden, 1971). The yield and quality of the fruits of *Zizyphus mauritiana* increased upon the application of *Sargassum wightii* extracts (Rama Rao, 1991). Seaweed liquid fertilizers improve the moisture holding capacity and also make favourable condition for growth of soil microorganisms. It enhances the soil health by increasing the micronutrients and microbial diversity. The seaweed *Ascophyllum nodosum* extract (0.2 %) sprayed in carrot plants showed significant defence against fungi diseases caused by *Alternaria radicina* and *Botrytis cinerea*. Extract of this seaweed enhances disease resistance in carrot, likely through induction of defence related enzymes, genes or proteins (Jayaraj et al., 2008). The application of seaweed extract as and/or seed treatment exhibited encouraging results on enhancement of vegetative growth and yield of several crops. The efficiency of SLF on seed germination varied between red, brown and green algae, may be because of the variability in the chemical nature between these species. The maximum germination was induced in root and shoot growth in Ragi by liquid fertilizers of *Ulva lactuca, Sargassum wightii* and *Gelidella acerosa*. The extract of *Enteromorpha intestinalis* enhanced the seed germination, root, shoot length and chlorophyll content of *Sesamum indicum*.

The extract of *Gracilaria edulis* (1 %) applied in the soil maximizes the germination, growth and development of *Zea mays* and *Phaseolus mungo. Hydroclathrtus* extracts showed considerable increase in the growth parameters of *Sorghum*. The foliar spray of *Gracilaria verucosa* and *Chaetomorpha linum* exhibited good results on vegetative growth of black gram, brinjal and tomato. The SLF of *Sargassum wightii* increases the height and number of branches in *Arachis hypogaea* when comparing to the chemical treatments. *Sargassum johnstonii* extract enhances the vegetative growth of *Lycopersicon esculentum*. The application of seaweed fertilizers with traditional fertilizers also yielded positive results on plant growth enhancement. The seaweed extracts not only enhance the plant vegetative growth but it also stimulates the early flowering and fruiting in crops. The biochemical properties of the plants also increased in the SLF treated plants. The pigment concentrations, protein, total soluble sugar, reducing sugar, starch, phenols, lycope, free aminoacids and vitamin C content were increased in several plants treated with seaweed liquid extracts. Seaweed fertilizers increase the fertility of soil by supplying various trace elements, by improving water holding capacity, by changing the soil structure. But, the chemical fertilizers have adverse effects such as increasing the acidic properties of the soil.

9.3.10 Bioplastics

The plastics are organic polymers and most of the plastics are derived from petrochemicals. The major challenge with plastics is their degradability and will take many centuries. It is one of the pollution causing agents in the earth. Hence, it is necessary to finding the new types of plastics with easily degradable and environmental friendly nature. The organic renewable biomass can be utilized for the production of plastics and these eco-friendly plastics are called as bioplastics. The advantages of the bioplastics are numerable and most important property is their quick degradability. Bioplastics utilization will help to reduce the massive emission of CO_2 from the fossil fuel and also preserve the fossil resource. Recently, the interest of feedstock for bioplastic is marine algae as an alternative raw material. The important factors include their rapid growing capabilities up to 20 % per day, easy cultivation and vast diversity. Development of raw material from marine algae can ensure a sustainable raw material for producing plastics. The moisture barrier content of plastic film derived from red algae agar displayed is better than cassava starch-based plastics and their mechanical properties were also similar to some low-density polyethylene plastics. The algae based bioplastic production is in infancy stage and rapid research is needed to commercialize and expand their utilizations.

9.4 Conclusion

The greatest diversity of marine algae provides the diverse amount of industrially important chemicals. Marine algal phycocolloids such as agar, alginates and carrageenan are important industrial products with several applications such as nutraceutical, pharmaceutical and biotechnological industries. In the process of novel drug development from the nature for treating various diseases, marine algal compounds are a significant source. Marine algae based functional food contains various essential nutrients and provide health benefits for human. Algal liquid fertilizers are having significant improvements in the plant growth, yield and soil fertility. Application of marine algae as the raw feedstock in biofuel and bioplastic productions has an immense opportunity to overcome the environmental pollution and produce enough bioenergy. Even though, several industrially important chemicals are obtained from the marine algae it is necessary to bring advancement in the production and its applications.

References

Abbott, I. (1996). Ethnobotany of seaweeds: Clues to uses of seaweeds. *Hydrobiologia*, 326/327, 15–20.

Armisen, R. (1995). World-wide use and importance of Gracilaria. *Journal of Applied Phycology*, 7, 231–243.

Bixler, H.J. and Porse, H. (2011). A decade of change in the seaweed hydrocolloids industry. *Journal of Applied Phycology*, 23, 321–335.

Blunden, G. (1971). The effect of aqueous seaweed extract as fertilizer additives. Proc. VII. Int. Seaw. Symp. Tokyo, pp. 584–589.

Demirbas, M.F. (2010). Microalgae as a feedstock for biodiesel. *Energy Education Science & Technology A*, 25, 31–43.

Draget, K.I., Smidsrød, O. and Skjåk-Bræk, G. (2002). Alginates from Algae. *In:* De Baets S, Vandamme, E.J., Steinbuchel, A. (eds) Biopolymers, vol 6. Polysaccharides II. Wiley, Weinheim, Germany, pp. 215–244.

FAO (2013). Fisheries and Aquaculture Information and Statistics Service – 18/12/2013.

FAO (2014). Fisheries and Aquaculture Information and Statistics Service- 16/03/2014.

Hu, C.C., Lin, J.T., Lu, F.J., Chou, F.P. and Yang, D.J. (2008). Determination of carotenoids in *Dunaliella salina* cultivated in Taiwan and antioxidant capacity of the algal carotenoid extract. *Food Chemistry*, 109, 439–446.

Jayaraj, J., Wan, A., Rahman, M., & Punja, Z. K. (2008). Seaweed extract reduces foliar fungal diseases on carrot. *Crop Protection, 27*(10), 1360–1366.

Lahaye, M. and Ray, B. (1996). Cell-wall polysaccharides from the marine green alga Ulva rigida (Ulvales, Chlorophyta)-NMR analysis of ulvan oligosaccharides. *Carbohydrate Research*, 283, 161–173.

Mc Hugh, D.J. (2003). A guide to the seaweed industry. FAO Fisheries Technical Papers T441.

Rama Rao, K. (1991). Effect of seaweed extract on *Zizyphus mauratiana* Lamk. *Journal of the Indian Botanical Society*, 71, 19–21.

Rinaudo, M. (2007). Seaweed polysaccharides. *In:* Kalmerling, J.P. (ed.), Comprehensive glycoscience from chemistry to systems biology, vol 2. Elsevier, London, pp. 691–735.

Tseng, C.K. (1987). *Laminaria* mariculture in China. *In:* Doty, M.S., Caddy, J.F., Santelices B. (eds), Case Studies of seven Commercial Seaweed Resources. FAO Fisheries Technical Paper 281. Food and Agriculture Organisation of the United Nations, Rome, Italy, pp. 239–263.

Chapter 10
Recent Developments on Algae as a Nutritional Supplement

Hari Niwas Mishra, Anupriya Mazumder, and P. Prabhuthas

10.1 Introduction

The increasing population and life threatening diseases are major growing concern of today's world. To meet the nutritional demand of the rising population, researchers are looking for alternative sources for food which are easy to cultivate, cost effective and produce large amount of bioactive compounds useful to prevent major diseases. Cultivation of algae is advantageous over other plant crops. Algae cultivation requires less water and terrestrial land. Furthermore, algae can be cultured using brackish or marine water resources. Algae are the source of many essential nutrients. They are a diverse group of autotrophic organisms ranging from unicellular to multicellular forms. Some of the algae are recognized as balance foods which offer sufficient quantity of proteins, carbohydrates, vitamins, and minerals for normal functioning of human body. Blue green algae, red algae and green algae are reported to have higher contents of dietary fibre.

The marine forms of algae are commonly called as seaweeds. They have been used as food and medicine for many centuries. The algal extracts are used as supplements in many food, dairy, and pharmaceutical industries. Algae are used as one of important medicinal sources due to its antioxidant, anticancer and antimicrobial properties.

In last few decades functional foods have emerged rapidly due to high market demands and promising health benefits. The new trend of research is inclined in development of algal fortified functional food items consumed popularly like pasta (Fradique et al., 2010), salad dressing (Gouveia et al., 2006), mayonnaises and

H.N. Mishra (✉) • A. Mazumder • P. Prabhuthas
Food Chemistry and Technology Laboratory, Agricultural and Food Engineering Department, Indian Institute of Technology Kharagpur, Kharagpur 721302, India
e-mail: hnm@agfe.iitkgp.ernet.in

D. Das (ed.), *Algal Biorefinery: An Integrated Approach*,
DOI 10.1007/978-3-319-22813-6_10

gelled desserts (Batista et al., 2012). Incorporation of algae to food products serves various purposes like natural food colour and also as a source of high value bioactive compounds. Different species of microalgae are used for natural compounds for food industry. Apart from food, algae have various other applications in feed, pharmaceutical and production of biofuel. The bioactive compounds present in algae include carotenoids, sulphated polysaccharides, essential fatty acids, polypeptides, amino acids, vitamins and minerals (Gouveia et al., 2006). Microalgal pigments are an excellent source of natural food colours which can replace synthetic food colours used in industry. European Commission has recently committed to replace 46 synthetic food colours as these are allergic and showing hyperactivity in children (McCann et al., 2007).

10.2 Major Nutritional Components of Algae

10.2.1 Lipids and Fatty Acids

Glycolipids are major lipids present in algal cell. The other lipids present in lesser quantities are neutral and phospholipids. Lipids present in microalgae are in monogalactosyl, diacylgylerols, digalactosyl and phosphatidyl glycorel form and exhibit anti-inflammatory and anti-thrombotic activity (Antonopoulou et al., 2005). Microalgae are rich sources of long chain fatty acids like linoleic (C18:2), α-linolenic acids (C18:3) and palmitic acid (C16:0) than vegetable oils like soya or sunflower oil (Draaisma et al., 2013). The most interesting fact of marine algae is that it contains large amount of ω-3 polyunsaturated fatty acids (PUFAs) of eicosapentaenoic acids (EPA, 20:5) and docossahexaenoic acid (DHA, 22:6). These functional fatty acids are traditionally extracted from marine fishes mainly from Salmon, Cod and Tuna. Marine algal oil rich in PUFA are an excellent alternative source of nutrition for vegetarian population. An EPA content of 39 % of total fatty acids is documented in *Phaeodactylum tricornutun* and *Nannochlorpsis sp* (Adarme-Vega et al., 2012). Docosahexaenoic acid (DHA) extracted from *Crypthecodinium cohnii* or *Schizochytrium* sp. for formulation of infant food are marketed by Martek Biosciences (Kyle et al., 1995).

The fatty acid profiling of *Spirulina platensis* reveals that it is a rich and active source of lauric, palmitoleic and oleic acids. The super critical extract of same algae does not possess a significant amount of γ-linoleic acid (GLA) in the fatty acid extracts (Mendiola et al., 2007). The marine seaweeds are also a very good source of ω-3 and ω-6 fatty acids. According to the studies conducted by MacArtain et al., *Palmaria* and *Sargassum* species have well balanced ω-3/ω-6 ratio along with highest lipid content of 3.8 % and 3.9 % respectively. Another polyunsaturated fatty acid known as oleic acid are present abundantly in *Gracilaria* sp., *Ulva* sp. and *Fucus* sp. (Ortiz et al., 2006).

10.2.2 Sugars and Polysaccharides

Algal cell wall has rigid structure due to accumulation of various sugars and polysaccharides. Algal cells are highly efficient for photo conversion and can accumulate upto 50 % carbohydrate in dry weight basis. Polysaccharides deposited in cell walls act as a shield to sustain in tides and ocean currents. Different algal species synthesize different sugars and polysaccharides as a storing material. In cyanobacteria, glycogens are produced whereas in green algae amylopectins are present in the cell wall.

Seaweeds contain sulphated polysaccharides, fucoidan, polyelectrolytes like alginates, carrageenans, agarans etc. The species rich in polysaccharides mainly belong to brown algae or pheophycea family and are prevalent in coastal area. *Ascophyllum*, *Porphyra* and *Palmaria* are few brown algal species which contain very high amount of hetero polysaccharides. The green seaweed *Ulva* also contains huge amount of hetero polysaccharides (65 % on dry weight basis). The nutrient availability and stress factors like light, pH and salinity around the residing environment of macro algae mostly determine the sugar content present in the algal cell wall. Macro algae contain large amount of fucose, mannose, galactose, glucose and uronic acid. *Sargassum vulgare* contains upto 70 % carbohydrates in dry weight basis. The polysaccharides present in *Sargassum* sp. exhibits potential antiviral action as it contains alginic acid, xylofucans and fucans.

Seaweeds are rich sources of dietary fibres. According to literature the dietary content in edible seaweed varies from 33 % to 62 % of the total fibres on dry weight basis. There are two varieties of dietary fibres present in seaweeds namely soluble and insoluble dietary fibres. The soluble dietary fibres include laminarin, porphyran, furonan and alginic acids. The insoluble dietary fibre include cellulose, hemicelluloses, mannans and xylans. Highest amount of total dietary fibres are present in Undaria (58 %) followed by Fucus (50 %), Porphyra (30 %) and Saccharina (29 %).

Another interesting species of microalgae is *Phorphiridium* sp. It is encapsulated with sulphated polysaccharides and composed of 10 different sugars. The prominent sugars present in these algae are xylose, glucose and galactose. Mannose, methylated galactose and pentoses are present in small quantities. The glucoronic acid and sulphur ester groups present in the cell of *Phohiridium* species negatively charge the cell wall.

10.2.3 Proteins and Essential Amino Acids

Microalgae are richest source of single cell proteins. *Spirulina* sp. which is the most investigated microalgae till date contains protein up to 70 % in dry weight basis (González-Benito et al., 2009). The amino acid present in *Spirulina* species also has high percentage of protein digestibility (Habib, 2008). The other high protein containing species are *Chlorella vulgaris* (55 %) and *Dunaliella* sp. (57 %). Samarakoon and Jeon (2012) did a comparative analysis of defatted soyabean flour and microalgal powder of *Nannochloropsis* spp., *Porphyridium cruentum* and *Phaeodactylum tricornutun* and found that the former two species are rich in hydrophobic and hydrophilic amino acid than soya flour.

Phycobiliproteins are the accessory pigment present in microalgal strains in enormous quantity. These proteins are derived during photosynthesis. *Spirulina platensis* contains a significant amount of phycobiliproteins in their cell matrix. The other pigments attached with phycobiliproteins are phycocyanin giving blue green colour to cyanobacteria: phycoerithrin and allophycocyanin (Viskari and Colyer, 2003). The phycobilin proteins have important pharmacological benefits to human health. It acts as therapeutic medicine to various cancer cells specially leukaemia treatment (Subhashini et al., 2004). The purified form of phycobilin proteins are used in cosmetic industry, natural food colour and as fluorescent labels in analytical chemistry (Kronick, 1986).

10.2.4 Vitamins

Both micro and macro algae accumulates essential vitamins and minerals due to their autotrophic nature. The algal biomass is richer in depositing water soluble vitamins (B_1, B_2, B_6, B_{12}, C) than Baker yeast and meat. Microalgae accumulates vitamin in larger quantities as compared to soyabeans and cereals. The rhodophycean and chlorophycean group of algae do not synthesize vitamin B_{12} but can derive it from other bacteria grown together with the algae (Becker, 2004). The brown seaweed species of *Ascophyllum*, *Fucaceae*, *Fucus* sp. contain large amount of α, β and γ tocopherol or vitamin E. Total 200–600 mg kg^{-1} of vitamin E is available in the above mentioned brown seaweed species in dry weight basis. The green algae contains only α-tocopherol. Vitamin E has high antioxidative capacity and prevents oxidation of algal lipids (Jayashree & Kamat, 1985).

10.3 Functional Properties of Algae

10.3.1 Antioxidant Activity

Antioxidant activities have been recognized in various forms of algae, especially from the marine origin like red algae, green algae and brown algae. Various forms of anti-oxidant assays, like 1,1-diphenyl-2-picrylhydrazyl (DPPH), superoxide anion (O_2), hydroxyl radical ($OH^{\cdot}$) and peroxyl radical (ROO•), were conducted by many researchers on different algal species to prove its antioxidant potential. Radical scavenging activity using DPPH have revealed antioxidant effects of a methanol extract of the blue-green algae *Anabaena* species. Ascorbate, iron and

H_2O_2 assays have also revealed antioxidant effects of phycobiliprotein phycocyanin in *Spirulina platensis* extract. The potent antioxidant properties of *Ulva lactuca* is associated mainly with its flavonoids contents. Radical scavenging activity of the green algae *Ulva fasciata* Delile was found to be associated mainly with its sesquiterpenoids. Animal studies of hot water extract of *Ulva reticulate* found it to reduce hepatic oxidative stress (Rao et al., 2004).

Sometimes, the antioxidant properties of algal extract are showing more specific reaction. For example, studies on methanol extracts of *Fucus vesiculosus* and *F. serratus* have shown to protect DNA damage of human epithelial colorectal adenocarcinoma cells (Caco-2) induced by H_2O_2. However, the same extract fails to protect DNA damage induced by tert-butyl hydroperoxide. A similar study conducted by O'Sullivan et al. (2011) using methanolic extracts of *Pelvetia canaliculata* reported that the extract have inhibited H_2O_2 induced superoxide dismutase depletion in Caco-2 cells. Solvent extracts from red algal species also showed antioxidant properties. Ethanol extracts of *Callophyllis japonica* suppressed H_2O_2 induced cellular apoptosis (Kang et al., 2005). Phlorotannins compounds from brown algae like *Eisenia bicyclis*, *Ecklonia cava* and *Ecklonia kurome* showed DPPH radical scavenging activities (Shibata et al., 2008).

10.3.2 Anti-Cancer Activity

Algal species are possessing variety of intra cellular compounds which showed anticancer properties. Various studies on algae showed that the polysaccharides are the major components that exhibit anti-cancer properties. *Spirulina platensis* extracts showed chemo preventive effect against carcinogenesis induced by dibutyl nitrosamine. The extract from the *Spirulina maxima* suppressed the expression of Bcl2 in A549 cells and inhibited viability of other human cancer cells. de Jesus Raposo et al. (2013) stated that polysaccharides from red microalgal species showed anticancer properties in two human tumour cell lines MCF-7 and HeLa. Cancer preventive action of the polysaccharides from the microalga *Porphyridium cruentum* was also reported by Gardeva et al. (2009). Sulphated polysaccharides from tropical green algae showed promising anticancer activities. Up to 58 % inhibition of HeLa cell proliferation was shown by sulphated polysaccharides isolated from *Caulerpa prolifera*. Two polysaccharide fractions obtained from the green alga *Caulerpa racemosa* showed anticancer properties against H22 tumour transplanted in mice. At a dose of 100 mg kg^{-1} d^{-1} these two fractions showed anti-tumour activity up to 84 % and 54 % in 48 h and 14 h, respectively (Ji et al., 2008).

Among marine algae, brown and red algae have been studied more as the sources of polysaccharides with anti-cancer and anti-tumour properties than green algae. The anti-cancer and anti-tumour properties of these marine algae are mainly associated with a specific group of polysaccharides called ulvans. Ulvans are water soluble sulphated polysaccharides normally found in the cell walls of marine green algae. These ulvans are characteristic of the plants, belonging to the genera *Ulva*,

Enteromorpha, Monostroma, Caulerpa etc. Water-soluble sulphated polysaccharide fractions of *Enteromorpha prolifera* showed *in vivo* and *in vitro* stimulation of immunity by significantly increasing ConA-induced splenocyte proliferation and induced the production of various cytokines via up-regulated m-RNA expression. Double the response was found with ulvan obtained from *Ulva rigida*. Increase in the expression of cytokines stimulated the activity of macrophages as well as induced an increase in COX-2 and NOS-2 expressions (Leiro et al., 2007). Ulvans from *Ulva pertusa* showed significant enhancement of immunity and cytokine production. They had little cytotoxicity against tumour cells. There are also several studies conducted on the antioxidant activities of ulvans in experimental rats against hepatitis.

In case of brown algae, extensive research on its crude extracts against different cancer cell lines showed promising anticancer potential. The *in vivo* studies on antitumour activity of brown algae showed the importance of these seaweeds for cancer therapy. The enzymatic extract as well as crude polyphenolic and polysaccharide fractions of *Ecklonia cava* showed antiproliferative activity against murine colon cancer cell line (CT-26), human leukemia (THP-1), mouse melanoma (B-16), and human leukemia (U-937) cells. The nuclear staining experiment showed that activity against CT-26 cells was due to apoptotic cell demise. The remarkable anticancer and antitumour properties of various brown algae are due to the presence of sulphated polysaccharide of fucoidan and carotenoid of fucoxanthin. They were found to be the most important active metabolites of brown algae as potential chemotherapeutic or chemopreventive agents. Some of the studies on brown algae that showed anti-cancer properties are listed in Table 10.1. In addition to fucoidan and fucoxanthin, alginic acid in brown algae also showed some cancer preventive properties. The alginic acid is an anionic polysaccharide distributed widely in the cell walls of brown algae. They have the ability to neutralize toxins and heavy metals by binding them in intestines and convert into a chemical which is less detrimental to human. Apart from brown algae, red algae also showed anti-cancer properties in research studies. Several well known polysaccharides from red algae have wide application in microbiology, biotechnology and food technology due to their gelling properties. Carrageenans are the best examples. Carrageenans indirectly showed anti-cancer properties by increasing antioxidant, antiviral and antitumour immunity.

10.3.3 Anti-Microbial Activity

Most of the studies on algal species showed that they are the actual producers of many biologically active secondary metabolites which are having antimicrobial effect. Some of this secondary metabolite includes polysaccharides, quinines, alkaloids, cyclic peptide, phlorotannins, diterpenoids and sterols. Several researchers made efforts to bring out these bioactive substances from algae using different methods. Screening of different classes of algae for their antibiotic value is recorded in many literatures.

Table 10.1 Some selected studies on brown algae that showed anti-cancer properties

Cell-lines showing positive response	Algal species used	Sample used for studies	Reference
Ehrlich carcinoma	*Sargassum ringgoldianum,*	Powdered tissue	Shevchenko et al. (2007)
	Lonicera japonica,		
	Lessonia nigrescens,		
	Scytosiphon lomentaria		
Mammary tumour cells	*Undaria pinnatifida*	Aqueous extract	Thinh et al. (2013)
Breast cancer,	*Undaria pinnatifida*	Aqueous extract	Synytsya et al. (2010)
Mammary carcinogenesis			
Human breast adenocarcinoma,	*Padina pavonica,*	Methanolic extract	Li et al. (2006)
Human prostate cancer cells	*Cystoseira mediterranea*		
Human leukaemic T cell lymphoblast,	*Sargassaceae sp.,*	Aqueous extract	Chizhov et al. (1999)
Human Burkitt's lymphoma,	*Dictyota dichotoma,*		
Human chronic myelogenous leukaemia	*Desmarestia ligulata*		

Of all classes of algae, seaweeds take up an important place as a source of antimicrobial compounds. Apart from antimicrobial properties, the biomolecules from seaweeds are also reported to have anticoagulant and antifouling properties (Marechal et al., 2004). Some of the antimicrobial compounds derived from algae are acrylin acid, chlorellin, aliphatic compounds, phenolic inhibitors, sesquiterpenes, diterpernoids, etc.

The aqueous extract of selected macroalgae (*Ulva fasciata, Gracilaria corticata, Sargassum wightii* and *Padina tetrastromatica*) from south-west coast of India showed significantly higher anti-bacterial activity. Gram negative bacterial species are highly sensitive than Gram positive bacterial species. The maximum zone of inhibition was noted with *G. corticata* against *Proteus mirabilis* and *P. tetrastromatica* against *Staphylococcus aureus* and *Vibrio harveyi*. The green algal extract of *Cladophora fascicularis* when tested against three pathogens viz. *Staphylococcus aureus*, *Escherichia coli* and *Bacillus subtilis* showed significant inhibitory effect which was due to an ether containing compound present in the extract. Dubber and Harder (2007) in their studies showed that methanolic and hexane extract from three different macroalgal species (*Mastocarpus stellatus, Laminaria digitata* and *Ceramium rubrum*) have anti-microbial activity against 12 fish pathogens.

Ulva lactuca, commonly called as Sea lettuce, was reported to have antimicrobial activity. These green algae have long been used as traditional food to treat helminthic infections and urinary diseases. Its antimicrobial activity is due to the presence of acrylic acid. Apart from *Ulva lactuca*, three different seaweeds which include *Padina gymnospora, Sargassum wightii* and *Gracilaria edulis* were also shown to have severe antimicrobial activity against common human bacterial pathogen. Methanolic extracts of 32 different macroalgae isolated from Atlantic and Mediterranean coastal area of Morocco were found to have antibacterial property against *S. aureus*, *E. coli*, *Enterococcus faecalis* and *Klebsiella pneumonia* (Ibtissam et al., 2009). Aqueous extract of *Dictyota dichotoma* and *Padina gymnosora* from Indian waters were effectual against *Bacillus megatherium* and *S. aureus*. Green algal extract of *Caulepra prolifera* showed significant anti-microbial activity against common marine bacterial strains. *Zandariania prototypus*, *Cystoseria sricata* and *Cymbula compressa* when extracted with ethanol showed inhibitory effect against bacteria and fungi. Similarly, El-Naggar (1987) reported that extracts of marine algae, *Dictyota dichotoma*, *Dilophus fasciola* and *Cystoseria barbata* isolated from Egyptian water were found to have significant antibacterial activities against common human pathogens.

In vitro studies on selected human upper respiratory tract pathogens treated with extracts of seaweeds showed significant positive effect. Various seaweed species belonging to the family *Chlorophyceae, Phaeophyceae* and *Rhodophyceae* that were isolated from coastal region of South Africa showed strong antibacterial activity. The level of antimicrobial activity of various marine algal extracts varies with the season during which it was isolated for studies at the same location. Another group of compounds that are more frequently reported for its anti-microbial properties is bromophenol compounds. They are found widely in many marine species of red and brown algae, and especially rich in the red algae of family *Rhodomaceae* (Oh et al., 2008). These bromophenol compounds are exhibiting a wide spectrum of pharmacological properties which includes enzyme inhibition, cytotoxic, antioxidant, antinflammatory and antimicrobial activities.

Al-Saif et al. (2013) studied antimicrobial activity of various solvent extracts of five different algal species (*Caulerpa occidentalis*, *Cladophora socialis*, *Dictyota ciliolata, Gracilaria dendroides* and *Ulva reticulata*) isolated from Red Sea coastal area in Saudi Arabia. The extracts obtained from the red alga *G. dendroides* showed better action against the tested common pathogens of human (*Escherichia coli, Enterococcus faecalis, Pseudomonas aeruginosa* and *Stapylococcus aureus*) followed by the green alga *U. reticulata*, and brown algae *D. ciliolata*. The reason for the better activity of *G. dendroides* was revealed by analyzing its chemical composition, where it showed highest percentages of fats, proteins and other flavonoid groups. Out of the four different extracts obtained from four different solvents (chloroform, ethanol, petroleum ether and water), chloroform extract showed significant antibacterial activity followed by ethanolic extract.

10.3.4 Anti-Inflammatory Activity

Oxidative stress plays a major role in the development of cancer, endothelial dysfunction, lung diseases, gastrointestinal disorder and atherosclerosis. All these diseases involve inflammatory reactions. Many naturally occurring food materials are

Table 10.2 Bioactive compounds from various algal sources (anti-inflammatory property)

Bioactive compounds	Organism	References
C-phycocyanin	*Spirulina platensis*	Shih et al. (2009)
Sulphated galactofucan	*Lobophora variegata*	Medeiros et al. (2008)
Alginic acid	*Sargassum wightii*	Sarithakumari et al. (2013)
Galactan	*Gelidium crinale*	Silva et al. (2010)
Methanol extract	*Neorhodomela aculeata*	Lim et al. (2006)
Fucans	*Lobophora variegate*	Siqueira et al. (2011)
Lectin	*Pterocladiella capillacea*	Silva et al. (2010)
Methanol extract	*Bryothamnion triquetrum*	Cavalcante-Silva et al. (2012)
Sulphated polysaccharide fraction	*Gracilaria caudate*	Chave et al. (2013)
Mucin-binding agglutinin	*Hypnea cervicornis*	Silva et al. (2010)

having variety of antioxidants that play a major role in anti-inflammatory activity. Algal species are the major sources of naturally occurring antioxidant, especially the marine algal species. Bowel inflammation are caused due to various reasons, one among them is increased formation of acetic acid, especially during fasting. Animal studies using rats showed that the acetic acid induced inflammation. This is effectively reduced by the crude extract of *Dunaliella bardawil*, a green algae rich in beta-carotene. Lycopene from *Chlorella marina* showed the anti-inflammatory activity against arthritis in rat model studies. Similarly, in a sheep model study conducted by Caroprese et al. (2012), extracts rich in phytosterols obtained from *Dunaliella tertiolecta* showed significant anti-inflammatory effects. The various bioactive compounds from different algal sources showing anti-inflammatory property are listed in Table 10.2.

10.4 Algal Nutraceuticals and Its Safety Issues

Algae are a diverse group of photo autotrophs that have ability to grow quickly by using light, CO_2, and produce more biomass per acre than plants. When referred to food supplement, microalgae plays an important role as a best source. Some of the important genus in microalgae that are more prevalent in food supplements are *Spirulina, Chlorella, Dunaliella, Nostoc, Botryococcus, Anabaena, Chlamydomonas, Scenedesmus* etc. Most of the microalgal groups are having almost balanced nutrient content which include vitamins, minerals, essential amino acids and fatty acids. Growing concern among consumers on proper nourishment along with increasing population, easy and cost-effective sources of food supplement are in demand. Due to plentiful production of beneficial compounds and easy for cultivation, the market for algal-based nutraceuticals are coming up with very profitable outcome. Since microalgae are commonly used as food and feed supplements, some of the microalgae of commercial importance are described in this chapter. Bishop and Zubeck (2012) report very well described the use of microalgae especially *Spirulina, Chlorella, Dunaliella, Haematococcus* and *Aphanizomenon*, as nutritional supplements and also explained about their challenges.

Spirulina is a rich source of vitamins, pigments, antioxidants and essential fatty acids along with numerous minerals. The major issue with consumption of *Spirulina* was its digestibility index and bioavailability of selected nutrients like vitamin B_{12}. This issue was also sorted out by using efficient post-harvest processing steps that improve the digestibility index and bioavailability of *Spirulina* biomass. Another issue is the identification of toxins called microcystin in the biomass. This microcystin was reported to cause liver cancer and other diseases. However, the traces of the toxin microcystin in biomass were due to invasion of other algae like *Microcystis* sp. in the *Spirulina* cultivation area. *Spirulina* as such was devoid of any toxin that is harmful for human consumption. Also *Spirulina* was reported to accumulate heavy metals in their biomass. But again this issue arises only in the species that was obtained from uncontrolled environment. Usually those species are employed for biofuel production and not for food purposes. In spite of the production of *Spirulina* sp. from the controlled environment there are some side effects viz. diarrhea, nausea and vomiting that were reported with the use of *Spirulina* products. The exact reason for that is still not clear.

Like *Spirulina*, the green algae *Chlorella* is also one of the nutrition-rich microalgae used as food supplement worldwide. It contains approximately 60 % protein along with other essential minerals and vitamins. The protein of *Chlorella* contains all essential amino acids required for the nutrition of heterotrophic organisms. Lutein and xanthophylls are also found abundantly in the *Chlorella* sp. *Chlorella* extracts were shown to have various properties like antitumor, antioxidant, anti-inflammatory, antimicrobial activities etc. It was also reported to control blood pressure, lower cholesterol levels, and enhance the immune system. Though benefits of *Chlorella* sp. are enormous, some investigations were also carried out to find out the possible side effects of *Chlorella* sp. Sometimes *Chlorella* products are labeled and sold in the market as low allergen which may be of clinical significance to certain types of people. Some consumers have reported few side effects after consumption of *Chlorella* biomass tablets and capsules which include low digestibility, nausea, vomiting, and other gastrointestinal related problems (Tiberg et al., 1995). This may be due to the improper processing of biomass before consumption. *Chlorella* tablets have been reported to cause acute tubule-interstitial nephritis often resulting in renal failure. Apart from some negative effects of *Chlorella* products, cultivation of *Chlorella* is also more challenging as well. Unlike *Spirulina*, *Chlorella* cultivation systems are expensive and often become contaminated by other algae.

Dunaliella is well known for its pigment beta-carotene, which accounts for up to 14 % of its dry weight. An ideal growth condition can yield up to 400 mg beta-carotene m^{-2} using *Dunaliella* sp. The other major pigment includes alpha-carotene, lutein and lycopene.

Dunaliella carotenoids are having potent antioxidant properties that reduce levels of lipid peroxidation and enzyme inactivation. Studies have shown beta-carotene of *Dunaliella* can prevent cancer of various organs like lungs, stomach, colon, rectum, breast, prostate and ovary by means of its antioxidant activity. Despite the advancement in the production of beta-carotene from natural sources like *Dunaliella*, more than 90 % of commercialized beta-carotene is produced synthetically which pose a challenge for extraction of beta-carotene from *Dunaliella* sp. However, natural beta-carotene has a higher bioavailability as compared to synthetically manufactured beta-carotene. Further, the activity and amount of the antioxidant enzymes were significantly greater in *Dunaliella* as compared to synthetic. In case of *Dunaliella* only little data are there to show its negative effects. Multigenerational studies with rats that consumes up to 10 % *Dunaliella* in their diets also showed no significant negative effects. Thus it was indicative on the safety of *Dunaliella* for the human consumption.

Haematococcus sp. is a green alga that has wide applications in nutraceuticals, pharmaceuticals, cosmetics and aquaculture industries. The major pigment produced by *Haematococcus* sp. is the red coloured astaxanthin, which makes *Haematococcus* cells to appear red. United States, India and Israel are said to produce approximately 300 tons dry weight of *Haematococcus*. Astaxanthin is sold in the market at $2500 per kilogram dry weight with an annual worldwide market estimated at $200 million. However, 95 % of this market is covered by synthetically derived astaxanthin (Lorenz and Cysewski, 2000). *Haematococcus* is cultivated commercially in large-scale outdoor systems and controlled photobioreactors, predominantly for astaxanthin. Some of the major players worldwide in the production of *Haematococcus* include Cynotech Corporation, Parry Nutraceuticals, BioReal, Inc., Alga Technologies, Fuji Health Science, Valensa International, and Aquasearch Inc. So far none of the reports showed its negative consequences upon ingestion (Marazzi et al., 2011). Animal studies also showed no adverse effects of consuming 5 to 18 g/kg. *Haematococcus* algal extract containing astaxanthin was also tested on humans for adverse clinical parameters which showed no evidence of adverse effects even upto 20 mg/day for four weeks (Satoh et al., 2009). Food and Drug Administration (FDA) has approved for the selling of *Haematococcus pluvialis* as a new dietary ingredient in the United States.

Aphanizomenon is a blue-green alga commonly found in freshwater systems. Approximately 500 tons of dried *Aphanizomenon* is estimated to be produced annually for use in food and pharmaceutical products. The major players in this field of Aphanizomenon production for food supplements are Cell Tech International Inc., Life Enthusiast Co-op, AquaSource and Klamath Valley Botanicals, Inc. *Aphanizomenon* sp. also showed no major side effects. Aphanizomenon flos-aquae, the dominant species used in production has not been shown to produce hepatotoxins (Carmichael et al., 2000).

10.5 Future Scope and Conclusion

Demand for nutritive food and health products increases parallel with the increase in human population. Algae are one of the best sources to meet nutritive demands due to their fast growth and better health benefits. The role of algae in human health and nutrition will increase with more research in this area. Currently algae are produced for food, aquaculture, colorants, cosmetics, pharmaceuticals and nutraceuticals purposes. But usage wise only small fraction is used for human food. Many algal species are still unexploited for its use as food and nutraceuticals purposes. So, there are more chances to enhance algal production and its usage. If the ability to use algae as functional food is developed by consumers on regular basis, undoubtedly there will be a continual interest on its value addition by researchers and manufacturers.

References

Adarme-Vega, T.C., Lim, D.K., Timmins, M., Vernen, F., Li, Y. and Schenk, P.M. (2012). Microalgal biofactories: A promising approach towards sustainable omega-3 fatty acid production. *Microb. Cell Fact*, 11, 96.

Al-Saif, S.S., Abdu-llah, Abdel-Raouf, N., El-Wazanani, H.A. and Aref, I.A. (2013). Antibacterial substances from marine algae isolated from Jeddah coast of Red Sea, Saudi Arabia, Saudi. *Journal of Biological Sciences*, 21, 57–64.

Antonopoulou, S., Nomikos, T., Oikonomou, A., Kyriacou, A., Andriotis, M., Fragopoulou, E. and Pantazidou, A. (2005). Characterization of bioactive glycolipids from *Scytonema julianum* (cyanobacteria*). Comparative Biochemistry and Physiology Part B: Biochemistry and Molecular Biology*, 140(2), 219–231.

Arad, S.M. and Levy-Ontman, O. (2010). Red microalgal cell-wall polysaccharides: Biotechnological aspects. *Current opinion in biotechnology*, 21(3), 358–364.

Batista, A.P., Nunes, M.C., Fradinho, P., Gouveia, L., Sousa, I., Raymundo, A. and Franco, J.M. (2012). Novel foods with microalgal ingredients–Effect of gel setting conditions on the linear viscoelasticity of *Spirulina* and *Haematococcus* gels. *Journal of Food Engineering*, 110(2), 182–189.

Becker, E.W. (2004). Microalage in human and animal nutrition. *In:* Handbook of microalgae culture. (Ed.) A. Ricjmond. Blackwell Publishing, Oxford, pp. 312–351.

Bishop, W.M. and Zubeck, H.M. (2012). Evaluation of Microalgae for use as Nutraceuticals and Nutritional Supplements. *Journal of Nutrition & Food Sciences*, 2, 5.

Carmichael, W.W., Drapeau, C. and Anderson, D.M. (2000). Harvesting of *Aphanizomenon flos-aquae* Ralfs ex Born. and Flah. var. flos-aquae (Cyanobacteria) from Klamath Lake for human dietary use. *Journal of Applied Phycology*, 12, 585–595.

Caroprese, M., Albenzio, M., Ciliberti, M.G., Francavilla, M. and Sevi, A. (2012). A mixture of phytosterols from Dunaliella tertiolecta affects proliferation of peripheral blood mononuclear cells and cytokine production in sheep. *Veterinary Immunology & Immunopathology*, 13(1-2), 27–35.

Cavalcante-Silva, L.H., da Matta, C.B., de Araujo, M.V., Barbosa-Filho, J.M., de Lira, D.P., de Oliveira Santos, B.V., de Miranda, G.E. and Alexandre-Moreira, M.S. (2012). Antinociceptive and anti-inflammatory activities of crude methanolic extract of red alga Bryothamnion triquetrum. *Marine Drugs*, 13(9), 1977–1992.

Chave, S., Nicolau, L.A., Silva, R.O., Barros, F.C., Freitas, A.L., Aragao, K.S., Ribeiro, A., Souza, M.H., Barbosa, A.L. and Medeiros, J.V. (2013). Anti-inflammatory and antinociceptive effects in mice of a sulfated polysaccharide fraction extracted from the marine red algae *Gracilaria caudata*. *Immunopharmacology and Immunotoxicology*, 13(1), 93–100.

Chizhov, A.O., Dell, A., Morris, H.R., Haslam, S.M., McDowell, R.A., Shashkov, A.S., Nifant, N.E., Khatuntseva, E.A. and Usov, A.I. (1999). A study of fucoidan from the brown seaweed *Chorda filum*. *Carbohydrate Research*, 320, 108–119.

de Jesus Raposo, M.F., de Morais, R.M. and de Morais, A.M. (2013). Bioactivity and applications of sulphated polysaccharides from marine microalgae. *Marine Drugs*, 11, 233–252.

Draaisma, R.B., Wijffels, R.H., Slegers, P.M., Brentner, L.B., Roy, A. and Barbosa, M.J. (2013). Food commodities from microalgae. *Current opinion in Biotechnology*, 24(2), 169–177.

Dubber, D. and Harder, T. (2007). Extracts of *Ceramium rubrum*, *Mastocarpus stellatus* and *Laminaria digitata* inhibit growth of marine and fish pathogenic bacteria at ecologically realistic concentrations. *Aquaculture*, 274, 196–200.

El-Naggar, M.E.E. (1987). Antimicrobial activity of some marine algae of Egypt. *Mansoura Science Bulletin*, 14 (2), 189–201.

Fradique, M., Batista, A.P., Nunes, M.C., Gouveia, L., Bandarra, N.M. and Raymundo, A. (2010). Incorporation of *Chlorella vulgaris* and *Spirulina maxima* biomass in pasta products. Part 1: Preparation and evaluation. *Journal of the Science of Food and Agriculture*, 90(10), 1656–1664.

Gardeva, E., Toshkova, R., Minkova, K. and Gigova, L. (2009). Cancer protective action of polysaccharide derived from microalga *Porphyridium cruentum* – biological background. *Biotechnology & Biotechnological Equipment*, 23, 783–787.

González-Benito, G., Barrocal, V., Bolado, S., Coca, M. and García-Cubero, M.T. (2009). Valorisation of by-products from food industry, for the production of single cell protein (SCP) using microalgae. *New Biotechnology*, 25, S262.

Gouveia, L., Raymundo, A., Batista, A.P., Sousa, I. and Empis, J. (2006). *Chlorella vulgaris* and *Haematococcus pluvialis* biomass as colouring and antioxidant in food emulsions. *European Food Research and Technology*, 222(3-4), 362–367.

Habib, M.A.B. (2008). Review on culture, production and use of Spirulina as food for humans and feeds for domestic animals and fish. FAO Fisheries and Aquaculture Circular. No. 1034.

Ibtissam, C., Hassane, R., Jose, M., Francisco, D., Antonio, G., Hassan, B. and Mohamed, K. (2009). Screening of antibacterial activity in marine green and brown macroalgae from the coast of Morocco. *African Journal of Biotechnology*, 8(7), 1258–1262.

Jayasree, V., Solimabi and Kamat, S.Y. (1985). Distribution of tocopherol (vitamin E) in marine-algae from Goa, west-coast of India. *Indian Journal of Marine Sciences*, 14(4), 228–229.

Ji, H., Shao, H., Zhang, C., Hong, P. and Xiong, H. (2008). Separation of the polysaccharides in *Caulerpa racemosa* and their chemical composition and antitumor activity. *Journal of Applied Polymer Science*, 110, 1435–1440.

Kang, K.A., Bu, H.D., Park, D.S., Go, G.M., Jee, Y., Shin, T. and Hyun, J.W. (2005). Antioxidant activity of ethanol extract of *Callophyllis japonica*. *Phytotherapy Research*, 13(6), 506–510.

Kronick, M.N. (1986). The use of phycobiliproteins as fluorescent labels in immunoassay. *Journal of Immunological Methods*, 92(1), 1–13.

Kyle, D.J., Reeb, S.E. and Sicotte, V.S. (1995). Production of Docosahexaenoic Acid from Dinoflagellates. US Patent #5, 407, 957.

Leiro, J.M., Castro, R., Arranz, J.A. and Lamas, J. (2007). Immunomodulating activities of acidic sulphated polysaccharides obtained from the seaweed *Ulva rigida* C. Agardh. *International Immunopharmacology*, 7, 879–888.

Li, B., Wei, X.J., Sun, J.L. and Xu, S.Y. (2006). Structural investigation of a fucoidan containing a fucose-free core from the brown seaweed *Hizikia fusiforme*. *Carbohydrate Research*, 341, 1135–1146.

Lim, C.S., Jin, D.Q., Sung, J.Y., Lee, J.H., Choi, H.G., Ha, I. and Han, J.S. (2006). Antioxidant and anti-inflammatory activities of the methanolic extract of *Neorhodomela aculeate* in hippocampal and microglial cells. *Biological & Pharmaceutical Bulletin*, 13(6), 1212–1216.

Lorenz, R.T. and Cysewski, G.R. (2000). Commercial potential for *Haematococcus* microalgae as a natural source of astaxanthin. *Trends in Biotechnology*, 18, 160–167.
MacArtain, P., Gill, C.I., Brooks, M., Campbell, R. and Rowland, I.R. (2007). Nutritional value of edible seaweeds. *Nutrition Reviews*, 65(12), 535–543.
Marazzi, G., Cacciotti, L., Pelliccia, F., Iaia, L. and Volterrani, M. (2011). Long-term effects of nutraceuticals (berberine, red yeast rice, policosanol) in elderly hypercholesterolemic patients. *Advances in Therapy*, 28, 1105–1113.
Marechal, J.P., Culoli, G., Hellioc, H., Thomas Guyon, M.E., Callow, A.S. and Ortaleo Magne, A. (2004). Seasonal variations in antifouling activity of crude extracts of the brown alga *Bifurcaria bifurcate* (Cystoseiraceae) against Cyprids of Balanus amphitrite and the marine bacteria *Cobertia marina* and *Pseudomonas haloplanktis. Journal of Experimental Marine Biology and Ecology*, 313, 47–62.
McCann, D., Barrett, A., Cooper, A., Crumpler, D., Dalen, L., Grimshaw, K. and Stevenson, J. (2007). Food additives and hyperactive behaviour in 3-year-old and 8/9-year-old children in the community: A randomised, double-blinded, placebo-controlled trial. *The Lancet*, 370(9598), 1560–1567.
Medeiros, V.P., Queiroz, K.C., Cardoso, M.L., Monteiro, G.R., Oliveira, F.W., Chavante, S.F., Guimaraes, L.A., Rocha, H.A. and Leite, E.L. (2008). Sulfated galactofucan from *Lobophora variegata*: anticoagulant and anti-inflammatory properties. *Biochemistry (Moscow)*, 13(9), 1018–1024.
Mendiola, J.A., Jaime, L., Santoyo, S., Reglero, G., Cifuentes, A., Ibanez, E. and Senorans, F.J. (2007). Screening of functional compounds in supercritical fluid extracts from *Spirulina platensis. Food chemistry*, 102(4), 1357–1367.
O'Sullivan, A.M., O'Callaghan, Y.C., O'Grady, M.N., Queguineur, B., Hanniffy, D., Troy, D.J., Kerrya, J.P. and O'Brien, N.M. (2011). In vitro and cellular antioxidant activities of seaweed extracts prepared from five brown seaweeds harvested in spring from the west coast of Ireland. *Food Chemistry*, 13, 1064–1070.
Oh, K.B., Lee, J.H., Chung, S.C., Shin, J., Shin, H.J., Kim, H.K. and Lee, H.S. (2008). Antimicrobial activities of the bromophenols from the red alga *Odonthalia corymbifera* and some synthetic derivatives. *Bioorganic & Medicinal Chemistry Letters*, 18, 104.
Ortiz, J., Romero, N., Robert, P., Araya, J., Lopez-Hernandez, J., Bozzo, C. and Rios, A. (2006). Dietary fiber, amino acid, fatty acid and tocopherol contents of the edible seaweeds *Ulva lactuca* and *Durvillaea antarctica. Food Chemistry*, 99(1), 98–104.
Rao, H.B.R., Sathivel, A. and Devaki, T. (2004). Antihepatotoxic nature of *Ulva reticulata* (Chlorophyceae) on acetaminophen-induced hepatoxicity in experimental rats. *Journal of Medicinal Food*, 13(4), 495–497.
Samarakoon, K. and Jeon, Y.J. (2012). Bio-functionalities of proteins derived from marine algae—A review. *Food Research International*, 48(2), 948–960.
Sarithakumari, C.H., Renju, G.L. and Kurup, G.M. (2013). Anti-inflammatory and antioxidant potential of alginic acid isolated from the marine algae, *Sargassum wightii* on adjuvant-induced arthritic rats. *Inflammopharmacology*, 21(3), 261–268.
Satoh, A., Tsuji, S., Okada, Y., Murakami, N. and Urami, M. (2009). Preliminary Clinical Evaluation of Toxicity and Efficacy of a New Astaxanthin-rich *Haematococcus pluvialis* Extract. *Journal of Clinical Biochemistry and Nutrition*, 44, 280–284.
Shevchenko, N.M., Anastiuk, S.D., Gerasimenko, N.I., Dmitrenok, P.S., Isakov, V.V. and Zvyagintseva, T.N. (2007). Polysaccharide and lipid composition of the brown seaweed *Laminaria gurjanovae. Russian Journal of Bioorganic Chemistry*, 33, 88–98.
Shibata, T., Ishimaru, K., Kawaguchi, S., Yoshikawa, H. and Hama, Y. (2008). Antioxidant activities of phlorotannins isolated from Japanese Laminariaceae. *Journal of Applied Phycology*, 13(5), 705–711.
Shih, C.M., Cheng, S.N., Wong, C.S., Kuo, Y.L. and Chou, T.C. (2009). Anti-inflammatory and antihyperalgesic activity of C-phycocyanin. *Anesthesia & Analgesia*, 13(4), 1303–1310.

Silva, L.M., Lima, V., Holanda, M.L., Pinheiro, P.G., Rodrigues, J.A., Lima, M.E. and Benevides, N.M. (2010). Antinociceptive and anti-inflammatory activities of lectin from marine red alga *Pterocladiella capillacea. Biological and Pharmaceutical Bulletin*, 13(5), 830–835.

Siqueira, R.C., da Silva M.S., de Alencar, D.B., Pires, A.F., de Alencar, N.M., Pereira, M.G., Cavada, B.S., Sampaio, A.H., Farias, W.R. and Assreuy, A.M. (2011). In vivo anti-inflammatory effect of a sulfated polysaccharide isolated from the marine brown algae *Lobophora variegate. Pharmaceutical Biology*, 49(2), 167–174.

Subhashini, J., Mahipal, S.V., Reddy, M.C., Mallikarjuna Reddy, M., Rachamallu, A. and Reddanna, P. (2004). Molecular mechanisms in C-Phycocyanin induced apoptosis in human chronic myeloid leukemia cell line-K562. *Biochemical Pharmacology*, 68(3), 453–462.

Synytsya, A., Kim, W.J., Kim, S.M., Pohl, R., Synytsya, A., Kvasnicka, F., Copikova, J. and Park, Y.I. (2010). Structure and antitumour activity of fucoidan isolated from sporophyll of Korean brown seaweed *Undaria pinnatifida. Carbohydrate Polymers*, 81, 41–48.

Thinh, P.D., Menshova, R.V., Ermakova, S.P., Anastyuk, S.D., Ly, B.M. and Zvyagintseva, T.N. (2013). Structural characteristics and anticancer activity of fucoidan from the brown alga *Sargassum mcclurei. Marine Drugs*, 11, 1456–1476.

Tiberg, E., Dreborg, S. and Björkstén, B. (1995). Allergy to green algae (*Chlorella*) among children. *Journal of Allergy and Clinical Immunology*, 96, 257–259.

Viskari, P.J. and Colyer, C.L. (2003). Rapid extraction of phycobiliproteins from cultured cyanobacteria samples. *Analytical Biochemistry*, 319(2), 263–271.

Chapter 11
Engineering *Spirulina* for Enhanced Medicinal Application

Chitrangada Das Mukhopadhyay

11.1 Introduction

Cyanobacteria are prokaryotes which can perform photosynthesis like higher plants. Their genomic organisation is very simple and thus is suitable for the study of detailed photosynthesis mechanism at a molecular level also for many other genomic manipulations relevant to benefit of living organisms. This unicellular alga, *Spirulina* has a thin thread like elongated structure and classified under *Cyanobacteriaceae* which is blue green in colour. Under microscope it looks like bunch of bright helical threads (Fig. 11.1).

This is also known as *Arthrospira platensis*. Their odour and taste is similar to seaweed. They grow well in fresh water like ponds, lakes and rivers. Pollution-free medium with abundant sunlight and moderate temperature is favourable for growth of *Spirulina*. However, they can also sustain and adapt harsher environmental condition. *Spirulina* can be referred as a complete nutritional food supplement, being a rich source of many important nutrients, including protein, carbohydrates, iron and vitamins like A, K and B complex. Most of the dry weight of this unicellular alga is protein which is very essential for growth and regeneration. It is a better substitute for meat and dairy products which are rich in fatty acid and cholesterol. A few grams of *Spirulina* intake can fulfill our everyday requirement of protein, iron and vitamins, especially vitamins A in form of beta carotene, B complex, D and K. *Spirulina* though contains no vitamin C, but it helps maintain potency of vitamin C. *Spirulina* has great antioxidant properties incurred by its yellow xanthophyll content. Thus, *Spirulina* can be a wise option to be used as a dietary supplement to maintain good health and resist several diseases.

C.D. Mukhopadhyay (✉)
Centre for Healthcare Science and Technology, IIEST, Shibpur 711103, India
e-mail: chitrangadadas@yahoo.com

D. Das (ed.), *Algal Biorefinery: An Integrated Approach*,
DOI 10.1007/978-3-319-22813-6_11

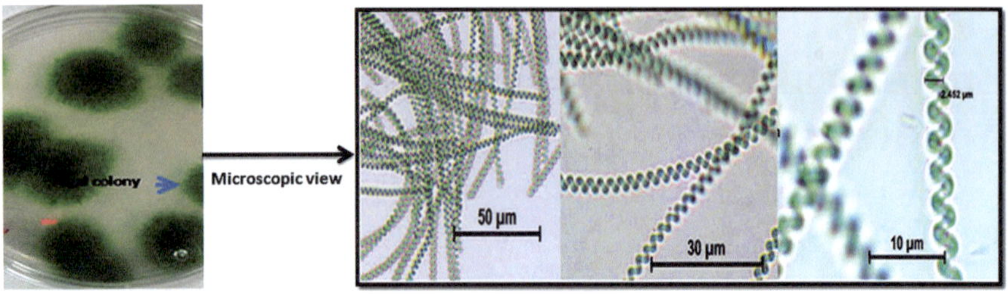

Fig. 11.1 *Spirulina platensis* under light microscope at different magnification.

Spirulina can be genetically engineered for many practical purposes, but neither any stable gene transfer system nor any suitable expression vector has been established so far. Salvi et al. (1994) are the pioneers of gene transfer studies in *Spirulina* and concluded that identification of a suitable restriction modification system can accelerate the progress in this direction. Kaneko et al. (1996) sequenced the whole genome of *Synechocystis* sp. PCC 6803, a unicellular cyanobacterium which was followed by whole genome sequencing over 40 strains of cyanobacteria. Some nitrogen-fixing filamentous species, and some unicellular species, which are either nitrogen fixing or non-nitrogen fixing are to name among them. Only recently in 2010, Fujisawa et al. first sequenced the filamentous non-nitrogen-fixing species viz, *A. platensis*.

Sequencing of whole genome opened up the way to use genes from higher plants as probes (Amao and Nakamura, 2006) to identify and clone cyanobacterial genes involved in photosynthesis. Subsequently, many cyanobacterial photosynthetic mutants were possible to create by genetic manipulations conducted by Ruengjitchatchawalya et al. (2002) and Cao et al. (1999). They created some genetically engineered strains which can efficiently produce organic substances. One advantage of cyanobacteria is that they use solar energy, CO_2, H_2O and inorganic substances as efficiently as higher plants. This feature can be exploited for production of useful organic products with minimum expenditure of resources and energy. A few decades ago Kuhlemeier et al. (1981) reported some expression vectors for *Anacystis nidulans* R2. Later, Kurdrid et al. (2011) developed cyanobacterial host vector systems to analyze the expression and structure function relation of the genes. One major advantage of *Spirulina* is its simpler growth requirements than other bacteria and is nonpathogenic. Therefore, the mass culture of *Spirulina* sp. can be performed efficiently (Vonshak, 1997). Because of these characteristics, many attempts to express foreign genes in *Spirulina* sp. have been made. One enzyme called human carbonic anhydrase, responsible for inter conversion of carbon dioxide and bicarbonate to maintain acid-base balance in blood and other tissues and removing extra carbon dioxide out of tissues and another called Mn superoxide dismutase from *E. coli* were cloned and expressed by Cao et al. in cyanobacteria. In this chapter an effort has been made to summarise ongoing global research on the most efficient conditions for expression of the valuable genes of *Spirulina* sp. in other organisms and also research on how genes from other organisms can be cloned and expressed in *Spirulina*. The problems in engineering *Spirulina* and its probable solution have also been discussed.

11.2 Amazing Potential of *Spirulina* sp.

Algae are important, innovative solution for many health problems. They are also known to promote regular functioning of major physiological systems, including the immune system, humeral system, cardiovascular system, respiratory system, and nervous system (Edwards, 2013). Algal anti-inflammatory properties have been explored to produce several natural products in indigenous cultures as a sun protectant, a moisturising agent and as a treatment for wounds, burns and bruises. *Spirulina* offers an endless number of possibilities for enhancing health and wellness. Recent research has even suggested this algae as a therapeutic solution for a number of serious conditions, including diabetes, arthritis, heart disease, autoimmune diseases, and cognitive decline, in the form of dementia and Alzheimer's. Some biomolecules of *Spirulina* have been shown to stall the spreading of the HIV and SARS virus (http://newhope360.com). It also enhances natural cleansing and detoxification, improves gastrointestinal and digestive health, reducing cancer risks with antioxidant protection. In conclusion, there are an endless number of medical and health applications for algae. The naturally derived substance has broad implications for the prevention and/or treatment of a wide range of diseases, conditions, and minor ailments. The other economic value of these algae includes use of technologies in the area of energy production and pollution abatement. Health Enhancement Products, Inc. (HEPI) recognizes the benefits of algae, and strives to optimise the value of this natural product by conducting research and development on its vast capabilities. The company promotes overall health and wellness by seeking effective solutions for the most pressing medical issues faced by people today.

11.2.1 Healthy Dieting with Spirulina *sp.*

Spirulina contains a very valuable compound called γ-linolenic acid or GLA. This is also found in human breast milk that contributes to generate strong immune system in new born babies. This microalga can be digested easily and is a rich source of several vitamins and nutrients. This is a great option to resist malnutrition in children and women. It helps the body absorb essential nutrients when it has lost its ability to absorb common forms of food. *Spirulina* favours growth of beneficial microorganisms like *Lactobacillus* and *Bifidobacteria* in human digestive tract to promote healthy digestion and proper bowel function while reducing harmful microorganisms like *E. coli*, *Candida* and yeast. Beneficial microorganisms increase absorption and assimilation of nutrients from the foods we eat, and helps to protect against infection. It also acts as a natural cleanser by removing mercury and other lethal toxins ingested by the body. *Spirulina* also increases stamina and immunity levels in athletes, and its high protein content helps to build muscle mass. *Spirulina* can effectively curb the appetite of athletes and soldiers during their vigorous training schedules helping them to maintain their right body weight and immunity.

11.2.2 Disease Prevention by Spirulina

Other than iron and beta-carotene, *Spirulina* contains many other micro-nutrients like copper, chromium, zinc, selenium and manganese. Human body generates a lot of toxic free radicals which react with these metals and subsequently eliminate from the body during excretion. A unique combination of phytonutrients viz. chlorophyll, phycocyanin and polysaccharides exists in *Spirulina* which makes it naturally suitable to fight many diseases. Some molecules which may harm human system and are absorbed by the body through pollution, poor diet, injury, or stress may also be eliminated with the help of these micronutrients. By removing free radicals, the nutrients help the immune system fight cancer and cellular degeneration. In some experimental findings, *Spirulina* sp. was found effective in regression of oral cancer in laboratory animals, and may pave way in cancer treatment. *Spirulina* sp. also can reduce low density lipoproteins (LDL) which are associated with severe cardiovascular disorders like atherosclerosis and strokes. LDL sticks to inner wall of the arteries and get hardened thus blocking the normal blood flow. Spirulina can lower the LDL accumulation in the body and helps prevent the many cardiovascular problems. It also helps lower blood pressure. *Spirulina* sp. also acts as antihistaminic component.

11.2.3 Anti-Aging with Spirulina *sp.*

Spirulina being rich in various antioxidants prevents aging. *Spirulina* contains concentrated form of beta-carotene which is essential for eyes. *Spirulina* is rich in iron, magnesium and trace minerals, and these minerals get easily absorbed by the body than common iron supplements found in the market. *Spirulina* is the highest source of vitamin B_{12}, which is essential for healthy nerves and tissue, especially for vegetarians. In Table 11.1, the nutritional values of *Spirulina* and its medicinal importance are summarised.

Table 11.1 Potential of *Spirulina* at a glance

	Spirulina a rich source of nutrition		Top 10 causes to engineer Spirulina for enhanced medicinal application
1	Protein content 65–71 % by dry weight, and contains all the essential amino acids except histidine.	1	Strengthens the immune system
2	It contain more than 13 types of minerals It has relatively high concentrations of K, Ca, Zn, Mg, Mn, Se, Fe and P.	2	Supports cardiovascular health and lowers cholesterol.
3	Vitamins B_1, B_2, B_6, B_{12}, E, biotin, pantothenic acid, folic acid, inositol and niacin are present. It is the richest vegetative source of B_{12} in the world.	3	Improves gastrointestinal and digestive health.
4	Carotenoids present in the form of alpha-carotene, beta-carotene, xanthophylis, cryptoxanthin, echinenone, zeaxanthin and lutein.	4	Enhances natural cleansing and detoxification.
5	Pigment such as chlorophyll is found in great abundance in *Spirulina* along with phycocyanin and porphyrin.	5	Reduces cancer risks with better antioxidant protection.
6	*Spirulina* contains very little carbo-hydrates, and about 3.9 Kcal g^{-1}. There is also very little Na, which is important for some people.	6	Provides complete daily nutrition need for the body.
		7	Promotes body metabolism.
		8	Neutralizes body acidity.
		9	Inhabits growth of cancerous cells.
		10	Reported to combat HIV attack.

11.3 Structural Features of *Spirulina* sp. Genome

A single closed circular genome of about 7.0 Mbp length is found in this cyanobacteria. However, no plasmid DNA has been sequenced so far. In general, the bacterial chromosome containing circular DNA sometimes contains uneven distribution of G and C bases on both strands; this can be detected by G-C skewing which help to locate ORFs and termination of DNA replication. In case of *Spirulina*, any such apparent shift was detected by this analysis unlike other cyanobacteria. *Spirulina* genome comprises 6630 potential protein coding genes, 49 RNA genes which consist of two sets of rRNA genes, 40 tRNA genes which code for tRNA, tmRNA, β subunit of RNAse P and signal-recognition particle RNA. Among the protein coding genes 3/4th are of known function but no functional characterisation was done for about 1500 genes. Most of these genes are responsible for general cyanobacterian metabolism but the most strikingly the biosynthesis genes for cyanotoxins such as alkaloid toxins (anatoxins and saxitoxins), non-ribosomal peptide toxins (microcystins), urea-derived toxins (cylindrospremopsin) and ribosomal depsipeptide toxins (microviridins) were absent in the genome, which logically explains the use of

Spirulina as a food since long. The genes for photosynthesis viz, psa X for photosystem I, cytochrome c550-like genes and genes for cytochrome b6, ATP synthase and NDH are detected as in other cyanobacteria. Although there are two copies of cytochrome c6 gene, any gene coding for electron carrier proteins in the thylakoid lumen such as plastocyanin and cytochrome Cm is not yet characterized.

11.3.1 Important Genes in Spirulina *sp. Genome*

One very unique feature of *Spirulina* is its ability to adapt in new or unfavourable environmental condition as in case of high salt concentrations. They have well equipped regulatory mechanism consisting of cAMP signalling cascade which helps them adapt in response to changes in the external environment (Ohmori et al., 2009). *Spirulina* develops a number of diverse cAMP- dependent signal cascades to adapt to different severe environmental conditions (Kashahara et al., 2001; Ohmori and Ohmori, 2002) as was revealed by comparative analysis of genes coding for adenylate cyclase. Amao and Nakamura (2006) highlighted the potential of *Spirulina* for production of biofuels due to presence of hydrogenase enzyme.

11.3.2 Lipid and Carotenoid Producing Genes

Undoubtedly *Spirulina* is the richest source of beta-carotene containing about ten different kinds of carotenoids. Most important of them are alpha, beta and gamma carotenes. They also contain diverse variety of yellow xanthophylls. They work together to protect our body from antioxidants and carcinogens. Synthetic beta-carotenes do not always show these benefits. The reason is that natural one is easily assimilated and contains cis isomer, which is lacking in synthetic beta-carotenes.

Kerfeld et al. (2003) and Wilson et al. (2006) did in-depth comparative analysis of all known cyanobacterial genes involved in carotenoid biosysnthesis and concluded that *Spirulina* contains all of them excluding genes for β-carotene ketolase. Interestingly they found a product of ketolase viz, 3-hydroxyechinenone, to bind to an orange carotenoid protein during energy dissipation from phycoliposome in photosystem II. Therefore, it is possible that an unknown type of ketolase may be present in the genome. All known cyanobacterial genes involved in biosysnthesis of fatty acids, vitamin E, lipoic acid, glycerolipids, lipo-polysaccharides and polyhydroxy alkanoates were detected by them. However, a gene for ω-3 fatty acid production desatyrase (desB) was consistently missing in *Spirulina* as evidenced by the biochemical analysis (Murata et al., 1992).

11.3.3 Prevention of Reactive Oxygen Species Formation by Spirulina

Spirulina harbours several genes encoding enzymes against reactive oxygen species. Genes for peroxiredoxin, thioredoxin peroxidase, and other putative peroxidases include five genes for bacterioferritin co-migratory proteins. However there was no evidence of gene for enzyme catalase. In a laboratory experiment by Pak et al. in 2012 using animal models it was revealed that several symptoms like increase in production of reactive oxygen species from liver mitochondria, activation of NF-kappa β, and imbalance in lymphocyte surface antigen ratio was significantly abated by administration of *Spirulina* and phycocyanin. It confirmed the therapeutic property of *Spirulina* sp. or phycocyanin that can inhibit the inflammatory response through anti-oxidative and anti-inflammatory mechanisms, breaking the crosstalk between oxidative stress and inflammation. Phycocyanin and phycocyanobilin from *Spirulina platensis* also protect against diabetic nephropathy by inhibiting oxidative stress. Zheng et al. (2012) have reported that bilirubin and its precursor biliverdin may have beneficial effects on diabetic vascular complications, including nephropathy, via its antioxidant effects. They investigated the role of chromophores viz. phycocyanin and phycocyanobilin obtained from *Spirulina platensis* and its chromophore in oxidative stress and renal dysfunction in diabetic rodent model.

11.3.4 Gene Transfer into Spirulina *sp.*

The biggest limitation in engineering any beneficial gene into *Spirulina* is lack of complete understanding of the restriction modification system of this cyanbacterium, high polysaccharide content of genomic DNA and excessive methylation. Genomic DNA of some blue green alga obtained through cesium chloride gradient purification and ultracentrifugation is poorly digested while using common restriction enzymes. But digestion with these enzymes along the whole length of genome at specific restriction sites as in other microbes cannot be assured due to interference of excessive DNA methylation and polysachharide content.

One alternative strategy was proposed successfully by Promega Inc. which involves no restriction digestion of the genomic DNA. DNA is first extracted by incubating the cyanobacterial cells at 37 °C for 1 h with 10 mg mL^{-1} lysozyme (Sigma) in 50 mM EDTA followed by addition of nuclei lysis solution. Briefly the procedure for genomic DNA library preparation involves sonication of g-DNA, blunting of the overhangs with exonuclease digestion, adding adenine nucleotide adapter at 3′ends using DNA-Taq polymerase enzyme and finally ligation into the pGEM ®-T Vector System. This method is called "T-vector method" where genomic DNA from *Spirulina platensis*, a typical filamentous cyanobacterium that is well known for the resistance of its DNA to restriction enzyme digestion has been used.

11.3.5 Vectors Used for Cloning in Cyanobacteria

One of the crucial criteria in identifying suitable vector construction is its stability of exogenous DNA expression in microalgal expression systems. With the rapid advances of genome sequencing technologies, gene sequences of various microalgae was published and large-scale approaches have been developed to utilize the sequence information into functional information. Generally, function of any gene can be derived using various approaches, such as analysis of gene expression pattern by promoter activity assay, gene silencing, ectopic expression, structure-function analysis, protein sub-cellular localization examination, and *in vitro* or *in vivo* biochemical assays (Kumar and Hirochika, 2001; Curtis and Grossniklaus, 2003). The conventional method for engineering an expression vector construct depends on the restriction digestion of a suitable DNA fragment followed by ligation. This method is time consuming and is dependent on availability of appropriate restriction sites. This is a significant technical barrier for large-scale functional gene analysis studies. Recently, the Gateway cloning system has been developed to facilitate large-scale production of gene constructs, and it is able to achieve rapid cloning of one or more genes into multiple destination vectors using site-specific recombination. The recombination cloning system is based on a two-step process (Earley et al., 2006). Firstly, the desired DNA fragment is cloned into a general donor vector by "GATEWAY" – BP reaction. Then, the DNA fragment flanked by two site-specific recombination sites (attB1 and attB2) in the donor plasmid can be transferred precisely into a variety of expression vectors by site-specific recombination reactions. Once the DNA product is targeted into a donor vector, the transfer of the DNA constructs into an expression destination vector which becomes simple and requires no traditional restriction enzyme/ligase cloning. This Gateway technology have been widely used in the research community, and many Gateway-compatible open reading frame clone collections and expression plasmids have been created for functional genomic analysis in many organisms.

Two plasmids were constructed by Kuhlemeier et al. (1981), recombining the *E. coli* vector pACYC184 and the cyanobacterial plasmid pUC1. These recombinants plasmids, designated as pUC104 and pUC105, could be transformed to *E. coli* K12 as well as to the cyanobacterium *Anacystis nidulans* R2 and in both hosts they expressed their antibiotic markers. pUC104 and pUC105 differ with respect to the location and the orientation of the pACYC184 segment in pUC1. pUC104 was tested to be more stable in different environmental conditions. Transformation of pUC105 to *Spirulina* was done successfully with antibiotic resistance marker for chloramphenicol.

Another shuttle vector was constructed by introducing multiple cloning site of plasmid pUC18 between EcoRI and HindIll sites of pBR322 to produce pBR322M. Then a 4.4-kb BamHI-Xhol fragment of pBA1 that originated from *A. nidulans* 6301 was ligated to the 2.5-kb Pvu II Eco47III fragment of plasmid pBR322M and the ends were filled with T4 DNA polymerase. An ampicillin resistant marker was also introduced to give rise pBAX18R.

11.3.6 Methods for Transformation into Microalgae

So far, microalgae have been successfully transformed through various transformation methods. The popular transformation methods are biolistic transformation and electroporation. In some cases, transformation resulted in the successful and stable expression of transgenes from either the nucleus or the plastid, but in most cases, only transient expression was observed. So far lot of research has been performed to increase the lipid yield of microalgae through optimization of growth and induction conditions, such as temperature, light intensity and duration, salt concentration, and nutrient requirement, while genetic modifications of microalgae to alter quantity, quality or composition of lipid are reported less. The main reason is probably the lack of a generally applicable transformation protocol for microalgae. Since microalgae are such a diverse group of organisms, it cannot be guaranteed that a method that works for one species could be applied to another one. Some antibiotics routinely used in the transformation of plants, such as kanamycin and zeocin, have been successfully used as markers in cyanobacterial gene manipulation. Also, heterologous gene expression in microalgae demands for preferential codon usage and appropriate promoter sequences to modulate each expression. This can be achieved by getting information on completely annotated and sequenced microalgal genomes. Any protocol for the genetic transformation of a microalgal strain needs to be customized to meet its specific requirements and overcome corresponding limitations.

11.4 Studies on Cloning, Structure and Function of Extrachromosomal DNA in *Spirulina* sp.

Cao Xue Cheng et al. (2005) described an efficient protocol for the extraction and purification of extrachromosomal DNA (exDNA) from *Spirulina platensis* strains, such as Sp-HO1, Sp-D, Sp-S, Sp-T etc. This method was commonly described as CTAB-Proteinase K method. The homology of several exDNAs, as well as between exDNA and chromosomal DNA was analyzed by Southern blotting. Partial sequences of exDNA from Sp-S were cloned and the primary and secondary structures, open reading frames (ORFs) of the cloned fragments, and function of proteins encoded by ORFs were analyzed through bioinformatics analyses. Briefly the CTAB method includes traditional method of cell lysis while adding Proteinase K during crude DNA extraction and purification by phenol-chloroform-isoamyl alcohol. In this way proteinase K helps removing protein contaminants and improves DNA quality. This method was successfully used to extract extrachromosomal DNA from more than ten strains of *Spirulina*. The results suggested that different strains of *Spirulina* contains exDNA of different size, e.g. strain HO1 exDNA has a length of 0.75 kb whereas strain D, J and Z has that of approximately 1.1 kb. Sp-S has two exDNA: shorter one is 1.8 kb long and longer one is 3.6 kb long. Same is the case for Sp-T. Although the longer one is double the length of shorter exDNAs in above two strains 2D gel electrophoresis revealed no new bands and confirmed that they were two independent strains. However, these DNAs were not abundant and difficult to isolate and purify from total DNA. The results revealed that three methods viz. freeze-thaw extraction, DNA gel extraction kit and alkaline method, can be used for exDNA purification. However, the freeze-thaw extraction method is the best one. It is with recovery efficiency over 90 %. Molecular characteristics of exDNA in *Spirulina* through studies on the genetic stability, probable molecular conformation and restriction enzyme patterns of the exDNA, it was found that these DNAs were stable.

11.4.1 Analysis of Restriction Endonuclease Digestion of Genomic DNA from **Spirulina** *sp.*

The modification of the CTAB extraction procedures (Poberski et al., 1997) includes adding proteinase K to lyse cells, high NaCl concentrations in the buffer to remove polysaccharides and an additional proteinase K digestion of crude DNA extracts to remove any excess proteins. The CTAB-proteinase K method is efficient to prepare digestible genomic DNA from *S. platensis*. Subsequently it was established that the genomic DNA can be digested by more than 12 restriction enzymes. The proper reaction condition involves an enzyme concentration of 5 U g^{-1} DNA, and a digestion time of 4 h.

11.4.2 **Spirulina** *Cloning and Expression Analysis of the Serine/Threonine Kinase Gene Family*

Serine/threonine kinases (STKs) have been found by Qin Song et al. (1993) in an increasing number of cyanobacteria, showing its important roles in signal transduction. Their work aims at molecular cloning and functional elucidation of a putative STK sequence in *Spirulina*. Ongoing *Spirulina platensis* genome project offers us a wealth of information concerning the sequences and organisation of STK gene family in *Spirulina platensis*. Thirty three putative STK homologues were identified in *Spirulina platensis* draft map. Motifs and invariant amino acids typical in eukaryotic STKs were conserved well in these proteins. These STK proteins were classified into three major families according to their domain structures. Their research provides a fundamental clue for further study of signal transduction system in *Spirulina platensis*.

11.5 Important *Spirulina* Genes Cloned in other Organisms

Besides this, several genes obtained from *Spirulina platensis* were cloned in other organisms by several researchers. Tiboni et al. (1984) cloned the genes for the large and small subunits of ribulose-1,5-bisphosphate carboxylase from *S. platensis* in *E. coli*. Sanangelantoni et al. (1990) also cloned the gene for ribosomal S2protein and a part of the gene responsible for the peptide elongation from *S. platensis* in *E. coli*. Riccardi et al. (1991) constructed the genomic library of *S. platensis* DNA using lambda EMBL3 vector. After that, they cloned the acetohydroxy acid synthase gene from this recombinant library in *E. coli*.

Bini et al. (1992) cloned the gene for β-isopropylmalate dehydrogenase of *S. platensis* (*leuB*) from a λEMBL3 genomic library by heterologous hybridization using the *Nostoc* UCD 7801 *leuB* gene as a probe. Salvi et al. (1994) identified, cloned, characterized and expressed the gene encoding serine esterase from *S. platensis* in *E. coli*. Kawata et al. (1998) used TA cloning vector for construction of a genomic DNA library. In their studies, the researchers cloned the gene coding phytoene synthase from *S. platensis* in *Synechococcus* and *Synechocystis*. Lui et al. (2005) cloned one operon named C-phycocyanin from *Spirulina platensis* into pMD18-T followed by its sequencing and genomic characterization. Zhang et al. (2005) cloned and characterized the partial *hoxH* genes encoding large subunit of nickel-iron hydrogenase of two cyanobacterial genera, including five strains of *Arthrospira* and two strains of *Spirulina* in *E. coli*.

Several investigations have been studied on the acyl-lipid desaturases genes. Meesapyodsuk et al. (2001) cloned the *desC* gene from *S. platensis* in baker's yeast and thus, the cyanobacterial gene product appeared to be functional in yeast. Apiradee et al. (2004) cloned and successfully expressed for the first time the genes encoding the acyl-lipid desaturases (desC, desA, desD) which are involved in γ-linolenic (GLA) synthesis in *E. coli*. Later, Kurdrid et al. (2005) cloned, expressed and characterized the *desD* gene in *Sachharomyces cerevisiae*.

Buttarelli et al. in 1989 sequenced the 5.3 kb DNA segment containing the *str* operon of *S. platensis*. Kasahara et al. (1997) sequenced the *cyaC* gene encoding an adenylate cyclase of *S. platensis*. Kawamura et al. (1986) have purified three restriction endonucleases from *S. platensis* subspecies *siamese* and named them SplI, SplII and SplIII respectively. Milano et al. (1992) demonstrated the two isoenzymatic forms of the enzyme acetohydroxy acid synthase (AHS), which catalyse the first common step in the biosynthesis of isoleucine, leucine and valine in *S. platensis* and they sequenced the genes *ilvX* and *ilvW* encoding these two enzymes. Tanioka et al. (2010) characterized cobalamin-dependent methionine synthase to study the physiological function of pseudovitamin 12 B or adeninyl-cobamide in *Spirulina platensis* NIES-39 and finally cloned the full-length *Spirulina* MS. Linjawi (2011) investigated the protective effect of *Spirulina* against mitomycin C (MMC)-induced genotoxic damage in male rats and suggested that *Spirulina* exerts its anti-mutagenic properties by inhibiting alterations in the gene expression.

11.5.1 Cloning, Sequencing and Over-Expression of Spirulina *Phycocyanin Gene*

Phycobiliprotein is an important light-harvesting pigment-protein available in cyanobacteria (*Cyanophyceae*), red algae (*Rhodophyceae*), the hidden of algae (*Cryptophyceae*) and a few dinoflagellates (*Pyrrophyceae*). According to the different composition and absorption spectra, the phycobiliproteins are generally divided into three categories: the phycocyanin (phycocyanin, referred to as the PC, the absorption spectrum in the 610–640 Nm), the phycoerythrin (phycoerythrin, referred to as PE, absorption spectra in the 500 to 570) and allophycocyanin (allophycocyanin, referred to as the APC, absorption spectra in the 650–671 Nm). These proteins can be used as industrial raw materials or food additives. Phycobiliprotein is widely used in the food industry, chemical industry, medical diagnosis, and clinical medicine. In addition, many kinds of phycobiliproteins, fluorescent, high quantum yield, easy combination of a number of proteins with biotin antibodies, phycobiliprotein acts as a fluorescent probe. Phycobiliprotein can be isolated and purified from *Spirulina. Spirulina* farming difficulties, its outdoor large-scale cultivation of the lack of systematic and comprehensive study, and high farming costs are expensive. So, phycobilisome protein from *Spirulina* sp. is not only the high cost but also complex process.

The genetic engineering methods for the production of phycobiliprotein has played an important role. Qin Song (2008) showed that the *Spirulina maxima* phycocyanin gene is 1119 bp having 99 % homology with *Spirulina platensis* phycocyanin gene. The β subunit gene sequences in the alpha subunit gene sequences upstream connection between 111 bp fragment of the gene, β subunit and α-subunit gene contained 519 bp and 489 bp, encoding 172 and 162 amino acid residues. Possible ribosome binding sites in the upstream of the gene sequence of the β subunit 8–11 bp exists at Gaga the binding site and a prokaryote ribosome binding sites usually ggaa similar, but not identical. *Spirulina maxima* phycocyanin three chromophore binding sites are the α subunit amino acids-Cys 84 and β subunits of amino acids-Cys82 and Cys153. The third chromophore binding site Cys153 located in the carboxy terminus of the β-subunit, has proved that this binding site is due to the subunit 12 amino acid residues (146–157). The phycocyanin contains many hydrophobic amino acid residues which play an extremely important role in phycocyanin gathering process.

11.5.2 Cloning of RuBisCO

Tiboni et al. (1984) cloned the genes for two subunits of ribulose-1,5-bisphosphate carboxylase from filamentous cyanobacterium, *S. platensis* into *E coli*. Those genes were found located very close having a total length of 4.6 kb. The amount of large subunit produced in the bacterial host represents at least 10 % of the total extracted protein.

11.5.3 Cloning and Characterization of Spirulina *Esterase*

Sergio et al. (1994) identified a 23 kDA gene known to code for a serine esterase from the cyanobacterium *Spirulina platensis*. It was cloned and expressed in *Escherichia coli* before functional characterization. DNA sequencing studies revealed the primary structure of the esterase which had partial similarity with the carboxyl-esterase (esterase II) encoded by *estB* of *Pseudomonas fluorescens*. Notably, the highest degree of homology was found in a stretch of 11 identical or highly conserved amino acid residues corresponding to the GXSXG consensus motif found in the catalytic site of many other esterases, lipases and serine proteases.

11.5.4 Generation of Mutants in Spirulina

Mingyue Fang et al. in 2013 tried to improve the carbohydrate yield of *Spirulina platensis* by generation of mutants with increased growth rate and carbohydrate content. They constructed a mutant library of *S. platensis* with diverse phenotypes using a new high throughput rapid mutagenesis technique at room temperature. The screening of the mutants was performed in the 96-well microplate and identified with microplate reader. The mutants which were stable even after several subcultures were considered ideal and selected. The mutants in mutant library showed diverse phenotypes in terms of cell growth rate, carbohydrate content and flocculation intensity. This mutagenesis method was established to be an effective tool to generate the mutant library for multicellular microalgae.

11.5.5 Use of Transposons for Cloning in Spirulina *sp.*

The oxidative photosynthesis of microalga is well known for its ability to reduce carbon dioxide emissions. This feature is employed industrially for wastewater treatment. Mass cultivation and extraction of useful substances from the microalgae are also in practice. Kawata et al. in 2004 developed artificial transposon systems by extracting essential elements from natural transposons. They mutated transposase and transposon complex system which improved the transformation efficiency by electroporation. They used Tn5, a natural transposon, transposase and cation liposome complex by electroporation to improve transformation efficiency for *Spirulina platensis* -C1 (Arthrospira sp. PCC9438) and selected the cloned cells growing in 2.0 μg mL^{-1} chloramphenicol containing medium. This genetically modified *Spirulina* have the potential for industrial wastewater treatment and production of useful chemical materials.

11.6 Challenges in Engineering *Spirulina* sp.

Because of its several benefits, studies on cyanobacteria have been increased all over the world. Especially during the past few decades, they have been used to understand photosynthesis and genetic expression controlled by photoregulation, cell differentiation, N_2 fixation, metabolism of carbon and hydrogen, resistance to environmental stress and molecular evolution (Koksharova and Wolk, 2002). Recently some cloning vectors and other genetic tools have been developed for cyanobacteria. Transformation, electroporation and conjugation techniques are being used for gene transfer studies. Mutant strains for specific genes have been developed by mutagenesis also. Complete genomic sequences of some strains have been obtained and some genomic sequence projects are under way. Condition of efficient electroporation into *Spirulina* was studied by Toyomizu et al. (2001b). The plasmid pHSG399 was transferred into *Spirulina* using electroporation technique and suggested the best electroporation condition is 5.0 ms pulse duration with an electric field of 4–8 kV per cm. On the other hand, Kawata et al. (2004) transferred Tn5 transposon, transposase and cation liposome complex into *Spirulina* and suggested 5.0 ms pulse duration with an electric field of 7.5 kV per cm.

11.7 Some Harmful Effect of *Spirulina* sp.

Although *Spirulina* is a suitable organism for producing recombinant proteins, there are not much researches going on gene transfer studies on *Spirulina platensis*. Despite of having very high nutritional value, this alga may also cause some side effects to some individuals including allergic reactions like rashes, hives, and difficulty in breathing. Some commercial *Spirulina* supplements may contain toxic substances as additives. It is, therefore, absolutely necessary to purchase *Spirulina* only from authentic sources. People with the serious metabolic condition phenylketonuria (PKU) cannot metabolize phenylalanine, and therefore, should not consume *Spirulina*. Similarly, this is not advisable for people with autoimmune diseases such as rheumatoid arthritis and multiple sclerosis. The *Spirulina* may also antagonise the effectiveness of certain medicines, such as prednisone, which are commonly used for treating asthma and other inflammatory diseases. Although *Spirulina* is apparently safe at larger doses, but doctors may advice the right dosage.

11.8 Conclusion

Use of seaweeds and other algae in oriental countries is an old practice as a remedy to cure or prevent various physical ailments. Several researchers have tried to establish a connection between these nutrient-rich sea plants and the body's immune system response. Intensive studies have started to identify potential sources of pharmacologically active agents. The mechanisms through which these algae have a preventive role include reduction of plasma cholesterol, binding of bile steroids, lipids, antioxidant activity, prevention of carcinogenic effect, binding and complement the significant pollutant trace elements in the diet. Since this microalga are non-toxic and have immense nutritious value to human and live stocks, *Spirulina platensis* is considered one of the most commercially important species among microalgae. It contains protein, carbohydrates, minerals, chlorophyll *a*, phycocyanin, vitamin 12B, β-carotene and essential fatty acids like γ-linoleic acid for human and animal nutrition. So, its production and consumption are increasing every year. The production is over 2000 tons per year. Some countries such as China, South Africa, Japan, Mexico, Australia and Chile are leading countries on production of *Spirulina*.

The research on stem cells is in vogue because of its potential application in repair and regenerative medicine. Some researchers have investigated the effects of *Spirulina* on stem cell growth and proliferation and revealed that this cyanobacteria has the potential to stimulate endogenous stem cells leading to healing and regeneration. Sequencing studies on *Spirulina* genome have helped to decipher functional role of many genes which can be exploited to produce useful neutraceuticals. The lack of suitable restriction modification system and expression vectors limits gene transfer into *Spirulina*; however several alternate strategies are coming up and construction of complete genomic library will help to understand structure function relationship at genomic level leading to engineering *Spirulina* for medicinal purposes. Recently, Verseux et al. (2015) indicated the potential of *Spirulina* in International Space Research stations. *Spirulina* can adapt itself in unfavourable environmental conditions like high salt concentration, adverse temperature and pH of medium where not much bacterial species can survive. So, pure cultures of *Spirulina* can be maintained easily. Cultivation of *Spirulina* will also help to enable environmental purification during its photosynthetic process. Thus beneficial effect of this cyanobacterium can be utilized in space too as in earth.

Spirulina can be obtained from commercial sources for human consumption but most of them are cultivated in laboratory. Some unavoidable toxic contamination may happen during preparation of tablet or powder form of *Spirulina* and unfortunately that cannot be removed during downstream processing of large scale production. Genetically improved variety of this cyanobacterium may eventually overcome this challenge in future.

References

Amao, Y. and Nakamura, N. (2006). Biohydrogen production with the light harvesting function of grana from *Spirulina* and colloidal platinum. *International Journal of Hydrogen Energy*, 31, 39–42.

Apiradee, H., Kalyanee, P., Pongsathon, P., Patcharaporn, D., Matura, S., Sanjukta, S., Supapon, C. and Morakot, T. (2004). The expression of three desaturase genes of *Spirulina platensis* in *Escherichia coli* DH5α. *Molecular Biology Reports*, 31, 177–189.

Bickford, P.C., Tan, J. and Shytle, R.D. (2006). Nutraceuticals synergistically promote proliferation of human stem cells. *Stem Cells and Development*, 15, 118–123.

Bini, F., Rossi, E., Barbierato, L. and Riccardi, G. (1992). Molecular cloning and sequencing of the b-isopropylmalate dehydrogenase gene from cyanobacterium *Spirulina platensis. Jl Gen. Microbiol.*, 138, 493–498.

Buttarelli, F.R., Calogero, R.A., Tiboni, O., Gualerzi, C.O. and Pon, C.L. (1989). Characterization of the str operon genes from *Spirulina platensis* and their evolutionary relationship to those of other prokaryotes. *Molecular and General Genetics*, 217, 97–104.

Cao, J., Zengfu, X., Guohua, Q. and Baojian, L. (1999). Studies on the sensitivity of *Spirulina platensis* to antibiotics and herbicide: Relationship with selectable markers for genetic transformation. *Bioresource Technology*, 70(1), 89–93.

Cao, X., Wang, Z., Yang, L., & Xu, B. (2007). High Efficient Extra-genomic DNA Extraction and Purification from Spirulina. *Oceanologia Et Limnologia Sinica*, *38*(3), 198.

Chi, X., Yang, Q., Zhao, F., Qin, S., Yang, Y., Shen, J., & Lin, H. (2008). Comparative analysis of fatty acid desaturases in cyanobacterial genomes. *Comparative and functional genomics*, 2008.

Curtis, M.D. and Grossniklaus, U. (2003). A gateway cloning vector set for high-throughput functional analysis of genes in planta. *Plant Physiology*, 133, 462–469.

Earley, K.W., Haag, J.R., Pontes, O., Opper, K., Juehne, T. et al. (2006). Gateway-compatible vectors for plant functional genomics and proteomics. *Plant Journal*, 45, 616–629.

Edwards, Mark (2013). Algae offer global nutrient deficiency solutions. Algae Industry Magazine. com. 16 Sept. 2013 <http://www.algaeindustrymagazine.com/algae-medical-solutions-part-2/>.

Fujisawa, T., Narikawa, R. and Okamoto, S. et al. (2010). Genomic structure of an economically important cyanobacterium, *Arthrospira platensis* NIES-39. *DNA Research*, 17, 85–103.

Gennero, L., Mortimer, P. and Sperber, K. (2006). Stem cells an alternative to organ transplantation in chronic, degenerative and infectious disease? *New Microbiology*, 29, 151–167.

http://newhope360.com/search/results/allergen-free (Demonstrates growing market for allergen-free food, evidenced by the number of articles on allergen-free foods and links to allergen-free products)

Jeamton, W., Dulsawat, S., Laoteng, K., Tanticharoen, M. and Cheevadhanarak, S. (2011). Phycocyanin promoter of Spirulina platensis controlling heterologous expression in cyanobacteria. *Journal of Applied Phycology*, 23, 83–88.

Kaneko, T., Sato, S. and Katoni, H. (1996). Sequence analysis of the genome of the unicellular cyanobacterium *Synechocystis* sp. Sequence determination of entire genome and assignment of potential protein–coding regions. *DNA Research*, 3, 109–136.

Kasahara, M., Yashiro, K., Sakamoto, T. and Ohmori, M. (1997). The *Spirulina platensis* adenylate cyclase gene, cyaC, encodes a novel signal transduction protein. *Plant and Cell Physiology*, 38, 828–836.

Kashahara, M., Unno, T., Yashiro, K. and Ohmori, M. (2001). Cyan-G, a novel cyanobacterial adenylyl cyclase and a possible ancestor of mammalian guanylyl cyclases. *Journal of Biological Chemistry*, 276, 10564–10569.

Kawamura, M., Sakakibara, M., Watanabe, T., Kita, K., Obayashi, A., Takagi, M. and Yano, K. (1986). A new restriction endonuclease from *Spirulina platensis. Nucleic Acids Research*, 14, 1985–1989.

Kawata, Y., Shin-ichi Yano, M.S., Thankappan, A.K. and Kojima, H. (2002). Constructing Genomic Libraries Using the pGEM ® -T Vector, Proemga notes 73.

Kawata, Y., Yano, S. and Kojima, H. (1998). Efficient library construction with a TA vector and its application to cloning of the phytoene synthase gene from the cyanobacterium *Spirulina platensis. Current Microbiology*, 37, 289–291.

Kawata, Y., Yano, S., Kojima, H. and Masaaki, T. (2004). Transformation of *Spirulina platensis* strain C1 (Arthrospira sp. PCC9438) with Tn5 transposase-transposon DNA-cation liposome complex. *Marine Biotechnology*, 6, 355–363.

Kerfeld, C.A., Sawaya, M.R. and Brahmandam, V. (2003). The crystal structure of a cyanobacterial water soluble carotenoid binding protein. *Structure*, 11, 55–65.

Koksharova, O.A. and Wolk, C.P. (2002). Genetic tools for cyanobacteria. *Applied and Environmental Microbiology*, 58, 123–137.

Kuhlemeier, C. J., Borrias, W. E., Van den Hondel, C. A. M. J. J., & Van Arkel, G. A. (1981). Vectors for cloning in cyanobacteria: construction and characterization of two recombinant plasmids capable of transformation to Escherichia coli K12 and Anacystis nidulans R2. *Molecular and General Genetics MGG, 184*(2), 249–254.
Kumar, A. and Hirochika, H. (2001). Applications of retrotransposons as genetic tools in plant biology. *Trends Plant Science*, 6, 127–134.
Kurdrid, P., Subudhi, S., Hongsthong, A., Ruengjitchatchawalya, M. and Tanticharoen, M. (2005). Functional expression of *Spirulina*-Δ6 desaturase gene in yeast, *Saccharomyces cerevisiae. Molecular Biology Reports*, 32, 215–226.
Kurdrid, P., Senachak, J., Sirijuntarut, M., Yutthanasirikul, R., Phuengcharoen, P., Jeamton, W., & Hongsthong, A. (2011). Comparative analysis of the Spirulina platensis subcellular proteome in response to low-and high-temperature stresses: uncovering cross-talk of signaling components. *Proteome Sci, 9*(1), 39.
Lin, Z. and Yu Ping, K. (2002). Cloning and Sequencing of the Phycocyanin Gene from Spirulina Maxima and Its Over-expression in *Pichia pastoris. Chinese Journal of Molecular Genetics*, 27(2), 236–239.
Linjawi, S.A. (2011). Protective effect of *Spirulina* against mitomycin c-induced genotoxic damage in male rats. *Journal of American Science*, 7, 922–931.
Lui, J., Zhang, X., Sui, Z., Zhang, X. and Mao, Y. (2005). Cloning and characterization of c-phycocyanin operon from the cyanobacterium *Arthrospira platensis* FACHB341. *Journal of Applied Phycology*, 17, 181–185.
Meesapyodsuk, D., Reed, D.W., Cheevadhanarak, S., Deshnium, P. and Covello, P.S. (2001). Probing the mechanism of a cyanobacterial D9 fatty acid desaturase from *Spirulina platensis* C1 (*Arthrospira* sp. PCC 9438). *Comparative Biochemistry and Physiology*, Part B, 129, 831–835.
Milano, A., Rossi, E., Zanaria, E., Barbierato, L., Ciferri, O. and Riccardi, G. (1992). Molecular characterization of the genes encoding acetohydroxy acid synthase in the cyanobacterium *Spirulina platensis. Journal of General Microbiology*, 138, 1399–1408.
Mingyue, F., Jin, L., Chong, Z., Yinyee, Tan, Peixia, J., Nan, G., Heping Li and Xinhui, X. (2013). Rapid Mutation of *Spirulina platensis* by a New Mutagenesis System of Atmospheric and Room Temperature Plasmas (ARTP) and Generation of a Mutant Library with Diverse Phenotypes. *Plos one*, 8,10 I e77046.
Murata, N., Wada, H. and Gombos, Z. (1992). Modes of fatty acid denaturation in cyanobacteria. *Plant Cell Physiology*, 33, 933–941.
Ohmori, K. and Ohmori, M. (2002). cAMP stimulates Na+dependent ATP formation in the lakophylic cyanobacterium *Spirulina platensis. Microbes and Environments*, 17, 144–147.
Ohmori, K., Ehira, S., Kimura, S. and Ohmori, M. (2009). Changes in the amount of cellular trehalose, the activity of maltooligosyl trehalose hydrolase and the expression of its gene in response to salt stress in the cyanobacterium *Spirulina platensis. Microbes and Environments*, 24, 52–56.
Pak, W., Takayama, F., Mine, M., Nakamoto, K., Kodo, Y., Mankura, M., Egashira, T., Kawasaki, H. and Mori, A. (2012). Phycocyanin from *Spirulina* shown to have similar effect to Bilirubin (a well-known antioxidant) in Kidney Cell Protection. *Journal of Clinical Biochemistry and Nutrition*, 51(3), 227–234.
Poberski, S., Bailey, G. and Baum, B. (1997). A modification of CTAB DNA extraction protocol for plants containing high polysaccharide and polyphenol components. *Plant Molecular Biology Reporter*, 15(1), 8–15.
Riccardi, G., Rossi, E., Milano, A., Forlani, G. and Felice, M. (1991). Molecular cloning and expression of *Spirulina platensis* acetohydroxy acid synthase genes in *Escherichia coli. Archives of Microbiology*, 155, 360–365.
Ruengjitchatchawalya, M., Chirasuwan, N., Chaiklahan, R., Bunnag, B., Tanticharoen, M. and Deshnium, P. (2002). Photosynthetic characterization of a mutant of *Spirulina platensis. Journal of Applied Phycology*, 14(2), 71–76.

Salvi, S., Trinei, M., Lanfaloni, L. and Pon, C.L. (1994). Cloning and characterization of the gene encoding an esterase from *Spirulina platensis. Molecular and General Genetics*, 243, 124–126.

Sanangelantoni, A.M., Calogero, R.C., Buttarelli, F.R., Gualerzi, C.O. and Tiboni, O. (1990). Organization and nucleotide sequence of the genes for ribosomal protein S2 and elongation factor Ts in *Spirulina platensis. FEMS Microbiology Letters*, 66, 141–146.

Sergio, S., Mirella, T., Lanfaloni, L., Cynthia, L. and Pon, C.L. (1994). Cloning and characterization of the gene encoding an esterase from *Spirulina platensis. Molecular and General Genetics, 243(1), 124–126.*

Song, Q., Tong, S., Zhang, P. and Shen, C.K. (1993). Isolation of plasmid from the blue-green alga Spirulina platensis. Chinese J. of Oceanol. Limnol., 11(3), 285–288.

Tanioka, Y., Miyamoto, E., Yabuta, Y., Ohnishi, K., Fujita, T., Yamaji, R., Misono, H., Shigeoka, S., Nakano, Y., Inui, H. and Watanabe, F. (2010). Methyladeninylcobamide functions as the cofactor of methionine synthase in a Cyanobacterium, *Spirulina platensis* NIES-39. *FEBS Letters*, 584, 3223–3226.

Tiboni, O., Pasquale, G. and Ciferri, O. (1984). Cloning and expression of the genes for ribulose-1,5-bisphosphate carboxylase from *Spirulina platensis. Biochimica et Biophysica Acta*, 783, 258–264.

Toyomizu, M., Suzuki, K., Kawata, Y., Kojima, H. and Akiba, Y. (2001b). Effective transformation of the cyanobacterium *Spirulina platensis* using electroporation. *Journal of Applied Phycology*, 13, 209–214.

Verseux, C., Baqué, M., Lehto, K., de Vera, J. P. P., Rothschild, L. J., & Billi, D. (2015). Sustainable life support on Mars–the potential roles of cyanobacteria. International Journal of Astrobiology, 1–28.

Vonshak, A. (1997). *Spirulina Platensis Arthrospira*: Physiology, Cell-Biology and Biotechnology. CRC Press. ISBN: 0748406743.

Wilson, A., Ajlani, G., Verbavatz, J. M., Vass, I., Kerfeld, C. A., & Kirilovsky, D. (2006). A soluble carotenoid protein involved in phycobilisome-related energy dissipation in cyanobacteria. *The Plant Cell, 18*(4), 992–1007.

Zhang, X., Shiraiwa, Y., Mao, Y., Sui, Z. and Liu, J. (2005). Cloning and characterization of *hoxH* genes from *Arthrospira* and *Spirulina* and application in phylogenetic study. *Marine Biotechnology*, 7, 287–296.

Zheng, J., Inoguchi, T., Sasaki, S., Maeda, Y., McCarty, M., Fujii, M., Ikeda, N., Kobayashi, K., Sonoda, N. and Takayanagi, R. (2012). New study shows that Spirulina protects against non-alcoholic liver cell inflammation. *American Journal of Physiology Regulatory, Integrative and Comparative*, 23(1), 522–531.

Chapter 12
Algae as a Source of Phycocyanin and Other Industrially Important Pigments

Chitralekha Nag Dasgupta

12.1 Introduction

Algae are morphologically diverse and widely distributed photosynthetic organism recognized as an excellent source of pigments including chlorophylls, carotenoids and phycobiliproteins. Algal pigments are in prime focus as they can be produced in large quantities in renewable manner. All these pigments are of great potential in biotechnological applications including nutraceuticals, pharmaceuticals and cosmetic industries as well as in biomedical research and clinical diagnostics. Specially, the use of phycocyanin as non-toxic and non-carcinogenic natural colourants is gaining importance worldwide in view of the potential toxicity and carcinogenicity of the synthetic colourants (Prasanna et al., 2007).

12.1.1 Source of Pigments

Three types of photosynthetic pigments are available in algae. They are chlorophylls, carotenoids and biliproteins. According to the diversity of pigments and other morphological characteristics, Fritsch (1944) classified the whole of the algae into eleven classes. They are Chlorophyceae (green algae), Xanthophyceae (yellow green algae), Chrysophyceae (golden-brown algae), Bacillariophyceae (diatoms), Cryptophyceae (brownish-green protozoa-like algae), Dinophyceae, Chloromonodineae, Euglinineae, Phaeophyceae (brown algae), Rhodophyceae (red algae) and Myxophyceae (Cyanophyceae, blue-green algae). While chlorophyll-*a* is universal in all algal classes, chlorophyll *b*, *c*, *d* and *e* are restricted to some classes

C.N. Dasgupta (✉)
DST Women Scientist-A, CSIR-National Botanical Research Institute,
Lucknow, Uttar Pradesh, India
e-mail: chitralekha.dasgupta@gmail.com

D. Das (ed.), *Algal Biorefinery: An Integrated Approach*,
DOI 10.1007/978-3-319-22813-6_12

of algae. The chlorophyll-*a* is found in all plant cells but others are commonly extracted from fresh water algae. The most important source of chlorophyll is Chlorophyceae and Cyanophyceae. Chlorophyll-*a* is abundantly found in *Chlorella* and *Spirulina. Chlorella* is called 'Emerald food' due to its amazingly high content of chlorophyll (Bewicke and Potter, 2009). It contains five times the amount of chlorophyll than *Sprirulina*. The chlorophyll content of *Chlorella* is about 7 % of the biomass. Chlorophyll-*a*, due to its stability properties, has been widely used as a colouring substance. The yellow, orange or red coloured pigments are called carotenoids. It includes the caroteins and the xanthophylls. All carotenoids are an antioxidant. It protects the body from damaging molecules called free radicals. β-carotene was the first carotenoid from algae to be commercialized. The average concentration of carotenoids in most algae is only 0.1–2 %, but *Dunaliella* when grown under the right conditions of high salinity and light intensity will produce up to 14 % β-carotene (Mimouni et al., 2012) . It is the precursor for vitamin A biosynthesis in the body. In addition to chlorophyll and carotenoids, certain types of microalgae especially *Cyanophyceae, Cryptophyceae* and *Rhodophyceae* contain phycocyanin and phycoerythrin. These photosynthetic accessory pigments, collectively known as phycobiliproteins, are deeply coloured (red or blue), water soluble, complex, proteinaceous compounds. These pigments have the potential as natural colourants for food, cosmetics and pharmaceuticals. R-phycoerythrin (R-PE) is originally isolated from red algae and has not been found in other taxa. Phycoerythrin is a major light-harvesting pigment of red algae and cyanobacteria that is widely used as a fluorescent probe and analytical reagent. Visually, the phycoerythrins appear red, the phycocyanins range from purple (phycoerythrocyanin, R-phycocyanin) to deep blue (C-phycocyanin), and the allophycocyanins are blue with a hint of green (Cook, 1945) (Table 12.1).

12.2 Phycocyanin

Phycocyanin is the part of phycobiliprotein complexes assembled into extremely large macromolecular complexes called phycobilisomes. Phycobiliproteins are brilliantly coloured, highly fluorescent, water-soluble protein components of the photosynthetic light-harvesting antenna. These proteins are classified into two large groups based on their colours, the phycoerythrin (red), and the phycocyanin (blue). The phycocyanins include C-phycocyanin (C-PC), R-phycocyanin (R-PC) and allophycocyanin (A-PC). Phycobilisomes consist of allophycocyanin cores surrounded by phycocyanin on the periphery. Phycocyanin is the major constituent while allophycocyanin functions as the bridging pigment between phycobilisomes and the photosynthetic lamella (Eisele et al., 2000). Phycocyanin pigment has a single visible absorption maximum between 615 and 620 nm and a fluorescence emission maximum at ~650 nm. Its molecular weight is between 70,000 and 110,000 Daltons. All phycocyanins with only phycocyanobilin chromophores, the most common PC type, were named C-phycocyanin (C-PC). The crystal structure of C-phycocyanin

Table 12.1 Different algal classes with their characteristic pigments and features

Algal classes	Common name	Pigments	Characteristic features
Chlorophyceae	Green algae	Chlorophylls-*a* and *b*, carotene and xanthophyll.	Dark to light green colouration due to chlorophyll-*a* and *b* similar as higher plants. Ranges from simple, one-celled organisms to complex, multi-celled organisms. They may also live in large colonies. There are both marine and freshwater.
Xanthophyceae	Yellow green algae	Chlorophyll-*a*, *c* and *e* but lack *b*. β-carotene and xanthophylls including neoxanthin and violaxanthin.	Excess of yellow xanthophylls gives the characteristic colour yellow green. Most of them are coccoid or filamentous, but some are siphonous.
Chrysophyceae	Golden brown algae	Chlorophyll-*a*, sometimes chlorophyll *c*, and perhaps chlorophyll *e*, carotenes, diadinoxanthin, and fucoxanthin.	Predominance of fucoxanthin gives the characteristic golden-brown colour. Mostly found in freshwater.
Bacillariophyceae	Diatom	Chlorophylls-*a* and *c*, β-carotene, fucoxanthin, diatoxanthin and diadinoxanthin.	Most common types of phytoplankton. Their cells are solitary or united into colonies of various kinds, which may be linked by siliceous structures; mucilage pads, stalks or tubes; amorphous masses of mucilage, or by threads of chitin, (polysaccharide) which are secreted through strutted processes of the cell.
Cryptophyceae	Crytomonads	Chlorophyll-*a* and *b*; many secondary pigments such as α-carotenes biliprotiens and unique xanthophylls.	Protozoa-like algae of various colours ranging from green, blue, brown, olive and red. Their phycobilin pigmentations do not form phycobilisomes. They are located on the interior of the thylakoids.

(continued)

Table 12.1 (continued)

Algal classes	Common name	Pigments	Characteristic features
Dinophyceae	Dinoflagellates	Chlorophylls *a* and *c*2 as well as peridinin (a type of carotenoid only found in dinoflagellates), β-carotene, small amounts of diadinoxanthin and dinoxanthin.	Dinoflagellates are unicellular protists which exhibit a great diversity of form. These species reproduce in such great numbers that the water may appear golden or red, producing a "red tide". They produce a neurotoxin which affects muscle function in susceptible organisms.
Chloromonodineae and Eugleniae		Very little is known about the pigmentation of the classes Chloromonodineae and Eugleniae, though the predominant green colour implies normal chlorophyll.	Freshwater, bright green in colour.
Pheophyceae	Brown algae	Chlorophyll-*a* and *c*, β-carotene and fucoxanthin.	They are mostly found in cold waters along continental coasts. Freshwater species are rare. Species colour varies from dark brown to olive green, depending upon the proportion of golden-brown pigment. Fucoxanthin masks the chlorophyll and carotenoid pigments.
Rhodophyceae	Red algae	Chlorophyll-*a* and *d* (no Chlorophyll *b*), β-carotene and a number of unique xanthophylls. They are lutein, fucoxanthin, myxoxanthin and violaxanthin. Also contain phycoerythrin and phycocyanin.	The pigment phycoerythrin and phycocyanin reflects the characteristic red colour of the algae; this masks the other pigments. Because blue light penetrates water to a greater depth than light of longer wavelengths, these pigments allow red algae to photosynthesize and live at somewhat greater depths.

(continued)

Table 12.1 (continued)

Algal classes	Common name	Pigments	Characteristic features
Myxophyceae	Cyanobacteria, blue green algae	They contain only one form of chlorophyll, chlorophyll-*a* (green pigment). In addition, they contain various yellowish carotenoids, the blue pigment phycobilin.	They are prokaryotic organism but resemble the eukaryotic algae in many ways, including their capability to photosynthesize, different morphological characteristics and using similar ecological niches. Phycobilins are water-soluble, brilliantly coloured and highly fluorescent protein-pigment complexes which include phycocyanin (blue pigment), allophycocyanin (bluish-green pigment) and phycoerythrin (red pigment). The combination of phycobilin and chlorophyll produces the characteristic blue-green colour from which these organisms derive their popular name.

belongs to the monoclinic crystal form, which has not been previously reported for phycobiliprotein structures. The pigment is composed of two subunits, α and β, which occur in equal numbers, but the exact number of α and β pairs which make up the molecule may vary among the species (Glazer, 1976). They are commercialized for fluorescent probes, biotechnology, cosmetic, nutraceutical and clinical and immunological purposes. Phycocyanin is a major light harvesting accessory pigment of red algae and blue green algae. They contain various bioactive components which can reduce inflammation, inhibit lipid peroxidation and have free radical scavenging activity, which can be beneficial for the protection against oxidative stress. The aforementioned effects of BGA can contribute to the prevention of metabolic and inflammatory diseases (Ku et al., 2013). There are 297 patents on phycobiliproteins in global patent databases. The majority of the patents are from USA, Japan and Europe. Patents are grouped into fluorescent applications, general applications and production aspects of phycobiliproteins and the features of each group are reported (Sekar and Chandramohan, 2008).

The blue green algae *Spirulina platensis* is the most popular algal source of this pigment. The principal phycobiliproteins present in *Spirulina platensis* are C-phycocyanin and allophycocyanin which are made up of dissimilar α and β polypeptide subunits (Wang et al., 2001). These two pigments were separated and purified from *Spirulina platensis* by precipitation with ammonium sulphate, ion exchange chromatography and gel filtration chromatography (Zhang and Chen, 1999). In the light of its many commercial applications in food and pharmaceutical industry, purity of the pigment plays a major role. Pharmaceutical industry demands a highly pure phycocyanin. Phycocyanin could be extracted in sodium phosphate buffer (pH 7) after macerating in liquid nitrogen, purified by dialysis and finally by gel filtration chromatography (Bhaskar et al., 2005). Ion-exchange chromatography has also been used for purification of phycocyanin from *Spirulina* (Patil et al., 2006). C-phycocyanin was extracted from fresh *Spirulina platensis* by deploying a species of non-pathogenic nitrogen-fixing bacteria, namely, *Klebsiella pneumoniae*. The extraction was clean and efficient, and the purity and concentration of C-PC proved to be of adequate quality (Zhu et al., 2007). The specific absorbance, fluorescence maxima, sub-unit make-up and amino acid composition of the biliproteins in *Spirulina platensis* resemble those reported for other blue-green algae. However, the minimum molecular weights (44,000 for C-phycocyanin and 38,000 for the allophycocyanin) and the specific extinction coefficients (73 and 58 for C-phycocyanin and allophycocyanin respectively) of these biliproteins were different from these values in other blue-green algae (Boussiba and Richmond, 1979). It has been observed that glucose and acetate enhanced cell growth and phycocyanin production of *S. platensis* (Chen et al., 1996).

Red algae, *Galdieria sulphuraria* 074G, was investigated for the production of phycocyanin and found that it produced significant amounts of pigment in darkness on glucose, fructose, sucrose and sugar beet molasses. Furthermore, the productivity of phycocyanin in the heterotrophic fed-batch cultures of *G. sulphuraria* was higher as compared to outdoor cultures of *Spirulina platensis* (Schmidt et al., 2005).

12.2.1 Applications

12.2.1.1 Colourant

Natural colourants are gaining importance over synthetic colours, as they are non-toxic and non-carcinogenic. The Japanese company Dai Nippon Ink and Chemical company extracts the blue phycocyanin from *Spirulina platensis* and sells it as a natural blue pigment "lina blue" commercially, for use in health foods and cosmetic products. With its increasing popularity, Dainippon Ink has carried out clinical studies on the uses of the colourant in uses of phycocyanin in foods which include the colouring of fermented milk products, ice creams, chewing gum, soft drinks, alcoholic drinks, desserts, sweet cake decoration, milk shakes and in fish feeds for Fancy Koi Corps. Other applications of the biliprotein include confectionaries, can-died ices and sherbets (Branen et al., 2001).

12.2.1.2 Fluorescent Probes

The most extensive use of algal pigments has been in diagnostics as fluorescent tags or "phycofluor probes" (Glazer and Stryer, 1984; Glazer, 1994). Three phycobiliproteins—R-phycoerythrin, B-phycoerythrin and allophycocyanin—serve as valuable fluorescent tags with numerous applications in flow cytometry, fluorescence activated cell sorting, histochemistry and to a limited degree in immunoassay and detection of reactive oxygen species. They are valuable candidates in the design and characterisation of light sensing elements in biosensors (Ayyagari et al., 1995).

When phycobilisomes are extracted into aqueous buffers, they disintegrate and the phycobiliproteins lose their natural acceptors of excitation energy and become highly fluorescent. Compared to other fluorophores, phycobiliproteins have high molar extinction coefficients and fluorescence quantum yields, and large Stokes shifts. The apo-protein chains contain amino and carboxyl groups that can form bonds to other molecules (Glazer and Stryer, 1984). Phycobiliproteins conjugated to immunoglobins, protein A and avidin were developed into fluorescent probes by Oi et al. (1982) and have since obtained wide usage in histochemistry, fluorescence microscopy, flow cytometry, fluorescence-activated cell sorting and fluorescence immunoassays (Glazer and Stryer, 1984; Sekar and Chandramohan, 2008). The property that phycobiliproteins retain their colour during electrophoretic movement can be an added advantage for their use as interval protein markers (Aráoz et al., 1998). Therefore, it can be used into the field of proteomics. A novel on-line fluorescence monitoring system for marine cyanobacterial cultivation based on phycocyanin fluorescence was developed by Sode et al. (1991). These chemically stabilised C-PC trimers can be used in fluorescent probes with spectral properties different from other phycobiliproteins. Also genetically stabilised C-PC fusion proteins fused to biospecific recognition domains have been used directly as biospecific fluorescent probes (Cai et al., 2001). *In vivo* fluorescence from PC has been used for on-line monitoring of growth in cyanobacterial cultures (Sode et al., 1991), detection of toxic cyanobacteria in drinking water (Izydorczyk et al., 2005) and remote sensing of cyanobacteria in natural waters (Simis et al., 2005).

12.2.1.3 Nutraceuticals and Pharmaceuticals

Purified C-PC has nutraceutical and pharmaceutical potentials. More attention has been paid on the use of C-PC as a nutraceutical, particularly in health foods in which dried *S. platensis* is the functional component. Beyond their nutritional value, whole cyanobacteria are suggested to stimulate the immune defence system and possess antioxidant, anti-inflammatory, anti-viral, anti-cancer and cholesterol-lowering effects, partly because of their C-PC contents. The largest route for human intake of C-PC is in non-purified form via *S. platensis* health food products (Spolaore et al., 2006). However, cyanobacteria may also contain a range of other biologically active compounds (Jensen et al., 2001; Singh et al., 2005), and it is, therefore, difficult to attribute health effects from consumption of whole cyanobacteria directly to the C-PC component.

Anti-Inflammatory

Phycocyanin enhances biological defense activity against infectious diseases through sustaining functions of the mucosal immune system and reduces allergic inflammation by the suppression of antigen-specific IgE antibody (Nemoto-Kawamura et al., 2004). Free bilirubin functions physiologically as a potent inhibitor of NADPH oxidase activity. The chromophore phycocyanobilin (PCB), found in blue-green algae also has been found to be a potent inhibitor of this enzyme complex, because it is rapidly reduced to phycocyanorubin in mammalian cells, a close homolog of bilirubin. Particularly in light of rodent studies demonstrating that orally administered *Spirulina* or phycocyanin (the *Spirulina* holoprotein that contains PCB) can exert a wide range of anti-inflammatory effects. Intake of two heaping table spoons daily would be likely to have important antioxidant activity in humans—assuming that humans and rodents digest and absorb *Spirulina*-bound PCB in a comparable manner. An intake of this magnitude can be clinically feasible if *Spirulina* is incorporated into "smoothies" featuring such ingredients as soy milk, fruit juices, and whole fruits. Such a regimen should be evaluated in clinical syndromes characterized and in part mediated by NADPH oxidase overactivity in affected tissues (McCarty, 2007). Based on these results, it is suggested that the inhibition of NO and prostaglandin E(2) over-production through suppressing iNOS and COX-2 induction and attenuation of TNF-alpha formation and neutrophil infiltration into inflammatory sites by C-PC may contribute, at least in part, to its anti-hyperalgesic activity (Shih et al., 2009).

C-Phycocyanin (CPC) protects against lipopolysaccharide- (LPS-) induced acute lung injury (ALI) in rats. Rats were challenged with LPS (5 mg kg^{-1} body weight) intratracheally to induce ALI. After 3 h LPS instillation, rats were administrated with CPC (50 mg kg^{-1} body weight, i.p.) for another 3 h. Post-treatment with CPC significantly inhibited LPS-induced elevation of protein concentration, nitrite/nitrate level, release of proinflammatory cytokines, the number of total polymorpho nuclear cells in bronchoalveolar lavage fluid, and lung edema evidenced by decrease of lung wet/dry weight ratio accompanied by a remarkable improvement of lung histopathological alterations. Furthermore, CPC significantly attenuated LPS-induced myeloperoxidase activity, O_2 (–) formation, expression of inducible nitric oxide synthase, and cyclooxygenase-2 as well as nuclear factor-kappa B (NF-κ B) activation in lungs. Additionally, CPC significantly downregulated proapoptotic proteins such as caspase-3 and Bax, but upregulated antiapoptotic proteins such as Bcl-2 and Bcl-XL in lungs exposed to LPS. These findings indicate that C-PC could be potentially useful for treatment of LPS-related ALI by inhibiting inflammatory responses and apoptosis in lung tissues (Leung et al., 2013).

Anti-Oxidant

The antioxidant and radical scavenging activities of PC from different cyanobacteria are well documented. Antioxidative activities of free phycocyanobilin are comparable to phycocyanobilin bound in C-PC (Lissi et al., 2000) and antioxidative activity is increased by denaturing or trypsin digestion of C-PC (Zhou et al., 2005). A-PC and C-PC have been reported to be potent antioxidants. Apo-proteins of A-PC and C-PC from *Spirulina platensis* were cloned, and the recombinant proteins were produced in *Escherichia coli* to study their antioxidant effects. A-PC exhibited higher activity than C-PC in scavenging peroxyl radicals, whereas C-PC exhibited higher activity than A-PC in scavenging hydroxyl radicals. All of the apophycocyanin subunits possessed strong antioxidant. However, the selection of the most useful antioxidant should depend on the type of targeted free radical to obtain the highest efficiency (Cherdkiatikul and Suwanwong, 2014). Phycocyanorubin, a reduced form of phycocyanobilin, is the important antioxidant species *in vivo* based on its similarity to bilirubin.

Bilirubin is a natural antioxidant in plasma. It is synthesised from biliverdin by biliverdin reductase; it binds to albumin and protects lipids from oxidation, while it itself is reoxidised to biliverdin (Stocker et al., 1987). Phycocyanobilin is also synthesised from biliverdin and reduced to phycocyanorubin by biliverdin reductase (Terry et al., 1993). Phycocyanorubin is also a radical scavenger, being re-oxidised to phycocyanobilin (Bhat and Madyastha, 2000). Like bilirubin, its antioxidative capacity is regenerated by biliverdin reductase and can, therefore, exceed the antioxidative capacity of administered phycocyanin. In addition, bilirubin (Kwak et al., 1991) and also phycocyanorubin (McCarty, 2007) inhibit formation of superoxide radicals by NADPH oxidase and may, therefore, play additional protective roles by reducing the generation of reactive oxygen species in the body. C-PC is bleached during scavenging of peroxyl radicals (Atanasiu et al., 1998; Hirata et al., 2000; Bhat and Madyastha, 2000). C-PC could serve as an effective natural antioxidant for efficient management of tributyltin (TBT) (Gupta et al., 2011) and peroxynitrite (ONOO (–)) induced oxidative damage (Bhat and Madyastha, 2001).

Bleaching of PC fluorescence has actually been used to investigate properties of other antioxidants in competition assays with PC (Atanasiu et al., 1998). Enhanced radical scavenging activities have been found in selenium-enriched C-PC, so-called Se-PC (Huang et al., 2007) obtained from *S. platensis* grown in Se-enriched medium (Chen et al., 2006; Huang et al., 2007). The effects of Se-PC were studied on plasma cholesterol, early atherosclerosis, cardiac production of superoxide anions, and NAD(P)H oxidase expression in hamsters. Chronic consumption of Se-rich *Spirulina* phycocyanin powerfully prevents the development of atherosclerosis. The underlying mechanism is related mainly to inhibiting pro-oxidant factors and at a lesser extent improving the serum lipid profile (Riss et al., 2007). Organic as well as inorganic Se are present in purified Se-PC but Se amino acids have not been identified (Huang et al., 2007), and mass spectrometry has shown similar molecular weights of α- and β-subunits of Se-PC and C-PC, respectively (Chen et al., 2006). Therefore, it is not clear whether the increased radical scavenging capability of

Se-PC is a result of covalently incorporated Se or associated Se compounds. Accumulation of Se in mixotrophic culture of *S. platensis* was investigated and found that glucose was better than acetate as an organic carbon source for mixotrophic culture of *S. platensis.*

The application of mixotrophic culture *S. platensis* with stepwise addition of Se to the medium could offer an effective and economical way for the production of high Se-enriched algal products (Chen et al., 2006). The PC could also attenuate ischemia-reperfusion (I/R)-induced cardiac dysfunction through its antioxidant and antiapoptotic actions (Khan et al., 2006). Oxalate induced renal calculi formation and the associated renal injury is thought to be caused by free radical mediated mechanisms. PC acts as antioxidant against calcium oxalate urolithiasis. It significantly controlled the early biochemical changes in calcium oxalate stone formation (Farooq et al., 2004). C-PC has also capability to modulate the antioxidant enzyme status, thereby could retard sodium selenite-induced cataract incidence both *in vitro* and *in vivo* (Kumari et al., 2013). It was found that C-PC was also useful in overcoming the drug resistance in cellular systems by scavenging the reactive oxygen species (Roy et al., 2007).

Anti-Diabetic

The encouraging results from different experiments have established the potential of PC as a clinical measure in preventing diabetes. An alloxan-injured mice was investigated by oral administration of phycocyanin obtained from *Spirulina platensis*. It has been observed that alloxan-injured mice was having decreased fasting blood glucose and glycosylated serum protein (GSP); maintained total antioxidative capability (T-AOC); avert malondialdehyde (MDA) formation in the liver, kidney, and pancreas; decreased total cholesterol (TC) level and triglycerides (TG) level in serum and liver; increased the levels of hepatic glycogen level; maintained glucokinase (GK) expression in the liver; and decreased p53 expression in the pancreas at mRNA level. Acute toxicity study further shows that phycocyanin is relatively safe to use as drug (Ou et al., 2012). Furthermore, it was found that the antidiabetic effect of PC on mice is most likely due to its ability to enhance insulin sensitivity, amelioration of insulin resistance of peripheral target tissues and regulation of glucolipide metabolism. Therefore, PC may have a potential clinical utility in combating type-2 diabetes (Ou et al., 2013). Oral administration of phycocyanin and phycocyanobilin may offer a novel and feasible therapeutic approach for preventing diabetic nephropathy (Zheng et al., 2013).

Anti-Thromboembolic

A novel and very potent (in nanomolar concentrations) antiplatelet agent could be C-PC for treatment of arterial thromboembolism. It was involved in the following inhibitory pathways: (1) C-phycocyanin increases cyclic GMP/VASP Ser157 phosphorylation and subsequently inhibits protein kinase C activity, resulting in inhibition of both P47 phosphorylation and intracellular Ca^{2+} mobilization, and (2) C-PC may inhibit free radicals (such as hydroxyl radicals) released from activated platelets, which ultimately inhibits platelet aggregation (Hsiao et al., 2005). The thromboxane B2 formation caused by collagen or arachidonic acid was significantly inhibited by C-PC due to suppression of cyclooxygenase and thromboxane synthase activity. Similarly, the rise of platelet intracellular calcium level stimulated by arachidonic acid and collagen-induced platelet membrane surface glycoprotein expression were also attenuated. In addition it significantly increased the platelet membrane fluidity and the cyclic AMP level through inhibiting cyclic AMP phosphodiesterase activity (Chiu et al., 2006).

Anti-Hypertensive

Endothelial dysfunction is associated with hypertension, atherosclerosis and metabolic syndrome. PC has antihypertensive effect. Long-term administration of PC may ameliorate systemic blood pressure. Phycocyanin may be beneficial for preventing endothelial dysfunction-related diseases in metabolic syndrome (Ichimura et al., 2013).

Hypo-Lipidemic

Spirulina platensis concentrate (SPC) was identified as the novel hypocholesterolemic agent by Nagaoka et al. (2005). The hypocholesterolemic action of PC may involve the inhibition of both cholesterol absorption and bile acid reabsorption. The aqueous extract of *S. platensis* may inhibit the intestinal absorption of dietary fat by inhibiting pancreatic lipase activity. PC inhibited the pancreatic lipase activity in a dose-dependent manner. These results suggest that the inhibitory effects of *S. platensis* on postprandial triacylglycerolemia may be due in part to the inhibition of pancreatic lipase activity by glycolipid H-b2 and PC (Han et al., 2006). Hyperlipidemia and oxidation play major roles upon cardiovascular diseases (CVDs). C-PC was effective in lowering serum cholesterol, total cholesterol (TC), TG, LDL, GOT and GPT. C-PC is found to decrease the malondialdehyde (MDA) equivalents and delay the diene conjugation in the Cu^{2+}-mediated oxidation of LDL. The lipid-lowering and antioxidation effects of C-PC suggested its roles in prevention of CVD and atherosclerotic formation (Sheu et al., 2013).

Anti-Carcinogenic

Both selenium (Se) and PC have been reported to show potent cancer chemopreventive activities. Se-PC was identified as a potent antiproliferative agent against human melanoma A375 cells and human breast adenocarcinoma MCF-7 cells. Induction of apoptosis in both A375 and MCF-7 cells by Se-PC was evidenced by accumulation of sub-G1 cell populations, DNA fragmentation and nuclear condensation. Further investigation on intracellular mechanisms indicated that depletion of mitochondrial membrane potential ($\Delta\Psi$m) was involved in Se-PC-induced cell apoptosis (Chen and Wong, 2008). The regulatory effect of phycocyanin (PC) from Spirulina platensis on cluster of differentiation 59 (CD59) gene expression of HeLa cells and antitumoral mechanism of PC was investigated by Li et al. (2005). PC can promote the expression of CD59 protein in HeLa cells, hold back the reproductions of HeLa cells, and moreover, a dosage effect was found between them. Namely, with the ascendance of PC concentration, the expression quantities of CD59 protein and apoptosis-inducing Fas protein increased and the multiplication activity of HeLa cells declined, whereas PC was of no use to CD59 and Fas protein expression, and reproduction of normal CHO cells as well. Furthermore, there was a significant decrease in the number of cells that survived for HeLa cells treated with C-PC compared with control cells untreated with C-PC.

Electron-microscopic studies revealed that C-PC could induce characteristic apoptotic features, including cell shrinkage, membrane blebbing, microvilli loss, chromatin margination and condensation into dense granules or blocks. Agarose electrophoresis of genomic DNA of HeLa cells treated with C-PC showed fragmentation pattern typical for apoptotic cells. Flow-cytometric analysis of HeLa cells treated with different concentrations of C-PC demonstrated an increasing percentage of cells in sub-G0/G1 phase. It was found that C-PC could promote the expression of Fas and ICAM-1 (intercellular cell-adhesion molecule 1) protein, while it held back the Bcl-2 (B-cell lymphocytic-leukaemia proto-oncogene 2) protein expression. This suggested that C-PC could induce the activation of pro-apoptotic gene and downregulation of anti-apoptotic gene expression and then facilitate the transduction of tumoural apoptosis signals that resulted in the apoptosis of HeLa cells (*in vitro*) (Li et al., 2006). It was found that C-PC was also effective on hepatocellular carcinoma (Roy et al., 2007) and having antineoplastic effects in colon carcinogenesis (Saini et al., 2012; Saini and Sanyal, 2014). A significant down regulation of multi drug resistance was observed in C-PC treated human hepatocarcinoma cells (Nishanth et al., 2010). C-PC could significantly reduce the expressions of N-methyl D-aspartate receptor subunit 2B (NR2B), tumour necrosis factor-α (TNF-α), interleukin-1β (IL-1β), and cyclooxygenase type 2 (COX-2) genes in the cochlea and inferior colliculus (IC) of mice (Hwang et al., 2013).

12.3 Phycoerythrin

Phycoerythrin is a red pigment extracted from blue green algae, cryptomonads and red algae like *Porphyridium cruentum. Porphyridium cruentum* is the most commonly used species for phycoerythrin production. It is cultured in artificial seawater with added potassium nitrate. Optimum temperature of growth for *Porphyridium* is 21 °C. Visually, the phycoerythrins appear red, the phycocyanins range from purple (phycoerythrocyanin, R-phycocyanin) to deep blue (C-phycocyanin), and the allophycocyanins are blue with a hint of green. Pure phycobiliproteins from crude algal extracts are usually obtained by a combination of different and non-scaleable methods. Particularly, phycoerythrin is classically purified by a combination of several techniques such as ammonium sulphate precipitation, ion-exchange chromatography, gel filtration and chromatography on hydroxylapatite. Purification procedures are often long and complex. For this reason the use of this biliprotein has been somewhat limited by the tedious preparation of adequate amounts of the purified protein.

12.3.1 Applications

Phycobiliproteins are used as colourants in food (chewing gums, dairy products, ice sherbaths, gellies, etc.) and cosmetics such as lipstick and eyeliners in Japan, Thailand and China. Dainippon Ink and Chemicals produces a blue food colourant from *Spirulina*, called lina blue, that is used in chewing gum, ice slush, sweets, soft drinks and dairy products. Phycobiliproteins can be commercially produced from *Spirulina* and the red microalgae *Porphyridium* and *Rhodella* (Becker, 1994; Singh et al., 2005; Spolaore et al., 2006). Phycobiliproteins are widely used in clinical or research immunology because they are very powerful and have highly sensitive fluorescent properties (Spolaore et al., 2006). In 1997 the global market for phycobiliproteins colourants was estimated at US$ 50 million and prices vary from US$ 3 to US$ 25 mg^{-1} (Spolaore et al., 2006). Prominent players in the phycoerythrin market are Europa-bioproducts, Cambridge, UK and Invitrogen, NY, USA.

12.4 β-Carotene

Algal carotenoids are present in chloroplasts as a complex mixture characteristic of each class of algae. The genus *Dunaliella* is a halotolerant, unicellular, motile green algae with exceptional morphological and physiological properties belonging to family *Chlorophyceae* (Ben-Amotz and Avron, 1992). It is the best carotenoid providing organism among the algae and other organisms. Initial growth phase requires 12–14 days in nitrate-rich medium. Magnesium salt is essential as it is required for chlorophyll production. For the carotenogenesis phase, nitrate depletion along with salinity maintenance and high light intensity will give best result (Ben-Amotz and Avron, 1983). It has both cis- and trans-isomers of carotenoids for high bioavailability and bioefficacy (Yeum and Russell, 2002) and contains oxygenated carotenoids (xanthophylls), which have higher bioactivity and better anticancer properties (Roodenburg et al., 2000). Commonly cultivated species are *Dunaliella salina* and *D. bardawil.* On average, *Dunaliella salina* under ideal conditions can yield 400 mg β-carotene m^{-2} of cultivation area (Finney et al., 1984).

12.4.1 Applications

This pigment is used mainly as food colourant and it imparts a yellow-orange colour. It is a non-harmful colourant used in numerous cosmetic and body-care products and food colourings. It is one of the leading food colourants in the world applied to a different food and beverage products to improve their appearance to customers, including items such as margarine, cheese, fruit juices, baked goods, dairy products, canned goods, confectionery, health condiments etc. (Coma, 1991). Apart from its use as a colourant, it is used popularly as a nutraceutical additive. The human body converts β-carotene to vitamin A via body tissues as opposed to the liver, hence avoiding a build-up of toxins in the liver. Vitamin A is essential for the human body in that it assists the body's immune system and helps battle eye diseases, such as cataracts and night blindness, various skin ailments such as acne, signs of aging, and various forms of cancer (Agarwal and Rao, 2000). Other therapeutic applications include antihypertensive, bronchodilator, analgesic and muscle relaxant and antiedema activity (Villar et al., 1992). Antioxidant qualities could reduce the harmful effects of free radicals, which are implied in over 60 life-threatening diseases including various forms of cancer, coronary heart disease, premature aging and arthritis (Törnwall et al., 2004). Additionally, the antioxidant qualities assist the body in suppressing the effects of premature aging caused by UV rays. The antioxidant and precursory vitamin A properties increase the appeal and application of β-carotene in pet foods (Cantrell et al., 2003). *Dunaliella* has been marked as powder form for human use. Extracted purified β-carotene is sold mostly in vegetable oil in bulk concentrations from 1 to 20 % to colour various food products and for personal use. The purified natural β-carotene is generally accompanied by the other carotenoids of *Dunaliella*, predominantly: lutein, neoxanthin, zeaxanthin, violaxanthin, cryptoxanthin, α-carotene for a total of approximately 15 % of the carotene concentration and is marketed under the title 'carotenoids mix'. Several companies involved in marketing of *Dunaliella* major in India are Parry's Agro Ltd. and ABC Biotech Ltd., located in Tamil Nadu which produce *Dunaliella* as well as β-carotene for pharmaceutical purpose. Other than these two worldwide several companies are involved in cultivation of *Dunaliella* for commercial purposes. The complete biomass can be utilized for bread and other products (Finney et al., 1984), the whole cells can be utilized for animals, poultry and fish feed since they are safe (Mokady et al., 1989).

12.5 Fucoxanthin

Fucoxanthin is a carotenoid, which performs a limited form of photosynthesis in brown algae and other heterokonts. The pigment is produced in abundant quantities from edible brown algae like *Laminalia japonica, Undaria pinnatifida, Eisenia bicyclis, Sargassum fulvellum, Hizikia fusiformis,* etc. It is responsible for the brown or olive-green colour of these algae. Commercial-scale production of fucoxanthin is often found difficult because fucoxanthin is produced by only a few edible brown algae that are refractory to harvesting on a mass scale.

12.5.1 Applications

Fucoxanthin shows biological activities such as anti-obesity, anti-oxidant, antitumour, anti-diabetes and anti-inflammatory properties (Heo et al., 2010). The anti-carcinogenic activity of fucoxanthin is reported to be the strongest among xanthophylls and carotenoids and is shown to prevent liver and skin cancer due to its antioxidant activity, and breast and prostate cancer through induction of apoptosis (Muthuirulappan and Francis, 2013). Fucoxanthin exhibits anti-obesity effects through stimulating the expression of mitochondrial uncoupling protein in white adipose tissue (Maeda et al., 2009). The leading companies manufacturing fucoxanthin are AlgaNova International (China) and Leili Natural Products Co. Ltd. (China).

12.6 Astaxanthin

Astaxanthin is yellow to red ketocarotenoid mainly obtained from *Haematococcus pluvialis* (green alga). It can produce up to 0.2–2.0 % (on dry weight basis). The astaxanthin is contained in the aplanospore of the algae. It can grow both under autotrophic and heterotrophic conditions. It is a freshwater algae and thus open-air culture is extremely difficult due to contamination by many other undesirable algal species. It may require closed culture systems such as tubular photobioreactors and although some pilot-scale units have been tested (Borowitzka and Borowitzka, 1989).

12.6.1 Applications

It is a high value product with applications in pharmaceuticals, nutraceuticals and animal nutrition with a potent pigment with antioxidant property and tinctorial value (Lorenz and Cysewski, 2000). Astaxanthin from *Haemato-coccus* algae is under consideration for US Food and Drug Administration clearance and several European countries had approved of its marketing as dietary supplement ingredient for human consumption. Its applications in many degenerative diseases and cancer prevention show its potential for technology development. It is better than other carotenoids in the sense it is more stable as compared to other carotenoids. It has high antioxidant potential (10 times more than β-carotene and 500 times more than α-tocopherol); it can easily cross blood brain barrier and has high tinctorial property. It is also helpful in treating Alzheimer's and Parkinson's diseases. It is a powerful quencher of singlet oxygen as evident from in vitro studies (Di Mascio et al.,

1990). It also has strong activity as an inhibitor of lipid peroxidation mediated by active forms of oxygen (Kobayashi and Sakamoto, 1999). The strong antioxidative activities of astaxanthin suggest its potential as photoprotectant against UV irradiation (Savoure et al., 1994). Astaxanthin-containing preparations for prevention of light-induced aging of skin have been developed by Suzuki et al. (1996). It can be used to prevent arteriosclerosis, coronary artery disease and ischemic brain development. A number of astaxanthin health products are under study. Studies on rats had shown no toxicity of astaxanthin preparations (Nishikawa et al., 1997). Dietary administration of astaxanthin has proved to significantly inhibit carcinogenesis in the mouse urinary bladder, rat oral cavity and rat colon (Tanaka et al., 1995). In addition, it is reported to induce xenobiotic-metabolizing enzymes in rat liver (Gradelet et al., 1996).

Astaxanthin has been shown to enhance *in vitro* antibody production by mouse spleen cells stimulated with sheep red blood cells and in human blood cells *in vitro* (Jyonouchi et al., 1991). Furthermore it has not exhibited any mutagencity *in vitro* study at doses up to 14.4 mg d^{-1} for two weeks (Miki et al., 1998). Fassett and Coombes (2012) has reported anti-inflammatory effect of astaxanthin which reduces cardiovascular disease caused by atherosclerosis. An oral preparation has been developed by Alejung and Wadstroem (1998) for the treatment of *Helicobacter* infections of the mammalian gastrointestinal tract. There was strong evidence to suggest that astaxanthin was shown to modulate the humoral and non-humoral immune systems. It enhances the release of interleukin-1 and the tumour necrosis factor in mice to a greater extent than canthaxanthin or β-carotene and has the highest cytokinine inducing activity (Okai and Higashi-Okai, 1996). Astaxanthin has a significant enhancing action on the production of immunoglobulin A, M and G and on T-helper cell antibody production, even when suboptimal amounts of antigen are present (Jyonouchi et al., 1994). It has also been used as salmon and trout feeds. Several companies are producing astaxanthin from *Haematococcus* worldwide: Mera Pharmaceuticals, Kailua–Kona, Hawaii 96740, USA; Cyanotech Corporation, Kailua–Kona, HI 96740, USA; BioReal Inc., Hawaii, USA; a subsidiary of Fuji Chemical Industry, Toyama, Japan; and Parry's Pharmaceuticals, Chennai, India.

12.7 Chlorophyll

Micro alga presents one of the highest chlorophyll contents found in nature, corresponding to 1.15 % of its biomass. It is mainly derived from green and blue green algae mainly *Chlorella* sp. and *Spirulina platensis*. In Brazil, the chlorophyll used as natural green colourant is obtained from spinach, which contains approximately 0.06 mg g^{-1} whereas the *Spirulina* sp biomass contains 1.15 mg g^{-1} of chlorophyll (El-Baky et al., 2007). Chlorophyll is generally produced by *Spirulina* using fermentation process. It has been shown that the composition of the cultivation medium, cellular age, and light intensity are the main factors influencing chlorophyll content in *S. platensis* biomass. Research studies have shown that *Spirulina* cultivated under poor light intensity present higher chlorophyll concentration in the biomass than that cultivated under highly illuminated conditions. This suggests an inverse relationship between light intensity and chlorophyll content. The concentration of chlorophyll in *S. platensis* also increases with the increase in nitrogen concentration of the medium (Vonshak et al., 1982). Chlorophyll production from *Chlorella* can be carried out in open ponds as well as fermenters. *Chlorella* cultivation is also carried out in open ponds, but concrete ponds are not economical.

12.7.1 Applications

Chlorophyll used as a food colourant is found to exhibit anti-mutagenic property. This is accomplished by inducing production of carcinogen detoxifying enzymes, and thereby reducing the risk of cancer. Chlorophyll and its derivatives are known to have antioxidant activity. Consumption of vegetables rich in chlorophyll and chlorophyll derivatives such as chlorophyllin, is associated with reduced risks of certain types of cancers. Consumption of chlorophyll-rich diet could prevent or delay the onset of certain diseases such as cancer that manifest with aging and are induced by free radicals (Koníčková et al., 2014). The function of chlorophyll in animals is suggested to be inhibition of lipid peroxidation and protection of mitochondria from oxidative damage induced by various free-radicals and other reactive oxygen species. Chlorophyllin has also been reported to inhibit radiation-induced DNA and mitochondrial membrane damage and it would also appear to be a potent protector of DNA with regard to oxidative damage (Plaza et al., 2010). Chlorophyll increases peristaltic action and thus relieves constipation, and also normalizes the secretion of digestive acids. It soothes the inflammation and reduces the excess pepsin secretion associated with gastric ulcers. Chlorophyll actually helps remove heavy metals from the body that have accumulated due to the ingestion of contaminated food products.

During World War II, the drying action of chlorophyll and its antiseptic qualities made it a common first-aid measure to prevent festering of wounds. In addition, chlorophyll soothes swelling and promotes granulation, the process that regenerates new tissue over injuries. Chlorophyll appears to promote regeneration of damaged liver cells, and also increases circulation to all the organs by dilating blood vessels. It is believed that if chlorophyll is ingested with sufficient iron, the magnesium can be displaced to yield a haemoglobin molecule (Chernomorsky and Segelman, 1988). Experiments in Japan have demonstrated that *Spirulina* has a marked positive effect on anaemia, possibly due to the conversion of chlorophyll into haemoglobin. In the food industry, chlorophyll is used as a natural pigment ingredient in processed foods. Because of its strong green pigment and consumers growing preference for natural foods, chlorophyll is gaining importance as food additive. Increasing number of researches are also reporting health benefits from consumption of high chlorophyll diet. This in turn is encouraging food processors to switch from artificial pigments to chlorophyll-based natural colouring. In the cosmetics industry, chlorophyll-*a* is used in soaps and cosmetics products (Dweck, 2002) (Table 12.2).

Table 12.2 Algal pigments with their industrial applications

Pigments	Main source	Industrial applications	Leading companies
Phycocyanin	*Spirulina platensis* (blue green algae)	Colourant, Fluorescent probes, Anti-inflammatory, Anti-oxidant, Anti-diabetic, Anti-thromboembolic, Anti-hypertensive, Hypo-lipidemic, Anti-carcinogenic.	Parry Nutraceuticals (India), EcoFuel Laboratories (Czech Republic), Dainippon Ink and Chemicals (Japan).
Phycoerythrin	*Spirulina* (blue green algae), *Porphyridium* and *Rhodella* (red algae)	Colourant-Lina blue, Food colourant, fluorescent properties.	Europa-bioproducts, Cambridge, UK and Invitrogen, NY, USA.
β-carotene	*Dunaliella salina* and *D. bardawil* (green algae)	Yellow-orange colourant, source of vitamin A, Anti-oxidant, Anti-hypertensive, Bronchodilator, Analgesic and Muscle relaxant and Anti-edema activity.	In India: Parry's agro Ltd., ABC biotech Ltd., located in Tamil Nadu.
Fucoxanthin	*Laminalia japonica, Undaria pinnatifida, Eisenia bicyclis, Sargassum fulvellum* and *Hizikia fusiformis* (brown algae)	Anti-obesity, Anti-oxidant, Anti-tumour, Anti-diabetes, Anti-inflammatory, Anti-carcinogenic.	AlgaNova International (China) and Leili Natural Products Co., Ltd. (China).
Astaxanthin	*Haematococcus pluvialis* (green algae)	Anti-oxidant property and tinctorial value, Treating Alzheimer's and Parkinson's diseases.	Mera Pharmaceuticals, Kailua–Kona, Hawaii 96740 (USA), Cyanotech Corporation, Kailua–Kona, HI 96740 (USA), BioReal Inc., Hawaii (USA), a subsidiary of Fuji Chemical Industry, Toyama (Japan) and Parry's Pharmaceuticals, Chennai (India).
Chlorophyll	*Chlorella* sp. (green algae) and *Spirulina platensis* (blue green algae)	Colourant, Anti-mutagenic, Anti-oxidant,	British Chlorophyll Ltd (UK)

12.8 Future Prospective

In the last 10–15 years, there has been increasing interest in the potential uses of C-PC in fluorescent applications, nutraceuticals and pharmaceuticals. C-PC has been stabilised chemically and by protein engineering, and new purification procedures allow very pure C-PC to be obtained at high yields. Positive health effects have been related to intake of purified C-PC. Although most of these are still not fully understood, recognition of functional resemblances between phycocyanorubin and the natural, physiological antioxidant bilirubin has turned C-PC into a prospective nutraceutical or pharmaceutical candidate with more well documented functionalities than health foods in general. Successful nutraceutical or pharmaceutical applications will depend on C-PC produced under well-controlled conditions that are difficult to accommodate in open cultures. Recombinant and heterotrophic production procedures seem more promising for novel C-PC synthesis at industrial scales. Recombinant expression of C-PC and holo-C-PC α-subunits in Anabaena sp. and E. coli has already demonstrated that protein engineering can generate C-PC with improved stability or novel functionalities. However, complete multimeric, recombinant C-PC complexes are still not made in heterotrophic cultures. Balanced co-synthesis and addition of chromophores combined with effective folding and complex assembly will remain major challenges to the productivity of C-PC in heterologous hosts. Instead G. sulphuraria is a promising candidate for heterotrophic synthesis of C-PC at large scale. Very high volumetric C-PC productivities have been obtained in this alga, and well controlled, axenic cultures can readily be scaled up.

References

Agarwal, S. and Rao, A.V. (2000). Tomato lycopene and its role in human health and chronic diseases. *Canadian Medical Association Journal*, 163(6), 739–744.

Alejung, P. and Wadstroem, T. (1998). Oral preparation for treatment of Helicobacter sp. Infections-comprises xanthophylls, especially astaxanthin esterified with a fatty acid and derived from the alga Haematococcus sp. World Patent, 9837874.

Aráoz, R., Häder, D.P. and Huss, V.A.R. (1998). DNA sequence and secondary structure of the small subunit rRNA (Accession No. Y14950) from the green alga *Chlorella* spec. Yanaqocha RA1 (Trebouxiophyceae, Chlorophyta). *Plant Physiol*, 116, 1604.

Atanasiu, R.L., Stea, D., Mateescu, M.A., Vergely, C., Dalloz, F., Briot, F. and Rochette, L. (1998). Direct evidence of caeruloplasmin antioxidant properties. *Molecular and Cellular Biochemistry*, 189(1-2), 127–135.

Ayyagari, M., Pande, R., Kamtekar, S., Gao, H., Marx, K., Kumar, J., Tripathy, S., Akkara, J. and Kaplan, D. (1995). Molecular assembly of proteins and conjugated polymers: Toward development of biosensors. *Biotechnol. Bioeng.*, 45, 116–121.

Becker, E.W. (1994). Microalgae. Biotechnology and Microbiology. Cambridge: Cambridge University Press. ISBN 978-0-521-06113.

Ben-Amotz, A. and Avron, M. (1983). On the factors which determine massive β-carotene accumulation in the halotolerant alga Dunaliella bardawil. Plant Physiology, 72(3), 593–597.

Ben-Amotz, A. and Avron, M. (1992). Dunaliella: Physiology, biochemistry, and biotechnology. CRC Press.

Bewicke, D. and Potter, B. (2009). Chlorella: The Emerald Food. Ronin Publishing, Berkeley, CA.

Bhaskar, S.U., Gopalaswamy, G. and Raghu, R. (2005). A simple method for efficient extraction and purification of C-phycocyanin from *Spirulina platensis* Geitler. *Indian J Exp Biol*, 43(3), 277.

Bhat, V.B. and Madyastha, K.M. (2000). C-Phycocyanin: A Potent Peroxyl Radical Scavenger in vivo and in vitro. *Biochemical and biophysical research communications*, 275(1), 20–25.

Bhat, V.B. and Madyastha, K.M. (2001). Scavenging of Peroxynitrite by Phycocyanin and Phycocyanobilin from *Spirulina platensis*: Protection against Oxidative Damage to DNA. *Biochemical and Biophysical Research Communications*, 285(2), 262–266.

Borowitzka, L.J. and Borowitzka, M.A. (1989b). Industrial production: Methods and economics. *In:* R.C. Cresswell, T.A.V. Rees, N. Shah (Eds), Algal and Cyanobacterial Biotechnology. Longman Scientific, London. pp. 244–316.

Boussiba, S. and Richmond, A.E. (1979). Isolation and characterization of phyco-cyanins from the blue-green alga *Spirulina platensis. Archives of Microbiology*, 120(2), 155–159.

Branen, A.L., Davidson, P.M., Salminen, S. and Thorngate, J. (Eds). (2001). Food additives. CRC Press.

Cai, Y.A., Murphy, J.T., Wedemaye, G.J. and Glazer, A.N. (2001). Recombinant phycobiliproteins. Recombinant C-phycocyanins equipped with affinity tags, oligomerization, and biospecific recognition domains. *Anal Biochem*, 290, 186–204.

Cantrell, A., McGarvey, D.J., Truscott, T.G., Rancan, F. and Böhm, F. (2003). Singlet oxygen quenching by dietary carotenoids in a model membrane environment. *Archives of Biochemistry and Biophysics*, 412(1), 47–54.

Chen, F., Zhang, Y. and Guo, S. (1996). Growth and phycocyanin formation of *Spirulina platensis* in photoheterotrophic culture. *Biotechnology Letters*, 18(5), 603–608.

Chen, T. and Wong, Y.S. (2008). In vitro antioxidant and antiproliferative activities of selenium-containing phycocyanin from selenium-enriched *Spirulina platensis. Journal of Agricultural and Food Chemistry*, 56(12), 4352–4358.

Chen, T., Zheng, W., Wong, Y.S., Yang, F. and Bai, Y. (2006). Accumulation of selenium in mixotrophic culture of *Spirulina platensis* on glucose. *Bioresource Technology*, 97(18), 2260–2265.

Cherdkiatikul, T. and Suwanwong, Y. (2014). Production of the α and β Subunits of Spirulina Allophycocyanin and C-Phycocyanin in *Escherichia coli*: A Comparative Study of Their Antioxidant Activities. *Journal of Biomolecular Screening*, 19(6), 959–965.

Chernomorsky, S.A. and Segelman, A.B. (1988). Biological activities of chlorophyll derivatives. *New Jersey Medicine: The Journal of the Medical Society of New Jersey*, 85(8), 669.

Chiu, H.F., Yang, S.P., Kuo, Y.L., Lai, Y.S. and Chou, T.C. (2006). Mechanisms involved in the antiplatelet effect of C-phycocyanin. *British Journal of Nutrition*, 95(02), 435–440.

Coma (1991). Dietary Reference Values for Food Energy and Nutritions for the United Kingdom. Report of the panel on dietary reference values. Committee on Medical Aspects of Food and Nutrition Policy. HMSO, London. Containing drink. Patent application number 10155459. Japanese.

Cook, A.H. (1945). Algal pigments and their significance. *Biological Reviews*, 20(3), 115–132.

Di Mascio, P., Devasagayam, T.P., Kaiser, S. and Sies, H. (1990). Carotenoids, tocopherols and thiols as biological singlet molecular oxygen quenchers. *Biochemical Society Transactions*, 18(6), 1054–1056.

Dweck, A.C. (2002). Natural ingredients for colouring and styling. *International Journal of Cosmetic Science*, 24(5), 287–302.

Eisele, L.E., Bakhru, S.H., Liu, X., MacColl, R. and Edwards, M.R. (2000). Studies on C-phycocyanin from *Cyanidium caldarium*, a eukaryote at the extremes of habitat. *Biochimica et Biophysica Acta (BBA)-Bioenergetics*, 1456(2), 99–107.

El-Baky, H.H.A., El Baz, F.K. and El-Baroty, G.S. (2007). Enhancement of antioxidant production in Spirulina plantensis under oxidative stress. *American-Eurasian Journal of Scientific Research*, 2(2), 170–179.

Farooq, S.M., Asokan, D., Sakthivel, R., Kalaiselvi, P. and Varalakshmi, P. (2004). Salubrious effect of C-phycocyanin against oxalate-mediated renal cell injury. *Clinica Chimica Acta*, 348(1), 199–205.

Fassett, R.G. and Coombes, J.S. (2012). Astaxanthin in cardiovascular health and disease. *Molecules*, 17(2), 2030–2048.

Finney, K.F., Pomeranz, Y. and Bruinsma, B.L. (1984). Use of algae Dunaliella as a protein supplement in bread. *Cereal Chemistry* (USA), 61(5), 402–406.

Fritsch, F.E. (1944). Present-day classification of algae. *The Botanical Review*, 10(4), 233–277.

Glazer, A.N. (1976). Phycocyanins: Structure and function. *In:* Photochemical and Photobiological Reviews (pp. 71–115). Springer US.

Glazer, A.N. (1994). Phycobiliproteins—A family of valuable, widely used fluorophores. *J Appl Phycol*, 6, 105–112.

Glazer, A.N. and Stryer, L. (1984). Phycofluor probes. *Trends in Biochemical Sciences*, 9(10), 423–427.

Gradelet, S., Astorg, P., Leclerc, J., Chevalier, J., Vernevaut, M.F. and Siess, M.H. (1996). Effects of canthaxanthin, astaxanthin, lycopene and lutein on liver xenobiotic-metabolizing enzymes in the rat. *Xenobiotica*, 26(1), 49–63.

Gupta, M., Dwivedi, U.N. and Khandelwal, S. (2011). C-Phycocyanin: An effective protective agent against thymic atrophy by tributyltin. *Toxicology Letters*, 204(1), 2–11.

Han, L.K., Li, D.X., Xiang, L., Gong, X.J., Kondo, Y., Suzuki, I. and Okuda, H. (2006). Isolation of pancreatic lipase activity-inhibitory component of spirulina platensis and it reduce postprandial triacylglycerolemia. Yakugaku zasshi. *Journal of the Pharmaceutical Society of Japan*, 126(1), 43–49.

Heo, S.J., Yoon, W.J., Kim, K.N., Ahn, G.N., Kang, S.M., Kang, D.H. and Jeon, Y.J. (2010). Evaluation of anti-inflammatory effect of fucoxanthin isolated from brown algae in lipopolysaccharide-stimulated RAW 264.7 macrophages. *Food and Chemical Toxicology*, 48(8), 2045–2051.

Hirata, T., Tanaka, M., Ooike, M., Tsunomura, T. and Sakaguchi, M. (2000). Antioxidant activities of phycocyanobilin prepared from Spirulina platensis. *Journal of Applied Phycology*, 12(3-5), 435–439.

Hsiao, G., Chou, P.H., Shen, M.Y., Chou, D.S., Lin, C.H. and Sheu, J.R. (2005). C-phycocyanin, a very potent and novel platelet aggregation inhibitor from Spirulina platensis. *Journal of Agricultural and Food Chemistry*, 53(20), 7734–7740.

Huang, Z., Guo, B.J., Wong, R.N.S. and Jiang, Y. (2007). Characterization and antioxidant activity of selenium-containing phycocyanin isolated from Spirulina platensis. *Food Chemistry*, 100(3), 1137–1143.

Hwang, J.H., Chen, J.C. and Chan, Y.C. (2013). Effects of C-phycocyanin and Spirulina on salicylate-induced tinnitus, expression of NMDA receptor and inflammatory genes. *PloS one*, 8(3), e58215.

Ichimura, M., Kato, S., Tsuneyama, K., Matsutake, S., Kamogawa, M., Hirao, E. and Omagari, K. (2013). Phycocyanin prevents hypertension and low serum adiponectin level in a rat model of metabolic syndrome. *Nutrition Research*, 33(5), 397–405.

Izydorczyk, K., Tarczynska, M., Jurczak, T., Mrowczynski, J. and Zalewski, M. (2005). Measurement of phycocyanin fluorescence as an online early warning system for cyanobacteria in reservoir intake water. *Environ Tox*, 20, 425–430.

Jensen, G.S., Ginsberg, D.I. and Drapeau, C. (2001). Blue-green algae as an immuno-enhancer and biomodulator. *J Amer Nutraceut Assoc*, 3(4), 24–30.

Jyonouchi, H., Hill, R.J., Tomita, Y. and Good, R.A. (1991). Studies of immunomodulating actions of carotenoids. I. Effects of β-carotene and astaxanthin on murine lymphocyte functions and cell surface marker expression in in vitro culture system. *Nutr Cancer*, 16(2), 93–105.

Jyonouchi, H., Zhang, L., Gross, M. and Tomita, Y. (1994). Immunomodulating actions of carotenoids: Enhancement of in vivo and in vitro antibody production to T-dependent antigens. *Nutr Cancer*, 21(1), 47–58.

Khan, M., Varadharaj, S., Ganesan, L.P., Shobha, J.C., Naidu, M.U., Parinandi, N.L. and Kuppusamy, P. (2006). C-phycocyanin protects against ischemia-reperfusion injury of heart through involvement of p38 MAPK and ERK signaling. *American Journal of Physiology-Heart and Circulatory Physiology*, 290(5), H2136-H2145.

Kobayashi, M. and Sakamoto, Y. (1999). Singlet oxygen quenching ability of astaxanthin esters from the green alga Haematococcus pluvialis. *Biotechnology Letters*, 21(4), 265–269.

Konícková, R., Vanková, K., Vaníková, J., Vánová, K., Muchová, L., Subhanová, I. and Vítek, L. (2014). Anti-cancer effects of blue-green alga Spirulina platensis: A natural source of bilirubin-like tetrapyrrolic compounds. *Ann. Hepatol*, 13, 273–283.

Ku, C.S., Yang, Y., Park, Y. and Lee, J. (2013). Health benefits of blue-green algae: prevention of cardiovascular disease and nonalcoholic fatty liver disease. *Journal of Medicinal Food*, 16(2), 103–111.

Kumari, R.P., Sivakumar, J., Thankappan, B. and Anbarasu, K. (2013). C-phycocyanin modulates selenite-induced cataractogenesis in rats. *Biological Trace Element Research*, 151(1), 59–67.

Kwak, J.Y., Takeshige, K., Cheung, B.S. and Minakami, S. (1991). Bilirubin inhibits the activation of superoxide-producing NADPH oxidase in a neutrophil cell-free system. *Biochimica et Biophysica Acta (BBA)*—Protein Structure and Molecular Enzymology, 1076(3), 369–373.

Leung, P.O., Lee, H.H., Kung, Y.C., Tsai, M.F. and Chou, T.C. (2013). Therapeutic effect of C-phycocyanin extracted from blue green algae in a rat model of acute lung injury induced by lipopolysaccharide. Evidence-Based Complementary and Alternative Medicine. 2013, 916590 (eCAM2013).

Li, B., Gao, M.H., Zhang, X.C. and Chu, X.M. (2006). Molecular immune mechanism of C-phycocyanin from *Spirulina platensis* induces apoptosis in HeLa cells in vitro. *Biotechnology and Applied Biochemistry*, 43(3), 155–164.

Li, B., Zhang, X., Gao, M. and Chu, X. (2005). Effects of CD59 on antitumoral activities of phycocyanin from *Spirulina platensis. Biomedicine & Pharmacotherapy*, 59(10), 551–560.

Lissi, E.A., Pizarro, M., Aspee, A. and Romay, C. (2000). Kinetics of phycocyanine bilin groups destruction by peroxyl radicals. *Free Radical Biology and Medicine*, 28(7), 1051–1055.

Lorenz, R.T. and Cysewski, G.R. (2000). Commercial potential for Haematococcus microalgae as a natural source of astaxanthin. *Trends in Biotechnology*, 18(4), 160–167.

Maeda, H., Hosokawa, M., Sashima, T., Murakami-Funayama, K. and Miyashita, K. (2009). Anti-obesity and anti-diabetic effects of fucoxanthin on diet-induced obesity conditions in a murine model. *Molecular Medicine Reports*, 2(6), 897–902.

McCarty, M.F. (2007). Clinical potential of *Spirulina* as a source of phycocyanobilin. *Journal of Medicinal Food*, 10(4), 566–570.

Miki, W., Hosada, K., Kondo, K. and Itakura, H. (1998). Astaxanthin-containing drink. Japanese patent, 10155459.

Mimouni, V., Ulmann, L., Pasquet, V., Mathieu, M., Picot, L., Bougaran, G. and Schoefs, B. (2012). The potential of microalgae for the production of bioactive molecules of pharmaceutical interest. *Current Pharmaceutical Biotechnology*, 13(15), 2733–2750.

Mokady, S., Abramovici, A. and Cogan, U. (1989). The safety evaluation of Dunaliella bardawil as a potential food supplement. *Food and Chemical Toxicology*, 27(4), 221–226.

Muthuirulappan, S. and Francis, S.P. (2013). Anti-cancer mechanism and possibility of nano-suspension formulation for a marine algae product fucoxanthin. *Asian Pac. J. Cancer Prev*, 14(4), 2213–2216.

Nagaoka, S., Shimizu, K., Kaneko, H., Shibayama, F., Morikawa, K., Kanamaru, Y. and Kato, T. (2005). A novel protein C-phycocyanin plays a crucial role in the hypocholesterolemic action of *Spirulina platensis* concentrate in rats. *The Journal of Nutrition*, 135(10), 2425–2430.

Nemoto-Kawamura, C., Hirahashi, T., Nagai, T., Yamada, H., Katoh, T. and Hayashi, O. (2004). Phycocyanin enhances secretary IgA antibody response and suppresses allergic IgE antibody response in mice immunized with antigen-entrapped biodegradable microparticles. *Journal of Nutritional Science and Vitaminology*, 50(2), 129–136.

Nishanth, R.P., Ramakrishna, B.S., Jyotsna, R.G., Roy, K.R., Reddy, G.V., Reddy, P.K. and Reddanna, P. (2010). C-Phycocyanin inhibits MDR1 through reactive oxygen species and cyclooxygenase-2 mediated pathways in human hepatocellular carcinoma cell line. *European Journal of Pharmacology*, 649(1), 74–83.

Nishikawa, Y., Minenaka, Y. and Ichimura, M. (1997). Physiological and biochemical effects of carotenoid (beta-carotene and astaxanthin) on rat. *Kashien Daigku Kiyo*, 25, 19–25.

Oi, V.T., Glazer, A.N. and Stryer, L. (1982). Fluorescent phycobiliprotein conjugates for analyses of cells and molecules. *The Journal of Cell Biology*, 93(3), 981–986.

Okai, Y. and Higashi-Okai, K. (1996). Possible immunomodulating activities of carotenoids in in vitro cell culture experiments. *International Journal of Immunopharmacology*, 18(12), 753–758.

Ou, Y., Lin, L., Pan, Q., Yang, X. and Cheng, X. (2012). Preventive effect of phycocyanin from *Spirulina platensis* on alloxan-injured mice. *Environmental Toxicology and Pharmacology*, 34(3), 721–726.

Ou, Y., Lin, L., Yang, X., Pan, Q. and Cheng, X. (2013). Antidiabetic potential of phycocyanin: Effects on KKAy mice. *Pharmaceutical Biology*, 51(5), 539–544.

Patil, G., Chethana, S., Sridevi, A.S. and Raghavarao, K.S.M.S. (2006). Method to obtain C-phycocyanin of high purity. *Journal of Chromatography* A, 1127(1), 76–81.

Plaza, M., Santoyo, S., Jaime, L., Reina, G.G.B., Herrero, M., Señoráns, F.J. and Ibáñez, E. (2010). Screening for bioactive compounds from algae. *Journal of Pharmaceutical and Biomedical Analysis*, 51(2), 450–455.

Prasanna, R., Sood, A., Suresh, A., Nayak, S. and Kaushik, B.D. (2007). Potentials and applications of algal pigments in biology and industry. *Acta Botanica Hungarica*, 49(1), 131–156.

Riss, J., Décordé, K., Sutra, T., Delage, M., Baccou, J.C., Jouy, N. and Rouanet, J.M. (2007). Phycobiliprotein C-phycocyanin from Spirulina platensis is powerfully responsible for reducing oxidative stress and NADPH oxidase expression induced by an atherogenic diet in hamsters. J. *of Agricultural and Food Chemistry*, 55(19), 7962–7967.

Roodenburg, A.J., Leenen, R., van het Hof, K.H., Weststrate, J.A. and Tijburg, L.B. (2000). Amount of fat in the diet affects bioavailability of lutein esters but not of α-carotene, β-carotene, and vitamin E in humans. *The American Journal of Clinical Nutrition*, 71(5), 1187–1193.

Roy, K.R., Arunasree, K.M., Dhoot, A., Aparna, R., Reddy, G.V., Vali, S. and Reddanna, P. (2007). C-Phycocyanin inhibits 2-acetylaminofluorene-induced expression of MDR1 in mouse macrophage cells: ROS mediated pathway determined via combination of experimental and *in silico* analysis. *Archives of Biochemistry and Biophysics*, 459(2), 169–177.

Saini, M.K. and Sanyal, S.N. (2014). Piroxicam and c-phycocyanin prevent colon carcinogenesis by inhibition of membrane fluidity and canonical Wnt/β-catenin signaling while up-regulating ligand dependent transcription factor PPARγ. *Biomedicine & Pharmacotherapy*, 68(5), 537–550.

Saini, M.K., Vaiphei, K. and Sanyal, S.N. (2012). Chemoprevention of DMH-induced rat colon carcinoma initiation by combination administration of piroxicam and C-phycocyanin. *Molecular and Cellular Biochemistry*, 361(1-2), 217–228.

Savoure, N., Briand, G., Amory-Touz, M.C., Combre, A., Maudet, M. and Nicol, M. (1994). Vitamin A status and metabolism of cutaneous polyamines in the hairless mouse after UV irradiation: Action of beta-carotene and astaxanthin. *International Journal for Vitamin and Nutrition Research*. Internationale Zeitschrift fur Vitamin-und Ernahrungsforschung. Journal international de vitaminologie et de nutrition, 65(2), 79-86.

Schmidt, R.A., Wiebe, M.G. and Eriksen, N.T. (2005). Heterotrophic high cell-density fed-batch cultures of the phycocyanin-producing red alga *Galdieria sulphuraria. Biotechnology and Bioengineering*, 90(1), 77–84.

Sekar, S. and Chandramohan, M. (2008). Phycobiliproteins as a commodity: Trends in applied research, patents and commercialization. *Journal of Applied Phycology*, 20(2), 113–136.

Sheu, M.J., Hsieh, Y.Y., Lai, C.H., Chang, C.C. and Wu, C.H. (2013). Antihyper-lipidemic and Antioxidant Effects of C-phycocyanin in Golden Syrian Hamsters Fed with a Hypercholesterolemic Diet. *Journal of Traditional and Complementary Medicine*, 3(1), 41.

Shih, C.M., Cheng, S.N., Wong, C.S., Kuo, Y.L. and Chou, T.C. (2009). Antiinflammatory and antihyperalgesic activity of C-phycocyanin. *Anesthesia & Analgesia*, 108(4), 1303–1310.

Simis, S.G.H., Peters, S.W.M. and Gons, H.J. (2005). Remote sensing of the cyanobacterial pigment phycocyanin in turbid inland water. *Limnol Oceanogr*, 50, 237–245.
Singh, S., Kate, B.N. and Banerjee, U.C. (2005). Bioactive compounds from cyanobacteria and microalgae: An overview. *Critical Reviews in Biotechnology*, 25(3), 73–95.
Sode, K., Horikoshi, K., Takeyama, H., Nakamura, N. and Matsunaga, T. (1991). On-line monitoring of marine cyanobacterial cultivation based on phycocyanin fluorescence. *Journal of Biotechnology*, 21(3), 209–217.
Spolaore, P., Joannis-Cassan, C., Duran, E. and Isambert, A. (2006). Commercial applications of microalgae. *Journal of Bioscience and Bioengineering*, 101(2), 87–96.
Stocker, R., Yamamoto, Y., McDonagh, A.F., Glazer, A.N. and Ames, B.N. (1987). Bilirubin is an antioxidant of possible physiological importance. *Science*, 235(4792), 1043–1046.
Takaichi, S., Sandmann, G., Schnurr, G., Satomi, Y., Suzuki, A., & Misawa, N. (1996). The Carotenoid 7, 8-Dihydro-ψ end Group can be Cyclized by the Lycopene Cyclases from the Bacterium Erwinia Uredovora and the Higher Plant Capsicum Annuum. *European journal of biochemistry*, *241*(1), 291–296.
Tanaka, T., Makita, H., Ohnishi, M., Mori, H., Satoh, K. and Hara, A. (1995). Chemoprevention of rat oral carcinogenesis by naturally occurring xanthophylls, astaxanthin and canthaxanthin. *Cancer Research*, 55(18), 4059–4064.
Terry, M.J., Maines, M.D. and Lagarias, J.C. (1993). Inactivation of phytochrome-and phycobiliprotein-chromophore precursors by rat liver biliverdin reductase. *Journal of Biological Chemistry*, 268(35), 26099–26106.
Törnwall, M.E., Virtamo, J., Korhonen, P.A., Virtanen, M.J., Taylor, P.R. and Albanes, D. (2004). Effect of α-tocopherol and β-carotene supplementation on coronary heart disease during the 6-year post-trial follow-up in the ATBC study. *European Heart Journal*, 25, 1171–1178.
Villar, R., Laguna, M.R., Calleja, J.M. and Cadavid, I. (1992). Effects of Phaeodactylum tricornutum and Dunaliella tertiolecta extracts on the central nervous system. *Planta Medica*, 58, 405–409.
Vonshak, A., Abeliovich, A., Boussiba, S., Arad, S. and Richmond, A. (1982). Production of Spirulina biomass: Effects of environmental factors and population density. *Biomass*, 2(3), 175–185.
Wang, X.Q., Li, L.N., Chang, W.R., Zhang, J.P., Gui, L.L., Guo, B.J. and Liang, D.C. (2001). Structure of C-phycocyanin from *Spirulina platensis* at 2.2 A resolution: A novel monoclinic crystal form for phycobiliproteins in phycobilisomes. *Acta Crystallographica Section D: Biological Crystallography*, 57(6), 784–792.
Yeum, K.J. and Russell, R.M. (2002). Carotenoid bioavailability and bioconversion. *Annual Review of Nutrition*, 22(1), 483–504.
Zhang, Y.M. and Chen, F. (1999). A simple method for efficient separation and purification of c-phycocyanin and allophycocyanin from Spirulina platensis. *Biotechnology Techniques*, 13(9), 601–603.
Zheng, J., Inoguchi, T., Sasaki, S., Maeda, Y., McCarty, M.F., Fujii, M. and Takayanagi, R. (2013). Phycocyanin and phycocyanobilin from *Spirulina platensis* protect against diabetic nephropathy by inhibiting oxidative stress. *American Journal of Physiology-Regulatory, Integrative and Comparative Physiology*, 304(2), R110-R120.
Zhou, Z.P., Liu, I.N., Chen, X.L., Wang, J.X., Chen, M., Zhang, Y.Z. and Zhou, B.C. (2005). Factors that effect antioxidant activity of c-phycocyanins from *Spirulina platensis. Journal of Food Biochemistry*, 29(3), 313–322.
Zhu, Y., Chen, X.B., Wang, K.B., Li, Y.X., Bai, K.Z., Kuang, T.Y. and Ji, H.B. (2007). A simple method for extracting C-phycocyanin from *Spirulina platensis* using *Klebsiella pneumoniae. Applied Microbiology and Biotechnology*, 74(1), 244–248.

Chapter 13
Liquid Fuels Production from Algal Biomass

Shantonu Roy and Debabrata Das

13.1 Introduction

Energy crisis is looming the global economy and environment. The rate at which fossil fuels are depleting, a necessity of alternate fuel has been gaining importance. The use of fossil fuels for energy is unsustainable and causes build up of greenhouse gases in the atmosphere leading to global warming. Biofuels store energy chemically that can be harnessed easily. It can also be used in existing combustion engines after blending with petroleum diesel to various degrees. No separate transportation infrastructures would be required for such fuels (Amaro et al., 2011). In biorefinery concept, every component of the biomass material would be used to produce commercially important products. At present, first generation biofuels are produced using sucrose and starch crops. Second generation biofuels are produced using lignocellulosic biomass. Lignocellulosic biomass gained importance because of their abundant availability but need of pretreatment and saccharification processes has hindered their usage as feedstock. Moreover, bioenergy production using agricultural crops or agricultural wastes as feedstock is disadvantageous as resources for water and agriculture lands are limited (Li et al., 2008). Algal biomass has been considered as third generation feedstock for biofuel production (Metzger and Largeau, 2005). Many algal species having high lipid content thus could be explored for oleo-fuel generation. Similarly, algal species having high carbohydrate content can be exploited for bioethanol or biogas production.

Algae have superior annual productivity and oil content as compared to seed crops. Oil productivity of soybean, canola and palm are 450, 1200 and 3000 litres per hectare, respectively. Algae can yield approximately 90,000 litres per hectare (Chisti, 2007). Algae do not require arable land for cultivation and thus it does not

S. Roy • D. Das (✉)
Department of Biotechnology, Indian Institute of Technology Kharagpur,
Kharagpur 721302, India
e-mail: ddas.iitkgp@gmail.com

D. Das (ed.), *Algal Biorefinery: An Integrated Approach*,
DOI 10.1007/978-3-319-22813-6_13

necessitate agricultural space. In fact, algae cultivation facilities can be built on marginal land that has few other uses. The water resources available for algal cultivation are fresh water, saline and industrial wastewaters (Brown and Zeiler, 1993). Moreover, algae fix atmospheric CO_2 with greater competence than land plants.

13.2 Biodiesel

Biodiesel consists of monoalkyl esters that are derived from different oil-rich sources e.g. organic oils extracted from plants, animal, algae, etc. via transesterification process. The chemical reaction for biodiesel transesterification involves hydrolysis of ester bond between glycerol and fatty acid chain and then further esterification with methanol as shown in Eqn. (13.1).

$$\text{Triglyceride} + 3\text{Methanol} \xrightleftharpoons{\text{catalyst}} \text{Glycerine} + 3\text{Methylesters}\ \left(\textit{Biodiesel}\right) \quad (13.1)$$

The presence of a catalyst and alkali such as potassium hydroxide enhances transesterification. Since it's a reversible reaction, an excess of methanol could be used to force the reaction in forward direction. Solvent recovery could be done by soxhlet technique and can be reused. Elevating temperature to 60 °C could increase the kinetics of the reaction and the process could be completed in 90 minutes. The biodiesel production process involves following steps (Xu et al., 2006):

- Reactants such as triglycerides, methanol and catalyst were to be placed in a controlled reactor so as to initiate transesterification reaction.
- The initial products formed after transesterification are placed in a separator to remove the glycerine and other unreacted components
- Recovery of excess methanol from the methyl esters is done by evaporation.
- The final biodiesel needs to be rinsed with water, neutralized and dried.

Many reports are available suggesting that under stressed or unfavourable conditions, algae produce more oil as compared to optimal growth conditions. Under favourable growth conditions, fatty acids are synthesized principally for esterification into glycerol-based membrane lipids. It constitutes about 5–20 % w/w of their dry cell weight (DCW). Composition of fatty acids present under such conditions can be categorized into following species: medium-chain (C10–C14), long-chain (C16–C18) and very long chain (C20) and fatty acid derivatives. Under unfavourable or stress conditions algae redirect their lipid biosynthetic pathways towards the formation and accumulation of neutral lipids (20–50 % DCW). Major constituents of such lipids are in the form of triacylglycerol (TAG). The role of TAGs is not of forming cell membrane, rather it serves as storage energy (Hu et al., 2008). The accumulated TAGs are stored in the cytoplasm of the algal cell as densely packed lipid bodies. Lipid accumulation also takes place in the inter-thylakoid space of the chloroplast in certain green algae. However, the challenges in production cost of high grade algae oils is the constraint in operational conditions (low temperature, low light intensity and nitrogen deficiency) that leads to accumulation of high grade oil in microalgae. In current scenario, it is quite difficult to obtain cheap algae biomass with 20 % w/w lipid content. With the advent of recent advances in photobioreactors configurations (closed and open), production of algal biomass is still a costly affair.

13.2.1 Microbiology

Algae have a very simple cellular structure as compared to higher autotrophic photosynthetic. They have higher nutrient assimilation ability because of their large surface to volume body ratio. The mechanism of CO_2 fixation via photosynthetic pathway is similar to that of higher plants. Because of their simple cellular structure, algae are generally more efficient in harnessing solar energy (Ghirardi, 2000; Singh et al., 2008). Therefore, microalgae produce 30 times the amount of oil per unit area of land as compared to terrestrial oilseed. This makes microalgae as a potential source for biodiesel production, thus completely displacing dependence on fossil oils such as diesel (Table 13.1). Many microalgae have oil content up to 80 % w/w of dry cell biomass. Generally oil content are in the range of 20–50 % w/w of dry cell biomass (Mata et al., 2010; Khan et al., 2009). Under stressed conditions accumulation of lipids is induced in microalgal species. The average lipid contents in algal cells vary between 1 % and 70 % w/w of dry cell weight (Table 13.1). Even within the same genus, different lipid content has been reported. *Botryococcus braunii* is a well known microalgae reported to have highest oil content (75 % w/w of dry cell weight), but is associated with a low productivity and much of it is secreted into the cell wall (Banerjee et al., 2002). In general, lipid content of microalgae like *Chlorella, Dunaliella, Isochrysis, Nannochloris, Nannochloropsis, Neochloris*, *Nitzschia*, *Phaeodactylum* and *Porphyridium* sp. lies in between 20 % and 50 % w/w DCW. Each of the algae has distinct productivity and growth characteristics. Many reports were available on *Chlorella* sp. as promising option for biodiesel production. On other hand, marine microalgae have greater lipid productivities. Thus marine algae could be used for mass production. There are certain advantages in marine algae cultivation e.g. a high salinity environment prevents extensive exogenous and endogenous contamination and moreover, sea water can be directly used for cultivation instead of putting burden on fresh water resources.

The fatty acid composition of the microalgal cell is also important. The heating power of biodiesel depends on its composition. In general, it consists of saturated and unsaturated fatty acids containing 12–22 carbon atoms, often belonging to omega 3 and omega 6 types. The fatty acid compositions of many fresh water microalgae species consist of C14:0, C16:0, C18:1, C18:2 and C18:3 fatty acids. Many fatty acid residues were species-specific, e.g. *Ankistrodesmus* sp. contains C16:4 and C18:4, *Isochrysis* sp. contains C18:4 and C22:6, *Nannochloris* sp. contains C16:3 and C20:5, and *Nitzschia* sp. contains C16:3 and C20:5. The fatty acid composition

Table 13.1 Potential of different algal species for biodiesel production

Habitat	Microalgal species	Lipid content (%w/w DCW)	Lipid productivity (mg L^{-1} d^{-1})
Fresh water	*Botryococcus* sp.	25.0–75.0	–
	Chaetoceros muelleri	33.6	21.8
	Chaetoceros calcitrans	14.6–16.4/39.8	17.6
	Chlorella emersonii	25.0–63.0	10.3–50.0
	Chlorella protothecoides	14.6–57.8	1214
	Chlorella sorokiniana	19.0–22.0	44.7
	Chlorella vulgaris	5.0–58.0	11.2–40.0
	Chlorella pyrenoidosa	2	–
	Chlorella sp.	18.0–57.0	18.7
	Chlorococcum sp.	19.3	53.7
	Cylindrotheca sp.	16–37	
	Ellipsoidion sp.	27.4	47.3
	Haematococcus pluvialis	25	–
	Scenedesmus obliquus	11.0–55.0	–
	Scenedesmus quadricauda	1.9–18.4	35.1
	Scenedesmus sp.	19.6–21.1	40.8–53.9
	Schizochytrium sp.	50–77	–
Marine water	*Dunaliella salina*	6.0–25.0	116
	Dunaliella tertiolecta	16.7–71.0	-
	Dunaliella sp.	17.5–67.0	33.5
	Isochrysis galbana	7.0–40.0	-
	Isochrysis sp.	7.1–33	37.8
	Nannochloris sp.	20.0–56.0	60.9–76.5
	Nannochloropsis oculata	22.7–29.7	84.0–142.0
	Nannochloropsis sp.	12.0–53.0	60.9–76.5
	Neochloris oleoabundans	29.0–65.0	90.0–134.0
	Nitzschia sp.	30.9	30.9
	Pavlova salina	30.9	49.4
	Pavlova lutheri	35.5	40.2
	Phaeodactylum tricornutum	18.0–57.0	44.8
	Porphyridium cruentum	9.5	34
	Spirulina platensis	4.0–16.6	–
	Tetraselmis sp. F&M-M34	14–18	43

of microalgae is greatly influenced by different nutritional and processing factors, cultivation conditions and growth phases. Under nitrogen deficiency and salt stress, accumulation of C18:1 is induced in all species. In *B. braunii*, under above conditions, accumulation of C20:5 takes place. Even CO_2 assimilation also influences the fatty acid profile in algal cells. The biomass productivity and carbon dioxide fixation ability of *Scenedesmus* sp. is high at 10 % (v/v) CO_2. Similarly, *B. braunii* grown under 10 % (v/v) CO_2 gave higher biomass productivity and it is also suitable for biodiesel production due to its high proportion of oleic acid (Yoo et al., 2010).

Following factors are associated with different microalgae cultivation:

(i) Growth rate – it is measured as total amount of biomass accumulated per unit time and unit volume.
(ii) Lipid quantity and quality – in the harvested biomass, the actual distribution of fatty acid residues within acylglycerols.
(iii) Robustness of the process – biomass productivity influenced by variation in temperature, nutrient input and light. Competition with other microalgae and/or bacterial species also important.
(iv) Nutrient predilection and rate of substrate utilization – during growth, CO_2, nitrogen and phosphorus utilization varies species to species.
(v) Ease of biomass harvesting – e.g. efficiency of cell lysis, extraction and purification of lipids depends upon the ease of harvesting of biomass.

Photo-biological formulae needed to be optimized for individual species followed with cost-effective cultivation techniques so as to make this process commercially viable. Using existing technologies, the commercial viability of biodiesel production from native microalgae is trivial. A significant degree of development would be required in volumetric productivity of biomass so as to make the process lucrative for commercialization. This goal could be achieved by exploring genetic engineering of the microalgae cells. Redirection of metabolic pathway towards appropriate lipids by usage of molecular biology tools like synthetic biology, gene knockout etc. along with improvement of bioprocess parameters could propel the biodiesel production process to better yields.

13.2.2 Genetic Engineering of Microalgae

When the wild species available in formal collections and described in the literature are not feasible for commercial production of biodiesel, one may resort to genetic engineering so as to improve and tune the features of native microalgae, with the aim of enhancing productivity and yield (Hu et al., 2008). However, the pending concerns of biological contamination—materialized in restrictive legislation at large—have hindered a broader utilization of genetically engineered microalgae. The several shortcomings found at present include: very few strains of microalgae that underwent genetic modification (with the notable exception of *Chlamydomonas reinhardtii*—the genome of which has been fully elucidated, but which is a fresh water species); poor elucidation of the mechanisms underlying regulation of gene expression; and lack of specific molecular biology tools, e.g., efficient nuclear transformation, availability of promoter and selectable marker genes, and stable expression of transgenes. Although lengthy in time and cost-intensive, transformation of microorganisms to better respond to non-expensive operating conditions brings about a few advantages. Metabolic engineering may increase the yields of acylglycerols, or even lead to different molecules with better performance as biodiesel. In higher plants, several studies have explored the effects of over expressing enzymes associated with lipid synthesis, yet little change in oil content was achieved in species containing higher levels of acetyl-CoA carboxylase – the rate limiting step in fatty acid biosynthesis, possibly because of the complex regulation of this enzyme.

At present, one is indeed still far from globally understanding the detailed molecular pathways and forms of regulation of lipid metabolism in microalgae. Bioinformatics applied to already sequenced microalgal genomes has unfolded essentially similar biochemical routes. Therefore, little experimental validation of putative enzyme activities has so far been done. Lipid accumulation is easily induced in microalgae by nitrogen deprivation. This provides a useful experimental basis for observing changes in gene transcription, protein synthesis and metabolic activities that prevail during lipid accumulation in microalgae. Nitrogen depletion coupled with RNAi suppression on the changes in the lipid and protein qualitative and quantitative profiles in *C. reinhardtii* during lipid droplet formation has been investigated. In cultures transferred to N-depleted media, the total fatty acid content per cell basis, increased by 2.4-fold within 72 h, of which 65 % of the total fatty acids were esterified to triacylglycerols (Moellering and Benning, 2010). Proteomic analysis provided evidence of a 'major lipid droplet protein' (MLDP) which was rather abundant in said lipid bodies. The mRNA transcript abundance of this protein followed the observed increase in lipid droplets after N-depletion, and RNAi lines of *C. reinhardtii*, with a 55–60 % reduction of MLDP transcription, produced lipid droplets characterized by 40 % larger diameters than the control line – thus implying that this protein is involved in regulation of lipid droplet size. Another efficient way of increasing lipid yield is to delete 'redundant' pathways in the selected microorganism, thus leading to precursor metabolites more suitable for biofuel production. An increase in triacylglycerol contents of lipid droplets was observed in a *C. reinhardtii* starchless mutant which was deficient in ADP-glucose pyrophosphorylase (an essential enzyme in starch production) following 48 h of N-depletion. Wild-type cells had increased their lipid droplet content by 15-fold.

13.2.3 Chemistry Behind Biodiesel Production

Alcohols are the most frequently used acyl-acceptors, particularly methanol and, to a lesser extent, ethanol. Other alcohols can be used, e.g. propanol, butanol, isopropanol, tert-butanol, branched alcohols and octanol but the cost is much higher. Regarding the choice between methanol and ethanol, the former is cheaper, more reactive and the fatty acid methyl esters (FAME) are more volatile than those of the fatty acid ethyl esters (FAEE). However, ethanol is less toxic and it can be considered more renewable because it can be easily produced from renewable sources by fermentation. In contrast, methanol is mainly produced from non-renewable fossil sources, such as natural gas. Regarding their characteristics as fuels, FAME and FAEE show slight differences e.g. FAEE have slightly higher viscosities and slightly lower cloud and pour points than the corresponding FAME. The transesterification of triacylglycerols can be carried out by different catalytic processes, or in supercritical conditions (Marchetti et al., 2007). The catalyst used may be classified as: (1) alkaline-catalyst (sodium hydroxide, NaOH; Potassium hydroxide, KOH; sodium metoxide, NaOMe); (2) acid-catalyst (sulphuric acid, phosphoric acid, hydrochloric acid, sulphonic acid); (3) enzymatic-catalyst (lipases); and (4) inorganic heterogeneous catalyst (solid phase catalyst) (Fig. 13.1).

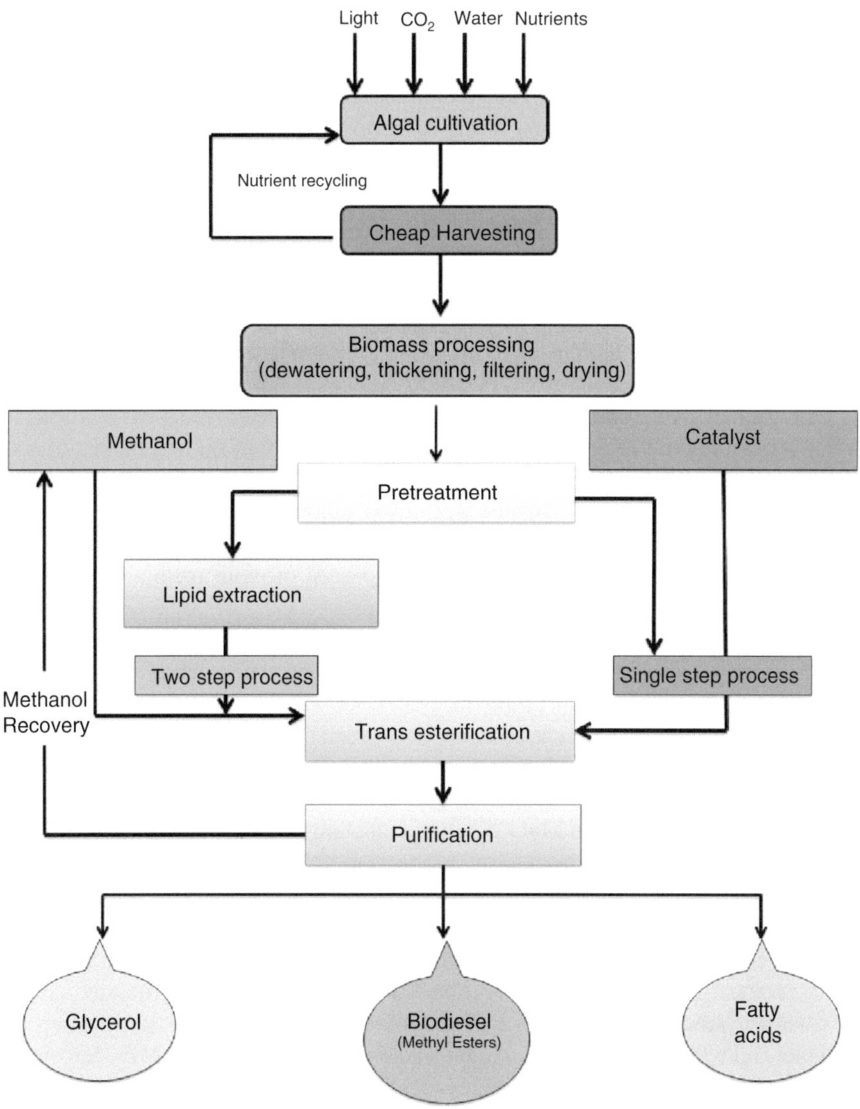

Fig. 13.1 Algal biodiesel production processes.

The transesterification using enzymes has been reported to be very expensive (the enzyme costs are very high), shows deactivation problems and requires a much longer reaction time (Vyas et al., 2010). On the other hand, acid and basic transesterification are widely used for biodiesel production. It is well known that basic catalyzed transesterification is faster than the acid catalyzed reaction (about 4000 times). However, acid catalysts can simultaneously promote esterification of free fatty acids (FFAs) and transesterification of triglycerides (Demirbas, 2007). Traditionally, homogeneous catalysts have been used for both acid and basic catalyzed reaction. Sulphuric acid is the main acid catalyst used for biodiesel production whereas NaOH, KOH and Na_2CO_3 mixed with alcohol, are commonly used for homogeneous basic catalysis (Helwani et al., 2009). However, one of the major disadvantages of homogeneous catalysts is that they cannot be reused or regenerated, because they are consumed in the reaction and separation of catalysts from products is difficult and requires more equipment, which could result in higher production costs. Besides, the process is not environmentally friendly because a large amount of wastewater is produced in the separation step. Based on the above premises, the use of solid catalysts seems to be an appropriate solution to overcome problems associated with homogeneous catalysts. Nevertheless, one of the major problems associated with heterogeneous catalysts is the formation of three phases with alcohol and oil, which leads to diffusion limitations thus lowering the rate of the reaction.

One way of overcoming mass transfer problem in heterogeneous catalysts is using certain amount of co-solvent to promote miscibility of oil and methanol and accordingly accelerate the reaction rate. Tetrahydrofuran (THF), dimethyl sulphoxide (DMSO), n-hexane and ethanol were used more frequently as co-solvent in transesterification of vegetable oils with methanol and solid catalysts. CaO as a solid base catalyst for transesterification of rapeseed oil with methanol and after 170 min of reaction time methyl ester gave improved yield of 93 % (Zabeti et al., 2010). However, by adding certain amount of THF into rapeseed oil/methanol mixture the same yields of 93 % were observed after 120 min of reaction time. Another way to promote mass transfer problems associated with heterogeneous catalysts is using structure promoters or catalyst supports which can provide more specific surface area and pores for active species where they can anchor and react with large triglyceride molecules.

Basic solids like CaO and MgO supported on alumina and hydrotalcites (Suppes, 2004) have been used for biodiesel production from vegetable oils. To avoid diffusion limitations, catalysts with higher surface area (porous silica-metal oxide composites) were tested in the transerification of vegetable and animal oils providing high conversion to (C10–C30) alkyl methyl esters and glycerin. On the other hand, zeolites, ion-exchange resins, mixed metal oxides or mesostructured solids have shown promising results in the acid esterification and transesterification of vegetable oil with high content of free fatty acids (FFAs) to obtain FAMEs. Recently, the transesterification of triglycerides contained in waste oilseed fruits with methanol has been studied using zeolites as strong acid catalysts (USY, BEA, FAU-X), together with weak acid catalysts (siliceous MCM-41 and ITQ-6) and base catalysts such as K-MCM-41 and K-ITQ-6 (Macario et al., 2010). Zeolites are microporous crystalline metallosilicates featured by exhibiting molecular sieve and shape selective properties, which have found widespread applications in catalytic, adsorption and ion exchange processes. Zeolites have usually been synthesized with crystal sizes in the micrometre range and, therefore, with negligible external surface area. These properties impose severe limitations for their use in the conversion of bulky compounds. A huge interest has emerged recently for the synthesis of new zeolitic materials with enhanced accessibility. In this sense, nanocrystalline hierarchical zeolites contain a bimodal porosity (micro- and mesopores) and high external surface area where active sites can catalyze reactions involving large molecules like triglycerides. The synthesis of hierarchical nanozeolites is based on the incorporation of organosilanes in the synthesis gel to prevent zeolite crystal growth and thereby to stabilize zeolitic particles with ultra small sizes.

In any case, acid catalyst is the recommended process when the starting materials are low grade or have a high concentration of free fatty acids. The fatty acids would deactivate the alkaline catalyst. Acid catalysts could be used in conjunction with base catalysts (two-stage process). This two-stage process allows the use of low-cost feedstock like waste oil with high content of free fatty acids. The acid catalysts are used in the primary stage to convert free fatty acid to methyl esters, followed by a base catalyst process to convert the remaining triglycerides to methyl esters. Acid catalysts should be the method when using oils extracted from microalgal biomass. For example, a maximum yield of 10 mg of FAME from 250 mg of lipids was observed under following conditions: 0.6 N hydrochloric acid–methanol catalysts for 0.1 h at 70 °C, using the lipids extracted from *Chaeotoceros mulleri* (Nagle and Lemke, 1990). In comparison, only 3.3 mg of FAME were produced when sodium hydroxide was used as the catalyst, at the same conditions that gave maximum FAME yield. A simultaneous extraction and transesterification method for microalgal fatty acids using an acid catalyst was also studied (Belarbi et al., 2000). Fatty acids were extracted either from freeze-dried microalgal biomass or from centrifugally harvested biomass paste that has been freeze stored. The biomass paste had a moisture content of 82 % by weight. The biomass belonged to either the diatom *Phaeodactylum tricornutum* or the green alga *Monodus subterraneus* and a maximum yield of 8.37 g of FAME was obtained from 10.8 g of lipids in the following conditions: biomass paste (500 g, 82 % moisture, 12 % of lipids by wt.) of *P. tricornutum*, methanol, 1 L and 50 mL of acetyl chloride. The resulting slurry was heated in a boiling water bath for 120 min at 2.5 atm. The most usual method to transform oil into biodiesel is transesterification. This consists of the reaction between triacylglycerols (contained in the oils) and an acyl-acceptor. The acyl group acceptors may be carboxylic acids (acidolysis), alcohols (alcoholysis) or another ester (inter esterification). Only alcoholysis and inter esterification are of interest to produce biodiesel. The starting esters in both are triacylglycerols (oils), and if the transformation is quantitative they yield a mixture of monoalkyl esters (biodiesel) and glycerol (alcoholysis) or another triacylglycerol (inter esterification).

13.2.4 *Quality of Biodiesel*

The physical and chemical properties of biodiesel such as density, viscosity, flash point, cold filter plugging point, solidifying point and heating value must be determined for assessment of the potential of biodiesel as a substitute of diesel fuel. One of the important interfering factor in biodiesel production process is bound glycerol. It is associated with the residual amount of triglycerides and partial glycerides in the biodiesel. Biodiesel fuel, in the form of FAME, is now manufactured in many countries. The relevant standard for biodiesel is the ASTM Biodiesel Standard D 6751 which is followed in the United States (Knothe, 2006). European Union follows a separate standard for biodiesel when it is used for internal combustion engines (Standard EN 14214) and for use as heating oil (Standard EN 14213). The acid number of biodiesel limit according to ASTM D 6751 standard (US) was synchronized with the European biodiesel value of 0.5. Algal oils are quite rich in polyunsaturated fatty acids unlike most of the vegetable oils. Biodiesel intended to be used as heating oil is not affected by PUFA content. For getting acceptance as heating oil, it must meet criteria relating to the extent of total unsaturation of the oil, which is indicated by its iodine value. According to European standard EN 14214 and EN 14213, the iodine value of biodiesel should exceed 120 and 130 g iodine/100 g biodiesel, respectively.

13.3 Bioethanol

In the global scenario, ethanol is produced from sugar plants (55 %), grain (37 %), synthetically (8 %) and other raw materials (2 %) (İçöz et al., 2009). Bioethanol for fuel purpose has certain characteristics viz. they are derived from carbohydrate-rich biomass. They are biodegradable and environment friendly. Different feedstock such as cellulosic biomass, agricultural waste, and wood waste were commonly used for bioethanol production. Complex sugars present in the feedstock are first converted to simple sugars (mainly hexose). Then these simple sugars are utilized by solventogenic organisms to produce ethanol by fermentation (Eq. 13.2):

$$C_nH_{2n}O_n \rightarrow \frac{n}{3}CH_3CH_2OH + \frac{n}{3}CO_2 + 227.0\ kJ/mol \tag{13.2}$$

The ethanol thus produced can be recovered by fractional distillation. The ethanol on combustion gives 1370.7 kJ mol^{-1} (Eq. 13.3) of energy that can be harnessed for cooking, automobile combustion engine etc.

$$CH_3CH_2OH + 3O_2 \xrightarrow{Combustion} 2CO_2 + 2H_2O + 1370.7\ kJ/mol \tag{13.3}$$

The exemplary characteristics of bioethanol are as follows (Yoon and Lee, 2011):

- Easy for storage and no need of separate infrastructure for distribution.
- Highly suitable automobile fuel when blended with naturally occurring fossil fuels.
- Emission of harmful unburned hydrocarbon and carbon monoxide are extremely low as compared to fossil fuel combustion.

The technologies and skills of ethanol production were once confined to handful countries around the globe. Its production and usage as fuel have started to show its presence globally. At present, the viability of bioethanol production from starch or sugar in a wide variety of crops is debatable as a replacement for fossil fuels. It has led to a debate of "food vs. fuel". There is the potential for rising food prices as food and fuel markets compete for scarce arable land. The requirement of large amount of arable land, the amount of energy to be spent and environmental pollution etc. are the major public concerns. Production of ethanol from lignocellulosic biomass promises to assuage the above concerns. Lignocellulosic biomass such as sugarcane bagasse, wheat straw, rice husk, rice straw, corn straw and other lignocellulosic biomass are explored as feedstock for ethanol production (Binod et al., 2010; Talebnia et al., 2010). To exploit the entrapped sugars from their complex polymeric forms, various pretreatment techniques have been developed viz. physical, chemical, physiochemical and biological pretreatment methods. The plant cell wall is rigid and complex structure composed of lignin, cellulose, hemicelluloses, etc. which makes them resistant to pretreatment techniques and thus leads to poor sugar yield.

So need of the hour is a source of carbohydrate-rich biomass that has simple cellular structure and doesn't compete for arable land. Algal biomass as feedstock holds the perfect future for bioethanol production because of their high carbohydrate content, cellulosic cell walls and starch based cytoplasm. Less harsh pretreatment techniques are required for saccharification for algal biomass. Moreover, lignin removal is a rate-limiting step for lignocellulosic biomass. Since there is an absence of lignin in algal biomass, it reduces the costs, time and difficulty of the saccharification process (Sarkar et al., 2012). The efficiency of fermentative bioethanol production strongly depends on the pretreatment and saccharification conditions. Under optimized conditions of pretreatment and saccharification it would lead to solubilization of carbohydrates and conversion of them to simple fermentable sugars. However, improper pretreatment and saccharification might lead to further degradation of fermentable sugars to undesirable products such as formic acid, acetic acid, and some furanic compounds (Lee et al., 2013).

13.3.1 Microbial Insight on Algae as a Source of Energy Crop

The algae are generally categorized into two groups viz. microalgae and macroalgae on the basis of their size and morphology. Algae are photosynthetic, eukaryotic organisms devoid of multicellular sex organs. It contains green chlorophyll along

Table 13.2 Ethanol yield from different feedstock (Mussato et al. 2010)

Feedstock	Ethanol yield (L ha^{-1})
Corn stover	1,050–1,400
Wheat	2,590
Cassava	3,310
Sweet sorghum	3,050–4,070
Corn	3,460–4,020
Sugar beet	5,010–6,680
Switch grass	10,760
Microalgae	46,760–140,290

with other photosynthetic pigments that give them characteristic colour which further helps in identification of key divisions. They fix atmospheric CO_2 to complex carbohydrates such as starch and cellulose via photosynthesis (Singh et al., 2011). Macroalgae also known as seaweeds are inhabitants of both intertidal and sub-tidal zone of coastal region where there is sufficient light penetration. They are composed of multiple cells organized into structures having analogy with roots, stems and leaves of higher plants. On the other hand, microalgae belongs to a large group of microscopic unicellular photosynthetic organisms. The yield of ethanol that could be achieved from microalgae is approximately 5000–15,000 gal of ethanol ac^{-1} yr^{-1} (46,760–140,290 L ha^{-1}). This yield is of higher magnitude as compared to other feedstock (Mussato et al., 2010) (Table 13.2).

Fermentative methods have been developed to utilize carbohydrate-rich microalgae for the production of bioethanol.

Major commercial advantages that are attracting researchers and entrepreneurs in the field of bioethanol production from algal biomass are:

(i) Countering the perception of "Food vs. Fuel", bioethanol production using algae wouldn't compete for either land or water.
(ii) As they are rich in carbohydrates, both marine and freshwater algae may be used for ethanol production (Singh et al., 2011) (Table 13.3).
(iii) Algae don't have lignin in their cellular ultra structure. Moreover, it has very low hemicelluloses content. This endorses for improved hydrolysis efficiency when subjected to pretreatment thus enhancing fermentation yields (Gouveia and Oliveira, 2009).
(iv) Rapid growth of algae and its versatility to grow in a variety of aquatic environments such as fresh water, saline water, or municipal wastewater are two most contrasting features that make them an ideal feedstock for bioethanol production.
(v) They have a high photosynthetic efficiency which is much higher than that of terrestrial biomass (Harun et al., 2009; Ross et al., 2008).

Table 13.3 List of algae rich in carbohydrates

Habitat	Algal source	% starch or biomass (g/dry weight)
Marine water	*Saccharina latissima*	50.0 (reserve food material)
	Green alga NKG 121701	>50.0 (starch)
	Laminaria hyperborean	55.0 (reserve food material)
	N. maculiforme TISTR 8406	30.1 (starch)
	Synechococcus sp.	15.0 (starch)
	Kappaphycus alvarezii	64 (starch)
Fresh water	*Spirogyra* sp.	43.3 (biomass after oil extraction)
	Oedigonium sp.	33.6 (biomass after oil extraction)
	Chlamydomonas reinhardtii UTEX 90	53.0 (starch)
	C. reinhardtii (UTEX2247)	45.0 (starch)
	C. reinhardtii	17.0 (starch)
	C. vulgaris	37.0 (starch)
	Chlorella sp. TISTR 8485	27.0 (starch)
	Chlorella sp. TISTR8593	22.0 (starch)
	Chlorococcum sp. TISTR8583	26.0 (starch)
	Scenedesmus sp. TISTR 8579	20.4 (starch)
	S. acutiformis TISTR 8495	16.4 (starch)
	S. acutus TISTR 8447	18.6 (starch)
	S. arcuatus TISTR 8587	12.9 (starch)
	S. armatus TISTR 8591	15.4 (starch)
	S. obliquus TISTR 8522	23.7 (starch)
	S. obliquus TISTR 8546	23.4 (starch)
	Nostoc sp. TISTR 8872	30.7 (starch)
	Nostoc sp. TISTR 8873	32.9 (starch)
	N. muscorum TISTR 8871	33.5 (starch)
	N. paludosum TISTR 8978	32.1 (starch)
	N. piscinale TISTR 8874	17.4 (starch)
	Oscillatoria sp. TISTR 8869	19.3 (starch)
	O. jasorvensis TISTR 8980	9.7 (starch)
	O. obscura TISTR 8245	12.6 (starch)
	Phormidium angustissimum	28.5 (starch)
	Spirulina fusiformis	37.3–56.1 (starch)

Steps involved in bioethanol production from algae are similar to that of cellulosic bioethanol production which requires four major unit operations viz. pretreatment, hydrolysis for saccharification, fermentation and distillation. Pretreatment process destroys the physical barriers in the cell wall. This enhances the accessibility of complex carbohydrate towards enzymatic hydrolysis. As compared to lignocellulosic biomass, algal biomass has a soft cellular organization and high moisture content rendering ease towards pretreatment. Different types of pretreatment used for algal biomass are physical, physico-chemical, chemical and biological.

13.3.2 Pretreatment and Saccharification of Algal Biomass

There are two ways of bioethanol production from algal biomass i.e. by either the sugar fermentation pathway or syngas pathway. Algal biomass on pretreatment and saccharification gives simple sugars that could be directly fermented to produce ethanol. In syngas pathway, hydrocarbons present in algal biomass are converted to syngas through gasification. The syngas thus generated could be subjected to fermentation to produce bioethanol. Such fermentation are carried out by strict autotrophic bacteria like *Clostridium ljungdahlii* (Younesi et al., 2005).

There are few important points to assess efficacy of pretreatment techniques which are as follows (Agbor et al., 2011):

(a) Quantitative and qualitative estimation of sugar and carbohydrate content in the liquid- and solid-fractions, respectively.
(b) Screening fermentation inhibitors like furfurals and neutralizing them prior to fermentation.
(c) Selecting the source from which bioethanol will be produced. It is judged on the basis of sugar and carbohydrate analyses whether liquid hydrolysates or WIS (water insoluble solids) to be taken for fermentation.
(d) Exploring pretreated samples for producing other value added products.

Crystalline structure of cellulose present in algal biomass is derived from β-D-glucopyranose units condensed by β-1,4-glycosidic bonds (Mittal et al., 2011). The initial degree of crystallinity is a crucial element to assess the pretreatment process. The distortion of crystallinity can be studied with the help of X-ray crystallography (XRD) analysis. The wide angle X-ray diffraction counts at 2θ angle close to 22° and 18° according to the Segal empirical method. The CrI can be calculated by the following equation:

$$CrI = \frac{I_{22} - I_{18}}{I_{22}} \times 100\% \tag{13.4}$$

where I_{22} is the peak intensity of the crystalline material (2θ = 22°) and I_{18} is the peak intensity of the amorphous material (2θ = 18°).

Disordered crystalline or amorphous cellulose shows higher hydrolysis rates as compared to partially disordered structure. In recent years, many studies have been performed on various pretreatment techniques to improve bioethanol fermentation.

Since algal cells are less rigid than plant cells, an extremely low acid pretreatment process is widely used. It obliterates the algal cell wall and releases carbohydrates to the liquid hydrolysates (Fig. 13.2).

Sugar recovery in such process can be maximized by cumulative optimization of three parameters viz. pretreatment time, temperature and acid concentration. It was reported that the temperature (50 to 180 °C) influences efficiency of saccharification at extremely low acid pretreatment (Lee et al., 2013). At 170 °C, extremely low acid pretreatment gave maximum glucan content of 32 % w/w using brown *Laminaria*

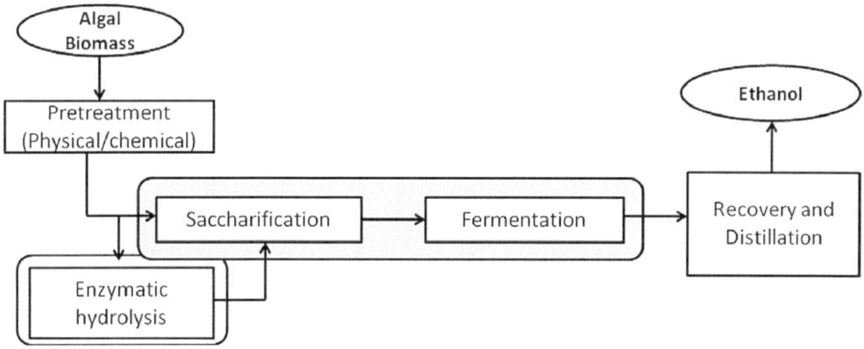

Fig. 13.2 Ethanol production process using algal biomass.

japonica biomass. In other reports, an increase in glucan content and enzyme digestibility was observed in ELA-treated *S. japonica* as compared to untreated algal biomass. Use of acids for pretreatment may lead to formation of furfural formation. Furfurals are growth inhibitors. Thus in some cases, the removal of furfurals was done prior to fermentation to avoid process inhibition. Dilute acid pretreatment of *Kappaphycus alvarezii* was treated by activated charcoal to remove hydroxyl methyl furfural (Hargreaves et al., 2013).

When acid treated algal biomass are neutralized using NaOH, it generates high concentrations of Na_2SO_4 that inhibit fermentation of *Ulvareticulate* sp. (Yoza and Masutani, 2013). Alkaline pretreatment was also explored for algal biomass pretreatment. Alkaline pretreatment of the green alga *Ulva lactuca* showed gelling property during alkaline pretreatment abolishing such pretreatment process for this algal species (Kim et al., 2011).

13.3.3 Enzymatic Hydrolysis of Pretreated Algal Biomass

Enzymatic hydrolysis has certain advantages as it does not generate or get affected by furfural content. The enzymatic liquefaction of starch in *Chlamydomonas reinhardtii* (UTEX90) could be used for dark fermentation. Hydrolytic enzymes such as α-amylase and amyloglucosidase are commercially available and are used for liquefaction and saccharification, respectively. The α-amylase disrupted the cell wall completely resulting in release of all carbohydrates. Ethanol yield was low as compared to acid pretreatment. Maximum enzymatic activity could be achieved by optimizing certain parameters like temperature, pH and time of exposure to enzyme. For *C. reinhardtii* and *U. pinnatifida,* the optimum enzymatic reaction conditions were 45 °C, pH 4.6 and 60 min for maximum extraction of glucose (Choi et al., 2010).

In biorefinery concept, the algal biomass left after extraction of lipid during biodiesel production could be a potential substrate for bioethanol production. The conversion of defatted microalgae biomass to bioethanol could be more economically viable than direct conversion of microalgae to bioethanol. After biodiesel production, the defatted biomass of *Dunaliella tertiolecta* and *Gracilaria verrucosa* were used for bioethanol production (Lee et al., 2013). The major bottleneck in using defatted algal biomass is the difficulty of removing solvents used during biodiesel production.

13.3.4 Microbial Fermentation of Algal Biomass

After hydrolysis, glucose and mannose are yielded from cellulose whereas xylose and arabinose are yielded from hemicelluloses. From commercial aspect, microorganisms are needed to convert hemicellulosic sugars viz. xylose and arabinose into bioethanol. One of the major hindrance towards effective production of ethanol is the inability of many microorganisms to utilize pentose sugars. There are plethora of microorganisms (mainly bacteria, fungi and yeasts) that are available in nature that can utilize pentose and hexose sugars into bioethanol. Each sugar fermenting microorganisms is different with respect to very narrow substrate range, ethanol tolerance, etc. These limitations can be overcome by development of recombinant strains which are tolerant to high ethanol concentrations and are capable of metabolizing various sugars. Industrially important prominent microorganism involved in bioethanol production is *Saccharomyces cerevisiae*. It is also capable of fermenting galactose. *Brettanomyces custersii* (KCCM11490) is also an ethanol producing mold that is preferred over *S. cerevisiae* during fermentation of red algae *G. amansii*. The hydrolysates of *G. amansii* are rich in galactose. The *B. custersii* (KCCM11490) can produce high bioethanol yields from galactose as compared to other sugars. Biochemical pathway involved in bioethanol production from galactose involves three routes (Goh and Lee, 2010; Park et al., 2012): (a) The D-galactose-6-phosphate pathway; (b) The Leloir pathway; and (c) The Entner-Deudoroff pathway.

When there is abundance of mannitol in the fermentation media, *Enterobacter* sp. (JMP3) and *Escherichia coli* (KO11) are reported to use it effectively for ethanol production (Kim et al., 2011). Co-culture technique was developed where two different microorganisms were used to ferment various saccharified products. One such example is fermentation of *L. japonica* hydrolyzates using *S. cerevisiae* and *E. coli* (KO11), sequentially giving high ethanol yields. Solid state fermentation of *S. japonica* hydrolyzates was fermented with cocktail of four different yeasts (*Pichia angophorae* [KCTC 17574], *Pichia stipitis* [KCTC 7228], *S. cerevisiae* [KCCM 1129] and *Pachysolen tannophilus* [KCTC 7937]) with *Bacillus* sp. (JS1) for ethanol production (Jang et al., 2012). Saccharification was performed by *Bacillus licheniformis* and the sugars thus generated were used by the four different types of yeasts to produce ethanol. Highest yield of ethanol is reported by *P. angophorae*. The oxidation of glucose via glycolytic pathway forms pyruvate along with NADH and ATP. Pyruvate is converted to ethanol and CO_2 under anaerobic conditions. The pyruvate is converted to acetaldehyde. The reaction is catalyzed by pyruvate decarboxylase enzyme. Acetaldehyde is then reduced to ethanol. This reaction is catalyzed by alcohol dehydrogenase. This conversion of pyruvate to ethanol can also be affected by intracellular electron balance i.e. $NADH/NAD^+$ ratio (Wang et al., 2013). *Zymomonas mobilis* is a well known bacterium capable of giving high ethanol yield. A strategy was developed to introduce pdc (pyruvate decarboxylase) and adhB (alcohol dehydrogenaseII) genes from *Zymomonas mobilis* into *E. coli*. This leads to development of an ethanologenic strain, *E. coli* (KO11).

13.3.5 Future Prospects of Algal Ethanol

With the encouraging trends in the field of algal ethanol production, many entrepreneurs are focusing on commercialization of this process. The major bottleneck for bioethanol production in commercial scale is availability of algae at very large quantities and at very low cost. Through breakthrough technological innovations and advancement, algal bioethanol can be produced in commercial scale. There are several roadblocks in the algae-to-ethanol technology. In open pond cultivation, the chance of external contamination arises. This may lead to a situation where the desired algae has been consumed by predators like *Paramecium* sp. or other protozoan or other stronger algal species have dominated the cultivation pond. Transgenic algae have commercially important traits but they are less fit for open cultivation. It is hypothesized that usage of transgenic extremophilic algae may be more robust and have better chances of survival in open pond culture. In such extreme conditions, the contaminants/competitors would be limited to minimum. The high cost of enzymatic hydrolysis of starch/cellulose also makes the cost of algal bioethanol several folds higher. Genetically modifying microalgal strain in such a way that it accumulates higher amount of starch/cellulose which could make them more lucrative commercially.

Application of synthetic biology could open a lead to paraphernalia of enzymes in microalgae that might help in saccharification of stored starch and cellulose. This would minimize the need of external enzymatic hydrolysis. The up regulation or over expression of biosynthesis pathway for starch/cellulose accumulation would definitely increase algal biofuel production potential. Dependence on large amount of fresh water for production of algal biomass might compete for fresh water requirements of crops and human consumption. By the year 2050, commercial bioenergy production is expected to consume 18–46 % of fresh water resources. Globally, world is facing shortage of fresh water recourses. Since 70 % of our earth is covered with marine water, development of high salt and temperature tolerant microalgae could be the potential candidate for mass production (Sheridan, 2009; Waltz, 2009). Places with greater availability of coastlines are the ideal regions for trapping sunlight and thus wouldn't compete for terrestrial lands. Thus the path of sustainable growth and energy generation can be realized truly by algal biofuel technologies.

13.4 Conclusion

Biodiesel from algal biomass has promising potential towards commercialization. The techniques that are currently used for biodiesel production are still trivial in terms of net energy balance. A careful assessment of the life cycle energy balances is required to study the sustainability of the process. Information regarding large scale demonstration of biodiesel production are scares which lead to incorrect economic assessments. Thus large scale pilot studies should be conducted under realistic setups which would include typical weather conditions. This would be useful for estimating biodiesel productivities. Innovation and breakthroughs are still required for development of design and technologies that would lead to costs reduction. Selection of strains for high lipid concentration which could be amicabily adapted to regional conditions along with genetic improvement could eventually make this process economically viable. A biofinery concept encompasses the vision where spent biomass after lipid extraction could be used for production of alternative bulk or fine chemical production. At present, emphasis is given on the production of bioproducts such as bioethanol or biomethane from lipid-extracted algal biomass. Since this approach utilizes complete waste resources, the overall energy conversion efficiency increases.

Bioethanol production from biomass is generally focused on utilization of lignocellulosic biomass. The lignocellulosic biomass for bioethanol production are cheap, easily available and renewable. But the production of the second-generation bioethanol is commercially not viable because of their recalcitrant nature. The algal biomass as a raw material for bioethanol production could become a sustainable and eco-friendly resource for renewable biofuel production. Currently, commercial production of bioethanol from algae is not feasible because of low product yield when compared to other conventional substrates. Issues pertaining to high costs of algae cultivation such as algae cultivation, harvesting, and biomass pretreatment are the major bottlenecks towards its commercialization. It is high time for human civilization that needs to take decisive steps on issues related to climate and environment and maintain a sustainable growth. Marine algae cultivation hold a promising avenue towards this endeavour.

References

Agbor, V.B., Cicek, N., Sparling, R., Berlin, A. and Levin, D.B. (2011). Biomass pretreatment: Fundamentals toward application. *Biotechnol. Adv.*, 29, 675–685.

Amaro, H.M., Guedes, A.C. and Malcata, F.X. (2011). Advances and perspectives in using microalgae to produce biodiesel. *Appl. Energy*, 88, 3402–3410.

Banerjee, A., Sharma, R., Chisti, Y. and Banerjee, U.C. (2002). *Botryococcus braunii*: A renewable source of hydrocarbons and other chemicals. *Crit. Rev. Biotechnol.*, 22, 245–279.

Belarbi, E.-H., Molina, E. and Chisti, Y. (2000). RETRACTED: A process for high yield and scaleable recovery of high purity eicosapentaenoic acid esters from microalgae and fish oil. *Process Biochem.*, 35, 951–969.

Binod, P., Sindhu, R., Singhania, R.R., Vikram, S., Devi, L., Nagalakshmi, S., Kurien, N., Sukumaran, R.K. and Pandey, A. (2010). Bioethanol production from rice straw: An overview. *Bioresour. Technol.*, 101, 4767–4774.

Brown, L.M. and Zeiler, K.G. (1993). Aquatic biomass and carbon dioxide trapping. *Energy Convers. Manag.*, 34, 1005–1013.

Chisti, Y. (2007). Biodiesel from microalgae. *Biotechnol. Adv.*, 25, 294–306.

Choi, S.P., Nguyen, M.T. and Sim, S.J. (2010). Enzymatic pretreatment of *Chlamydomonas reinhardtii* biomass for ethanol production. *Bioresour. Technol.,* 101, 5330–5336.

Demirbas, A. (2007). Importance of biodiesel as transportation fuel. *Energy Policy*, 35, 4661–4670.

Ghirardi, M. (2000). Microalgae: A green source of renewable H_2. *Trends Biotechnol.*, 18, 506–511.

Goh, C.S. and Lee, K.T. (2010). A visionary and conceptual macroalgae-based third-generation bioethanol (TGB) biorefinery in Sabah, Malaysia as an underlay for renewable and sustainable development. *Renew. Sustain. Energy Rev.*, 14, 842–848.

Gouveia, L. and Oliveira, A.C. (2009). Microalgae as a raw material for biofuels production. *J. Ind. Microbiol. Biotechnol.*, 36, 269–274.

Hargreaves, P.I., Barcelos, C.A., da Costa, A.C.A. and Pereira, N. (2013). Production of ethanol 3G from *Kappaphycus alvarezii*: Evaluation of different process strategies. *Bioresour. Technol.*, 134, 257–263.

Harun, R., Danquah, M.K. and Forde, G.M. (2009). Microalgal biomass as a fermentation feedstock for bioethanol production. *J. Chem. Technol. Biotechnol.*, 85(2), 199–203.

Helwani, Z., Othman, M.R., Aziz, N., Kim, J. and Fernando, W.J.N. (2009). Solid heterogeneous catalysts for transesterification of triglycerides with methanol: A review. *Appl. Catal. A Gen.*, 363, 1–10.

Hu, Q., Sommerfeld, M., Jarvis, E., Ghirardi, M., Posewitz, M., Seibert, M. and Darzins, A. (2008). Microalgal triacylglycerols as feedstocks for biofuel production: Perspectives and advances. *Plant J.*, 54, 621–639.

İçöz, E., Mehmet Tuğrul, K., Saral, A. and İçöz, E. (2009). Research on ethanol production and use from sugar beet in Turkey. *Biomass and Bioenergy*, 33, 1–7.

Jang, J.-S., Cho, Y., Jeong, G.-T. and Kim, S.-K. (2012). Optimization of saccharification and ethanol production by simultaneous saccharification and fermentation (SSF) from seaweed, *Saccharina japonica. Bioprocess Biosyst. Eng.*, 35, 11–18.

Khan, S.A., Hussain, M.Z., Prasad, S. and Banerjee, U.C. (2009). Prospects of biodiesel production from microalgae in India. *Renew. Sustain. Energy Rev.*, 13, 2361–2372.

Kim, N.-J., Li, H., Jung, K., Chang, H.N. and Lee, P.C. (2011). Ethanol production from marine algal hydrolysates using *Escherichia coli* KO11. *Bioresour. Technol.*, 102, 7466–7469.

Knothe, G. (2006). Analyzing biodiesel: standards and other methods. *J. Am. Oil Chem. Soc.*, 83, 823–833.

Lee, O.K., Kim, A.L., Seong, D.H., Lee, C.G., Jung, Y.T., Lee, J.W. and Lee, E.Y. (2013). Chemo-enzymatic saccharification and bioethanol fermentation of lipid-extracted residual biomass of the microalga, *Dunaliella tertiolecta. Bioresour. Technol.*, 132, 197–201.

Li, Y., Horsman, M., Wang, B., Wu, N. and Lan, C.Q. (2008). Effects of nitrogen sources on cell growth and lipid accumulation of green alga *Neochloris oleoabundans. Appl. Microbiol. Biotechnol.*, 81, 629–636.

Macario, A., Giordano, G., Onida, B., Cocina, D., Tagarelli, A. and Giuffrè, A.M. (2010). Biodiesel production process by homogeneous/heterogeneous catalytic system using an acid–base catalyst. *Appl. Catal. A Gen.*, 378, 160–168.

Marchetti, J.M., Miguel, V.U. and Errazu, A.F. (2007). Possible methods for biodiesel production. *Renew. Sustain. Energy Rev.*, 11, 1300–1311.

Mata, T.M., Martins, A.A. and Caetano, N.S. (2010). Microalgae for biodiesel production and other applications: A review. *Renew. Sustain. Energy Rev.*, 14, 217–232.

Metzger, P. and Largeau, C. (2005). *Botryococcus braunii*: A rich source for hydrocarbons and related ether lipids. *Appl. Microbiol. Biotechnol.*, 66, 486–496.

Mittal, A., Katahira, R., Himmel, M.E. and Johnson, D.K. (2011). Effects of alkaline or liquid-ammonia treatment on crystalline cellulose: Changes in crystalline structure and effects on enzymatic digestibility. *Biotechnol. Biofuels*, 4, 41.

Moellering, E.R. and Benning, C. (2010). RNA interference silencing of a major lipid droplet protein affects lipid droplet size in *Chlamydomonas reinhardtii*. *Eukaryot. Cell*, 9, 97–106.

Mussato, S.I., Dragone, G., Guimarães, P.M., Silva, J.P.A., Carneiro, L., Roberto, L.M., Vicente, A., Domingues, L. and Teixeira, J.A. et al. (2010). Technological trends, global market, and challenges of bio-ethanol production. *Biotechnol. Adv.*, 28, 817–830.

Nagle, N. and Lemke, P. (1990). Production of methyl ester fuel from microalgae. *Appl. Biochem. Biotechnol.*, 24–25, 355–361.

Park, J.-H., Hong, J.-Y., Jang, H.C., Oh, S.G., Kim, S.-H., Yoon, J.-J. and Kim, Y.J. (2012). Use of *Gelidium amansii* as a promising resource for bioethanol: A practical approach for continuous dilute-acid hydrolysis and fermentation. *Bioresour. Technol.*, 108, 83–88.

Ross, A.B., Jones, J.M., Kubacki, M.L. and Bridgeman, T. (2008). Classification of macroalgae as fuel and its thermochemical behaviour. *Bioresour. Technol.*, 99, 6494–6504.

Sarkar, N., Ghosh, S.K., Bannerjee, S. and Aikat, K. (2012). Bioethanol production from agricultural wastes: An overview. *Renew. Energy*, 37, 19–27.

Sheridan, C. (2009). Making green. *Nat. Biotechnol.*, 27, 1074–1076.

Singh, A., Nigam, P.S. and Murphy, J.D. (2011). Renewable fuels from algae: An answer to debatable land based fuels. *Bioresour. Technol.*, 102, 10–16.

Singh, S., Kate, B.N. and Banerjee, U.C. (2008). Bioactive Compounds from Cyanobacteria and Microalgae: An Overview. *Crit. Rev. Biotechnol.*, 25(3), 75–95.

Suppes, G. (2004). Transesterification of soybean oil with zeolite and metal catalysts. *Appl. Catal. A Gen.*, 257, 213–223.

Talebnia, F., Karakashev, D. and Angelidaki, I. (2010). Production of bioethanol from wheat straw: An overview on pretreatment, hydrolysis and fermentation. *Bioresour. Technol.*, 101, 4744–4753.

Vyas, A.P., Verma, J.L. and Subrahmanyam, N. (2010). A review on FAME production processes. *Fuel*, 89, 1–9.

Waltz, E. (2009). Biotech's green gold? *Nat. Biotechnol.*, 27, 15–18.

Wang, J., Kim, Y.M., Rhee, H.S., Lee, M.W. and Park, J.M. (2013). Bioethanol production from mannitol by a newly isolated bacterium, *Enterobacter* sp. JMP3. *Bioresour. Technol.*, 135, 199–206.

Xu, H., Miao, X. and Wu, Q. (2006). High quality biodiesel production from a microalga *Chlorella protothecoides* by heterotrophic growth in fermenters. *J. Biotechnol.*, 126, 499–507.

Yoo, C., Jun, S.-Y., Lee, J.-Y., Ahn, C.-Y. and Oh, H.-M. (2010). Selection of microalgae for lipid production under high levels carbon dioxide. *Bioresour. Technol.*, 101, 71–74.

Yoon, S.H. and Lee, C.S. (2011). Lean Combustion and Emission Characteristics of Bioethanol and Its Blends in a Spark Ignition (SI) Engine. *Energy & Fuels*, 25, 3484–3492.

Younesi, H., Najafpour, G. and Mohamed, A.R. (2005). Ethanol and acetate production from synthesis gas via fermentation processes using anaerobic bacterium, *Clostridium ljungdahlii*. *Biochem. Eng. J.*, 27, 110–119.

Yoza, B.A. and Masutani, E.M. (2013). The analysis of macroalgae biomass found around Hawaii for bioethanol production. *Environ. Technol.*, 34, 1859–1867.

Zabeti, M., Daud, W.M.A.W. and Aroua, M.K. (2010). Biodiesel production using alumina-supported calcium oxide: An optimization study. *Fuel Process. Technol.*, 91, 243–248.

Chapter 14
Gaseous Fuels Production from Algal Biomass

Shantonu Roy and Debabrata Das

14.1 Introduction

The most abundant renewable energy available to us is solar energy. It provides 178,000 TW energy to the Earth per year (Rupprecht et al. 2006). Entrapment of such consistent source of energy has been performed by photosynthetic organisms. The interest for microalgae is growing worldwide for their ability to harness solar energy and consequently providing biomass that can be used as feedstock for renewable energy generation. The rate of CO_2 fixation was up to 6.24 kg m^{-3} day^{-1} (Cheng et al., 2006). The productivity of algae could be 10 times higher (50 ton dry weight (DW) per hectare per year) when compared with conventional agricultural crops (Murphy and Power, 2009; Wijffels, 2008).

Hydrogen is considered as a clean and renewable source of energy for the future. It may be used as a potential alternative energy source in place of fossil fuels. It has the long-term potential to reduce the dependence on hydrocarbon-based fuels. Molecular H_2 has the highest energy content per unit weight among the known gaseous fuels (143 GJ ton^{-1}) and is the only carbon-free fuel which ultimately oxidizes to water as a combustion product. Therefore, burning hydrogen not only has the potential to meet a wide variety of end user applications but also does not contribute to greenhouse emission, acid rain or ozone depletion. Although hydrogen is the most abundant element in the universe, it must be produced from other hydrogen-containing compounds such as biomass, organic wastewater or water (Kotay and Das, 2008).

In a process called biophotolysis, green algae uses light energy to generate the energy carrier i.e. H_2 from water (Skjånes et al., 2007). Since water is considered as widely available resource, in recent times, many studies are focused on the

S. Roy (✉) • D. Das
Department of Biotechnology, Indian Institute of Technology, Kharagpur 721302, India
e-mail: shantonu20@gmail.com

D. Das (ed.), *Algal Biorefinery: An Integrated Approach*,
DOI 10.1007/978-3-319-22813-6_14

photobiological hydrogen production. The history of biological hydrogen production dates way back to three billion years ago when it developed photosynthesis process to convert CO_2, water and sunlight into hydrogen and oxygen. Although a comprehensive knowledge is available on biological H_2 production, a detailed emphasis is needed on basic and applied research for its application on commercial scale. A paramount requirement for understanding hydrogen production is in-depth knowledge of biochemistry and enzymology. Through understanding of these photobiological mechanisms, chemical mimicking of biophotolysis can be developed. This will, in addition, necessitate much more work on the isolation, characterization and stabilization of functional biological components.

For a stable and cost-effective hydrogen production, light conversion efficiency should be improved by at least 10 % so that it can compete with photoelectrical systems (Kruse et al., 2005; Prince and Kheshgi 2005). Metabolic engineering could prove a promising tool to increase conversion efficiencies and improve productivity. Fortunately, several advances in direct photolysis, indirect photolysis, and photo-fermentation have shown importance (Dickson et al., 2009). Although, hydrogen production from photolysis of water projects a promising future, the cumulative hydrogen production is quite low (Table 14.1). In case of *Chlamydomonas reinhardtii*, maximum cumulative hydrogen production of 3.1 ± 0.3 mL L^{-1} was observed which was quite low as compared to dark fermentative hydrogen production (Tamburic et al., 2011). Such low rate of production is the major bottleneck for large scale hydrogen production from photolysis of water.

For sustainable green energy production, biomass is considered as one of the most promising alternative. A model of decentralized power generation from hydrogen produced from biomass can be envisioned with the advent of cheap commercialization of fuel cells. There are many reports available on lab scale hydrogen production but commercial hydrogen production plants from biomass are not known in the world today. Algal biomass could prove to be a potential feedstock for biohydrogen production. This biomass is biodegradable, requires less harsher pretreat-

Table 14.1 Photobiological hydrogen production using algae

Microorganism	Photobioreactor	Rate of hydrogen production (mL L^{-1} h^{-1})
Chlamydomonas reinharditi strain 137 (mt$^+$)	Flat bottle	2
Chlamydomonas reinharditi	vials	2.1
Chlamydomonas reinharditi strain 137C+	vials	4.3
Chlorella sorokiniana Ce	-	1.35
Anabena cylindrica	-	0.74 μmol mg^{-1} Ch.l *a*
Synechocystis PCC6803	Flask	0.66 μmol mg^{-1} Chl *a*
Gloeobacter PCC 7421	Flask	1.35 μmol mg^{-1} Chl *a*
Nostoc muscorum	Flask	7.4 μmol mg^{-1} Chl *a*
Nostoc sp. PCC7422	Flask	100 μmol mg^{-1} Chl *a*

ment for saccharification and is produced in relatively less amount of time. Algal biomass mostly contains starch and cellulose and lacks lignin in their ultra structure. Suitable dark fermentative mixed consortia could be used to utilize such complex carbohydrates and produce hydrogen. During dark fermentation of organic material, energy carriers such as hydrogen and methane may be produced (Schink, 1997).

14.2 Photobiological Route for Hydrogen Production Using Algae

14.2.1 Hydrogen Production by Microalgae

Hydrogen production ability of green algae is coupled with photosynthesis. Water acts as direct source of electron. As the water splits to oxygen with the help of light energy and Photosystem II (PSII), the electron generated in this process is transferred to cytochrome b6f complex using plastoquinone (PQ) (Finazzi et al., 2002). Plastoquinone (PQ) then reduces plastocyanin (PC). Plastocyanin is a copper-containing electron transfer protein which then reduces Photosystem I (PSI). PSI can then reduce a ferredoxin (Eq. 14.1) or flavodoxin, which in turn reduces ($NADP^+$) and produces NADPH. The trans membrane potential of protons was used by membrane bound an ATPase to generate ATP and H^+ ions which enters into the lumen from stroma thylakoid. These protons are reduced by ferredoxin (reduced) to form molecular hydrogen via Fe-Fe hydrogenase enzymes (Eqn. 14.2).

$$\mathrm{Fd}_{(\mathrm{ox})} + \mathrm{e}^-_{(\mathrm{from\ PSI})} \xrightarrow{\text{ferredoxin oxidoreductase}} \mathrm{Fd}_{(\mathrm{Red})} \tag{14.1}$$

$$2\mathrm{H}^+ + 2\mathrm{e}^- \xrightarrow[\text{Ferredoxin}]{\text{Fe-FeH}_2\text{ genase}} \mathrm{H}_2 \tag{14.2}$$

The Fe-Fe-hydrogenase of green algae differs from most [FeFe]-hydrogenase as it lacks accessory Fe–S clusters at the N-terminus. It uses the electrons from ferredoxin (reduced) and combines it with the protons (H^+) to form molecular hydrogen.

14.2.1.1 Hydrogen Production by Cyanobacteria

Cyanobacteria, also called blue-green algae, belongs to the phylum of bacteria inhabiting watery environment. They obtain energy through oxygenic photosynthesis. This ability of cyanobacteria has converted early earth's reducing atmosphere in to an oxidizing one. These ancient life forms are harnessed in producing the eco-friendly energy of the future—hydrogen.

14.2.1.2 Non-heterocystous Cyanobacterial H_2 Production

Similar to microalgae, cyanobacteria also splits water and the electron thus generated is transferred to ferredoxin. Ferredoxin then channelizes these electrons to reduce $NADP^+$ to NADPH via FNR enzyme complex. NADPH then donates electron to bidirectional hydrogenase that converts proton to molecular hydrogen. In this process, bidirectional hydrogenase is Ni-Fe hydrogenase and thus plays a crucial role in conversion of solar energy and water into molecular hydrogen. These pigments system present in cyanobacteria are similar to microalgae. Cyanobacteria have large light-harvesting antenna complexes which give them an upper hand for survival in low-light conditions. Due to self-shading effect, the efficiency of large antenna systems decreases. Moreover, most of the energy from the incident light of the absorbed photons are dissipated as fluorescence and heat (Polle, 2002).

14.2.1.3 Heterocystous Cyanobacterial H_2 Production

Another category of cyanobacteria have filamentous morphology. These filamentous cyanobacteria have nitrogen fixing ability and have two types of cellular polymorphism. It has vegetative cells and heterocyst cells. Heterocystous cells are thick walled devoid of photosystem II (PS II). Thus photolysis of water doesn't take place inside the heterocysts thus anaerobic condition prevails inside the heterocyst. This anaerobic condition is very much required for nitrogenase and bidirectional hydrogenase activity. Nitrogenase is exclusively found in heterocysts. It fixes molecular nitrogen to ammonia and hydrogen is produced as a byproduct. Hydrogen is produced to maintain the redox potential inside the heterocyst. The turnover number of nitrogenase is significantly lower as compared to bi-directional [Ni-Fe] hydrogenase (Tamagnini et al., 2007). For substantial hydrogen production, larger amount of nitrogenase enzyme is required. Moreover, maintenance of nitrogenase enzyme is an energy demanding process as it consumes 16 moles of ATP to fix 1 mole of ammonia and to produce 1 mole of molecular hydrogen (Eq. 14.3).

$$N_2 + 8e^- + 8H^+ + 16ATP \rightarrow H_2 + 2NH_3 + 16(ADP + Pi) \qquad (14.3)$$

During nitrogen fixation, the electron allocation coefficient is defined as the ratio of electron used for nitrogen fixation to total electron consumed in the nitrogenase mediated reaction. The value of this coefficient remains in the range of 0.7 to 1. Therefore, in presence of nitrogen a large number of electrons are needed to be channelized to form hydrogen. This decreases the value of electron allocation coefficient. But scenario changes when there is a nitrogen deprived condition. In this case hydrogen is formed by expenditure of just 4 moles of ATP (Eq. 14.4).

$$2e^- + 2H^+ + 4ATP \rightarrow H_2 + 4(ADP + P_i) \qquad (14.4)$$

14.2.2 Molecular Approach for Improvement of H_2 Production in Green Algae

Photosynthetic efficiency of the antenna pigments are just 10 % which implies that the rate of electron transfer from PSII to PSI is about 10 times lower than the rate of photon capture by the antenna pigments. One approach to improve photosynthetic efficiency was performed in *Chlamydomonas reinhardtii*, by truncating the chlorophyll (Chl) antenna size of PSII. Most recent studies showed the application of RNA interference or RNAi to down regulate the genes encoding for light harvesting complexes (LHC). Silencing of LCH genes lead to occurrence of less tightly stacked grana with improved photosynthetic quantum yield (81 %) and it also reduced the sensitivity towards photo-inhibition (Rosenberg et al., 2008).

The [FeFe]-hydrogenases of green algae are oxygen sensitive. This hampers hydrogen production. Therefore, protein engineering strategies are envisioned to increase the oxygen tolerance of these enzymes. Use of random and site directed mutagenesis has led to strain improvement of *Chlamydomonas* sp. which showed 10-fold improvement of oxygen tolerance (Chen et al., 2003).

14.2.3 Biohydrogen Production Using Algal Biomass as Feedstock

Microorganisms generate hydrogen for two principle reasons, first to dispose of excess reducing equivalents and second as a byproduct in nitrogen fixation. Microbial H_2 production is an attractive process for supplying a significant share of the H_2 required for the near future (Fig. 14.1).

Obligate anaerobes, facultative anaerobes, methylotrophs, and photosynthetic bacteria are well known microorganisms that have H_2 production ability. Based on the pathway that these microbes follow to evolve H_2, they are broadly categorized as either dark fermentative or photo fermentative microorganisms. Operation of dark fermentative H_2 production has low energy demand as compared to other conventional production processes. Moreover, it requires moderate process conditions which results in low cost for operation and minimal pollution generation (Angenent et al., 2004; Valdezvazquez et al., 2005). Therefore, dark fermentation has attracted greater attention of the researchers. Among a large number of microbial species, strict anaerobes and facultative anaerobic chemoheterotrophs, such as *Clostridia* and *Enteric* bacteria are efficient H_2 producers while degrading various types of carbohydrates.

Anaerobic *Clostridia* are well known group of microorganisms having greater potential for H_2 production. Immobilized *C. butyricum* was reported to produce 2 mol H_2 mol^{-1} glucose at 50 % glucose conversion efficiency. The highest theoretical yield of 4 mole H_2 from 1 mole of glucose can be achieved on following acetic acid fermentation (Taguchi et al., 1996). Formation of other metabolites like etha-

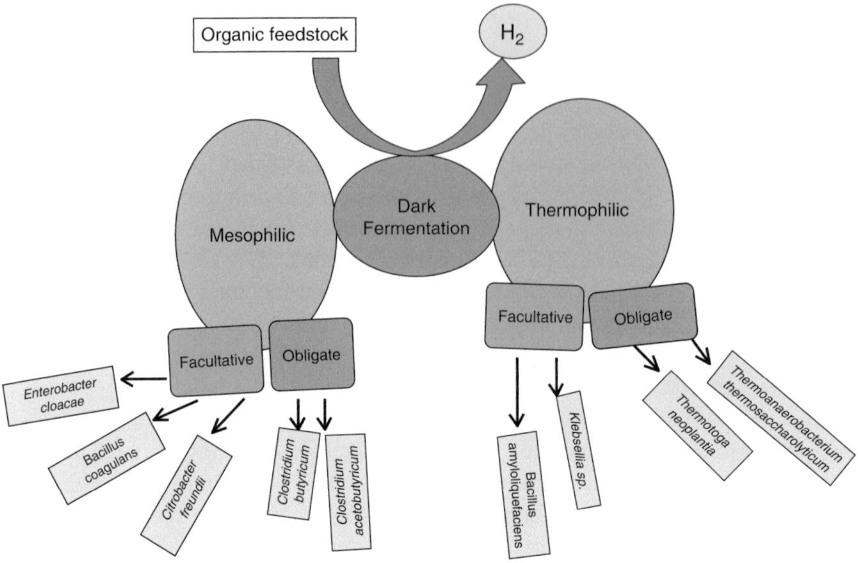

Fig. 14.1 Different domains of dark fermentative hydrogen production.

nol, propanol, butyrate, etc. leads to lower H_2 yields as formation of these metabolites compete for reducing equivalents inside the cell. For instance the conversion of one mole of glucose into butyrate is accompanied by the production of only 2 mol of H_2. *Clostridium* sp. such as *C. butyricum*, *C. welchii*, *C. pasteurianum and C. beijerinckii* are some newly isolated species that are potential H_2 producing microorganisms.

These organisms have plethora of hydrolytic enzymes which helps in degradation of organic matter to carbon dioxide and hydrogen. Such report on hydrogen production was available over 100 years ago by the biochemist Hoppe-Seyler (Stephenson and Stickland, 1932). The reduction of protons to H_2 serves to dissipate the excess electrons within the cell and generally permits additional energy generating steps in metabolism.

Plethora of microbial species is reported to produce hydrogen through dark fermentation viz. *Enterobacter, Citrobacter, Bacillus* and *Clostridium* sp. In recent times, apart from pure cultures, enriched mixed consortia are now gaining importance. By virtue of abundance of different array of hydrolytic enzymes, the mixed consortia or synthetic mixed consortia are ideal biocatalyst to utilize complex carbohydrates present in biomass for gaseous energy recovery. Nevertheless, the pursuit of ideal microbe(s) for H_2 production has attracted several researchers to screen exotic sources like hot spring, coal mine leachate, acid mine leachate, etc.

Anaerobic heterotrophic microorganisms can form hydrogen during the oxidation of organic substrates. A great advantage of fermentation is fast degradation of solids and other complex organics found in wastes and agricultural products. On the other hand, fermentation today converts only about 15 % of the energy to hydrogen.

On the other hand, biohydrogen production at thermophilic temperatures (60 °C) has many attractive advantages as compared to mesophilic temperatures (37 °C). Higher temperature condition leads to pathogenic destruction, lowers the risk of contamination by methanogenic archaea, higher rate of hydrolysis as well as H_2 yield.

The hyper thermophilic archaebacterium *Pyrococcus furiosus* produces H_2, organic acids and CO_2 from carbohydrates (Godfroy et al., 2000). Some cellulose degrading thermophiles that produce H_2 are *Anaerocellum, Caldi-cellulosiruptor, Clostridium, Dictyoglomus, Fervidobacterium, Spirocheta, Thermotoga* and *Thermoanaerobacter*. Hyperthermophilic *Thermotoga maritime* produces H_2 at 80 °C with yield of 4 mol mol^{-1} glucose which is equal to the maximal theoretical value (Schröder et al. 1994). Similar stoichiometries as of *T. maritima* were obtained for two more moderate thermophiles, *Acetothermus paucivorans* and *Acetomicrobium flavidum*, grown at 60 °C. These results showed that higher H_2 yields on hexose can be reached by extreme- and hyper-thermophiles as compared to mesophilic facultative and strict anaerobes.

Different pretreatments method could be employed to improve hydrogen production using algae as feedstock. Different physico-chemical pretreatments were employed to increase the accessibility of different complex sugars entrapped in algal biomass into simpler form. Carbohydrates in algal biomass are found as intracellular complex polymeric form bounded with rigid algal cell walls. Therefore, it is necessary to break the algal cell wall along with complex carbohydrate to facilitate the release of simple sugar. Cost of pretreatment of biomass adds significantly to overall hydrogen production process. Several methods of pretreatments such as physical (sonication, milling, grinding and pyrolysis), chemical (acid, alkali, thermal, H_2O_2) and biological methods (enzymatic, microbial) have been reported to break algal cell wall, hydrolyze the complex carbohydrates and release fermentable sugars. Each of the method has its own merits and demerits. Preference of chemical method such as acid treatment over others is because of higher conversion efficiency of polymeric carbohydrates into simpler sugars in lesser time. Physico-chemical methods are based on simpler technology but they are energy intensive processes limiting their use for commercial application. They also lead to formation of furfurals which inhibits the growth of the fermentative organisms. Biological based methods are developed to overcome these problems. It includes co-culture development where one organism will produce hydrolytic enzymes that would facilitate saccharification process.

Dark fermentative organisms can utilise simple sugars (which are released during saccharification) to produce hydrogen. Even crude enzymatic techniques were also studied for pretreatment of algal carbohydrates. But these processes are costly and time consuming with low reducing sugar yield. Very few reports were available on usage of algal biomass as feedstock for hydrogen production. In a study, algal biomass was used to produce H_2 using *Clostridium butyricum* and subsequent use of produced organic acids for H_2 production by photo fermentation using *R. sphaeroides* KD 131 (Kim et al., 2006). Tam et al. (2012) showed thermophilic hydrogen production using *Chalmydomonas reinharditti* (Nguyen et al., 2010). Nayak et al.

(2014), used amylase treated *Anabena variabilis* biomass to produce hydrogen via thermophilic dark fermentation. Similar study with *Chlorella sorokiniana* was also reported where acid-heat treated biomass was used by *Enterobacter cloacae* IITBT08 to produce hydrogen (Kumar et al., 2013). Thermophilic dark fermentation with acid-heat treated *Chlorella sorokiniana* biomass gave higher yield as compared to mesophilic process (Roy et al., 2014).

14.3 Biomethane Production Using Algae Biomass as Feedstock

Anaerobic digestion process (ADP) has been used since centuries to produce biogas. However, the first documented digestion plant was constructed in Bombay, India in 1859 (Lettinga, 2001). Utilization of biogas from a digester plant was first introduced in 1895 in Exeter, England where biogas was used for street lighting. Currently almost 15 million digesters, including small farm-based digesters, are operated in China and roughly 12 million digesters are established in India (Pathak et al., 2009).

An increasing awareness on greenhouse gas emissions and global warming has led to the global demand for alternate renewable fuels and has attracted researchers to promote ADP for industrial applications. Apart from conventional methods of treatment of sewage biosolids, livestock manure, and concentrated wastes from food industry, ADP can also serve as a significant source of renewable fuel. For instance, the anaerobic digestion (AD) of algal biomass to biogas possesses several advantages as compared to other biofuel sources and conversion techniques, such as:

- Uncritical water quality. Wastewater, brackish water and even seawater can be used for algae culturing in addition to fresh water.
- Reduced energy consuming steps.
- Maximal algal biomass utilization possible.
- Partial recycling of nutrients with AD effluent. AD releases nutrients in a potentially usable and recyclable form. The supernatant liquid rich in nitrogen and phosphorus content can be used as a fertilizer for algae culturing while the solid phase can be used as a fertilizer in agriculture or as a livestock nutrient.
- Possible integration with other technologies. ADP can be used as a co-technology for algal residues utilization after biodiesel, green diesel, bioethanol, and hydrogen production. Also, a variety of organic wastes and by-products can be co-digested with algae to produce biogas.

Nevertheless, methane production from algae suffers from certain limitations that need to be addressed prior to its industrial application, some of which are (Gujer and Zehnder 1983):

- High capital cost of algae production and AD units.

- Low algae productivity. Algae growth rate is relatively limited by low efficiency of photosynthesis, photoinhibition and carbon assimilation.
- Incomplete digestibility of algal cells. The algal biomass partially contains recalcitrant organic matter that cannot be hydrolyzed by the conventional ADP.
- Slow conversion rate. Biomass residence time in the ADP varies between 10 and 30 days.
- Unbalanced C:N ratio. A low C:N ratio can lead to the accumulation of NH_4^+ in an anaerobic digester to inhibitory levels while lack of nitrogen can limit anaerobic conversion and methane production.
- High sensitivity of the ADP. Methanogenic organisms are sensitive to fluctuations of environmental and operational parameters.

The main structural elements of macroalgae are composed of polysaccharides and their cell wall lack lignin. Based on the cell envelope components, they are classified into three major phyla: red, green and brown. Green algae contain ulvan and xylan in their cell envelopes whereas red algae contain carrageen, agar and xylose. Alginate and fucoidan are found in brown algae. In many algal classes cellulose is main structural element of cell wall. The main storage polysaccharides also vary among the three phyla of algae. Floridean starch is found in red algae; chlorophycean in green macroalgae while laminarin and mannitol in brown macroalgae. Remarkable variations are observed in the biochemical composition of microalgae, cyanobacteria and macroalgae which usually depends on several environmental factors such as temperature, salinity, light intensity and nutrient availability. Proteins and lipids account for the bulk of microalgal dry weight while carbohydrates contribute to minor component of cell dry weight. This discrepancy in the biochemical composition among different phylum or genera and also among similar species challenges the ADP of algae.

14.3.1 Principles of the Anaerobic Digestion Process

AD is a complex biological process in which organic matter is degraded into methane by naturally occurring microbial consortium comprising anaerobic bacteria and archaea. This section specifically deals with the detailed description of ADP biochemistry and microbiology, the influence of environmental and physicochemical parameters on process performance, and the importance of biogas composition and its application.

14.3.1.1 Biochemistry and Microbiology of Anaerobic Digestion

Algal biomass is a blend of organic and inorganic matter in which the organic part comprises complex polymeric macromolecules like proteins, polysaccharides, lipids, and nucleic acids that appear in particulate or colloidal form. During ADP, this organic matter is converted to final products (methane and carbon dioxide), new biomass, and/or inorganic residues by the action of diverse groups of microorganisms. The overall process comprises multiple stages with numerous intermediate products. Generally, ADP constitutes four key steps: (1) hydrolysis, (2) fermentation or acidogenesis, (3) acetogenesis and (4) methanogenesis. The overall transformation can be described by six distinct biological processes as mentioned in Fig. 14.2.

1. Hydrolysis of colloid and particulate biopolymers to monomers.
2. Fermentation or acidogenesis of amino-acids and sugars to intermediary products (propionate, butyrate, lactate, ethanol, etc.), acetate, hydrogen, and formate.

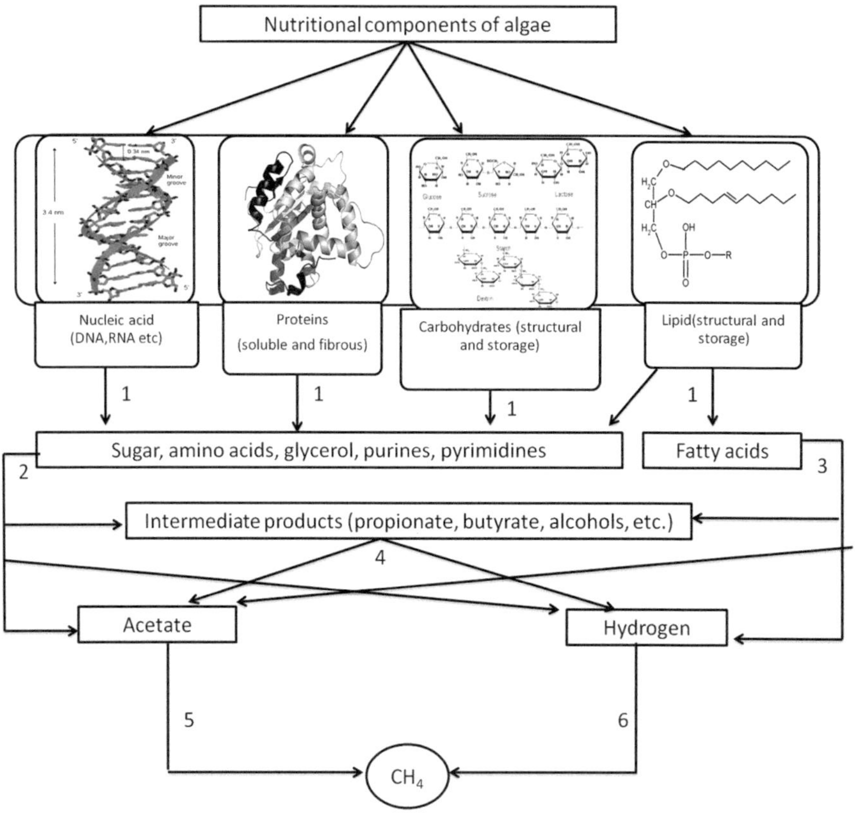

Fig. 14.2 Fate of organic components in anaerobic digestion.

3. β-oxidation of long-chain fatty acids and alcohol fermentation to volatile fatty acids (VFA) and hydrogen.
4. Anaerobic oxidation or acetogenesis of intermediary products, such as VFAs to acetate, carbon dioxide and hydrogen by obligate and facultative hydrogen producing species.
5. Transformation of acetate into methane by acetoclastic methanogens.
6. Transformation of molecular hydrogen and carbon dioxide into methane by hydrogenophilic methanogens.

Similar fermentative bacteria perform the first three steps which are sometimes referred to as acidogenesis or the acid-phase (Chen et al., 2008). Other important biological processes in AD are:

- Conversion of a variety of monocarbon compounds (e.g., formate, methanol) into acetic acid by homoacetogenic bacteria.
- Reduction of sulphur compounds to hydrogen sulphide by sulphur reducing bacteria.

14.3.2 Operational Parameters, Physicochemical Factors, and Inhibition of the Anaerobic Process

Among the microorganisms involved in AD, archaea possess the slowest growth rate and attain lowest energy during methanogenesis. Thus, for effective methane production, proper maintenance of environmental and operational parameters for archaea is the key factor. The main environmental factors for ADP are temperature, pH, alkalinity and redox potential. Operational parameters include C:N:P ratio, presence of essential micronutrients, organic loading rate (OLR), hydraulic (HRT) and solids retention time (SRT), and incoming salts and toxicants concentration. The accumulation of certain intermediates or byproducts, such as VFAs, ammonia and hydrogen sulphide, can lead to inhibition of methane production.

14.3.2.1 Temperature

The most important environmental factor for methanogenesis is temperature. The optimal temperature for growth of psychrophilic organisms is between 10 and 15 °C, mesophilic between 35 and 40 °C, and thermophilic between 58 °C and 68 °C. Compared to mesophiles and thermophiles, the rate of methane production is slowest with psychrophiles and, therefore, psychrophilic regime is rarely used for large-scale methane production. Methanogens are highly sensitive to temperature variation. Even a minute temperature change of 2–3 °C can cause accumulation of VFAs thereby decreasing methane generation rate especially at thermophilic conditions (Speece, 1983). The activity of all anaerobic microorganisms is drastically affected by considerable temperature drop and ceases methane production. However, these microorganisms are able to recover with appropriate temperature stabilization.

14.3.2.2 pH and Alkalinity

The pH is another important environmental factor for ADP as the optimum pH varies for different groups of methanogens. The acidogens exhibit maximum activity at pH 5.5–6.5 while methanogens exhibit at pH 7.8–8.2. Thus anaerobic digesters are usually maintained in the range of 7–8 to enhance methanogenesis. At pH higher than 8, dissociation of NH_4^+ to neutral NH_3 occurs that inhibits growth of methanogens (Hansen et al., 1998). An important marker of pH persistence in anaerobic digesters is alkalinity. Usually the pH fluctuations caused by the generation of VFAs and carbon dioxide is controlled by bicarbonate alkalinity buffers at pH close to neutral. A stable ADP is characterized by the bicarbonate alkalinity in the range of 1,000 to 5,000 mg L^{-1} as $CaCO_3$. The ratio between VFAs to alkalinity should be in the range of 0.1–0.25. Further increase of VFAs to alkalinity ratio indicates possible process deterioration and requires the OLR to decrease in order to lower the VFA formation rate.

14.3.2.3 Nutrients

Stable growth of anaerobic microorganisms is dependent on availability of appropriate micro and macro nutrients. The approximate ratio of carbon to nitrogen and phosphate should be in the range of 75:5:1 to 125:5:1. Chynoweth reported the nitrogen limiting conditions when C/N ratio exceeded 15 during digestion of *Macrocystis*. The most important element for growth is nitrogen as it serves dual role: it helps in synthesis of protein and nucleic acids, and when reduced to ammonia, serves as a base to maintain neutral pH. Certain macronutrients such as iron, nickel, cobalt, molybdenum, zinc, calcium, copper and boron, are necessary for stable AD in the ppm level (Kida et al., 2001). Most of these metals serve as the principal component of enzymes' active site. For example, copper and cobalt are constituents of B_{12}-enzyme which catalyze the methanogenesis; nickel is part of factor F 430 found only in methanogenic bacteria, and molybdenum and selenium are subcomponents of formate dehydrogenase, which is part of the active site of hydrogenase and acetyl-CoA synthase.

14.3.2.4 Oxidation–Reduction Potential

A substance's affinity to either gain or lose electrons is measured by its oxidation-reduction potential (ORP). In AD, it is a measure of available oxidants (oxygen or nitrate ions) or reductants (hydrogen). A high ORP (>50 mV) indicates presence of free oxygen in the anaerobic environment, ORP between 50 and −50 mV depicts an anoxic environment with nitrates and nitrites, and an ORP lower than −50 mV implies reducing environment. At an ORP range of −50 to −100 mV, sulphate reducing microorganisms surpass methanogens for hydrogen and acetate since sulphate is a more thermodynamically favourable electron acceptor. Acid production during fermentation occurs in the ORP range of −100 to −300 mV while methanogenesis require ORP $\leq$ 300 mV when carbon dioxide is used as an electron acceptor and methane is formed (Gerardi, 2003).

14.3.2.5 Organic Loading Rate, Hydraulic and Solids Retention Times

The organic loading rate (OLR), hydraulic and solids retention times (HRT and SRT) also play a significant role in ADP. With rapid increase in OLR, accelerated acid formation takes place that leads to alkalinity depletion and a subsequent pH drop. The volume and capital cost of AD system depend upon HRT while SRT influences volatile solids (VS) reduction and, thus the methane yield from biomass. Any deviation in OLR, HRT and SRT can disturb ADP and inhibit methane production.

14.3.2.6 Toxicants

Methanogens are strict anaerobes and even the lowest oxygen concentrations (as low as 0.1 mg L^{-1}) can inhibit methane production. Methanogenesis using marine algae is sensitive to high salts concentration (e.g. NaCl). Nevertheless, comparable methane yield was reported from green macroalgae diluted with seawater and fresh water. In another report, methanogens were observed to be adapted to 65 g L^{-1} NaCl concentration. Moreover, desalination of macroalgae by heat and pressure resulted in less methane yield possibly due to loss of easily digestible organic matter as compared to untreated algae (Braun et al., 1981). Some algae have a tendency to accumulate heavy metals which can be detrimental to methanogenic bacteria. However, their negative effect can be decreased by precipitation with sulphide compounds. Nevertheless, by-products, including ammonia and hydrogen sulphide, at high concentrations can be toxic for bacteria. Although the nitrogen and sulphur source in AD is proteins, certain seaweeds have a high amount of sulfated carbohydrates. The neutral forms of ammonia and hydrogen sulphide are more toxic to bacteria as they can easily penetrate the cell membrane.

This toxicity of ammonia and sulphide is dependent upon the presence of metals, temperature, and pH in digesters. However, some authors have reported increasing sulphide toxicity with increasing pH possibly due to different mechanisms of sulfide toxicity on different species which is usually associated with several factors: competition between methanogens for hydrogen and acetate; denaturation of native proteins through the formation of sulphide and disulphide cross linkage between polypeptide chains; interference with the assimilatory metabolism of sulphur; and the ability to remove essential metals (nickel, iron, cobalt) from the solution. Disruption of intracellular pH, potassium deficiency, and inhibition of a specific enzymatic reaction are associated with ammonia toxicity. The reported lethal concentration for ammonia toxicity for methanogenic microorganisms is 1.5–1.7 g L^{-1} at pH 7.4 and above (Sterling et al., 2001). However, methanogens can tolerate ammonia concentration up to 3–4 g L^{-1} at lower pH. Moreover, ADP levels of microorganisms are stable at high nitrogen concentrations (5–7 g L^{-1}). AD is inhibited by accumulation of non-ionic form of organic acids which is usually caused by decline in hydrogen utilization by acetoclastic methanogens and, therefore, leads to drop in pH. The mechanism by which organic acids inhibits ADP is probably through denaturation of cell proteins.

14.4 Anaerobic Digestion of Algae

Research on production of biogas from seaweeds and microalgae by AD started in mid-sixties. The first detailed large-scale study of seaweed cultivation and AD was performed in the Institute of Gas Technology, USA (Chynoweth et al., 2001). The composition of biogas obtained from ADP depends on the type of substrate and operational parameters. The theoretical yield of biogas can be estimated by the Bushwell equation (Eq. 14.5) (Buswell and Mueller, 1952).

$$C_cH_hO_oN_n + yH_2O \rightarrow xCH_4 + nNH_3 + (c-x)CO_2 \tag{14.5}$$

where

$$x = (4c - h - 2o - 3n - 2s)/8$$
$$y = (4c - h - 2o + 3n + 3s)/4$$

Methane yield depends on the average carbon oxidation state of the substrate. The theoretical methane yield from different algae is presented in Table 14.2. As observed from the table, lipids have lowest oxidation state and largest theoretical methane yield as compared to proteins, glycerol, and carbohydrates. Thus, macroalgae with high carbohydrate content and cyanobacteria with high protein content are theoretically poorer feedstock for methane production while microalgae with high lipid content have higher potential methane yield.

Table 14.2 Potential of different algae for methane production

Feedstock	CH_4 yield (L g^{-1}VS)	Purity of CH_4 (%)
Macrocystis pyrifera	0.64	59.5
Chlorella sp.	0.59	58.9
Chlorella pyrenoidosa	0.97	66.2
Anabaenopsis sp.	0.49	52.8
Chlamydomonas sp.	0.46	51.9
Laminaria sp.	0.45	51.9
Gracillaria verrucosa	0.44	49.3
Spirulina maxima	0.43	49.9
Ulva sp.	0.42	60.6
Fucus vesiculosus	0.41	50.1
Gracillaria tikvahiae	0.42	52.7

14.4.1 Anaerobic Digestion of Macroalgae

The biochemical composition varies considerably in different macroalgae species. However, cellulose is the main primary component among all algal groups. Extensive research has been conducted to study biological degradation of cellulose in recent years and the degradation mechanisms differ in anaerobic and aerobic bacteria. Anaerobic bacteria comprises multi enzyme complex—cellulosome in their cell envelope and consists of up to 11 different catalytic enzymes carried by scaffold-proteins. The enzymatic hydrolysis of algal cellulose is relatively slow and is inhibited by close association with other structural materials like polyphenols, fucoidan, protein and alginate. Thus, other species-specific sulphonated, methylated or carboxylated polysaccharides, mannitol, proteins and lipids usually determine the readily biodegradable fraction of algal biomass (Rees and Welsh, 1977).

14.4.1.1 Rhodophyta (Red Algae)

The main biochemical components of red algae consist of two sulphated polysaccharides-agars and carrageenans which are responsible for main part of cell dry weight. Both of them have an agarose backbone composed from α (1→3) linked galactose disaccharide units, connected by β (1→4) linkages. Galactose units on agars can be methylated (up to 20 %) and have few sulphate ester groups, while carrageenans can have from one to three sulphate ester groups, one for every disaccharide unit. Polysaccharides have non-uniform structure that depends on algal source, life stage, and season. Gamma Proteobacteria class, Bacteroidetes and

Planctomycetes phyla are usually able to degrade the red algal cell wall by secreting specific glycoside hydrolases, diverse agarases and carrageenases. Biodegradation of agar by a consortium of microorganisms from an anaerobic digester was studied on the biodegradation of *Gracilaria tikvahiae* and the results suggested that more than 50 % of the agar could be fermented over the course of three weeks. The rest of the agar, however, degraded slowly and complete fermentation (more than 90 %) required more than nine weeks. Overall agar was degraded faster than the rest of the organic matter in algae. This conclusion was supported by results of another experiment on *G. tikvahiae* batch digestion, where less than 50 % of the agar was fermented during 11 weeks. AD of carrageenan was studied on the biodegradation of red alga—*Eucheuma cottonii*, which has about 61.1–72.9 % of carrageenan content from dry weight. About 40 % of the biomass remained after 10 weeks of digestion but carrageenan accounted for 49.8–59.6 % of the fraction remaining (King et al., 1985). Therefore, it was concluded to be less biodegradable than rest of the algal organic matter.

Biogas production was observed to be highest during first four weeks when methane concentration ranged from 10 to 25 %. During later stages, however, methane production ceased probably due to accumulation of VFAs, out performance of the methanogens by sulphate reducing bacteria, or inhibition by sulphides. According to chemical composition, the theoretical methane yield for *G. tikvahiae* is 0.42 L g^{-1} VS and for *Gracilaria verrucosa* is 0.44 L/g VS. Biomethane potential (BMP) assays provided an experimental methane yield from *G. tikvahiae* of 0.35–0.4 L g^{-1} VS added, which corresponds to about 70–95 % of the theoretical methane yield. The methane yield from *G. verrucosa* was in the range 0.28–0.35 L $(g\ VS)^{-1}$, which corresponds to 58–77 % of the theoretical yield. A 60-d residence time was required for *Gracilaria* conversion to biogas.

14.4.1.2 Green Algae

Most frequently used feed stocks for AD consist of Ulvales, particularly *Ulva* (sea lettuce), *Enteromorpha* and *Cladophora* which are widespread all over the world specially in eutrophic ecological systems. However, *Ulva* sp. is considered as a potential feedstock for biomethane production since it has a low fraction of poorly biodegradable polyphenol materials (varies from 1 to 1.9 %). The major components of *Ulva* sp. are carbohydrates (about 40–60 % of cell dry weight), proteins (10–17 %), and lipids (1.8–3.5 %). Carbohydrates accumulate mostly as the cell envelope and consist of ulvan, cellulose, xyloglucan and glucuronan which are responsible for up to 38–54 % of the algal dry matter. In contrast to cellulose, ulvan is a water-soluble sulphated polysaccharide. The average composition of ulvan is rhamnose (16.8–45.0 % dw), xylose (2.1–12.0 %), glucose (0.5–6.4 %), uronic acid (6.5–19.0 %) and sulphate (16.0–23.2 %) but the main repeating disaccharide is ulvanobiouronic acid that is composed of aldobiouronic acid and 4-*O*-b-d-glucuronosyl-l-rhamnose. Despite the high ratio of sugars, only 8.9 % of ulvan and 16.6 % of *Ulva* organic matter is fermented by biota from the human. In another study degradation of 32, 25.9 and 50.9 % of sugars from *Ulva* was reported after 24 h of fermentation. It was concluded that the complex chemical structure of ulvan cell polysaccahrides makes them poorly accessible for enzymatic attack. The recommended OLR and HRT for *Ulva* AD fall in the range of 0.8–1.2 g VS $L^{-1}d^{-1}$ and 15–20 d, respectively. Differences in *Ulva* composition among strains and even seasonal changes for one strain account for this observed variability (Lee, 2013).

14.4.1.3 Phaeophyceae (Brown Algae)

Alginic acid, laminarin, mannitol and fucoidan form the major organic components of brown algae. Among these, mannitol lacks polymeric structure, is soluble, and can be easily transferred into the cell. Anaerobic microorganisms can utilize mannitol effectively with the formation of acetate and hydrogen as the major products, and minor production of ethanol, formate, lactate and succinate. AD of laminarin was reported using microbial consortium isolated from the human gut where almost complete (>90 %) laminarin was used up within 24 h along with the formation of butyrate and other VFA. Alginate has a more complex molecular structure and usually forms a gel in algae. The alginate lyases are enzymes found to be responsible for alginate depolymerization. Biological degradation of soluble Na-alginate gel is 6–8 times faster as compared to Ca-alginate gel due to calcium cross bridging in the polysaccharides. Alginate depolymerization leads to formation of mixture of oligosaccharides with different length, which are further degraded to 4-deoxy-l-erythro-5-hexoseulose uronic acid. The final products of alginate degradation are glyceraldehyde-3-phosphate and pyruvate AD of fucoidan is still not reported. However, several fucoidan-degrading marine bacteria were isolated and characterized and the possibility of AD of fucoidan containing waste sludge from alginate extraction has been explored. The molecular structure of particular strains makes the AD of fucans difficult. The AD of algal proteins and polyphenols and their impact on overall digestibility still needs to be addressed. It is assumed that polyphenols associate with proteins and polysaccharides in the cell envelope that decreases their availability for biological degradation.

Brown algae are one of the most prominent algal feedstock for AD and several species have been examined including *Macrocystis pyrifera*, *Ascophyllum nodosum*, *Durvillea antarctica*, *Sargassum* spp. and *Laminaria* spp. According to the chemical composition, the theoretical methane yield of 0.52 L g^{-1} VS and 0.49 L g^{-1} VS were predicted for *M. pyrifera* and *Laminaria* sp. respectively. Researchers have reported an experimental methane yield for *M. pyrifera* of 0.43 L g^{-1} VS (82 % VS reduction) but only 0.24–0.3 L g^{-1} VS (50–60 % VS reduction) for *Laminaria saccharina*. Variability in chemical composition between these genera signifies the difference in yield. *Laminaria* has a higher content of fucoidan, laminarin and alginate but lower content of mannitol. The ratio between experimental and theoretical methane yield by *M. pyrifera* species is highly correlated with the mannitol content, in contrast to polysaccharides. Mannitol can be easily and completely degraded by anaerobic microorganisms (*Oceanography and Marine Biology, An Annual Review*, 2003).

14.4.2 Anaerobic Digestion of Cyanobacteria and Microalgae

14.4.2.1 Cyanobacteria

In contrast to algae, cyanobacteria consists of proteins as their biochemical component and lack of hard polysaccharide-based cell wall which makes them suitable for ADP. Two genera, *Arthrospira* (*Spirulina*) and *Anabaena* have been used extensively as potential feedstock for ADP. The BMP assay for a cyanobacterium mixture collected from Lake Dian resulted in a higher methane yield of 0.37 L g^{-1} VS (HRT 35 days) with a methane content of 60–65 % in the biogas. The methane yield during batch digestion of *Arthrospira platensis* and *Arthrospira maxima* species varied from 0.29 to 0.33 L g^{-1} VS corresponding to 68–77 % of the theoretical methane yield (Mussgnug et al., 2010). When municipal anaerobic sewage sludge was used as inoculum with cyanobacterium as feedstock, methane yield of 0.26 L g^{-1}VS was observed at an OLR of 0.97 g VS $L^{-1}d^{-1}$, HRT of 33 d, and temperature at 30 °C. Despite high ammonia and fatty acids concentrations (2.5 and 2 g L^{-1}, respectively) methane production was stable possibly due to high alkalinity (8 g L^{-1}) and pH of 7.55. AD stability and methane yield was highly influenced by VS concentration, OLR and HRT (Samson and LeDuyt, 1986). The methane yield and methane volumetric production rate were 0.04–0.36 L g^{-1} VS and 0.17–0.8 L $L^{-1}d^{-1}$, respectively.

14.4.2.2 Microalgae

Microalgae comprise diverse group of organisms and their predominant organic constituent vary from carbohydrates to proteins or lipids. Due to the widespread nature, fast growth rate and robustness, green microalgae serve as potential substrates for ADP. The BMP determined for *Chlamydomonas reinhardtii*, *Chlorella kessleri* and *Scenedesmus obliquus* was 0.387, 0.218 and 0.178 L g^{-1} VS, respectively (Samson and LeDuyt, 1986). The amount of biogas production correlated well with the extent of algal degradation. *C. reinhardtii* exhibited a higher cell disintegration rate in comparison to *C. kessleri* and *S. obliquus*. Number of the *Chlorella* and *Scenedesmus* species as well as several other algae (e.g. *Nannochloropsis*) have resistant trilaminar membrane-like structure containing an hydrolysable sporopollenin-like biopolymer—algaenan. The overall cell wall structure consists of four distinct layers: rigid internal, micro fibrillar, medial trilaminar and external columnar (for green algae *Coelastrum*). The major algaenan functions are protection from parasites and desiccation. *Chlorella* and *Scenedesmus* have internal rigid cell walls, either glucose-mannose type or glucosamine-type. In contrast, *C. reinhardtii* has a cell wall composed of proteins and glycoproteins. Resistant cell wall retained *S. obliquus* cells undamaged after six months of digestion (Takeda, 1988).

14.5 Current and Prospective Methods for Algae to Methane Process Enhancement

The biological production of methane from algae or cyanobacteria is generally a two-step process. First, biomass is produced where sunlight energy is captured and converted to new algal cells. Second, the energy stored in biomass is transformed into a more applicable form, such as methane gas, through the ADP. Thereafter, methane can be easily stored, transported, and used later for the production of heat or electricity. Methane can also be used as a motor fuel. The overall efficiency of methane production depends on performance of these coupled steps.

14.5.1 Anaerobic Digestion Improvement

The main objective for improving ADP is increasing the conversion efficiency while simultaneously decreasing capital and operational costs.

14.5.1.1 Inoculum Source for Anaerobic Digestion of Algae

As discussed earlier, the biochemical composition of algal biomass varies with the type of species which consists of unique compounds, such as algin, laminarin and fucoidan. Moreover, marine algal biomass has a high salt concentration that can affect anaerobic microorganisms. Thus isolation and application of microorganisms adapted for digestion of specific algae is required which is labour-intensive but, if successful, has the potential to improve algal ADP. Typically anaerobic sludge from a domestic sewage plant or marine anaerobic sediment has been used in literature for startup of the algal ADP. Several authors have reported that anaerobic organisms adapt readily to algal biomass as a sole substrate, and the inoculum source has a minor or no effect on the final methane yield and VS reduction. Application of an inoculum adapted to high ammonia concentration is a possible solution to overcome the problem of ammonia inhibition.

14.5.1.2 Process Parameters and Reactor Design

VS reduction, OLR and HRT are the most important operating parameters that affect the methane yield in ADP. Generally, the ratios of actual to theoretically calculated methane yield and VS reduction are low (typically from 0.4 to 0.6) as the degradation ability is dependent on the ability of anaerobic bacteria to hydrolyze complex organic compounds as well as by the slow rate of acetogenesis and methanogenesis stages of AD. To increase the AD efficiency, appropriate adjustments are made with process parameters and reactor design. Feedstock pretreatment and conditioning and choice and source of specific anaerobic microorganisms can contribute in improving ADP. The general principles of digestion with non-algal feedstock can be applied to the AD of algae. Also environmental parameters have a significant impact on AD performance. With optimum process design methane yield is enhanced with increasing OLR and decreasing HRT.

The main goal of optimal reactor design is to reduce reactor volume and capital costs without compromising the maximum methane yield at high OLR and low HRT. Over the past few decades several high-rate digester configurations and their characteristics, advantages and disadvantages are described elsewhere. Decoupling HRT from SRT by anaerobic sludge immobilization is one of the strategies employed along with granulation and floc formation, biomass recycling, or membrane retention. The methane yield and methane production in a non-mixed vertical flow reactor (NMVFR) are larger and more stable at higher OLR as compared to CSTR. Another promising approach for biosolids digestion is separation of the hydrolysis and acetogenesis steps from methanogenesis, a process called a two-stage system. A two-stage anaerobic reactor system achieved stable methane production from *M. pyrifera* and *D. antarctica* with an HRT of one day for each stage (Vergarafernandez et al., 2008).

A special case of the two-stage system is preliminary treatment of macroalgae in percolation reactors with natural hydrolysis and acidogenesis processes. In percolation reactors, algae are stored in a tank yielding a drained liquid product containing VFAs and ethanol as substrates for methanogenesis. The AD of hydrolysis juices is more economically efficient as compared to digestion of whole macroalgae due to lower reactor size, energy for substrate heating, grinding and pumping. For example, the volume of digester with fixed bacteria for digestion of hydrolysis juices is 25 times smaller as compared to a CSTR digester required for whole algae (Nizami et al., 2009).

The relevant C/N ratios for microalgae are in the range of 4–6 and it was observed that the addition of carbon-rich cellulosic materials can balance the high nitrogen content. Addition of 25 and 50 % of waste paper to a mixture of *Scenedesmus* sp. and *Chlorella* sp. resulted in a 1.59- and 2.05-fold increase in the methane yield (Yen and Brune, 2007). The optimal ratio between algal biomass (*Scenedesmus* sp. and *Chlorella* sp.) and waste paper was found to be 40 % algae and 60 % paper with corresponding C:N ratio equal to 22.6. The authors also reported that paper addition stimulated cellulase activity in the anaerobic digester from 1.26 ± 0.14 mg L^{-1} min^{-1} to 3.02 ± 0.09 mg L^{-1} min^{-1} (50 % paper, C:N is equal to 18). Addition of nutrient-rich algal and cyanobacterial biomass to nutrient limited waste products that cannot be digested as sole substrate but can help to increase the process efficiency.

14.6 Conclusion

Biological production of hydrogen through dark fermentation process appears to be more suitable as compared to all other biological routes of hydrogen production. It has higher rate of hydrogen production and less energy input is required when compared with other biological routes. Renewable feedstock like algal biomass could make dark fermentative hydrogen production truly renewable. But there are certain bottlenecks regarding dark fermentative hydrogen production viz. low substrate conversion, influence of process parameters, accumulation of metabolites, etc. The spent media after dark fermentation is rich in volatile fatty acids which could be further used for biomethanation. This leads to enhanced gaseous energy recovery. Moreover, cyanobacteria and green algae could also be a promising feedstock for biogas production through anaerobic digestion process. The anaerobic digestion of algae can be applied for biofuel production.

References

Angenent, L.T., Karim, K., Al-Dahhan, M.H., Wrenn, B.A. and Domíguez-Espinosa, R. (2004). Production of bioenergy and biochemicals from industrial and agricultural wastewater. *Trends Biotechnol.*, 22, 477–485.

Braun, R., Huber, P. and Meyrath, J. (1981). Ammonia toxicity in liquid piggery manure digestion. *Biotechnol. Lett.*, 3, 159–164.

Buswell, A.M. and Mueller, H.F. (1952). Mechanism of Methane Fermentation. *Ind. Eng. Chem.*, 44, 550–552.

Chen, H.-C., Yokthongwattana, K., Newton, A.J. and Melis, A. (2003). *SulP*, a nuclear gene encoding a putative chloroplast-targeted sulfate permease in *Chlamydomonas reinhardtii. Planta*, 218, 98–106.

Chen, Y., Cheng, J.J. and Creamer, K.S. (2008). Inhibition of anaerobic digestion process: A review. *Bioresour. Technol.*, 99, 4044–4064.

Cheng, L., Zhang, L., Chen, H. and Gao, C. (2006). Carbon dioxide removal from air by microalgae cultured in a membrane-photobioreactor. *Sep. Purif. Technol.*, 50, 324–329.

Chynoweth, D.P., Owens, J.M. and Legrand, R. (2001). Renewable methane from anaerobic digestion of biomass. *Renew. Energy*, 22, 1–8.

Dickson, D., Page, C. and Ely, R. (2009). Photobiological hydrogen production from *Synechocystis* sp. PCC 6803 encapsulated in silica sol–gel. *Int. J. Hydrogen Energy*, 34, 204–215.

Finazzi, G., Rappaport, F., Furia, A., Fleischmann, M., Rochaix, J.-D., Zito, F. and Forti, G. (2002). Involvement of state transitions in the switch between linear and cyclic electron flow in *Chlamydomonas reinhardtii. EMBO Rep.*, 3, 280–285.

Gerardi, M.H. (2003). The Microbiology of Anaerobic Digesters (Google eBook). John Wiley & Sons, USA, ISBN 0-471-20693-8.

Gibson, R.N. and Atkinson, R.J.A. (eds) (2003). Oceanography and Marine Biology, An Annual Review, Volume 41. CRC Press, USA.

Godfroy, A., Raven, N.D.H. and Sharp, R.J. (2000). Physiology and continuous culture of the hyperthermophilic deep-sea vent archaeon *Pyrococcus abyssi* ST549. *FEMS Microbiol.* Lett., 186, 127–132.

Gujer, W. and Zehnder, A.J.B. (1983). Conversion Processes in Anaerobic Digestion. *Water Sci. Technol.*, 15(8–9), 1276–1277.

Hansen, K.H., Angelidaki, I. and Ahring, B.K. (1998). Anaerobic digestion of swine manure: Inhibition by ammonia. *Water Res.*, 32, 5–12.

Kida, K., Shigematsu, T., Kijima, J., Numaguchi, M., Mochinaga, Y., Abe, N. and Morimura, S. (2001). Influence of Ni^{2+} and Co^{2+} on methanogenic activity and the amounts of coenzymes involved in methanogenesis. *J. Biosci. Bioeng.*, 91, 590–595.

Kim, M., Baek, J., Yun, Y., Junsim, S., Park, S. and Kim, S. (2006). Hydrogen production from *Chlamydomonas reinhardtii* biomass using a two-step conversion process: Anaerobic conversion and photosynthetic fermentation. *Int. J. Hydrogen Energy*, 31, 812–816.

King, G.M., Guist, G.G. and Lauterbach, G.E. (1985). Anaerobic Degradation of Carrageenan from the Red Macroalga *Eucheuma cottonii. Appl. Envir. Microbiol.*, 49, 588–592.

Kruse, O., Rupprecht, J., Bader, K.-P., Thomas-Hall, S., Schenk, P.M., Finazzi, G. and Hankamer, B. (2005). Improved photobiological H_2 production in engineered green algal cells. *J. Biol. Chem.*, 280, 34170–34177.

Kumar, K., Roy, S. and Das, D. (2013). Continuous mode of carbon dioxide sequestration by *C. sorokiniana* and subsequent use of its biomass for hydrogen production by *E. cloacae* IIT-BT 08. *Bioresour. Technol.*, 145, 116–122.

Lee, J.W. (ed.) (2013). Advanced Biofuels and Bioproducts. Springer New York, New York, NY.

Lettinga, G. (2001). Challenge of psychrophilic anaerobic wastewater treatment. *Trends Biotechnol.*, 19, 363–370.

Meher Kotay, S. and Das, D. (2008). Biohydrogen as a renewable energy resource—Prospects and potentials. *Int. J. Hydrogen Energy*, 33, 258–263.

Murphy, J.D. and Power, N. (2009). Technical and economic analysis of biogas production in Ireland utilising three different crop rotations. *Appl. Energy*, 86, 25–36.

Mussgnug, J.H., Klassen, V., Schlüter, A. and Kruse, O. (2010). Microalgae as substrates for fermentative biogas production in a combined biorefinery concept. *J. Biotechnol.*, 150, 51–56.

Nayak, B.K., Roy, S. and Das, D. (2014). Biohydrogen production from algal biomass (*Anabaena* sp. PCC 7120) cultivated in airlift photobioreactor. *Int. J. Hydrogen Energy*, 39, 7553–7560.

Nguyen, T.-A.D., Kim, K.-R., Nguyen, M.-T., Kim, M.S., Kim, D. and Sim, S.J. (2010). Enhancement of fermentative hydrogen production from green algal biomass of *Thermotoga neapolitana* by various pretreatment methods. *Int. J. Hydrogen Energy*, 35, 13035–13040.

Nizami, A.S., Korres, E.N. and Murphy, J.D. (2009). Review of the Integrated Process for the Production of Grass Biomethane. *Environ. Sci. Technol.*. 43, 8496–8508.

Pathak, H., Jain, N., Bhatia, A., Mohanty, S. and Gupta, N. (2009). Global warming mitigation potential of biogas plants in India. *Environ. Monit. Assess.*, 157, 407–418.

Polle, J. (2002). Truncated chlorophyll antenna size of the photosystems? A practical method to improve microalgal productivity and hydrogen production in mass culture. *Int. J. Hydrogen Energy*, 27, 1257–1264.

Prince, R.C. and Kheshgi, H.S. (2005). The Photobiological Production of Hydrogen: Potential Efficiency and Effectiveness as a Renewable Fuel. *Crit. Rev. Microbiol.*, 31(1), 19–31.

Rees, D.A. and Welsh, E.J. (1977). Secondary and Tertiary Structure of Polysaccharides in Solutions and Gels. Angew. *Chemie Int. Ed. English*, 16, 214–224.

Rosenberg, J.N., Oyler, G.A., Wilkinson, L. and Betenbaugh, M.J. (2008). A green light for engineered algae: Redirecting metabolism to fuel a biotechnology revolution. *Curr. Opin. Biotechnol.*, 19, 430–436.

Roy, S., Kumar, K., Ghosh, S. and Das, D. (2014). Thermophilic biohydrogen production using pre-treated algal biomass as substrate. *Biomass and Bioenergy*, 61, 157–166.

Rupprecht, J., Hankamer, B., Mussgnug, J.H., Ananyev, G., Dismukes, C. and Kruse, O. (2006). Perspectives and advances of biological H_2 production in microorganisms. *Appl. Microbiol. Biotechnol.*, 72, 442–449.

Samson, R. and LeDuyt, A. (1986). Detailed study of anaerobic digestion of *Spirulina maxima* algal biomass. *Biotechnol. Bioeng.*, 28, 1014–1023.

Schink, B. (1997). Energetics of syntrophic cooperation in methanogenic degradation. *Microbiol. Mol. Biol. Rev.*, 61, 262–280.

Schröder, C., Selig, M. and Schünheit, P. (1994). Glucose fermentation to acetate, CO_2 and H_2 in the anaerobic hyperthermophilic eubacterium *Thermotoga maritima*: Involvement of the Embden-Meyerhof pathway. *Arch. Microbiol.*, 161, 460–470.

Skjånes, K., Lindblad, P. and Muller, J. (2007). BioCO_2—A multidisciplinary, biological approach using solar energy to capture CO_2 while producing H_2 and high value products. *Biomol. Eng.*, 24, 405–413.

Speece, R.E. (1983). Anaerobic biotechnology for industrial wastewater treatment. *Environ. Sci. Technol.*, 17, 416A–427A.

Stephenson, M. and Stickland, L.H. (1932). Bacterial enzyme liberating molecular hydrogen. *Biochem. J.*, 26, 712–724.

Sterling, M., Lacey, R., Engler, C. and Ricke, S. (2001). Effects of ammonia nitrogen on H_2 and CH_4 production during anaerobic digestion of dairy cattle manure. *Bioresour. Technol.*, 77, 9–18.

Taguchi, F., Yamada, K., Hasegawa, K., Taki-Saito, T. and Hara, K. (1996). Continuous hydrogen production by *Clostridium* sp. strain no. 2 from cellulose hydrolysate in an aqueous two-phase system. *J. Ferment. Bioeng.*, 82, 80–83.

Takeda, H. (1988). Classification of *Chlorella* strains by cell wall sugar composition. *Phytochemistry*, 27, 3823–3826.

Tamagnini, P., Leitão, E., Oliveira, P., Ferreira, D., Pinto, F., Harris, D.J., Heidorn, T. and Lindblad, P. (2007). Cyanobacterial hydrogenases: Diversity, regulation and applications. *FEMS Microbiol. Rev.*, 31, 692–720.

Tamburic, B., Zemichael, F.W., Maitland, G.C. and Hellgardt, K. (2011). Parameters affecting the growth and hydrogen production of the green alga *Chlamydomonas reinhardtii*. *Int. J. Hydrogen Energy*, 36, 7872–7876.

Valdezvazquez, I., Riosleal, E., Esparzagarcia, F., Cecchi, F. and Poggivaraldo, H. (2005). Semi-continuous solid substrate anaerobic reactors for H_2 production from organic waste: Mesophilic versus thermophilic regime. *Int. J. Hydrogen Energy*, 30, 1383–1391.

Vergarafernandez, A., Vargas, G., Alarcon, N. and Velasco, A. (2008). Evaluation of marine algae as a source of biogas in a two-stage anaerobic reactor system. *Biomass and Bioenergy*, 32, 338–344.

Wijffels, R.H. (2008). Potential of sponges and microalgae for marine biotechnology. *Trends Biotechnol.*, 26, 26–31.

Yen, H.-W. and Brune, D.E. (2007). Anaerobic co-digestion of algal sludge and waste paper to produce methane. *Bioresour. Technol.*, 98, 130–134.

Chapter 15
Integrating Microalgae Cultivation with Wastewater Treatment for Biodiesel Production

S. Venkata Mohan, M.V. Rohit, P. Chiranjeevi, R. Hariprasad, and Somya Arora

15.1 Introduction

Microalgae are responsible for more than half of the world's primary production of oxygen. They are the simplest and most abundant form of plant life on the earth (Energy from algae (2010). These photosynthetic organisms are categorized under third generation biofuels and are known to have high oil and biomass yields, can be cultivated with wastewater, do not need arable land for cultivation, do not compete with common food resources and very efficiently use water and nutrients for growth (Hannon et al. 2010). There are various routes of metabolism which microalgae have adopted for their growth and survival viz., autotrophic, heterotrophic and mixotrophic. They are capable of shifting their metabolism in response to changes in the environmental conditions (Devi et al., 2012). Algal cultivation for biodiesel production is considered more amenable a technology than the cultivation of oil crops (Chisti, 2007) because the yields of algae-derived oils are much higher (Abou-Shanab et al., 2010). Autotrophically algae gain energy through light by fixing atmospheric CO_2 (Devi and Venkata Mohan, 2012). However, low biomass yields, requirement of cultivation systems with large surface area and shallow depth for better access of light are some of the disadvantages associated with autotrophic mode of nutrition. In the absence of light, the photosynthetic process gets suppressed and algae gain energy from alternative organic processes using heterotrophic pathways that convert sugar into lipids (Perez-Garcia et al. 2010). This pathway leads to significantly denser biomass, facilitating greater lipid yields.

Integrating biodiesel production with CO_2 mitigation from industrial flue gases and wastewater treatment is considered as a viable strategy in algal cultivation and gives an additional offset towards waste remediation. Algae can be cultivated in

S. Venkata Mohan (✉) • M.V. Rohit • P. Chiranjeevi • R. Hariprasad • S. Arora
Bioengineering and Environmental Sciences (BEES), CSIR-Indian Institute of Chemical Technology (CSIR-IICT), Hyderabad 500007, India
e-mail: vmohan_s@yahoo.com

D. Das (ed.), *Algal Biorefinery: An Integrated Approach*,
DOI 10.1007/978-3-319-22813-6_15

both open raceways and closed systems. Closed systems reduce the chances of contamination with other bacteria/microalgae and thus are more suitable for pure strains. Various bioprocesses and downstream processing technologies are used extensively for recovery of lipids and various co-products from microalgae. They also include physical, thermo-chemical, biochemical and biological treatments to create energy-rich products from the source biomass (Demirbas, 2011). Integration of the biorefinery concept with wastewater treatment will ensure efficient utilization of algae biomass and reduces overall residual waste component of biomass favouring sustainable economics. Algae biodiesel research and development is gaining pace in biofuels markets and has vast scope for advancements in future (Bracmort, 2013). In this chapter, an attempt has been made to summarize the basic and applied aspects of nutritional modes employed by microalgae, algal-cultivation strategies, downstream processing technologies and biorefinery concepts by analyzing the contemporary literature in concurrence with recent developments.

15.2 Nutritional Modes

Nutritional modes significantly influence the carbon assimilation and lipid productivity of microalgae (Devi et al., 2013). Three types of nutritional modes—autotrophic, heterotrophic and mixotrophic—are reported to produce algal fuel in the presence of light. Autotrophic organisms can convert physical (light) and chemical (CO_2 and H_2O) energy into carbohydrates, which forms the basis for all other carbon containing biomolecules (Yoo et al., 2010). Autotrophic organisms are relatively self-sufficient because they obtain their energy from sunlight (Nelson et al., 1994). Photosynthetically fixed CO_2 in the form of glucose serves as the sole energy source for all the metabolic activities of the algal cells (Chang et al., 2011). Major advantage of the autotrophic nutritional mode is that the algal oil production occurs at the expense of atmospheric CO_2. Large scale microalgae cultivation systems (such as open/raceway ponds) are usually operated under photoautotrophic conditions (Mata et al., 2010). However, low biomass yields, requirement of cultivation systems with large surface area and shallow depth for optimum light availability are some of the inherent disadvantages. In the absence of light, the photosynthetic process gets suppressed and algae gain energy from alternative organic processes that convert sugar into lipids (Venkata Mohan et al., 2014a). Heterotrophic organisms utilize organic carbon produced by autotrophs as energy for their metabolic functions because they cannot utilize atmospheric CO_2.

Microalgae use organic molecules as primary sources of energy and carbon through heterotrophic nutritional mode and facilitate high biomass productivities, which are economically feasible for large scale production (Behrens, 2005; Perez-Garcia et al., 2011a). The growth of algae can be significantly denser resulting in greater yield because light is not a limiting factor. Heterotrophic nutrition takes place both in the presence and absence of light. This unique ability is shared by several species of microalgae (Perez-Garcia et al., 2011a). Light and carbon act as an energy source in photo-heterotrophic nutrition mode whereas the sole source of energy during dark conditions is organic carbon. Photo-heterotrophic nutritional mode avoids the limitations of light dependency which is the major obstruction for gaining high cell density in large scale photo-bioreactors. The major advantage of heterotrophic nutritional mode is the facilitation of wastewater treatment along with lipid production, which gives this nutritional mode an edge over the other two (Venkata Mohan et al., 2014b). Moreover, cost effectiveness, relative simplicity of operation, and easy maintenance are the main attractions of the heterotrophic growth approach (Perez-Garcia et al., 2011a, b). However, heterotrophic systems suffer from contamination problems (Olguin, 2012).

Microalgae can also function under mixotrophic nutrition by combining both the autotrophic and heterotrophic mechanisms thus assimilating available organic compounds as well as atmospheric CO_2 as carbon source (Chandra et al., 2014). Since mixotrophs can utilize organic carbon, light energy is not a limiting factor for biomass growth (Chang et al., 2011). Mixotrophism is often observed in the ecological water bodies where the homeostatic structure and function of a living system is supported by chemical, physical and organic activity of the biota, which balances the ecological status (Venkata Mohan, 2010). Depending on the species, growing conditions and growth stages, microalgae have been shown to produce various types of lipids including triacylglycerides, phospholipids, glycolipids and betaine lipids (Greenwell et al., 2010). The intracellular lipid granules stored under stress conditions act as precursors for fatty acid biosynthesis.

15.3 Wastewater Treatment vs Lipid Production

Algae based wastewater treatment involves photosynthetic conversion of solar energy into useful biomass utilizing nutrients such as nitrogen and phosphorus along with organic carbon. This wastewater, if discharged into water bodies, would lead to eutrophication. Algae are the main biocatalysts in oxidation ponds, raceway ponds and high rate algal ponds where they are used to remove concentration of nutrients, especially for polishing purpose. Heterotrophic/mixotrophic algae cultivation particularly requires carbon, water and inorganic salts. The availability of the nutrients affects the growth of algae as well as the lipid profile. Along with the ability to grow heterotrophically in nutrient-rich and organic environment, algae also remove toxins and heavy metals from wastewater. Levels of several contaminant heavy metals are significantly reduced by the cultivation of microalgae (Munoz and Guieysse, 2006). Moreover, the presence of photosynthetic component in treatment setup makes the whole process eco-friendly (Stephenson et al., 2010; Venkata Mohan et al., 2014a). The following studies were performed with different wastewaters for enhancement of lipid productivities along with nitrogen and phosphate removal.

Cultivation of *Botryococcus braunii* in secondarily treated sewage as treatment operation showed 17 % lipid accumulation (Orpez et al. 2009). Hu et al. (2012)

reported lipid productivity of 20 % with *Auxenochlorella protothecoides* on municipal wastewater. *Scenedesmus* sp. cultivation in fermented swine wastewater yielded good lipids and other value added products in association with nutrient removal (Kim et al., 2007). Organic content-rich wastewaters like diary wastewater are highly amenable for heterotrophic cultivation. The biomass and lipid productivities of 2.5 and 5.8 g L^{-1} were obtained for soya whey and ethanol thin stillage in heterotrophic mode using *Chlorella vulgaris* (Mitra et al., 2012). In the study conducted by Beevi and Sukumaran (2014), diary wastewater was supplemented with 4 % and 6 % of glycerol as carbon source. Lipid productivities of 39 % and 42 % were reported using *Chlorococcum* sp. RAP13 in heterotrophic mode (Beevi and Sukumaran, 2014). Acidogenic effluents rich in volatile fatty acids from fermentative hydrogen producing process were evaluated as substrate for lipid accumulation by heterotrophic cultivation (Venkata Mohan and Devi, 2012). Acetate gets easily assimilated by the algal cell, as a part of the acetyl- coenzyme A (acetyl-CoA) metabolism in a single-step reaction catalyzed by acetyl-CoA synthetase (Boyle and Morgan, 2009). TAG accumulation in response to environmental stress is likely to occur as a means of providing an energy that can be readily assimilated on return of favourable environment, to allow rapid growth (Devi et al., 2012). Thin layer culture systems are well-known for high biomass productivities. Attached mode growth of *Chlorella* sp. with dairy manure wastewater showed high biomass growth as well as fatty acid yield (Johnson and Wen, 2010).

15.4 Cultivations Methods

Cultivation of microalgae influences both biomass growth and lipid productivity. Culturing of algae requires the input of light as an energy source for photosynthesis with a sufficient supply of macronutrients (carbon, nitrogen and phosphate) and micronutrients (sulphur, potassium, magnesium) in dissolved form (Mata et al., 2010). These nutrients play a critical role in biomass enrichment by facilitating cell growth, maintenance and synthesis of different kinds of lipids (Table 15.1). The main options for algae cultivation on a commercial scale are open-ponds or closed systems called photobioreactors (Chisti, 2007; Robert et al., 2012). There are also hybrid configurations that include a combination of the two growth options. Innovations in algae cultivation aim towards increasing lipid productivity of algae while consuming resources that would otherwise be considered waste (Campbell, 2008).

15.4.1 Open Ponds/Raceway

Open ponds mimic the ecological niche of algae and can be categorized into natural waters (lakes, lagoons, ponds, etc.) and artificial ponds or containers (Pearson, 1996; Chisti, 2007). Artificial systems include shallow ponds (large in size),

Table 15.1 Effect of macro and micronutrients on algae growth.

Carbon	Major source of energy for heterotrophic algae (i.e. auxotrophs, mixotrophs, etc.). Higher rates of growth and respiration are obtained with glucose.
	High concentrations can lead to product inhibition and reduced growth rate
Nitrogen	Major constituent of protein, nucleic acids, vitamins, amino acids, purines, pyramidines and porphyrins
	Low nitrogen causes increase in polyunsaturated fatty acids (PUFAs)
Phosphorous	Helps in biosynthesis of nucleic acids and phospholipids and TAGs at low concentration. Constituent of cell membrane and is required for phosphorylation reactions
	Deficiency leads to decrease in phospholipids of cell membranes
Potassium	Helps in osmoregulation, ion exchange across intracellular membranes, protein synthesis and also helps in activation of certain enzymes.
	Deficiency leads to increase in respiration
Calcium	Major component for cell wall, important in spindle formation during mitotic cell division. Helps in regulation of certain metabolic activities, improves growth and maintains membrane stability
Magnesium	Helps in activation of enzymes related to photosynthesis and respiration. It's the central atom in chlorophyll. Assists in DNA and RNA synthesis
	Deprived conditions results in low turnout in photosynthesis.
Sulphur	Main component of amino acids like cysteine and methionine. Constituent of several cofactors, vitamins and ferredoxin
Iron	Plays a major role in formation of protein like ferrodoxin and cytochrome. Major component responsible for assimilation of N_2 also important in activation of enzymes of photosynthesis and respiration.
	High concentration leads in inhibition of algae growth
Manganese	Activates many enzymes related to photosynthesis, respiration and nitrogen metabolism. Actively participates in splitting water molecule into oxygen during photosynthesis.
	Decreases the efficiency of photosynthesis
Zinc	Essential for activation of carboxylases for fatty acids synthesis
Molybdenum	Active component in nitrogenase and nitrate reductase which help in nitrogen metabolism
	Deficiency results in low nitrogen absorption

raceway ponds, tanks and circular ponds. These ponds are usually constructed in shallow dimensions as light penetration is a major limiting factor that can slow down biomass growth of algae. Moreover, when these systems are operated in a continuous mode, CO_2 and nutrients have to be fed continuously to the pond (Chisti, 2007). Ponds are divided by a series of baffles, and water is moved through the ponds for proper mixing of nutrients thus ensuring uniform algae growth. Commercial production of single-cell protein, health food, and beta-carotene from algal biomass is one of the oldest industrial large open-pond cultivation systems since 1950s (Chisti, 2007; Perez-Garcia et al., 2011b). Open-pond cultivation incurs low construction and operating costs, which invariably results in low production costs (Stephenson et al., 2010). Open-pond cultivation inherits some drawbacks such as poor light diffusion, not-so-efficient mixing, losses due to evaporation, poor diffusion of atmospheric CO_2, uncontrolled pond temperature, contamination by predators and other fast-growing heterotrophs and the requirement of large areas of land (Harun et al., 2010; Perez-Garcia et al., 2011b). Uncontrolled environments in and around the pond pose a multitude of problems that can directly or indirectly stunt algae growth (Kazamia et al., 2012; Mata et al., 2010).

15.4.2 Photobioreactors

Photobioreactors (PBRs) provide a more controlled environment than open ponds, due to pre-set conditions. Everything that the algae need to grow (carbon dioxide, water and light) can be supplied within the system (Weissman, 1987; Pulz, 2001). PBRs facilitate better control of culture environment, such as carbon supply, water supply, optimal temperature, efficient exposure to light, pH levels, gas supply rate, mixing regime, etc. and can achieve high growth rates (Mata et al., 2010; Sierra et al., 2008). Higher biomass productivity can be obtained in closed cultivation systems where contamination can also be prevented to major extent (Ramanathan et al., 2011). High mass transfer is one of the important criteria for PBR design, especially for CO_2 sequestration (Ugwu et al., 2008).

Various types of closed cultivation systems or PBRs have been reported for algae cultivation. Tubular photobioreactors with diverse configurations are being used viz., horizontal, serpentine, vertical, near-horizontal, conical and inclined with suitable illuminated surfaces. Ventilation and mixing is performed by pump or ventilation systems. Sparger is attached at the bottom of the reactor to diffuse small bubbles of gas. Good mixing, mass transfer of CO_2 and effective removal of the O_2 produced during photosynthesis are major advantages with this configuration. Vertical column photobioreactors are low cost, easily constructed and compact systems which are suitable for large scale production. Air-lift photobioreactors comprise two interconnecting zones called the riser, where the gas mixture is sparged, and the down comer, which does not receive the gas. Mixing in the system is done by bubbling the gas through a sparger in the riser tube, with no physical agitation. Gas held up in the down comer significantly influences the fluid dynamics of the airlift reactor. Airlift reactors have the characteristic advantage of creating circular mixing patterns in which liquid culture passes continuously through dark and light phases, giving a flashing-light effect to algal cells. Tubular photobioreactors made from polyethylene tubes (plastic bags) are being used commercially for algal cultivation which provide light ranging in the visible and near-infra red region (low UV transmission) of spectrum and is associated with low cost. Stirred-tank photobioreactors are the conventional reactor setup in which agitation is provided mechanically with the help of impellers or baffles by providing illumination externally. Internally illuminated photobioreactor incorporates both solar and artificial lighting systems and switches to the artificial lighting system in the absence of solar light.

Flat-plate or flat panel photobioreactors comprise transparent flat plates made of transparent materials for maximum utilization of solar energy to achieve high photosynthetic efficiencies. Accumulation of dissolved oxygen concentrations is low compared to horizontal tubular photobioreactors. Lack of temperature control and gas engagement zones are some of the disadvantages observed with these systems. A coiled transparent and flexible tube of small diameter with separate or attached degassing unit is the basis for the helical type of bioreactor. A centrifugal pump is used to drive the culture through a long tube to the degassing unit. CO_2 gas mixture and feed can be circulated from either direction, but injection from the bottom gives higher photosynthetic efficiency. A degasser facilitates removal of photosynthetically produced oxygen and residual gas of the injected gas stream. The energy required by the centrifugal pump in recirculating the culture and associated shear stress and fouling on the inside of the reactor are some of the challenges encountered while working with this system.

15.5 Downstream Processing

Downstream process of algal biomass includes harvesting, dewatering, drying, cell disruption and extraction of product followed by biodiesel preparation by transesterification (Fig. 15.1).

15.5.1 Dewatering and Drying

Dewatering and drying of algae are most important steps in downstream processing of microalgae. Dewatering involves separating the extracellular water from the algal suspension. A subsequent drying stage removes all water from the biomass (~95 % by mass). The water content must be substantially reduced especially while preparing the algae for storage, in order to avoid rapid decay and decomposition (Bruton et al., 2009). The intracellular water remaining in the algal suspension can only be removed by thermal processes. Reducing the mass delivers considerable savings on transport costs, especially over long distances. Different drying techniques such as rotary drying, spray drying, solar drying, cross-flow drying, vacuum shelf drying, flash drying and freeze drying are used (Shelef et al., 1984). Drying by solar energy is the simplest and most economical method of drying although it depends on climatic conditions, and involves the risk of algal paste decay during the process at high light intensities (Becker, 1994). Drying is performed either by direct sunlight or by means of a circulating air flow heated by solar energy. Spray drying is frequently used in the production of algae for food purposes, because a large number of constituents are retained. Similar to flash dryers, it is a continuous process wherein the paste is dried in a few seconds (Becker, 1994).

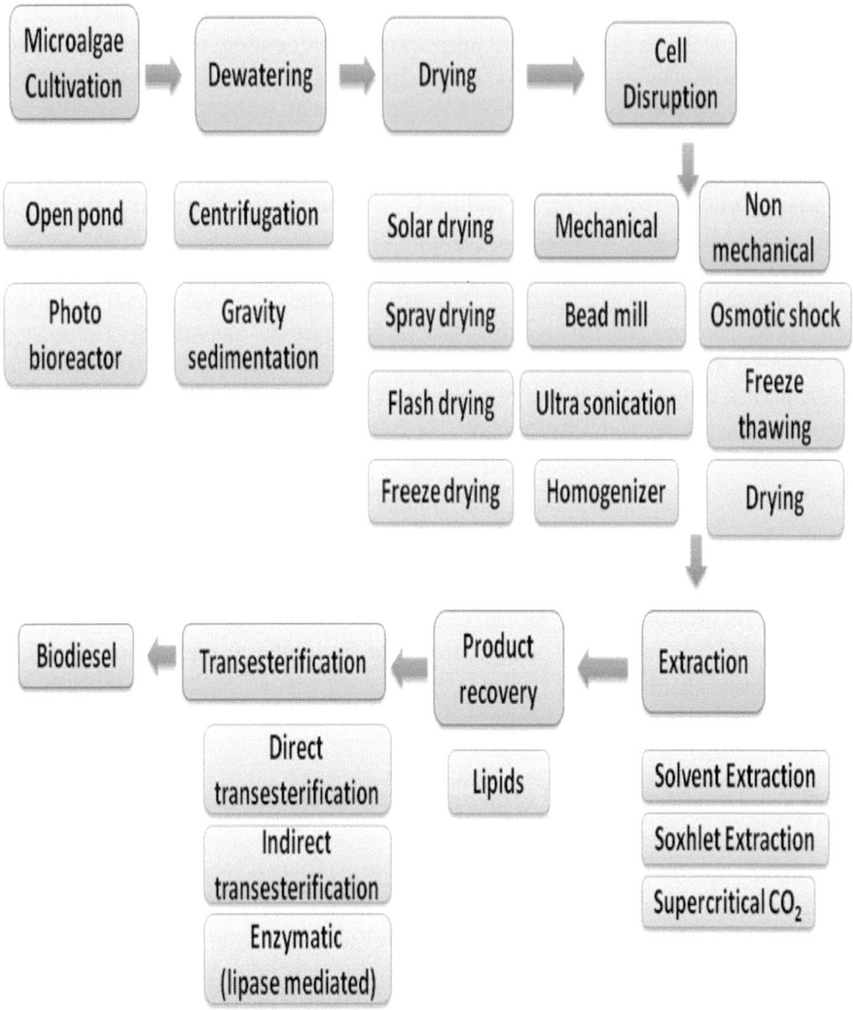

Fig. 15.1 Flow diagram of different unit operations during downstream processing of microalgae.

15.5.2 Cell Disruption

The disruption of algae cells prior to extraction is of particular importance because the contents of the extracted lipids are determined according to the disruption method and equipment employed. The selection of appropriate device for disruption is the key factor for enhancing the lipid extraction efficiency (Lee et al., 2010). Cell disruption methods are classified as mechanical and non-mechanical. Mechanical cell disruption includes bead mill, high pressure press, homogenizers, ultrasonication, autoclave, lyophilization and microwave while non-mechanical methods often involve lysing the microalgae cells with acids, alkalis, enzymes, or osmotic shocks.

Expeller press (or oil pressing) is a mechanical method applied for the disruption of algae cell membranes by squeezing the cells under high pressure (Mercer and Armenta, 2011) which can recover nearly 75 % of the oil from algae cells in a single step.

The advantages of this method include elimination of a solvent requirement and easy operation. This method involves the application of beads for the disruption of the algal cell wall. Continuous exposure of biomass to beads leads to cell-wall rupture, resulting in the release of intracellular contents into the solvent medium. Similar to expeller pressing, this method can also be applied for both disruption and extraction. Non-mechanical cell disruption uses chemical solvents, enzymes, solvent fluids and osmosis. Extraction solvents used are mainly hexane, acetone, chloroform and methanol. Organic solvents and supercritical carbon dioxide is used extensively to extract lipids from microalgal biomass. Both technologies have their own merits and demerits. Despite having low reactivity with lipids and being directly applicable to wet biomass, organic solvent extraction is slow and uses a large amount of toxic solvents.

15.5.3 Extraction of Lipids

Several methods have been employed for extracting microalgae lipids (Table 15.2). Among the processes described, solvent extraction is suitable for extracting lipids from mass cultures but requires large volumes of solvent. The existing methods of lipid extraction usually involve selective solvent extraction, and the biomass material may be subjected to drying prior to extraction (Lee et al., 2010). Lipids are soluble in organic solvents but sparingly soluble or insoluble in water. Selection of solvent systems is an important criterion for lipid extraction and typically depends on the type of lipid present (total/neutral lipids) and the proportion of non-polar (neutral) lipids (commonly known as triacylglycerols) and polar lipids (mainly phospholipids and glycolipids) in the sample (Huang et al., 2010). Recovery and reusability of the solvent are possible with this method. Ultrasonication extraction method is specifically used when dealing with small volumes of biomass, but can perform well when coupled with the enzymatic treatment, but both methods lack cost effectiveness and feasibility for large-scale applications. The combination of 'ultrasono-enzymatic treatment' causes faster extraction and facilitates higher oil yields as compared to ultrasonication and enzymatic extractions individually (Fajardo et al., 2007). Supercritical carbon dioxide extraction (SC-CO_2), pulse electric field procedure, osmotic shock, hydrothermal liquefaction, and wet lipid extraction require more optimization efforts for large-scale applications.

Each method has its own advantages and disadvantages for practical applicability. High cost, power consumption and difficulty involved in scaling up are some of the persistent limitations of many methods. Supercritical carbon dioxide extraction is a green technology that can be potentially used for large-scale microalgal lipid extraction. However, this process has high capital cost and energy requirement for supercritical fluid compression. For extraction of high valued products, bead mill is

Table 15.2 Various extraction techniques employed for oil extraction from microalgae.

Extraction Method	
Solvent (selective)	Biomass subjected to drying prior to extraction
	Selection of solvent systems is an important criterion for lipid extraction
	Hexane is used with Soxhlet and Goldfish methods
	Chloroform/methanol or chloroform/methanol/water with Folch or modified Bligh and Dyer procedures
	Best suited to extract non-polar lipids
Soxhlet	A semi-continuous process that allows the buildup of a solvent in the extraction chamber
	Solvent surrounding the sample is recycled back
	Provides a soaking effect and does not permit channelling
	Polar and membrane bound lipids are not recovered
Wet lipid	Uses wet algae biomass by using solvent proportionately
	Elimination of drying step
Hydrothermal liquefaction	Biomass converted in hot compressed water to a liquid biocrude.
	Processing temperatures range 200–350 °C with pressures of around 15–20 MPa
	Ideal for the conversion of high-moisture-content biomass (microalgae) as drying step is not necessary
Ultrasonic	Ultrasonic-assisted extractions recover lipids through cavitation
	Ultrasound breaks the cell wall by cavitation shear forces
Supercritical carbon dioxide (SC-CO_2)	Relatively low temperature and the stability of CO_2 allow most compounds to be extracted with little damage or denaturing
Pulse electric field (PEF)	Short pulses of a strong electric field enlarges the pores of the cell membranes and expels lipid contents
Enzymatic treatment	Enzymes degrade the cell walls by water acting as the solvent
	Oil fraction much easier to recover
Osmotic shock	Osmotic shock causes a release in the cellular contents of microalgae
	It is also induced to release cellular components for biochemical analysis

used along with chemical solvents (Chisti and Moo-young, 1986). Hexane and methanol are used in combination for high extraction efficiency and with economic viability (Neelma et al., 2013).

15.5.4 Biodiesel Preparation

After lipids are extracted, constituents of the lipids (TAG's) are subjected to trans-esterification. The triglyceride composition of algae upon transesterification with an alcohol can produce algae-derived biodiesel (alkyl esters). Different types of trans-esterification reactions viz., acid/base catalyzed, direct and enzymatic were used.

The transesterification process consists of the reaction of triglyceride molecules with alcohol in the presence of a catalyst to produce glycerol and mono-alkyl fatty acid esters also known as biodiesel (Harrison et al., 2012). The fatty acids react with methanol to form diacyl glycerides, monacyl glycerides, and finally, fatty acid methyl esters (FAMEs) (Gong and Jiang, 2011). In this process glycerol is formed as by-product. The transesterification process reduces the viscosity of the FAME as compared to the parent oil, whereas the fatty acid composition will not be altered. Alcohols are the key substrates in transesterification. The commonly used alcohols are methanol, ethanol, propanol, butanol and amyl alcohol, but methanol is widely applied in the transesterification of microalgae oils because of its low cost and physical and chemical advantages.

15.5.4.1 Direct Transesterification

Direct transesterification or in-situ transesterification is receiving much attention in current industrial processes. This technique combines lipid extraction and transesterification in a single step, further reducing the overall number of downstream processes required for biodiesel production (Wahlen et al., 2011). In this process, acid catalyst and pure methanol are added simultaneously to microalgal biomass (dried powder) wherein, methanol extracts the lipids which in the presence of acid catalyst get transesterified to produce fatty acid methyl esters (FAMEs). The lipid extraction and transesterification reaction takes place in a single step in this process (Sathish and Sims, 2012; Ehimen et al., 2010). The algae biomass can be effectively converted to fatty acid methyl esters through this process in relatively less time. Minimization of solvents and requirement of less time for reactions are the advantages of this method whereas the lipid productivity and success rate of the reactions are the associated drawbacks. Direct transesterification is feasible at pilot and large scale systems.

15.5.4.2 Acid and Base Catalyzed Transesterification

Acid catalysis (H_2SO_4/HCl) is usually performed at high alcohol-to-oil-molar ratios, low to moderate temperatures and pressures, and high acid-catalyst concentrations (Zhang et al., 2003). Base-catalyzed transesterification of microalgae oil is used most frequently and involves the presence of a base catalyst (hydroxides/carbonates) to precede the reaction (Meher et al., 2006; Vargha and Truter, 2005). In this reaction, the triglycerides are readily transesterified in the presence of the catalyst at an atmospheric pressure and temperature of 60–70 °C in the presence of excess methanol (Srivastava and Prasad, 2000). The main drawback with this process is the formation of soap at high free fatty acid concentrations (Furuta et al., 2004). Compared to base catalysts, acid catalysts are less susceptible to the presence of free fatty acids in the source feedstock (Helwani et al., 2009), but the reaction rates of converting triglycerides to methyl esters are too slow (Gerpen, 2005). Repeated application of catalyst in the reactions increases the acid value of the microalgae oil.

15.5.4.3 Enzyme-mediated Transesterification

This kind of reaction is catalyzed by the enzyme lipase, whereby total triacylglycerides (both extracellular and intracellular) can be converted to biodiesel (Bisen et al., 2010). The conversion process requires complex processing instruments and the expensive nature of enzymes makes the process limiting. As a solution to overcome the limitations, immobilization was employed but the low technical feasibility of the process makes the reaction complex (Helwani et al., 2009; Watanabe et al., 2001). A novel method has been established for microalgal lipid extraction (Origin Oil, 2010) which performs three simultaneous functions (dewatering, cell disruption, and lipid extraction) in a single downstream step. This method substantially reduces the energy expenditure required to produce biodiesel from microalgal biomass. The downstream technologies required for industrial-scale production of microalgal biodiesel helps overcome the economic hindrances.

15.6 Algal Biorefinery

Extensive studies on microalgae biomass have revealed that there is a huge potential for co-products which can be recovered using biochemical and thermo-chemical technologies. The extraction of more than one type of product from single biomass source increases the value of biomass and offers additional uses to end products (Mussgnug et al., 2010; Yang et al., 2011; Ehimen et al., 2011). The major process after biomass growth is extraction of lipids and conversion to biodiesel. The residual biomass (deoiled cake) can be subjected to a range of bio/thermo chemical processes like fermentation, anaerobic digestion or pyrolysis (Fig. 15.2). The microalgae biomass residues after biodiesel production can be used to generate biomethane which can be burnt to produce electricity. Recovery of methane and biohydrogen by using lipid extracted microalgae pulp as fermentative feedstock after proper pretreatment by biomethanization or acidogenic process accounts for renewability of biomass and increases economic outcomes (Subhash and Venkata Mohan, 2014). Thermo-chemical conversion of algae biomass can also be performed for synthesis of bio-oil and biochar (Sarkar et al., 2014). This will make the production of biodiesel from algae more competitive by reducing the overall production costs and energy needed for downstream processes (Harun et al., 2010). The high input costs will be the major limitation but integrated biorefinery concept can overcome the technical and economic constraints.

Polysaccharides are also produced by microalgae which are used as feedstocks for biofuels production. Carbohydrates derived from photo-synthesis are either accumulated in the plastids as reserve materials (e.g. starch) or become the main component of cell walls (e.g., cellulose, pectin and sulphated polysaccharides) (Ho et al., 2011). Microalgae that contain glucose-based carbohydrates are the most feasible feedstock for bioethanol production. On the other hand, the microalgae can improve the environmental load with respect to pollution and achieve more sustain-

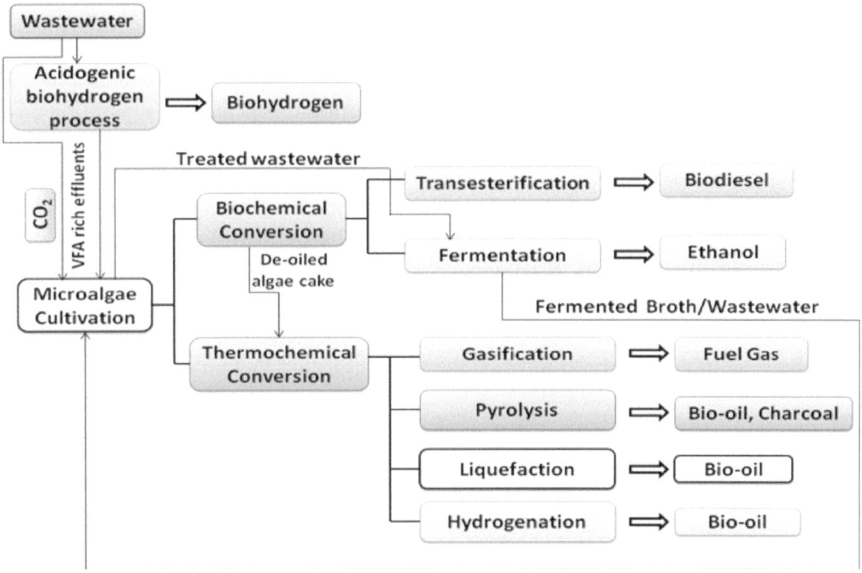

Fig. 15.2 Integrated algal biorefinery.

able lifestyles using CO_2 and wastewater (Sivakumar et al., 2012). Wastewater treatment integration with biofuel production is gaining attention in the current generation of sustainable fuels synthesis (Venkata Mohan et al., 2011). Integrating algal biorefinery concept with wastewater treatment will provide efficient utilization of algae biomass and reduces overall residual waste component of biomass and favours sustainable economics (Venkata Mohan et al., 2014b). A conceptual design of Green Wisdom Inc., USA, using microalgae for integrated bioremediation and biofuel production depicted the potential of microalgae for rural communities leading to economic acceleration and sustainability.

15.7 Future Scope and Challenges

In the future, microalgal biodiesel can be used directly or blended in appropriate ratios for internal combustion engines. The improved screening and selection enables the cultures to grow rapidly in wastewater containing elevated nutrient loads and flue gases. By osmosis technology, cleaned water will be separated from microalgae that will help in feasible production of biodiesel and alleviation of air and water pollution. Innovative cultivation systems and modification of the biochemical composition by simple changes in their cultivation conditions (nutrients, light intensity, temperature, pH, mixing etc.) can lead to higher productivities of the targeted products. On the other hand, their ability to synthesize and accumulate various high value products (e.g. biopolymers, proteins, polysaccharides, pigments) will help to reduce the production costs.

Intensive research needs to be focussed on advanced downstream and bioprocess technologies to reduce the scale up and production costs. Novel separation and extraction systems with higher efficiencies can provide promising results for future applications. The advances in material science and technology will also help producers utilize the proper equipment with lower cost. Employing photo bioreactors can increase the productivities of various species by improved process control and elimination of contaminants. Real time projections for prediction of cost per volume models are required for accurate extrapolation of lab scale data for development of marketable biodiesel technologies. Microalgae based biofuels should be deployed sensitively with the ability to forecast the trend of future technologies.

Acknowledgements The authors wish to thank Director, CSIR-IICT, Hyderabad for encouragement. Authors acknowledge funding from CSIR in the form of 12th five-year plan project-BioEn (CSC-0116). MVR/SA and PC acknowledge University Grants Commission (UGC) and Council of Scientific and Industrial Research (CSIR) for providing research fellowship respectively.

References

Abou-Shanab, R.A.I., Jeon, B.H., Song, H., Kim, Y. and Hwang, J. (2010). Algae-Biofuel: Potential use as sustainable alternative green energy. *Journal on Power and Energy Engineering*, 1, 4–6.

Bracmort, K. (2013). Algae's Potential as a Transportation Biofuel. Congressional Research Service Report for Congress. USA.

Becker, E.W. (1994). Microalgae Biotechnology and Microbiology. Cambridge University Press, Cambridge. Volume 10 of Cambridge Studies in Biotechnology. ISBN 0521350204, 9780521350204.

Beevi, U.S. and Sukumaran, R.K. (2014). Cultivation of microalgae in dairy effluent for oil production and removal of organic pollution load. *Bioresour. Technol.*, 165, 295–301.

Behrens, P.W. (2005). Photobioreactor and fermentors: The light and the dark sides of the growing algae. *In:* Andersen, R.A. (Ed.), Algal Culturing Techniques. Elsevier Academic Press, 189–204.

Bisen, P.S., Sanodiya, B.S., Thakur, G.S., Baghel, R.K. and Prasad, G.B.K.S. (2010). Biodiesel production with special emphasis on lipase-catalyzed transesterification. *Biotechnol. Lett.*, 32, 1019–1030.

Boyle, N.R. and Morgan, J.A. (2009). Flux balance analysis of primary metabolism in *Chlamydomonas reinhardtii. BMC Syst. Biol.*, 3, 4.

Bruton, T., Lyons, H., Lerat, Y., Stanley, M. and Rasmussen, M.B. (2009). A review of the potential of marine algae as a source of biofuel in Ireland. Sustainable Energy Ireland, Dublin, Ireland.

Campbell, M. (2008). Biodiesel: Algae as a Renewable Source for Liquid Fuel. *J. Guelph. Engineering*, 1, 2–7.

Chandra, R., Rohit, M.V., Swamy, Y.V. and Venkata Mohan, S. (2014). Regulatory function of organic carbon supplementation on biodiesel production during growth and nutrient stress phases of mixotrophic microalgae cultivation. *Bioresour. Technol.*, 165, 279–287.

Chang, R.L., Ghamsari, L., Manichaikul, A., Hom, E.F.Y., Balaji, S., Fu, W., Shen, Y., Hao, T., Palsson, B., Salehi-Ashtiani, K. and Papin, J.A. (2011). Metabolic network reconstruction of *Chlamydomonas* offers insight into light-driven algal metabolism. *Mol. Syst. Biol.*, 7, 518.

Chen, C.Y., Yeh, K.L., Aisyah, R., Lee, D.J. and Chang, J.S. (2011). Cultivation, photobioreactor design and harvesting of microalgae for biodiesel production: A critical review. *Bioresour. Technol.*, 102, 71–81.

Chisti, Y. and Moo-Young, M. (1986). Review: Disruption of microbial cells for intracellular products. *Enzyme Microb Technol.*, 8, 194–204.

Chisti, Y. (2007). Biodiesel from microalgae. *Biotechnol Adv.*, 25, 294–306.

Demirbas, A. (2011).Competitive liquid biofuels from biomass. *Appl Energy*, 88, 17–28.

Devi, M.P. and Venkata Mohan (2012). CO_2 supplementation to domestic wastewater enhances microalgae lipid accumulation under mixotrophic microenvironment: Effect of sparging period and interval. *Bioresour. Technol.*, 112, 116–123.

Devi, M.P., Venkata Subhash, G. and Venkata Mohan, S. (2012). Heterotrophic cultivation of mixed microalgae for lipid accumulation and wastewater treatment during sequential growth and starvation phases. Effect of nutrient supplementation. *J. Renew Energy*, 43, 276–283.

Devi, M.P., Swamy, Y.V. and Venkata Mohan, S. (2013). Nutritional mode influences lipid accumulation in microalgae with the function of carbon sequestration and nutrient supplementation. *Bioresour. Technol.*, 142, 278–286.

Ehimen, E. A., Sun, Z. F., & Carrington, C. G. (2010). Variables affecting the in situ transesterification of microalgae lipids. *Fuel*, *89*(3), 677–684.

Ehimen, E.A., Sun, Z.F., Carrington, C.G., Birch, E.J. and Eaton-Rye, J.J. (2011). Anaerobic digestion of microalgae residues resulting from the biodiesel production process. *Appl. Energ.*, 88, 3454–3463.

Energy from Algae (2010). Technical Summary. Scott Maden.

Fajardo, A.R., Cerdán, L.E., Medina, A.R., Gabriel, F., Fernández, A. and Moreno, P.A. (2007). Lipid extraction from the microalga *Phaeodactylum tricornutum*. *Eur. J. Lipid. Sci. Technol.*, 109, 120–126.

Furuta, S., Matsuhashi, H. and Arata, K. (2004). Biodiesel fuel production with solid superacid catalysis in fixed bed reactor under atmospheric pressure. *Catal. Commun.*, 5, 721–723.

Gerpen, V.J. (2005). Biodiesel processing and production. *Fuel Process Technol*, 86, 1097–1107.

Gomez, J.R.O., Aberasturi, G.J., Lopez, C.R. and Belsue, M. (2012). A brief review on industrial alternatives for the manufacturing of glycerol carbonate, a green chemical. *Org Process Res Dev.*, 16, 389–399.

Gong, Y. and Jiang, M. (2011). Biodiesel production with microalgae as feed stock: From strains to biodiesel. *Biotech Letters*, 33, 1269–1284.

Greenwell, H.C., Laurens, L.M.L., Shields, R.J., Lovitt, R.W. and Flynn, K.J. (2010). Placing microalgae on the biofuels priority list: A review of the technological challenges. *J. Royal Soc. Interface.*, 7, 703–726.

Hannon, M., Gimpel, J., Tran, M., Rasala, B. and Mayfield, S. (2010). Biofuels from algae: Challenges and potential. *Biofuels.*, 1, 763–784.

Harrison, B.B., Marc, E.B. and Anthony, J.M. (2012). Chemical and physical properties of algal methyl ester biodiesel containing varying levels of methyl eicosapentaenoate and methyl docosahexaenoate. *Algal Res*, 1, 57–69.

Harun, R., Danquah, M.K. and Forde, G.M. (2010). Microalgal biomass as a fermentation feedstock for bioethanol production. *J. Chem. Technol. Biotechnol.*, 85, 199–203.

Helwani, Z., Othman, M.R., Aziz, N., Fernando, W.J.N. and Kim, J. (2009). Technologies for production of biodiesel focusing on green catalytic techniques: A review. *Fuel Process Technol*, 90, 1502–1514.

Ho, S.H., Chen, C.Y., Lee, D.J. and Chang, J.S. (2011). Perspectives on microalgal CO_2-emission mitigation systems – A review. *Biotechnol. Adv.*, 29, 189–198.

Hu, B., Min, M., Zhou, W., Li, Y., Mohr, M. and Cheng, Y. (2012). Influence of exogenous CO_2 on biomass and lipid accumulation of microalgae *Auxenochlorella protothecoides* cultivated in concentrated municipal wastewater. *Applied Microbiol and Biotechnol.*, 166, 1661–1673.

Huang, G., Chen, F., Wei, D., Zhang, X. and Chen, G. (2010). Biodiesel production by microalgal biotechnology. *Appl. Energ.*, 87, 38–46.

Johnson, M.B. and Wen, Z.Y. (2010). Development of an attached microalgal growth system for biofuel production. *Appl. Microbiol. Biotechnol.*, 85, 525–534.

Kazamia, E., Czesnick, H., Van Nguyen, T.T., Croft, M.T., Sherwood, E. and Sasso, S. (2012). Mutualistic interactions between vitamin B_{12}-dependent algae and heterotrophic bacteria exhibit regulation. *Environ. Microbiol.*, 14, 1466–1476.
Kim, M.K., Park, J.W., Park, C.S., Kim, S.J., Jeune, K.H. and Chang, M.U. (2007). Enhanced production of Scenedesmus spp. (green microalgae) using a new medium containing fermented swine wastewater. *Bioresour. Technol.*, 98, 2220–2228.
Lee, J.Y., Yoo, C., Jun, S.Y., Ahn, C.Y. and Oh, H.M. (2010). Comparison of several methods for effective lipid extraction from microalgae. *Bioresour. Technol.*, 101, 75–77.
Mata, T.M., Martins, A.A. and Caetano, N.S. (2010). Microalgae for biodiesel production and other applications: a review. *Renew. Sust. Energ. Rev.*, 14, 217–232.
Meher, L.C., Sagar, D.V. and Naik, S.N. (2006). Technical aspects of biodiesel production by transesterification: A review. *Renew Sust Energ Rev*, 10, 248–268.
Mercer, P. and Armenta, R.E. (2011). Developments in oil extraction from microalgae. *Eur. J. Lipid. Sci. Technol.*, 113, 539–547.
Mitra, D., Van Leeuwen, J. and Lamsal, B. (2012). Heterotrophic/mixotrophic cultivation of oleaginous *Chlorella vulgaris* on industrial co-products. *Algal Res.*, 1, 40–48.
Munoz, R. and Guieysse, B. (2006). Algal-bacterial processes for the treatment of hazardous contaminants: A review. *Water Res.*, 40, 2799–2815.
Mussgnug, J.H., Klassen, V., Schluter, A. and Kruse, O. (2010). Microalgae as substrates for fermentative biogas production in a combined biorefinery concept. *J Biotechnol*, 150, 51–56.
Neelma, M., Nadia, S., Shagufta, N., Faiza, S. and Farkhanda, M. (2013). Harvesting and processing of microalgae – A review. *Sci. Tech. and Dev.*, 32, 235–243.
Nelson, J.A., Savereide, P.B. and Lefebvre, P.A. (1994). The CRY1 gene in Chlamydomons reinhardtii: Structure and use as a dominant selectable marker for nuclear transformation. *Mol. Cell Biol.*, 14, 4011–4019.
Olguin, E.J. (2012). Dual purpose microalgae–bacteria-based systems that treat wastewater and produce biodiesel and chemical products within a biorefinery. *Biotechnol. Adv.*, 30, 1031–1046.
Origin Oil (2010). Algae harvesting, dewatering and extraction a breakthrough technology to transform algae into oil. Worldbiofuels Markets, Amsterdam, The Netherlands. 15–17.
Orpez, R., Martinez, M.E., Hodaifa, G., El Yousfi, F., Jbari, N. and Sanchez, S. (2009). Growth of the microalga Botryococcusbraunii in secondarily treated sewage. *Desalination*, 246, 625–630.
Pearson, H.W. (1996). Expanding the Horizons of Pond Technology and Application in an Environmentally Conscious World. *Water Sci Technol.*, 33, 1–9.
Perez-Garcia, O., de-Bashan, L., Hernandez, J. and Bashan, Y. (2010). Efficiency of growth and nutrient uptake from wastewater by heterotrophic, autotrophic, and mixotrophic cultivation of Chlorella vulgaris immobilized with azopirillum brasilense. *J. Phycol.*, 46, 800–812.
Perez-Garcia, O., Escalante, F. M., de-Bashan, L. E., & Bashan, Y. (2011a). Heterotrophic cultures of microalgae: metabolism and potential products. *Water research*, *45*(1), 11–36.
Perez-Garcia, R.O., Bashan, Y. and Puente, M.E. (2011b). Organic carbon supplementation of municipal wastewater is essential for heterotrophic growth and ammonium removing by the microalgae Chlorella vulgaris. *J. Phycol.*, 190–199.
Pulz, O. (2001). Photobioreactors: Production systems for phototrophic microorganisms. *Appl. Microbiol. Biotechnol.*, 57, 287–293.
Ramanathan, G., Rajarathinam, K., Boothapandi, M., Abirami, D., Ganesamoorthy, G. and Duraipandi (2011). Construction of vertical tubular photobioreactor for microalgae cultivation. *J. Algal Biomass Utln.*, 2, 41–52.
Robert, M.H., Christina, E.C., Tom, N.K., Stephen, L.F., Oybek, K. and David, R.S. (2012). Evaluation of environmental impacts from microalgae cultivation in open-air raceway ponds: Analysis of the prior literature and investigation of wide variance in predicted impacts. *Algal Research*, 1, 83–92.
Sarkar, O., Agarwal, M., Kumar, A.N. and Venkata Mohan, S. (2014). Retrofitting hetrotrophically cultivated algae biomass as pyrolytic feedstock for biogas, bio-char and bio-oil production encompassing biorefinery. *Bioresour Technol* DOI: 10.1016/j.biortech.2014.09.070.

Sathish, A. and Sims, R.C. (2012). Biodiesel from mixed culture algae via a wet lipid extraction procedure. *Bioresour. Technol.*, 118, 643–647.

Shelef, G., Sukenik, A. and Green, M. (1984). Microalgae harvesting and processing: A literature review. Report prepared for the US Department of Energy, Technion Research and development Foundation Ltd., Haifa, Israel.

Sierra, E., Acien, F.G., Fernandez, J.M., Garcia, J.L., Gonzalez, C. and Molina, E. (2008). Characterization of a flat plate photobioreactor for the production of microalgae. *Chem Eng J.*, 138, 136–147.

Sivakumar, G., Xu, J., Thompson, R.W., Yang, Y., Smith, P.R. and Weathers, P.J. (2012). Integrated green algal technology for bioremediation and biofuel. *Bioresour Technol.*, 107, 1–9.

Srivastava, A. and Prasad, R. (2000). Triglycerides-based diesel fuels. *Renew Sus Energ Rev.*, 4, 111–133.

Stephenson, A.L., Kazamia, E., Dennis, J.E., Howe, C.J., Scott, S.A. and Smith, A.G. (2010). Life-cycle assessment of potential algal biodiesel production in the United Kingdom: A comparison of raceways and air-lift tubular bioreactors. *Energ. Fuels*, 24, 4062–4077.

Ugwu, C.U., Aoyagi, H. and Uchiyama, H. (2008). Photobioreactors for mass cultivation of algae. *Bioresour. Technol.*, 99, 4021–4028.

Vargha, V. and Truter, P. (2005). Biodegradable polymers by reactive blending transesterification of thermoplastic starch with poly (vinyl acetate) and poly(vinyl acetate-co-butyl acrylate). *Eur. Polymer J.*, 41, 715–726.

Venkata Mohan, S. (2010). Waste to renewable energy: A sustainable and green approach towards production of biohydrogen by acidogenic fermentation. *In:* Om, Singh, Steve, Harvey (Eds.), Sustainable Biotechnology: Renewable Resources and New Perspectives. Springer. 129–164

Venkata Mohan, S., Devi, M.P., Mohanakrishna, G., Amarnath, N., Lenin Babu, M. and Sarma, P.N. (2011). Potential of mixed microalgae to harness biodiesel from ecological water-bodies with simultaneous treatment. *Bioresour. Technol.*, 102, 1109–1117.

Venkata Mohan, S. and Devi, M.P. (2012). Fatty acid rich effluent from acidogenic biohydrogen reactor as substrate for lipid accumulation in heterotrophic microalgae with simultaneous treatment. *Bioresour. Technol.*, 123, 627–635.

Venkata Mohan, S., Devi, M.P., Subhash, G.V. and Chandra, R. (2014). Algae oils as fuels (Chap 8). *In:* Pandey, A., Lee, D.J., Chisti, Y. (Eds), Biofuels from Algae. Elsevier, pp. 155–187 (ISBN: 9780444595584).

Venkata Subhash, G. and Venkata Mohan, S. (2014). Deoiled algal cake as feedstock for dark fermentative biohydrogen production: An integrated biorefinery approach. *Int J Hydrogen Energy*, 39, 9573–9579.

Venkata Mohan, S., Rohit, M.V., Chiranjeevi, P., Chandra, R. and Navaneeth, B. (2014). Heterotrophic Microalgae Cultivation to Synergize Biodiesel Production with Waste Remediation: Progress and Perspectives. *Bioresour Technol.* DOI:10.1016/j.biortech.2014.10.056.

Wahlen, B.D., Willis, R.M. and Seefeldt, L.C. (2011). Biodiesel production by simultaneous extraction and conversion of total lipids from microalgae, cyanobacteria, and wild mixed cultures. *Bioresour Technol.*, 102, 2724–2730.

Watanabe, Y., Shimada, Y., Sugihara, A. and Tominaga, Y. (2001). Enzymatic conversion of waste edible oil to biodiesel fuel in a fixed-bed bioreactor. *J. Am. Oil Chem. Soc.*, 78, 703–707.

Weissman, J. (1987). Photobioreactor Design: Comparison of Open Ponds. *Bioeng. Biotech.*, 31, 336–344.

Yang, Z., Guo, R., Xu, X., Fan, X. and Luo, S. (2011). Fermentative hydrogen production from lipid-extracted microalgal biomass residues. *Appl. Energ.*, 88, 3468–3472.

Yoo, C., Jun, S.Y., Lee, J.Y., Ahn, C.Y. and Oh, H.M. (2010). Selection of microalgae for lipid production under high level of carbon dioxide. *Bioresour. Technol.*, 101, 71–74.

Zhang, Y., Dub, M.A., McLean, D.D. and Kates, M. (2003). Biodiesel production from waste cooking oil: 2 Economic assessment and sensitivity analysis. *Bioresour. Technol.*, 90, 229–240.

Chapter 16
Significance of Cyanobacteria as Inoculants in Agriculture

Dolly Wattal Dhar, Radha Prasanna, Sunil Pabbi, and Rashi Vishwakarma

16.1 Introduction

The utilization of beneficial microbes as biofertilizers has become vital in agriculture sector for their prospective role in food safety and sustainable crop production. Biofertilizers keep the soil environment rich in all kinds of micro- and macronutrients via nitrogen fixation, phosphate and potassium solubilisation or mineralization, release of plant growth regulating substances, production of antibiotics and biodegradation of organic matter in the soil (Sinha et al., 2014). When biofertilizers are applied as seed or soil inoculants, they multiply and participate in nutrient cycling and benefit crop productivity (Singh et al., 2011). Generally, 60 % to 90 % of the total applied fertilizer is lost and the remaining 10 % to 40 % is taken up by the plants. In this regard, microbial inoculants have supreme significance in integrated nutrient management systems to sustain agricultural productivity and healthy environment (Adesemoye and Kloepper, 2009).

The deployment of non-toxic monoculture of cyanobacteria and green algae as biofertilizer input for increasing soil nutrient availability reduces the dependence on costly chemical fertilizers in integrated nutrient management systems (Grzesik and Romanowska-Duda, 2014). Application of various forms of cyanobacteria and microalgae including *Microcystis aeruginosa*, *Anabaena* sp. PCC 7120, and *Chlorella* sp. can be beneficial to growth, development and metabolic activity of corn seedlings. Some of them have the remarkable ability to form intimate symbiotic associations with a wide range of eukaryotic hosts belonging to different plant groups. Cyanobacteria also are a biogeochemically important component of diverse

D.W. Dhar (✉) • S. Pabbi • R. Vishwakarma
Centre for Conservation and Utilisation of BGA, Indian Agricultural Research Institute (IARI), New Delhi, India
e-mail: dollywattaldhar@yahoo.com

R. Prasanna
Division of Microbiology, Indian Agricultural Research Institute (IARI), New Delhi, India

D. Das (ed.), *Algal Biorefinery: An Integrated Approach*,
DOI 10.1007/978-3-319-22813-6_16

ecosystems that play a significant role in carbon and nitrogen cycling. Non-toxic cyanobacterial and microalgae cultures can be used for ecological and integrated corn cultivation and will facilitate environmental protection by reducing the need to use toxic artificial fertilizers. *Anabaena* sp. and *Calothrix* sp. can also be used for the stimulated production of ammonia and indolic compounds.

Cyanobacteria constitute the largest, most diverse and widely distributed group of prokaryotes that perform oxygenic photosynthesis and many are capable of fixing atmospheric nitrogen and can convert insoluble phosphorus into soluble form (De, 1936, 1939; De and Sulaiman, 1950; Singh, 1961; Venkataraman, 1979; Roger et al., 1993; Kaushik, 1998; Irisarri et al., 2001). In the last few decades, cyanobacteria are also being increasingly explored as novel sources of pigments, bioactive metabolites and nutritional supplements. However, the major research output using these ubiquitous organisms has been for their significant role in enhancing soil fertility and plant productivity. They are important components of soil especially in flooded rice fields, which are known to have a diverse flora of morphologically distinct forms, which exhibit a seasonal and crop stage-specific diversity (De, 1939; Singh, 1961; Watanabe, 1959; Roger and Kulasooriya, 1980).

Rice (*Oryza sativa* L.) is one of the most important crops of South East Asia with acreage expanding by nearly 16 % in Kharif, 2010 to 42 million hectares. It ranks among the highest, in terms of fertilizer usage, consuming more than 60 million tonnes of nitrogen in India. It is one of the most important cereals for more than half of the world's population and speculated demand for five billion rice consumers by the year 2030 is to be met from available land and water resources (Khush, 2005). The successful production of rice depends upon efficient and economic supply of nitrogen, an element required in the largest quantity in comparison with other essential ones. However, the nitrogenous fertilizer use efficiency is low because of its loss from soils through various chemical and biochemical processes. Besides, increasing application of nitrogenous fertilizer is neither environmental friendly nor economically viable (Cassman and Pingali, 1994). It has, therefore, become necessary to look for alternative renewable sources to meet at least a part of the nitrogen demand of rice crop (Swaminathan, 1982). In this context, biological nitrogen fixation is especially important in rice fields, where cyanobacteria are recognized as significant contributors to the overall nutrient balance and improving the productivity of rice fields (Mandal et al., 1999).

The fact that wetland rice systems can produce 2–4 t ha^{-1} of grain yield without any fertilizer input (De Datta, 1987) highlights the significance of microbial biomass (including cyanobacteria/*Azolla*) in its dual role as a labile sink and source of nutrients in submerged rice systems (Inubushi et al., 1997a, b; Prasanna et al., 2010a, b). Further, unlike chemical N fertilizers, cyanobacteria neither contaminate the environment nor consume the photosynthesis of rice plants. Submerged rice soils are known to provide an ideal condition for their growth with respect to the requirement for light, water, temperature, humidity and atmospheric nitrogen fixation of cyanobacteria. These organisms are able to withstand extremes of temperature and drought and show remarkable variation in growth, nitrogen fixation and stress compatibility (Whitton and Potts, 2012).

16.2 Cyanobacterial Diversity in Rice Based Ecosystems

In India, rice was grown on the same land for long periods without the addition of manure to the soil and the ability of rice to grow in the absence of manure was suggested to be either the result of fixation of nitrogen in the soil or the rice plants themselves harbouring a nitrogen-fixing symbiont. Later work demonstrated that the N fertility of soil is sustained better under flooded conditions than under dry land conditions (Watanabe and Roger, 1984). Evidence was obtained that cyanobacteria are the main agents of nitrogen fixation, thus, maintaining the natural fertility of low land Indian paddy fields; and the part played by bacteria was thought to be relatively unimportant and possibly nil (De, 1939). A number of experiments carried out in India suggested the possibility of the role of cyanobacteria in enriching soil during water logged periods of rice growth (Venkataraman, 1972, 1981). Thereafter, quantification of N fixed biologically by free-living cyanobacteria in paddy floodwater and soil was undertaken globally, especially in South East Asia (Roger and Kulasooriya, 1980).

Application of cyanobacteria was shown to increase the productivity of crops like rice (Goyal, 1997; Yanni, 1992), wheat (Karthikeyan et al., 2007) and leguminous plants. *Prosopis juliflora* (Rai et al., 2004) and the technologies for inoculation of paddy fields with nitrogen fixing algae were developed (Venkataraman, 1972, 1981). Under favourable conditions, a good algal bloom in rice fields yield on average about 6–8 t of fresh biomass (Roger and Kulasooriya, 1980; Roger et al., 1987). The continued existence of such biomass in soils as organic matter depends upon its decomposability and some algae may be decomposed quickly while others may last longer (Watanabe and Kiyohara, 1960).

The need for algal inoculation arose from an earlier belief that N_2-fixing cyanobacteria strains were not widely present in rice fields. Occurrence of cyanobacteria was reported in only 5 % of 911 soil samples (Watanabe and Yamamoto, 1971), 33 % of 2213 samples (Venkataraman, 1975), and in 71 % of Japanese soil samples (Okuda and Yamaguchi, 1952). Khan et al. (1994) have analyzed 38 soil samples from 11 districts of Dhaka division (Bangladesh) for cyanobacterial flora. Of the 84 species documented, 50 % were heterocystous diazotrophic cyanobacteria, predominantly *Fischerella, Nostoc* and *Calothrix*, occurring in about 53 %, 47 % and 26 % of the soil samples respectively. A number of floristic studies have shown the dominance of heterocystous species (*Anabaena*, *Nostoc, Calothrix, Cylindrospermum, Gloeotrichia* and *Scytonema*) especially akinete formers in many rice fields. These species positively correlated with pH and available P content of the soils (Roger et al., 1993). The best pH range for fixation of N by them is between 7.0 and 8.5. Among the physical and chemical factors influencing the growth and establishment of these organisms, light exerts the most significant effects. This is manifested in terms of periodicity with respect to season and the cultivation cycle of the crop. Algal succession in rice fields is manifested in terms of differing populations of unicellular/colonial, non-heterocystous, followed by heterocystous—initially unbranched and by late harvest stage, branched heterocystous forms (Singh and

Bisoyi, 1989). Nitrogen fixers do not dominate either in the beginning or late stages of crop growth and the phase between tillering and panicle development leads to production of maximum biomass (Roger and Kulasooriya, 1980). Various microorganisms such as protozoans, fungi and grazers such as copepods, ostracods, snails and larvae significantly diminish cyanobacterial populations (Venkataraman, 1972). Management practices such as tillage, transplantation and weeding or water management also influence the proliferation of these organisms in the field. Application of insecticides or fungicides at recommended doses also do not significantly affect the growth of cyanobacteria (Venkataraman, 1975), as a wide range of tolerance is exhibited by these organisms.

The species possessing biofertilizer potential are generally the hetero-cystous filamentous forms belonging to the orders Nostocales and Stigo-nematales, in which the nitrogenase activity and oxygenic photosynthesis are separated spatially and nitrogenase activity is usually light dependent. Cyanobacterial strains, especially the *Nostoc* and *Anabaena* strains isolated from deepwater rice soils can also provide interesting evidence to their adaptations in the predominantly anoxygenic surroundings and their interactions with other flora/fauna in this niche.

Species of *Nostoc*, *Anabaena*, *Tolypothrix*, *Aulosira*, *Cylindrospermum*, *Scytonema*, *Westiellopsis* and several other genera are profuse in Indian rice field soils and are known to contribute significantly to their fertility in India. On an average, cyanobacteria accounts to about 33 % (Venkataraman, 1975) whereas reports show that up to 50 % of the total algae was cyanobacteria in some of the southern and eastern states. Common genera found in Indian rice soils include *Anabaena*, *Nostoc*, *Aulosira*, *Calothrix* and *Tolypothrix* etc. (Gupta and Khajuria, 1996; Swarnalakshmi et al., 2006).

Among the soil properties, pH is undoubtedly the most important factor determining the flora and fauna composition. In culture media, the optimal pH for the growth of cyanobacteria ranges 7.5–10, with a lower limit of 6.5–7.0. However, in soil-culture experiments, soils having slightly alkaline pH are more favourable, while in natural environments cyanobacteria prefer neutral to alkaline pH (De, 1936; Roger, 1996). Prasanna and Nayak (2007) have evaluated soil samples collected from nine locations, differing in their EC and pH values for cyanobacterial abundance and generic diversity. A total of 166 forms, including 130 heterocystous and 36 heterocystous isolates were recorded. Average population counts (measured as MPN) of various locations clearly suggested the tremendous diversity among the locations sampled. Soil samples from Jeypore (Orissa state) recorded highest diversity and 20 cyanobacterial forms, across nine genera were isolated. *Nostoc* and *Anabaena* were found to be the dominant genera in all the locations, in terms of their abundance and displayed highest diversity indices (Prasanna and Kaushik 2006; Prasanna and Nayak, 2007). A predominance of heterocystous forms (68–95 %) was observed at all locations, while non-heterocystous forms exhibited 5–32 % abundance in the various locations. Heterocystous forms represented highest % abundance at pH of 8.1, followed by 7.9 (83 and 80 % respectively). Soil samples with pH of 7.4 and 9.3 recorded highest % abundance of non-heterocystous forms. Nayak and Prasanna (2007) have observed significant positive correlation between EC and pH. Shannon's diversity index was maximum at a pH of 6.9, followed by pH of 7.4, indicative of the higher number of genera recorded in these soil samples. Richness and evenness (J and E) indices were highest in soil samples of pH of 9.3. Cyanobacteria belonging to 12 genera were isolated which included eight heterocystous forms—*Anabaena*, *Nostoc*, *Westiellopsis*, *Calothrix*, *Scytonema*, *Aulosira*, *Hapalosiphon* and *Cylindrospermum*—and four non-heterocystous forms—*Phormidium*, *Oscillatoria*, *Lyngbya* and *Aphanocapsa*. *Nostoc* and *Anabaena* recorded maximum number of isolates i.e. 10 at pH of 6.5. Isolates belonging to the genera *Anabaena* and *Nostoc* also showed the highest relative abundance of 100 %, while lowest values were recorded for *Aulosira* and *Lyngbya*, as they were isolated only from one of the nine locations. The genus *Nostoc* was most abundant (30 %, 51 isolates), followed by *Anabaena* (28 %, 46 isolates) and *Phormidium* (12 %, 20 isolates).

Diversity analyses of a set of 70 *Anabaena* strains isolated from diverse rice agro ecologies of India, using morphological and molecular datasets generated useful information for region specific abundance patterns which aided in the development of a comprehensive database based on the distribution of *Anabaena* strains in diverse agro ecologies of India. Useful primers based on repeat/palindromic sequences for PCR based differentiation of *Anabaena* isolates (Nayak et al., 2007, 2009) were also identified which are being utilized for evaluating the establishment of these strains in soil. Analyses of the species-wise distribution in different soil types and soil pH exhibited that *Anabaena iyengarii* was present at pH ranging from 5.5 to 8.5 and alluvium soils harboured all the species of *Anabaena* except *A. oscillarioides*.

Studies exploring the rhizosphere of rice and wheat revealed the morphological and functional diversity of these facultative prokaryotes. Interestingly, the predominance of *Nostoc* and *Anabaena* was observed and many strains exhibited nitrogen fixing and IAA producing potential in dark (Prasanna et al., 2009a; Karthikeyan et al., 2007, 2009).

Previous studies on the distribution pattern of cyanobacteria in soils of Andhra Pradesh, Haryana, Delhi, Rajasthan, Uttar Pradesh and Punjab have revealed that although recurrent combination of forms were distinguishable, there appeared to be a localized distribution of cyanobacteria depending upon the soil pH, electrical conductivity and exchangeable sodium. The predominant forms recorded species of *Nostoc*, *Calothrix*, *Scytonema*, *Hapalosiphon* and *Westiellopsis* were observed in soils of all the locations and seems to be tolerant to salt fluctuations (Kaushik, 2005). In salt-affected soils of Maharashtra, the species of *Nostoc* and *Calothrix* were predominant, although the pH of the soil varied from 6.0 to 7.2. Mucilage producing cyanobacterial species of *Scytonema*, *Lyngbya* and *Tolypothrix* are also common. On scrutinizing the total cyanobacterial flora, it was observed that out of the total 37 species, 50 percent were nitrogen- fixing strains, including the non-heterocystous nitrogen fixers. Successive cultivation of BGA makes the environment more favourable and after a few years it may facilitate to produce a reasonably good yield of crops, as observed by Singh (1961) for sugarcane after three years of reclamation with BGA. Although sporadic at pH below 6.0, their ability to grow in diverse pH ranges and modify their environment makes them successful in any niche. Acidic soils usually do not support their growth, although a few reports on their presence in soils with pH values between 5 and 6 are available (Aiyer et al., 1972; Prasad et al., 1978). The relative abundance of cyanobacteria in rice soils and biofertilizer inocula from four countries revealed that significant correlation could be made with respect to pH and available P content of soils (Roger et al., 1987). pH of the location sampled also showed a positive correlation for *Calothrix*, *Phormidium* and *Hapalosiphon. Nostoc* and *Anabaena* exhibited superior adaptive traits, although in terms of numbers, they showed an uneven distribution at different pH. Among the non-heterocystous cyanobacteria, *Phormidium* was the most cosmopolitan—20 isolates as against 9 and 5 belonging to *Oscillatoria/Lyngbya.*

The Centre for Conservation and Utilization of cyanobacteria at Indian Agricultural Research Institute (IARI) houses a germplasm facility wherein isolates from diverse agro-ecological regions of India are maintained.

16.3 Technologies for Production of Microalgal Biofertilizer

Microbial biofilms are communities of microorganisms adhering to abiotic/biotic surfaces and implanted in an organic matrix of biological origin which provides structure and stability to the community (Webb et al., 2003). Biofilms comprise layers of prokaryotic or eukaryotic cells, which can play a key role in plant-microbe interactions. Microbes in a biofilm are sedentary, sheathed in extracellular polysaccharide (EPS) matrix (Mah Thien-Fah and O'Toole, 2001) and establish intimate contact with other microbial cells in the environment. The enclosed polymer of biofilm provides protection from environmental stress such as extreme pH, osmotic shock, and desiccation. Moreover, microbes in biofilms show an increased resistance to antimicrobial agents (Stewart, 2002) and thus, survive in competitive environments. There is a great scope for developing efficient microbial biofilms (EMB) biofertilizer technology which can overcome the poor survival of the microbial inoculants, especially under harsh environmental conditions. Phototrophic biofilms include matrix enclosed microbial communities, mainly driven by light energy. The extracellular mucilage/polysaccharide layer of cyanobacteria symbolize a nutrient rich environment for the growth of associated bacteria (both photosynthetic/non-photosynthetic types) which can be exploited as a carrier for inoculants. Such biofilms exemplify assemblages of bacterial cells anchored to organic matrix of algae which guards microorganisms against adverse environmental conditions and improves their survival. Enrichment of useful native/potential PGPR (plant growth promoting rhizobacteria) in such biofilms can provide a viable alternative to carrier-based biofertilizers. Such rhizobacteria may directly stimulate growth by nitrogen fixation, solubilization of nutrients, production of growth hormones, 1-aminocyclopropane, 1-carboxylate (ACC) deaminase, and suppress by antagonizing pathogenic fungi by production of siderophores, chitinase, ß-1-3-glucanase, antibiotics, fluorescent pigments and cyanide. They can also cause substantial changes in plant gene expression (Glick, 1995; Glick et al., 1999). Swarnalakshmi et al. (2013) have evaluated novel biofilmed preparations, using Cyanobacteria (*Anabaena torulosa*) as a matrix for agriculturally useful bacteria (*Azotobacter*, *Mesorhizobium*, *Serratia* and *Psuedomonas*) in wheat crop. Their results have shown existence of synergism among phototroph-heterotroph partners.

Prasanna et al. (2011a, b) have created biofilmed preparations using the cyanobacterium *Anabaena* sp. as a matrix for agriculturally useful bacteria such as *Azotobacter, Mesorhizobium, Serratia* and *Pseudomonas*. Such biofilms demonstrated enhanced survival of bacteria, and improved plant growth-promoting traits under *in vitro* conditions. The development of cyanobacterial biofilmed biofertilizers (CBBs) by bacterial colonization (N-fixer, phosphate solubilizer and PGPR) has been found to enhance metabolic activities and survival, compared to monocultures (Prasanna et al., 2011a, b). Jayasinghearachchi and Seneviratne (2004) had earlier showed the increase of nodulation and N accumulation in soybean with fungal biofilm (*Bradyrhizobium-Penicillium* spp) under green house conditions. It is apparent that biofilms can play a central role in plant-microbe interactions in rhizospheric soil (Fujishige et al., 2006). Nonetheless, success of introduced strains is based on their survival and colonization. Swarnalakshmi et al. (2013) have hypothesized that colonization in the rhizosphere and other traits associated with soil fertility may be enhanced through the use of biofilmed inoculants.

Algalization has been recognized as an important input in rice cultivation as it forms a perpetually renewable source of nutrients and improves soil health (Singh, 1961; Venkataraman, 1981). Inoculation of rice fields with cyanobacteria was ushered in Japan in the early 1950s by Watanabe et al. (1951) and it was further developed in India, Burma, Egypt and China. They are often referred to as ‘Paddy Organisms’ because of their abundance in the paddy fields and play a considerable role in rice culture. Considering the fact that 87 % of the rice area in our country accounts for holdings of 1–4 ha and 13 % of the holdings even less than 1 ha, an inexpensive rural oriented algal biofertilizer technology for rice was developed (Venkataraman, 1981). The method was simple, inexpensive, and effectively adopted by farmers and the inoculum comprised of several species of the genera *Aulosira*, *Tolypothrix*, *Scytonema*, *Nostoc*, *Anabaena* and *Plectonema*. It was propagated by farmers in shallow trays or tanks with 5–15 cm water, about 4 kg soil m^{-2}, 100 g triple superphosphate m^{-2} and insecticide (2 mL of malathion/tray, 50 % EC). In one to three weeks when a thick mat developed on the soil or water surface, the trays were dried, and algal flakes scraped and stored in bags for use in the fields @ 10 kg ha^{-1}. However, the production capacity of cyanobacteria flakes in India was only 0.01 % of the total inoculum requirement for the country. Also, the wider application of this technology was not possible because of the fluctuations in environmental parameters as well as the variation in the nature of soil used as the source of nutrient, presence of aerial and soil-based contamination which leads to deterioration of the quality of the inoculum. Also, the bulky nature of soil-based inoculum poses transportation problems. Intensive efforts were undertaken to improve the mass production technologies for increasing the production and effectiveness. Alternative carriers (Multani mitti, straw) were employed and polyhouse-based production technology, which permitted year round production and contamination-free products (Goyal et al., 1997; Goyal, 1997; Kaushik, 1998; Kaushik and Prasanna, 1998; Prasanna and Kaushik, 1995). To overcome these problems, an alternative production technology was developed by Kaushik (1998) in which soil has been replaced by a cheaper (US $ 0.28 per 1000 litre) formulated medium (Prasanna et al., 1998) and the production done under controlled conditions by constructing multiplication units in glass or poly-houses to ensure continuous production of quality inoculum throughout the year.

The production technology has now been considerably improved with the introduction of new and cheap carrier materials such as multani mitti (Fuller's earth) and wheat straw that support higher algal load with longer shelf life, thus noticeably reducing the quantity of inoculum per unit area (Kaushik, 1998; Goyal et al., 1997; Prasanna et al., 1998). A comparative appraisal of the technologies is provided in Table 16.1.

In spite of the utility of BGA biofertilizers as low cost input which are renewable and pollution-free, the farmers' acceptance of this practice has been found far from satisfactory, because the crop response is generally not as spectacular as observed with mineral fertilizers (Fig. 16.1).

Reluctance among the farmers to adopt BGA biofertilizer production technology could be attributed to various reasons, including difficulty in its use, unavailability of the inputs required, market uncertainty, lack of adequate mass production units and non-availability of quality material, price fluctuations or preferences for very low management crop technology. However, as regards the cyanobacterial biofertilizer, the wide variation in agro climatic conditions under which the rice is grown in India, a single product cannot be effective throughout the country. The constraints may be physical, biological, environmental or technical. The establishment of inoculated strains vis a vis native strains is a major problem encountered. The present-day soil conditions such as acidity, alkalinity, pesticide application and high nitrate level may also limit the nitrogen fixing capacity of the inoculants, for which agronomic practices need to be incorporated such as liming, etc. The presence of antagonists or grazers, agrochemicals, and/or certain toxic elements in the rhizosphere can also play a significant role. The most important problem faced has been the mushrooming of small units selling substandard inoculants and lack of quality control measures as prescribed for other microbial inoculants. Inaccessibility of proper transportation and storage facilities and lack of the publicity is one major limitation in popularizing biofertilizers among the farming community.

Sugarcane wastes and paper wastes solid matrices was also used for the immobilization of blue green algae along with rice husks and soil for improving the quality of BGA biofertilizer (Kannaiyan, 1999). Shanmugasundaram (1996) prepared the biofertilizer inoculum "cyanostraw" by growing the desired algal strains together with the carrier in polybags under controlled conditions to get a product of high quality free from soil fungi, protozoans or contaminating algae. An almost similar method was used in Egypt, but only the floating algal flakes, relatively free of soil, were collected and dried. About 250 g ha^{-1} of dried algal flake was inoculated a week after transplanting rice (Watanabe, 1986).

Table 16.1 Comparative evaluation of different carrier-based cyanobacterial biofertilizer production technologies

	Types		
Parameters	Open air technology using soil	Straw	Multani mitti
Production site	Diverse field, any open area	Indoor using polyhouse/ glasshouse	Indoor using polyhouse/ glasshouse
Multiplication units	Shallow polythene lined pits; trays, troughs, brick/ cement lined, RCC tanks	Shallow RCC tanks in one/two tiers	Shallow RCC tanks
Nutrient source	Soil (10 kg/ unit + 250 g SSP)	Formulated medium	Formulated medium
Energy source	Sunlight	Sunlight or artificial light	Sunlight
Temperature control	Variable, dependent on+ambient temperature	Partial control	Partial control
Aeration	Natural	Can provide through cloth filters	Can provide through cloth filters
Carrier qualities	Present everywhere, but contamination cannot be reduced	Easily available, cheap and lightweight	Available in most areas and moderately priced
Harvesting cycle	3–4 weeks/ harvest	7–10 days/harvest	7–10 days/harvest
Population	10^2–10^3 propagules/g soil	106–107 propagules/g straw	106–107 propagules/g carrier
Distribution of genera in product	Generally 1/2 strains only	All the strains inoculated	All the strains inoculated
Quantity of inoculum per hectare	10 kg	1 kg	1 kg
Transportation	Expensive due to bulky nature	Less expensive, light weight	Less expensive, light weight
Storage	More space	Less space	Less space
Shelf life	More than 12 months	More than 12 months	More than 12 months
Benefits	Rural oriented technology, can be undertaken by farmers and can be scaled up/ down as per needs	Contaminant-free inoculants with high titre; Requires simple infrastructure and suitable for commercialization	Contaminant-free inoculants with high titre; Requires simple infrastructure and suitable for commercialization

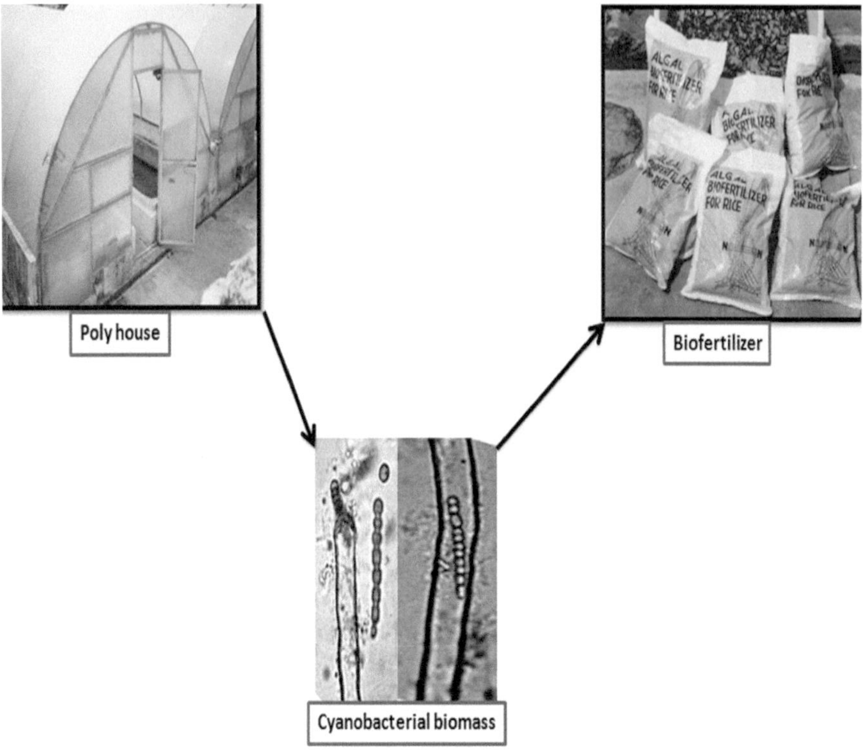

Fig. 16.1 Strategy for biofertilizer production

16.4 Cyanobacteria and Soil Fertility with Emphasis to Biological Nitrogen Fixation (BNF)

The colonization of plants by cyanobacteria is known to trigger the release of an assortment of biologically active metabolites in rhizosphere, which elicit induced systemic responses in plant, and enhance plant growth, even under conditions of abiotic or biotic stress (Prasanna et al., 2013a, b, c). The employment of cyanobacteria as a matrix for developing biofilmed biofertilizers is an economically appealing option (Prasanna et al., 2011a, b), providing multiple benefits, especially in terms of nutrient savings. Most cyanobacteria show independence for carbon and nitrogen and can support useful heterotrophic bacteria and fungi as consortia or biofilms. Carrier-based preparations using these biofilms have been favourable as PGP inoculants for wheat crop, tomato and legumes (Prasanna et al., 2013b, c; Swarnalakshmi et al., 2013), but need to be explored in other crops.

Cyanobacteria have been mainly employed as inoculants for rice (Prasanna et al., 2012), and recently, in wheat and other crops, as both consortia or biofilms (Chaudhary et al., 2012; Karthikeyan et al., 2007; Manjunath et al., 2011; Nain et al., 2010; Prasanna et al., 2013b; Rana et al., 2012). They can grow as biofilms in soil and aquatic bodies, playing a vital role in soil aggregation and metal binding/biosorption (Caire et al., 1990). Biofilms as inoculants attempt to replicate microbial communities, and are attracting attention because they can colonise and endure in diverse niches (Prasanna et al., 2013a, b; Seneviratne et al., 2009; Swarnalakshmi et al., 2013). Seneviratne et al. (2009) found noticeable increase in the growth of rice and dry weight of plants through the inoculation of biofilmed fertilizers as compared to the conventional method of monoculture inoculation.

Ayers and Edwards (2013) have provided system for inoculating soil with microalgae. The system can be used with many different water sources to culture microalgae and form an inoculant that is added to irrigation water used for watering crops. The system provides improved crop production matrics as compared to crops grown without the microalgae-based inoculation system. The system can be integrated into existing irrigation systems to add macronutrients and micronutrients into the water, thereby, providing highly bioavailable nutrients to crops.

Soil nitrogen pool is believed to be maintained through biological nitrogen fixation (Roger and Ladha, 1992; Kundu and Ladha, 1995) and fertilizer nitrogen. Nitrogen is brought into organic farming systems through the inclusion of nitrogen fixing crop in rotation or use of biofertilizers/cyanobacteria in rice crop. Jeffries et al. (1992) suggested that soil algae play a crucial role in nutrient cycling in desert and semi-desert ecosystem, particularly in cycling of nitrogen. Consequently, nitrogen balance studies under such systems are usually positive (Nguyen et al., 1995). A blue green algal bloom usually corresponds to less than 10 kg N ha^{-1} and a dense bloom may contain 10–20 kg N ha^{-1} (Roger, 1991).

Active extracellular release of fixed algal nitrogen and N release upon decomposition have been projected as likely sources of N for higher plants (Mayland and McIntosh, 1966; Stewart, 1967; Jones and Stewart, 1969). It is still uncertain whether extracellular products are taken up directly or after microbial degradation (Jones and Wilson, 1978). It has been reported that only one third of field population of cyanobacteria was decomposed and absorbed by the rice crop in the first year and the rest remained as residual soil nitrogen (Saha and Mandal, 1980). Decomposability of *Anabaena* in soil is reported to be faster than other cyanobacteria and algal biomass rich in akinetes does not degrade easily as compared to other vegetative cells (Mandal et al., 1999).

Lot of information has been generated in tropics regarding improvement in the fertility status of rice soils to sustain rice yields by utilizing diazotrophic cyanobacteria as the biological input (De and Sulaiman, 1950; Venkataraman, 1972, 1981; Fogg et al., 1973; Singh and Bisoyi, 1989). These organisms play a positive role in the sustenance of the nitrogen status of rice fields and enhancement in the soil available nitrogen (De and Mandal, 1956; Singh, 1961, 1978; Venkataraman, 1975; Saha et al., 1982; Goyal, 1989; Santra, 1991, 1993) and a considerable build up of nitrogen fertility in rice soil has been reported (Roger and Kulasooriya, 1980; Saha and Mandal, 1980; Roger and Reynaud, 1982). These were considered to be of great consequence in the N economy of rice cultivation under water logged conditions (Mahajan et al., 2003). Extensive literature is available on the ability of many species of cyanobacteria to fix N_2 and its subsequent addition to the ecosystem (Roger and Watanabe, 1986). Saha and Mandal (1979) isolated some species of cyanobacteria from alluvial low-land rice soils of West Bengal (India), which showed strong nitrogen fixing capacity in liquid culture medium. Subsequently, it was observed that inoculation of these species in soils could bring about considerable increase in the nitrogen content of soils through fixation.

Judicious use of the cultures of cyanobacteria as a biofertilizer could provide the country's entire rice acreage as much N as obtained from 15–17 lakh tonnes of urea and at farm level, it can contribute about 25–30 kgN ha^{-1} (Hegde and Dwivedi, 1993). Cyanobacterial biofertilizer is recommended only as a supplement to nitrogenous fertilizers and the supplementation effect may remain perceptible even in the presence of high levels of fertilizer nitrogen (Venkataraman and Goyal, 1969). Pronounced additive effect of algal application at lower levels of fertilizer nitrogen becomes important in extensive agriculture (Goyal, 1982, 1989) which envisages use of less fertilizer nitrogen on larger areas for reducing loss of fertilizer nitrogen and ensuring maximum utilization of the natural process. Multilocational trials conducted under varying agroclimatic conditions using different rice varieties indicated that algal inoculation can result in an addition of 30 kg N ha^{-1}. This, however, depends upon agroecological condition, which would regulate the activity and establishment of introduced algae (Venkataraman and Goyal, 1969; Venkataraman, 1979), though Roger and Kulasooriya (1980) and Singh and Singh (1987) recorded 30 kg N ha^{-1} $year^{-1}$ as a satisfactory value when environmental factors are favourable. The use of blue green algal biofertilizer is also reported to contribute about 25–50 kg N ha^{-1} to rice crop (Singh, 1961; Venkataraman, 1972; Singh 1989; Singh and Singh, 1992; Adil and Karte, 1992; Venkataraman and Shanmugasundaram, 1992). Experiments conducted at CRRI, Cuttack indicated that inoculation of soil with *Aulosira* sp. at the rate of 60 kg ha^{-1} (fresh weight) registered significant changes of soil nitrogen content. Cyanobacteria incorporated to soil increased 13–14 % of N content under field conditions and cyanobacteria amended soil released 50 % of ammonium N at 50 days of flooding (Singh et al., 1981). The rate of N released by cyanobacteria was 12 and 35 % after 7 and 35 days of flooding (Saha et al., 1982). Ghosh and Saha (1997) also reported that the inoculation of soil with soil based mixed culture of four diazotrophic cyanobacteria namely *Aulosira fertilissima, Nostoc muscorum, Nostoc commune* and *Anabaena* species radically increased the release of inorganic nitrogen in soil. Nitrogen content of soil was higher in exposed light incubated soil than unexposed soil due to N gain by cyanobacteria (Singh and Singh, 1987).

Chopra and Dube (1971) reported that the pots inoculated with *Tolypothrix tenuis* showed appreciable increase in total and organic nitrogen. Release of nitrogen from rapid decomposition of fresh or dry mass incorporated into the soil has been reported (Saha et al., 1982; Tirol et al., 1982; Miam and Stewart, 1985). Rao and Burns (1990) investigated the influence of cyanobacterial inoculation in reconstituted flooded soil columns and observed a significant enhancement in soil aggregation, microbial activity and microbial biomass. Inoculation with *Nostoc muscorum* also had a marked effect on soil chemical and biological properties, with total C increasing by 50–63 % and total N increasing by 111–120 %. Increases in the soil indigenous microbial population were recorded, with numbers of bacteria 500, fungi 16, and actinomycetes 48 times the non-inoculated values on day 300 in the high-rate soil (Rogers and Burns, 1994). Inoculation with *Anabaena variabilis, Aulosira fertilissima, Nostoc muscorum* and *Tolypothrix tenuis* and in combination recorded significant enhancement in soil microbial biomass carbon and available nitrogen, besides improving the N uptake and carbon content of rice plants. Analyses revealed a positive correlation between available nitrogen and soil microbial biomass carbon with N uptake and C content in plants (Swarnalakshmi et al., 2007a, b).

Field experiments were conducted at IARI, New Delhi to compare the efficiency of the three carrier based cyanobacterial biofertilizers (wheat straw, multani mitti and the traditional soil based cyanobacterial biofertilizer) on the grain yield of rice for a period of three years (Dhar et al., 2007). Treatments included five levels of nitrogenous fertilizer urea and their interaction with the three types of cyanobacteria biofertilizer on the grain yield of rice variety PNR 381. Highest grain yields were achieved with the application of multani mitti based biofertilizer along with 90 kg N/ha, although maximum percent increase in yield over control (37.97 %) was recorded, when applied along with 60 kg N/ha. The straw based and soil based biofertilizer treatments demonstrated highest yields when supplemented with 90 and 120 kg N/ha respectively. This study emphasized the utility of the multani mitti based and straw based cyanobacterial biofertilizers in rice cultivation for maximizing crop productivity, reducing inputs of chemical fertilizers and sustaining soil fertility.

Selection of promising strains which establish and proliferate in soil and promote the growth of crop and soil micro flora, leading to enhanced soil fertility and crop yields is an ongoing process for developing effective inoculants. Assessment of population dynamics of cyanobacterial inoculants in pot experiments revealed that several strains established in soil and persisted up to harvest stage of rice crop in soil and roots (Prasanna et al., 2009b). Analyses of harvest stages samples revealed a significant enhancement in soil microbial biomass carbon, available N and related soil microbiological parameters and increase in grain yields and 1000-grain weight as a result of inoculation of selected cyanobacterial strains.

Vertical migration of burrowing invertebrates promotes losses of NH_3 by increasing the rate of diffusion from reduced to oxidizable layers (Fenchel and Blackburn, 1979). Consequently, greater losses of mineralized N are likely from tubificid presence with both surface applied and buried algal treatments, as a result of NH_3 volatilization and nitrification-denitrification in the former case and primarily by NH_3 diffusion in the latter. In the absence of tubificids, a significantly larger proportion of algal N from the soil N pool was therefore utilized by the rice. The amount of algal N utilized was related to the rate of algal death and decomposition which is quicker in an anaerobic than aerobic environment (Jones and Wilson, 1978).

Incorporation of fresh organic materials into the soil offers an energy source for non-rhizospheric, nitrogen fixing bacteria normally limited by carbon and O_2 supply (Yoshida and Rinaudo, 1982) and it stimulates N_2 fixation in paddy soil and tubificid activity through irrigation of its burrows (Matsuguchi, 1979). Promotion of O_2 and N_2 diffusion downwards could intensify the micro-aerobic environment suitable for

N_2 fixation, which is otherwise largely restricted to the rice rhizosphere. Unlike the dry land soils of Wilson et al. (1980), those of Tirol et al. (1982) taken from flooded rice fields were neither dried nor sieved and would have contained a tubificid population. With a C:N ratio of approx 5, cyanobacteria is rapidly mineralized (Watanabe and Kiyohara, 1960) despite some species being more resistant to decay (Gunnison and Alexander, 1975).

16.5 Nitrogen Fixing Potential of Cyanobacteria—Tools and Analyses

Nitrogen fixation is carried out in specialized cells known as heterocysts in cyanobacteria, which have thick walls that offer necessary anaerobic conditions for the activity of nitrogenase enzyme. These cells have a high rate of respiration that scavenges the diffused oxygen, and they lack photosystem II, because of which there is no splitting of water and evolution of oxygen during photosynthesis (Thomas, 1970). Heinze (1906) claimed to have obtained evidence of nitrogen fixation in a culture of *Nostoc* contaminated with a *Streptothrix*, which by itself exhibited no nitrogen fixation (Bristol and Page, 1923). Two strains of *Nostoc* isolated from a state of symbiosis grew well in a nitrogen-free solution, and from this observation Molisch (1926) concluded that they were able to fix nitrogen. Aeration with 1 % CO_2 or the addition of sucrose to the medium stimulated nitrogen fixation. Allison and Hoover (1935), using a species of *Nostoc* isolated from soil, found it to possess the capacity of fixing as much as 10–20 mg of nitrogen per 100 ml of medium containing neither fixed nitrogen nor a source of energy in 50–80 d. When supplied with a suitable source of energy such as glucose, the alga fixed nitrogen in the dark. Copeland (I932) reported the nitrogen fixing potential of *Oscillatoria princeps*, *O. formosa*, *Spirulina labyrinthiformis* and *Phormidium laminosum* from nitrogen-deficient warm springs.

Till the findings of Wyatt and Silvey (1969), only heterocystous forms were considered to be capable of fixing nitrogen. Since then, many of the non-heterocystous forms have also been reported to fix nitrogen under anaerobic or microaerophilic conditions except *Gloeocapsa*, which is an aerobic nitrogen fixer (Stewart et al., 1979). Prasanna and Kaushik (1994) evaluated the nitrogen-fixing ability in a three non-heterocystous cyanobacteria in aerobic and microaerobic (argon/hydrogen) environments. *Plectonema* sp. showed significant nitrogenase activity in both types of environments, while for *Lyngbya* sp. and *Synechocystis* sp., hydrogen enriched environment was more stimulatory. Stewart and Lex (1970) suggested that all the vegetative cells of trichomes of diazotrophic cyanobacteria possess nitrogenase but enzyme gets inactivated in the presence of oxygen. Once nitrogen is fixed, many organisms are known to incorporate ammonia into amino acids by the enzyme glutamate dehydrogenase and this enzyme has been found to be lacking or present in small amounts in cyanobacteria (Hoare et al., 1967; Pearce et al., 1969). Under the conditions of ammonia limitation, most prokaryotes use a pathway consisting of glutamine synthetase and glutamate synthase to assimilate ammonia (Miflin and Lea, 1976). In cyanobacteria, this pathway is known to be the major ammonia assimilatory route under nitrogen fixing conditions (Thomas et al., 1975; Wolk et al., 1976).

Tropical conditions ensure increased occurrence of cyanobacteria in rice field soils because high humidity, temperature and shade provided by crop canopy favour the luxuriant growth of these organisms (Roger and Reynaud, 1979). Ecological studies of cyanobacteria in submerged rice fields are often hindered by problems in methodology, primarily in estimating algal biomass (quantitatively and qualitatively) and in measuring biological nitrogen fixation by cyanobacteria and their associations. This has been quantified using several methods (Watanabe et al., 1977). The evidence for nitrogen fixation has come from long-term fertility trials (Watanabe et al., 1981), N balance studies (Ventura and Watanabe, 1983) and experiments involving acetylene reduction assay (Yoshida and Ancajes, 1971). Accurate reproducible measurement of N fixing potential of cyanobacteria as an estimate of N increase are extremely important in field studies especially in rice cultivation (Watanabe et al., 1977; Roger and Ladha, 1992; Roger, 1996). Although a number of investigations have been undertaken to evaluate the N-fixation rates (Venkataraman, 1981), very few consistent data are available and there is a definite need for *in situ* measurements of nitrogenase activity (Watanabe and Cholitkul, 1979).

Significance of phototrophic N fixation in tropical rice fields was established using acetylene reduction assay (ARA) and total N measurements (Rice and Paul, 1971). ARA is a sensitive means to detect nitrogenase activity but its accuracy for quantification studies has been much debated (Lee and Watanabe, 1977; Roger, 1996). Nitrogen fixation by cyanobacteria has been predominantly estimated and data published before 1980 varied from a few to 80 kg N ha^{-1} $crop^{-1}$ (Roger and Kulasooriya, 1980). About 200 crop cycle measurements in experimental plots at IRRI showed the activities as 0–1200 μmole C_2H_4 m^{-2} h^{-1} for daily values and 20–500 μmole C_2H_4 m^{-2} h^{-1} for average ARA during a crop cycle. Extrapolated values (assuming $C_2H_2/N_2=4$) ranged from 0.2 to 50 kg N ha^{-1} $crop^{-1}$ and averaged 20 kg in no-N control plots, 8 kg in plots with broadcast urea and 18 kg in plots where N was deep placed. ARA was negligible in 75 % of 60 plots (Roger et al., 1988) where urea was broadcast. Watanabe and Cholitkul (1979) have described different methods for field estimation of biologically fixed nitrogen and reported an addition of 18–45 kg N ha^{-1} due to the activity of these diazotrophic organisms using ARA technique. They observed that the advancement and modifications in the methodology for ARA were needed for accurate measurements. Watanabe and Cholitkul (1979) inferred that N-fixation rates associated with algae in wetland rice may not exceed 10 kg ha^{-1} where no chemical nitrogenous fertilizers had been used. Other studies (Dommergues and Rinaudo, 1979) have reflected that cyanobacteria might become dominant anytime during the cultivation cycle, exhibiting one or several peaks of ARA. Lee et al. (1977) observed that ARA in the rhizosphere of the rice cv. IR 26 increased to reach a maximum value at the heading stage, followed by a rapid decrease at 4–6 weeks after transplanting. The synthesis and activity of nitrogenase in cyanobacteria is generally inhibited at high concentrations of NH_4^+-N (>300 mg/l), especially in paddy fields where the organic matter content is extremely high and nitrogenous fertilizers are applied in large doses (Rowell et al., 1977). Also, the symbiotic association of *Azolla-Anabaena* spp. has been known to retain significant nitrogenase activity when grown with nitrate or urea (Peters and Mayne, 1974).

A modified method for measurement of nitrogen fixation by cyanobacteria and *Azolla* sp. under field conditions using intact, moist soil cores was developed by Prasanna et al. (2003) in which chemical fertilizers, cyanobacteria and *Azolla* biofertilizers had been used. This method involved collection of fresh and moist soil cores (0–30 mm) using a soil auger, incubation with 10 % acetylene in airtight glass vials under field conditions for three hours, and measurement of the ethylene produced using a gas chromatograph. Experiments carried out over a period of three consecutive years revealed that the estimates of nitrogen fixation were comparable with those of other researchers obtained through the use of soil water columns, sophisticated chambers and automated sampling devices. The earlier techniques involving soil-water-column sampling and the use of portable chambers and automated samplers were tedious and prolonged (Lee and Watanabe, 1977; Watanabe et al., 1977; Roger and Ladha, 1992).

Nitrogen losses alone cannot resolve the disparities between treatments in the recovery of algal ^{15}N by rice. Isotopic dilution with ^{14}N fixed by growing algae may lead to underestimations of algal N recovery by plants in surface applied treatments. Besides, the greater uptake of ^{15}N observed from buried alga can be a function of its proximity to the rice roots (Grant and Seegers, 1985). In absence of fertilizers, lowland rice uses NH_4-N present at the time of flooding and N mineralized from soil organic matter. Availability of nitrogen fixed by cyanobacteria to the rice plant has been shown with the help of ^{15}N studies (Reynault et al., 1975; Inubushi and Watanabe, 1986). Using ^{15}N, Mayland and McIntosh (1966) and Stewart (1967) have shown the possible contribution of blue green algal nitrogen fixation. Miam and Stewart (1985) observed that about 50 % of total N fixed by cyanobacteria is released to the surroundings. Recently, contribution of N_2 fixing cyanobacteria to rice production and availability of nitrogen using ^{15}N labelled material to obtain more direct information on the dynamics of N by rice plants has been studied (Fernandez Valiente et al., 2000) in microplot experiment. In this study, the recovery of blue green algal nitrogen was compared with the recovery of same amount of labelled ammonium sulphate under field conditions. The availability of nitrogen to the rice plant was similar to that of chemical fertilizer even at the tillering stage indicating a fast mineralization of organic nitrogen in the soil followed by a rapid and fast transfer of fixed nitrogen to rice crops. The amount of blue green algal nitrogen recovered in plants was however lower in another study (Tirol et al., 1982). Quantitative measures over longer periods of the release of ^{15}N upon mineralization of cyanobacteria and its subsequent uptake by rice plants was also given by Wilson et al. (1980). However, the recoveries of ^{15}N recorded were not consistent. The inconsistency was supposed to be due to the failure to consider the role of aquatic invertebrates in rice ecosystem.

16.6 Influence of Cyanobacteria on Crop Improvement

Smith and Crews (2014) assert that mixed-species assemblages of cyanobacteria and eukaryotic microalgae may have a number of potentially important benefits with regards to total biomass production; crop protection against grazing losses; and crop protection against disease losses. They also suggest that the role of species diversity in reducing crop losses to pathogens or herbivores in agriculture is readily translatable to microalgal cultivation. By reducing the concentration of the pathogen or herbivore host populations, and by chemically or physically disrupting mechanisms of host location with non-host species, interspecific and intraspecific diversity can play an important role in crop protection. Research directed at developing diverse, pest-resistant algal communities holds considerable promise.

Nitrogen and phosphorus are major limiting factors to plant growth and addition of fertilizers has thus, become standard to maintain the healthy growth and persistence of crops (Saleque et al., 2004; Lin et al., 2006). In intensified rice systems, N use and N uptake efficiency decreases as application of N fertilizer increases (Carreres et al., 1996). Several workers have asserted that inoculation of soil with nitrogen fixing cyanobacteria increases rice yields, although the efficiency of particular inoculums may vary with the variation of soil characteristics and agroclimatic conditions (Singh, 1961; Venkataraman, 1975; Watanabe, 1965, 1973; Yamaguchi, 1976). The supplementation of chemical fertilizer with cyanobacteria could conserve up to 30 % of commercial chemical fertilizers (Venkataraman, 1972, 1981; Nayak et al., 2004) which has been also shown in field trials undertaken in different states of India (Kaushik, 2005). In rice ecosystems, these organisms are known to fix an average of 27 kg N ha^{-1} in N-free plots (Roger and Ladha, 1992; Carreres et al., 1996) and the rate is reduced to 8 kg N ha^{-1} when urea is broadcast (Roger and Ladha, 1992). Arora et al. (1989) reported that blue green algal application to rice grown in pots could save 50 % of the recommended dose of chemical nitrogen being applied to rice. It is generally believed that the nitrogen fixed by these organisms is made available to the rice plants through exudation or autolysis and microbial decomposition (Roger and Kulasooriya, 1980; Querijero-Palacpac et al., 1990). Direct evidence of transfer of blue green algal nitrogen to rice plant is, however, scarce (Mac Rae and Castro, 1967; Roger, 1996).

The outcome of algalization on grain yield under field conditions has been reported from China (Ley, 1959), Egypt (El-Nawawy and Hamdi, 1975), Japan (Watanabe, 1965) and India (Jha et al., 1965; Venkataraman, 1972, 1979, 1981; Singh, 1978, 1985, 1988; Chopra et al., 1985; Kaushik, 1985). The positive effect of blue green algal inoculation on growth and grain yield was attributed to the availability of fixed N by these organisms (Venkataraman, 1972, 1981; Roger and Kulasooriya, 1980; Querijero-Palacpac et al., 1990). Similar results regarding influence of cyanobacteria on grain yield have been reported by others (Sprent and Sprent, 1990; Yanni, 1992) and have been attributed to the favourable conditions for biological nitrogen fixation by cyanobacteria under flooded conditions of rice fields. The results as reviewed by Roger and Kulasooriya (1980) revealed an average increase of 14 % in rice yield over control which was equivalent to the application of 25–30 kg N ha^{-1} as biofertilizer. The potential for practical use of cyanobacteria as biofertilizer in rice farming was extensively discussed by Roger and Watanabe (1986). However, their inoculation studies failed to increase rice yields consistently (Watanabe, 1986; Roger et al., 1993).

Field experiments at Cuttack during 1978–79 indicated that mixed inoculation of *Aulosira, Aphonothece* and *Gloeotrichia* increased grain yield by 10, 17, 32 and 35 % during first, second, third and fourth crop (Singh, 1978). Singh and Singh (1987) also reported an increase of 15–23 % grain yield by cyanobacteria inoculation. Under pot culture conditions, unialgal and composite cultures of cyanobacteria were able to increase the rice yield considerably as compared to control uninoculated conditions (Dhaliwal et al., 1995). Increased rice yield due to algal inoculation have also been reported under control pot culture and field experiments by other workers (Singh, 1961; Relwani, 1963; Sundara Rao et al., 1963; Venkataraman and Goyal 1968). However, the amount of nitrogen fixed by cyanobacteria and its effect on growth and yield of rice crops varies with different doses of nitrogenous fertilizer (Venkataraman, 1972). Pandey et al. (1993) reported that the combination of 10 kg ha^{-1} cyanobacteria biofertilizer with 90 kg nitrogen as urea ha^{-1} resulted in better growth and maximum yield of the rice crop. Inoculation with soil based mixture of four heterocystous species led to a trivial increase in grain (8 %) and straw (11 %) yield which was, however, accompanied by significant increase in nitrogen uptake by the grain (30 %) and an increase in total uptake of 15.3 kg N ha^{-1} (Ghosh and Saha, 1997).

A field experiment conducted on loamy soil at Bilaspur, Madhya Pradesh (India) in rice-wheat sequence revealed that the rice yields under treatments of cyanobacteria (soil based) at the rate of 12 kg ha^{-1} and FYM at the rate of 5 t ha^{-1} were statistically *at par* with each other. Organic manures and cyanobacteria combined with 60 kg N + 37.5 kg P_2O_5 + 22.5 kg K_2O ha^{-1} as chemical fertilizers (Rathore et al., 1995) proved superior to other treatments in rice yields and their residual effect on wheat yields. Ahmed (2001) reported that the deployment of cyanobacterial strains together with chemical nitrogen fertilizers increased grain yield of rice in the fields of Nagaon-Sub division of Assam (India). It has also been shown that the combined application of cyanobacteria and N fertilizer is more effective in increasing the number of tillers and crop yield than the application of cyanobacteria alone in rice fields (Aiyer et al., 1972; Roychoudhury et al., 1983; Patel et al., 1984). Sometimes, the effect of cyanobacterial inocula on the yield of crops in the presence of N fertilizer has been ascribed to production of growth promoting substances which can accelerate root growth (Brown et al., 1956; Venkataraman and Neelakantan, 1967; Kopteva, 1970; Tupik, 1973). This in turn enables the crop plant to take up more nitrogen from the soil (Sundara Rao et al., 1963). N fertilization along with the algalization stimulated plant growth and produced more photosynthetic area, which probably helped in increasing the crop yield (Watanabe et al., 1951; Subramanyan and Manna, 1966; Jalapathi et al., 1977; Yadav et al., 1988). The application of cyanobacteria enhanced the aggregation status of rice field soils under different levels of irrigation and nitrogen fertilizer (urea) and increased the grain yield significantly (Roychoudhury et al., 1983).

Despite several studies on the rice-wheat cropping system, concerted efforts to evaluate the role of cyanobacteria on wheat crop are limited (Kaushik, 2005). At IARI, a survey of wheat fields revealed the presence of a number of heterocystous genera including *Calothrix* and *Westiellopsis* besides *Anabaena* and *Nostoc* (Prasanna et al., 2008a, b; Jaiswal et al., 2008; Karthikeyan et al., 2009) which were also prevalent in the rhizosphere/roots. Such strains exhibited the ability to fix nitrogen, secrete IAA and a number of amino acids and sugars. Pot experiments undertaken with a set of such strains using unsterile soil (in glass house conditions) and sterile soil (controlled conditions of National Phytotron Facility) revealed their promise as diazotrophs and plant growth promoting inoculants. Significant enhancement in soil microbiological parameters was also recorded revealing their positive interactions with the microflora (Karthikeyan et al., 2007). Further, studies were undertaken to screen, select and evaluate the interactive effect of a combination of bacterial and cyanobacterial isolates from the wheat rhizosphere for their role as biofertilizers in wheat (Nain et al., 2010). Significant enhancement in the soil microbiological (dehydrogenase activity, FDA, alkaline phosphatase and microbial biomass carbon) and plant growth/yield parameters were obtained in selected combinations of bacteria and cyanobacteria. Observations also revealed a two-fold increase in panicle weight in selected combinations, revealing the synergistic effects of these microbes on soil fertility and crop yields and their promise in integrated nutrient management of wheat crop.

In addition, cyanobacteria have also been recognized as important agents in the stabilization of soil surfaces (Bailey et al., 1973) primarily through the production of extracellular polysaccharides, which are chief agents of aggregate formation, and stabilization (Molope et al., 1985; Burns and Davies, 1986). The significance of polysaccharides in soil aggregation may be the direct result of binding soil particles into micro aggregates (Cheshire et al., 1983, 1984), although Tisdall and Oades (1982) considered polysaccharides as only transient adhesives. Aiyer et al. (1972) reported a considerable reduction in the oxidisable organic matter, total S^{2-} and Fe^{2+} content of soil after successive rice cropping along with cyanobacterial inoculation.

16.7 Cyanobacterial Mediated Bioremediation

Application of cyanobacteria to the salt affected soils has led to the enrichment of soils with nitrogen content (Kaushik, 1985, 1994). The increase in soil nitrogen varied from 17.9 to 28.5 % over the initial soil nitrogen (Kaushik, 1985), with concomitant enhancement in grain yields. Requirement of gypsum for amelioration of sodic soils was reported to reduce when Cyanobacteria inoculation was used in paddy fields (Kaushik, 1994).The availability of the phosphorus ranged 68–100 % in soils of Andhra Pradesh and about 11 % increase in available phosphorus in saline-sodic soils of Delhi (Kaushik, 1994). Repeated cultivation of *Anabaena torulosa* in saline soils (as a result of bad farm management practices) led to a significant reduction (Thomas, 1997) in soil salinity (12–35 %). The rice yield in saline soils is also affected when N is supplemented in the form of algal inocula vis-à-vis urea (Antarikanonda and Amarit, 1991). Successive cultivation of cyanobacteria makes the environment more favourable for the crop cultivation and after a few years it may help to produce a reasonably good yield of crops, as observed by Singh (1961) for sugarcane after three years of reclamation with cyanobacteria. Although such biological soil amelioration is a tedious process, it is considered to be a sustainable approach as compared to reclamation by chemical amendments (Olkarinen, 1996). Coal-based power plants generate a variety of pollutants along with a huge quantity of fly-ash (FA) that is usually dumped in nearby areas. According to some estimates, in areas situated close to National Thermal Power Plant (NTPC), Unchahar, Raebareli (UP), FA is deposited at a considerable rate (Tripathi et al., 2008). Increasing pollution of land by different contaminants including FA is reducing the cultivable area, thereby affecting the growth and yield of the rice plants (Dwivedi et al., 2007). Moreover, the threats associated with accumulation of heavy metals in rice grains are a major concern to humans (Meharg, 2004; Tripathi et al., 2007). Fly-ash is often used as soil modifiers (Sikka and Kansal, 1995; Gupta et al., 2002; Tripathi et al., 2004; Mittra et al., 2005; Jala and Goyal, 2006) due to its beneficial properties. Fly-ash is rich in boron and deficient in nitrogen. Boron has been verified to be essential for nitrogen fixation by heterocystous cyanobacteria strains such as *Anabaena* (Mateo et al., 1986; Blevins and Lukaszewski, 1998). Therefore, cyanobacteria appear to be a likely candidate for improving nitrogen status in FA contaminated paddy fields; however, long term trials need to be undertaken in order to understand the metal accumulation in soil and plant. Addition of FA at lower doses also enhances physical, chemical and biological properties of soil and has been demonstrated to result in increased growth of a number of plants (Gupta et al., 2002). However, its usage in agriculture and agronomy sector is still limited (<10 %) due to issues about the presence of toxic elements viz., Cd, As and Ni (Carlson and Adriano, 1993; Gupta et al., 2002; Jala and Goyal, 2006).

Recent investigations suggest that FA can find better application if combined with organic supplements, nitrogenous fertilizers (NF) and blue green algal biofertilizer (Rautaray et al., 2003; Tripathi et al., 2004; Rai et al., 2004). In this context a study was undertaken to evaluate the application of cyanobacterial biofertilizer and recommended/modified dose of nitrogenous fertilizer for safe utilization of FA in paddy cultivation and to develop an integrated technology for the farmers to cultivate rice crops in FA affected areas. The field experiments were conducted to analyze the result of different doses of FA @ 10 and 100 t ha^{-1} denoted as FA_{10} and FA_{100} mixed with garden soil (GS) respectively with and without cyanobacterial biofertilizer (@ 12.5 kg ha^{-1} as cyanobacteria) and NF @ 90 and 120 kg ha^{-1} (NF_{90} and NF_{120}) on growth, yield, phytotoxic and stress tolerance responses of rice (*Oryza sativa* L.) var. Saryu-52. Metal composition (Fe, Si, Zn, Mn, Cu, Ni, Cd and As) of various plant tissues (roots, leaves, seed husk and grain) was also investigated for securing health safety related to rice consumption. Significant enhancement of growth was observed in the plants growing on modified soils as compared to GS and best response was obtained with $FA_{10}+NF_{90}$+cyanobacteria. Arsenic accumulation was detected only in FA_{100} and its amendments. Inoculation of cyanobacteria caused slight reduction in Cd, Ni and As content of plants as compared to NF_{120} amendment. Cyanobacteria have a mucilaginous sheath and can show significant adsorption of the metals like Cd and metalloids like As (Tien, 2002) in addition to their absorption, thus, reducing the toxic effects of these metals (Rai et al., 2004). The high levels of stress inducible non-protein thiols (NP-SH) and cysteine in FA_{100} were reduced by application of NF and Cyanobacteria indicating stress amelioration. Increase in NP-SH content is suggestive of increase in thiols *viz*., cysteine, glutathione and phytochelatins (PCs). The levels of both cysteine and NP-SH was found to be maximum in FA_{100}, which decreased in $FA_{100}+NF_{90}$+cyanobacteria. This suggested that stress caused by accumulation of Cd and As was diminished by application of cyanobacteria, which also improved the growth of plants. Thiolic compounds show positive correlation with accumulation of metals. These thiols play a key role in the maintenance of redox status of the cell as well as in the binding of metal ions for their detoxification. This enhancement is credited to an increase in the whole S assimilation pathway (Rausch and Wachter, 2005; Herbette et al., 2006). This study recommended the integrated use of FA, cyanobacteria and NF for improved growth, yield and mineral composition of the rice plants besides reducing the high demand of nitrogen fertilizers (Tripathi et al., 2008). Application of cyanobacteria also improved N status of paddy soil through fixation of atmospheric nitrogen and NF requirement was reduced to 90 kg ha^{-1} from 120 kg ha^{-1} when cyanobacteria was applied emphasizing the use of cyanobacteria as an economical strategy in paddy cultivation. Besides enhancing the growth, it also augmented composition of essential micronutrients while the level of toxic metals was maintained under safe limit in FA amendments. It may be prudent to cultivate Saryu-52 variety of rice in FA contaminated areas with balanced doses of cyanobacteria with low exogenous NF input for improved yield and grain fortification with trace elements.

16.8 Role of Micro and Macrofauna vis a vis Cyanobacteria in Rice Fields

It is established that cyanobacteria are consumed by grazers and the development of zooplankton populations, especially those of cladocerans, copepods, ostracods, and mosquito larvae prevent the establishment of algal blooms in agriculture within 1–2 weeks (Venkataraman, 1961). A general trend among N_2-fixing cyanobacteria is that the strains forming mucilaginous colonies (such as *Nostoc* and *Gloeotrichia*) are less efficient N fixers and less susceptible to grazing than the non-colonial strains (Antarikanonda and Lorenzen, 1982; Grant et al., 1985). Plate counts indicated the preponderance of grazer-resistant mucilaginous cyanobacteria in 90 % of 102 soils studies (Roger et al., 1987). The application of neem cake or neem oil can offer a cheap means to control micro-crustaceans and snails that graze on cyanobacteria (Roger and Kulasooriya, 1980; Grant et al., 1985). *Cypris* sp., when present in large numbers, prevented the development of *Tolypothrix tenuis* inocula and suppressed the growth of inocula of *Anabaena* sp. added to flooded soil. The waterflea *Daphnia magna,* when present in large numbers, had only a minor effect on the growth rate and extent of development of algae. *Anabaena* sp. was less susceptible compared to *T. tenuis* to two microcrustaceans (Wilson et al., 1980). Therefore, it can be stated that there exists a close relationship between cyanobacteria and other organisms in flooded rice-field ecosystems.

Uptake of blue green algal nitrogen (N) and total N uptake by low land rice (*Oryza sativa*) was influenced by tubificid (Oligochaeta) presence in submerged soils. Recovery of algal ^{15}N by the first crop was 24–43 % but only 4–7 % for the second. Tubificid activities reduced the recovery of algal N by rice, increased its total N content and doubled losses of ^{15}N to the atmosphere. Soil N mineralization, measured as NH_4^+ production was doubled over seven days by their activities and algal mineralization was also enhanced. The NH_4^+ release of *Limnodrilus* sp. was 4.11 ± 0.06 ng NH_4^+-N mg ash free dry wt^{-1} h^{-1} (Grant and Seegers, 1985).

As part of an integrated pest management project to examine the role of blue-green algae (cyanobacteria) in the food web of rice-field ecosystems, ^{14}C labelled filamentous and mono-cellular cyanobacteria were used as food for fish, zooplankton and benthic fauna in artificial rice fields. It was ascertained that fish consumed the cyanobacteria at the fastest rates and in the largest amounts, followed by the benthic species and zooplankton. It was also found that filamentous cyanobacteria were consumed in higher amounts than monocellular cyanobacteria. The importance of grazing in nutrient recycling was highlighted (Rakshit et al., 1999) and the results sustain the theory mentioned earlier that grazing leads to the dominance of mucilaginous blue green algal strains which are usually less active in biological N fixation (Antarikanonda and Lorenzen, 1982; Grant et al., 1985). Nutrient recycling in rice-field floodwater is brought about by microorganisms, zooplankton, benthic and certain other invertebrate groups (Roger, 1996).

The results of a number of studies (Ganf and Blazka, 1974; Gardner and Miller, 1981; Chaturvedi and Agarwal, 1983; Sherr et al., 1983) have indicated that NH_4^+ is generally the excretion product of these groups, but small amounts of urea is sometimes detected. Quantitative data of Roger et al. (1987) and Osa-Afiana and Alexander (1981) indicated that rapid algal successions, recurrent at the beginning of the crop cycle, may indicate a rapid turnover rate of N and P pools. Without recycling by fauna, N fixed in cyanobacteria will gather at the soil surface and partly be lost by volatilization to ammonia. Also, without faunal activity, the N from cyanobacteria will become available to the plant late in the crop cycle and, thus, may have little impact on the yield of the current crop. Fish have been reported to have a number of beneficial effects in rice fields, including an increase in the rate of nutrient cycling and nutrient availability resulting in increased N uptake and higher rice yields (Lightfoot et al., 1990). Sevilleja and Cagauan (1992) attributed a pH near neutral levels and a subsequent reduction in N loss through volatilization to grazing by fish.

16.9 Genomics and Proteomics of Nitrogen Fixation

The dichotomy in *nif* gene organization in cyanobacteria, with respect to non-heterocystous and heterocystous forms, and in heterocyst and vegetative cells of heterocystous forms has been a subject of intensive research globally and in India at IARI, New Delhi and BARC, Trombay. The heterocystous cyanobacteria are known to exhibit a rearrangement of their structural genes for nitrogenase- *nif* H, D and K in the vegetative cells when forming heterocysts. On the other hand, non-heterocystous forms exhibit a similar contiguous cluster of *nif* H, D and K genes in all their cells. Using *xis* A (encodes gene involved in excision of 11 kb element between *nif* K and DH) as a marker, a set of branched and unbranched filamentous heterocystous and non-heterocystous cyanobacteria from the germplasm of CCUCyanobacteria, IARI, New Delhi were analyzed for the organization of *nif* HDK gene operon (Prasanna et al., 1993; Prasanna and Kaushik, 1995). Hybridization was not detected in *Mastigocladus* and *Scytonematopsis* sp. indicative of no rearrangement of *nif* HDK genes in these organisms, as reported earlier for another branched form—*Fischerella* sp. Interestingly, falsely branched heterocystous members, *Tolypothrix* sp. and *Scytonema*, showed hybridization. Among non-heterocystous forms, for the first time reports were generated in filamentous forms—*Lyngbya* and *Plectonema* sp. regarding the presence of *xis* A gene, while unicellular genus *Synechocystis* sp. did not exhibit hybridization.

The filamentous non-heterocystous strain *Plectonema boryanum* (UTEX 544) is known to exhibit diazotrophy when incubated in an anaerobic environment, in the presence of light (Misra and Tuli, 1994). Analyses of expression of genes involved in photosynthesis and nitrogen fixation revealed a differential pattern in the two phases. The onset of nitrogen fixation was preceded by a depression in photosynthesis, accompanied by a several-fold reduction in the level of *cpc*BA (encodes phycocyanin) and *psb* A and C, although *psb* C and D were not co-transcribed. In the nitrogen fixing phase, a dramatic increase in the level of *nif* H transcripts was recorded; however, levels of *gln* A were not altered. Reciprocal regulation of gene expression was observed to be well orchestrated with alternating cycles photosynthesis and nitrogen fixation in *Plectonema boryanum* (Misra and Tuli, 2000).

Cyanobacteria, with their ancient ancestry and ubiquity, exhibit in general, considerable tolerance to abiotic stress. In-depth analyses have been done in this context, regarding the gene regulation, protein expression and their relationship to nitrogen fixation by Apte and co-workers from BARC, Trombay. The genetic mechanisms involved are of great interest as cyanobacteria can provide, not only suitable options for reclamation of such soils for agriculture, but also serve as a valuable source of genes for developing transgenics.

Soil salinity is a principal deterrent to plant growth and a definite exists to employ biological strategies to overcome this problem and convert such inhospitable barren habitats to cultivable land. The response of cyanobacterial protein synthesis to salinity has been investigated by Apte and co-workers (Apte et al., 1987; Apte and Thomas, 1983, 1984, 1985; David and Thomas, 1979; Fernandes et al., 1993; Reddy et al., 1989; Thomas and Apte, 1984). They observed that curtailment of sodium influx is a major mechanism responsible for salt tolerance in *Anabaena* strains. The presence of certain inorganic nitrogenous compounds was found to enhance the salt tolerance of freshwater and brackish water strains of *Anabaena* and the mechanism elucidated revealed the primary role of combined nitrogen in protection of cyanobacteria against salt stress, by facilitating maintenance of low intracellular concentrations of Na^+.

Salinity mediated expression of specific proteins was found to be very rapid and sensitive to rifampicin in *Anabaena* strains. Bhagwat and Apte (1989) demonstrated a commonality in the protein synthesis regulation in response to heat, salinity or osmotic stress, which may play an important role in the maintenance of vital cellular functions, including nitrogen fixation. They also identified stress specific proteins which may constitute a second level of regulation in these organisms. Employing a subtractive hybridization procedure, Apte and Haselkorn (1990) found that a substantial part of the *Anabaena torulosa* genome, probably in excess of 100 kb, responds to salt. Ballal and Apte (2005) showed that *Anabaena* sp. L-31 has two distinct *kdp* operons, involved in K^+ scavenging and transport, among which *kdp*2 was observed to be expressed more and implicated in adaptation to K^+ limitation and desiccation stress and contribute to its survival and metabolic capabilities, including nitrogen fixation in saline environments.

Alahari and Apte (1998) recorded pleiotropic effects of K^+ deficiency in *Anabaena torulosa*, revealing that K^+ plays vital specific roles in cyanobacterial growth and metabolism, including regulation of photosynthesis and nitrogen fixation.

Apte and co-workers (Rajaram and Apte, 2003, 2008; Rajaram et al., 2001) investigated the heat shock response in *Anabaena* sp. L-31 and found that Cpn60 has a major role in carbon and nitrogen metabolism in *Anabaena* sp. A differential response of the cyanobacterial culture growing under nitrogen-fixing and nitrate supplemented cultures was found to be correlated with the loss of Cpn60, inactivation of photosynthetic machinery and nitrate reduction under the latter conditions. However, the over expression of Cpn60 enhanced the thermal stability of these vital metabolic processes.

The global orchestration of genes in various metabolic processes, including biotic stress, uptake of ions etc. in cyanobacteria, thus play key roles in the process of fixation and assimilation of nitrogen from the environment.

16.10 Conclusions

Depletion of soil fertility, low fertilizer use efficiency and increasing environmental pollution are of major concern to agriculture in terms of crop productivity. Amongst the array of biofertilizers developed for different crops, cyanobacteria constitute the most important inputs in rice cultivation, which are now gaining importance in other crops including wheat. They form an inexpensive farm grown input which helps in better crop nutrient management, while working in perfect harmony with nature. There is a definite need to utilize these biofertilizers together with organic composts and minimal doses of chemical fertilizers for reaping 'cleaner' and healthy harvests and securing food production, human health, protecting the environment and saving scarce natural resources.

Future research needs to be executed for field level testing of these cyanobacterial strains which have shown tremendous prospective in terms of their plant growth promoting activity through the dissection of traits involved in plant growth promotion, use of radio-labelled compounds, fluorescence microscopy and micro level analyses of exudates involved in the "cross-talk" between cyanobacteria-other microflora and plants. In future, there is also a definite need to tap the cyanobacterial diversity in terms of their metabolites and explore them for their use novel inputs in integrated pest and nutrient management practices for environmentally safe "organic" and "green" agriculture.

Extensive field level training should form an integral component for its popularization among farmers and extension workers. More emphasis should be given on practical aspects pertaining to the method of application, economic benefits, environmental benefits, long term and short term gains. Popularization through press and media and material such as leaflets etc. in local languages should be provided. Formulation of standards and criteria for quality control are very essential which will go a long way in widening the scope and use of cyanobacterial biofertilizers.

References

Adesemoye, A.O. and Kloepper, J.W. (2009). Plant-microbes interactions in enhanced fertilizer-use efficiency. *Appl Microbiol Biotechnol*, 85, 1–12.

Ahmed, S.U. (2001). Studies on application of some potential cyanobacteria strains in rice cultivation under unsterilized soil condition in pot culture. *Phykos*, 40, 149–153.

Aiyer, R.S., Salahudeen, S. and Venkataraman, G.S. (1972). Long-term algalization field trial with high yielding varieties of rice (*Oryza sativa* L.). *Ind. J. Agr. Sci.*, 42, 380–383.

Alahari, A. and Apte, S.K. (1998). Pleiotropic effects of potassium deficiency in a heterocystous, nitrogen fixing cyanobacterium *Anabaena torulosa. Microbiology*, 144, 1557–1563.

Allison, F. E., & Hoover, S. R. (1935). Conditions which favour nitrogen fixation by a blue-green alga. *Trans. 3rd Inter. Cong. Soil Sci*, 1, 145–147.

Antarikanonda, P. and Lorenzen, H. (1982). N_2-fixing blue-green algae (Cyanobacteria) of high efficiency from paddy soils of Bangkok, Thailand: Characterization of species and N_2-fixing capacity in the laboratory. *Arch. Hydrobiol. Suppl.*, 63, 53–70.

Antarikanonda, P. and Amarit, P. (1991). Influence of blue-green algae and nitrogen fertilizer on rice yield in saline soils. *Kasetsart Journal of Natural Sciences*, 25, 18–25.

Apte, S.K. and Haselkorn, R. (1990). Cloning of salinity stress-induced genes from the salt tolerant nitrogen-fixing cyanobacterium *Anabaena torulosa. Plant Mol Biol*, 15, 723–733.

Apte, S.K. and Thomas, J. (1984). Effect of sodium on nitrogen fixation in *Anabaena torulosa* and *Plectonema boryanum. J Gen Microbiol.*, 130, 1161–1168.

Apte, S.K. and Thomas, J. (1985). Effect of sodium on membrane potential, uptake of phosphate, nucleoside phosphate pool and synthesis of expression of nitrogenase in *Anabaena torulosa. Indian J Exp Biol.*, 23, 518–522.

Apte, S.K. and Thomas, J. (1983). Impairment of photosynthesis by sodium deficiency and its relationship to nitrogen fixation in the cyanobacterium *Anabaena torulosa*. FEMS *Microbiol Lett*, 16, 153–157.

Apte, S.K., Fernandes, T., Badran, H. and Ballal, A. (1998). Expression and possible role of stress-responsive proteins in *Anabaena*. *J Biosci.*, 23, 399–406.

Apte, S.K., Reddy, B.R. and Thomas, J. (1987). Relationship between sodium influx and salt tolerance of nitrogen fixing cyanobacteria. *Appl Environ Microbiol.*, 53, 1934–1939.

Arora, J.K., Garcha, H.S., Pandher, M.S. and Gupta, R.P. (1989). Evaluation of the most prevalent cyanobacteria for N_2-fixing capacity and thermo-tolerance in Punjab. *Ann. Biol.*, 5, 77–80.

Ayers, A. D., & Edwards, M. R. (2013). *U.S. Patent Application 14/069,932*.

Bailey, D., Mazurak, A. P., & Rosowski, J. R. (1973). AGGREGATION OF SOIL PARTICLES BY ALGAE1. *Journal of Phycology*, 9(1), 99–101.

Ballal, A. and Apte, S.K. (2005). Differential expression of the two *kdp* operons in the nitrogen-fixing cyanobacterium *Anabaena* sp. strain L-31. *Applied and Environmental Microbiology*, 71(9), 5297–5303.

Bhagwat, A.A. and Apte, S.K. (1989). Comparative analysis of proteins induced by heat shock, salinity, and osmotic stress in the nitrogen-fixing cyanobacterium *Anabaena* sp. strain L-31. *J Bacteriol*, 171, 5187–5189.

Blevins, D. G., & Lukaszewski, K. M. (1998). Boron in plant structure and function. *Annual review of plant biology*, 49(1), 481–500.

Bristol, B. M., & Page, H. J. (1923). A critical enquiry into the alleged fixation of nitrogen by green algae1. *Annals of Applied Biology*, 10(3–4), 378–408.

Burns, R. G., & Davies, J. A. (1986). The microbiology of soil structure. *Biological Agriculture & Horticulture*, 3(2–3), 95–113.

Caire, G.Z., Cano, M.S., Mule, M.C.Z. and Halperin, D.R. (1990). Antimycotic products from the cyanobacterium *Nostoc muscorum* against *Rhizoctonia solani*. *Phytopathology*, 51, 1–4.

Carlson, C. L., & Adriano, D. C. (1993). Environmental impacts of coal combustion residues. *Journal of Environmental quality*, 22(2), 227–247.

Cassman, K.G. and Pingali, P.L. (1994). Extrapolating trends from long-term experiments to farmers fields: The case of irrigated rice systems in Asia. *In:* Agricultural sustainability in economic environmental and statistical considerations. Barnett, V., Payne, R. and Steiner, R. (Eds) Wiley, New York. pp. 63–84.

Chaturvedi, M. L., & Agarwal, R. A. (1983). Ammonia excretion in snails Viviparus bengalensis (LAMARCK) and Pila globosa (SWAINSON) during active and dormant periods. *Internationale Revue der gesamten Hydrobiologie und Hydrographie*, 68(4), 599–602.

Chaudhary, V., Prasanna, R., Nain, L., Dubey, S.C., Gupta, V., Singh, R., Jaggi, S. and Bhatnagar, A.K. (2012). Bioefficacy of novel cyanobacteria-amended formulations in suppressing damping off disease in tomato seedlings. *World Journal of Microbiology and Biotechnology*, 28, 3301–3310.

Cheshire, M.V., Sparling, G.P. and Mundie, C.M., 1983. Effect of periodate treatment of soil on carbohydrate constituents and soil aggregation. J. Soil Sci., 34:105–112.

Cheshire, M.V., Sparling, G.P. and Mundie, C.M., 1984. Influence of soil type, crop and air drying on residual carbohydrate content and aggregate stability after treatment with periodate and tetraborate. Plant Soil, 76: 339–347.

Chopra, T.S. and Dube, J.N. (1971). Changes of nitrogen content of rice soil inoculated with *Tolypothrix tenuis*. *Plant Soil*, 35, 453–462.

David, K.A.V. and Thomas, J. (1979). Extracellular peptides of *Anabaena* L-31: Evidence of their role in regulation of heterocyst formation . *J Biosci*, 1, 447–445.

De Datta, S.K. (1987). Nitrogen transformation processes in relation to improved cultural practices for lowland rice. *Plant Soil*, 100, 47–69.

De, P.K. (1936). The problem of the nitrogen supply of rice. I. Fixation of nitrogen in the rice soil under waterlogged condition. *Indian J. Agric. Sci.*, 6, 1237–1242.

De, P.K. (1939). The role of blue-green algae in nitrogen fixation in rice fields. *Proc. Roy. Soc. London*, 127B, 129–139.

De, P.K. and Mandal, L.N. (1956). Fixation of nitrogen by algae in rice soils. *Soil Sci.*, 81, 453–458.

De, P.K. and Sulaiman, M. (1950). The influence of algal growth in the yield of crops. *Ind. J. Agric. Sci.*, 20, 327–342.

Dhaliwal, M.K., Pandher, M.S., Gupta, R.P., Garcha, H.S. and Gagneja, M.R. (1995). Effect of chemical nitrogen on the growth and nitrogen fixation by blue-green algae in Basmati rice. *Indian J. Ecol.*, 22(1), 7–10.

Dhar, D.W., Prasanna, R. and Singh, B.V. (2007). Comparative performance of three carrier based blue green algal biofertilizers for sustainable rice cultivation. *J. Sust .Agric*, 30(2), 41–50.

Dommergues, Y. R. and Rinaudo, G. (1979). Factors affecting N_2 fixation in the rice rhizosphere. *In:* Nitrogen and Rice Philippines, IRRI. pp. 241–260.

Dwivedi, S., Tripathi, R.D., Srivastava, S., Mishra, S., Shukla, M.K.,Tiwari, K.K., Singh, R. and Rai, U.N. (2007). Growth performance and biochemical responses of three rice (Oryza sativa L.) cultivars grown in fly-ash amended soil. *Chemosphere*, 67, 140–151.

El-Nawawy, A.S. and Hamdi, Y.A. (1975). Research on blue-green algae in Egypt, 1958–1972. In: *Nitrogen fixation by free living microorganisms*. Stewart, W.D.P. (ed.). Cambridge Univ. Press, London. pp. 219–228.

Fenchel, T., & Blackburn, T. H. (1979). *Bacteria and mineral cycling*. Academic Press, Inc. (London) Ltd.

Fernandes, T.A., Iyer, V. and Apte , S.K. (1993). Differential effects of salt and osmotic stress on growth and nitrogen fixation in *Anabaena* sp. strain L-31. *Appl Environ Microbiol*, 59, 899–904.

Fernandez Valiente, E., Ucha, A., Quesada, A., Leganes, F. and Carreres, R. (2000). Contribution of N_2 fixing cyanobacteria to rice production: Availability of nitrogen from ^{15}N-labelled cyanobacteria and ammonium sulphate to rice. *Plant Soil*, 221, 107–112.

Fogg, G.E., Stewart, W.D.P., Fay, P. and Walsby, A.E. (1973). The blue-green algae. Academic Press, London.

Fujishige, N.A., Kapadia, N.N., DeHoff, P.L., Hirsch, A.M. (2006). Investigations of *Rhizobium* biofilm formation. *FEMS Microbiol. Ecol.*, 56, 195–206.

Ganf, G. G., & Blazka, P. (1974). Oxygen uptake, ammonia and phosphate excretion by zooplankton of a shallow equatorial lake (Lake George, Uganda).*Limnol. Oceanogr*, 19(2), 313–325.

Gardner, W. S., & Miller III, W. H. (1981). Intracellular composition and net release rates of free amino acids in Daphnia magna. *Canadian Journal of Fisheries and Aquatic Sciences*, 38(2), 157–162.

Ghosh, T.K. and Saha, K.C. (1993). Effects of inoculation with N_2-fixing cyanobacteria on the nitrogenase activity in soil and rhizosphere wetland rice (*Oryza sativa* L.). *Biol. Fertil. Soils*, 16, 16–20.

Ghosh, T.K. and Saha, K.C. (1997). Effects of inoculation of cyanobacteria on nitrogen status and nutrition of rice (*Oryza sativa* L.) in an Entisol amended with chemical and organic sources of nitrogen. *Biol. Fertil. Soils*, 24, 123–128.

Glick, B.R., Patten, C.L., Holguim, G. and Penrose, D.M. (1999). Biochemical and Genetic Mechanisms used by Plant Growth Promoting Bacteria. Imperial College Press, London. pp. 180–248.

Glick, B.R. (1995). The enhancement of plant growth of free living bacteria. *Can. J. Microbiol*, 41, 109–117.

Goyal, S.K., Singh, B.V., Nagpal, V. and Marwaha, T.S. (1997). An improved method for production of algal biofertilizer. *Indian Agric. Sci.* 67, 314–315.

Goyal, S.K. (1982). Blue-green algae and rice cultivation. *Proc. Natl. Symp.*, BNF, IARI, New Delhi. p. 346–347.

Goyal, S.K. (1989). Stress compatibility in cyanobacteria. *Phykos*, 28, 267–273.

Goyal, S.K. (1997). Algae and the soil environment. *Phykos*, 36, 1–12.

Grant, I.F. and Seegers, R. (1985). Tubificid role in soil mineralization and recovery of algal nitrogen by low land rice. *Soil Biol. Biochem.*, 17, 559–563.

Grant, I.F., Tiroi, A.C., Aziz, T. and Watanabe, I. (1983). Regulation of invertebrate grazers as a means to enhance biomass and nitrogen fixation of Cyanophycep in wetland rice fields. *Soil Science Society of America Journal*, 47, 669–675.

Grant, I.F., Roger, P.A. and Watanabe, I. (1985). Effects of grazer regulation and algal inoculation on photodependent nitrogen fixation in a wetland rice field. *Biol. Fert. Soils*, 1, 61–72.

Grzesik, M. and Romanowska-Duda, Z. (2014). Improvements in germination, growth and metabolic activity of corn seedlings by grain conditioning and root application with cyanobacteria and microalgae. *Pol J Environ Stud*, 23, 1147–1153.

Gunnison, D. and Alexander, M. (1975). Resistance and susceptibility of algae to decomposition by natural microbial communities. *Limnol. Oceanogr.*, 20, 64–70.

Gupta, D.K., Rai, U.N., Tripathi, R.D. and Inouhe, M. (2002). Impacts of flyash on soil and plant responses. *J. Plant Res*, 115, 401–409.

Gupta, R.D. and Khajuria, M.R. (1996). Prospects and constraints of biofertilizers for rice with special reference to Jammu region of Jammu and Kashmir State. *Farmers and Parliament*, 31, 21–22.

Harrison, W.H. and Aiyer, S. (1914). *Mem. Dep. Agric. India Chem.*, 4, 1.

Hegde, D.M. and Dwivedi, B.S. (1993). Integrated nutrient supply and management as a strategy to meet nutrient demand. *Fert. News*, 38, 49–59.

Herbette, S., Taconnat, L., Hugouvieux, V., Piette, L., Magniette, M. L., Cuine, S., … & Leonhardt, N. (2006). Genome-wide transcriptome profiling of the early cadmium response of Arabidopsis roots and shoots. *Biochimie*, 88(11), 1751–1765.

Hoare, D.S., Hoare, S.L. and Moore, R.B. (1967). The photoassimilation of organic compounds by autotrophic blue-green algae. *J. Gen. Microbiol.*, 49, 351–370.

Inubushi, K. and Watanabe, I. (1986). Dynamics of available N in paddy soils. II. Mineralized N of chloroform-fumigated soil as a nutrient source for rice. *Soil Sci. Plant Nutr.*, 32, 561–577.

Inubushi, K., Hori, K., Matsumoto, S. and Wada, H. (1997a). Anaerobic decomposition of organic carbon in paddy soil in relation to methane emission to the atmosphere. *Water Sci. Technol.*, 36, 523–530.

Inubushi, K., Shibahara, F., Hasegawa, K. and Yamamura, S. (1997b). Effect of added organic matter on microbial biomass nitrogen dynamics and plant uptake in paddy soils. *In:* Plant nutrition for sustainable food production and environment. Ando et al. (eds). Kluwer Academic Publishers. Dordrecht (Netherlands). pp. 777–778.

Irisarri, P., Gonnet, S. and Monza, J. (2001). Cyanobacteria in Uruguayan rice fields: Diversity, nitrogen fixing ability and tolerance to herbicides and combined nitrogen. *J. Bacteriol.*, 91, 95–103.

Jaiswal, P., Prasanna, R., Nayak, S., Sood, A. and Suseela, M.R. (2008). Characteriza-tion of rhizo-cyanobacteria and their associations with wheat seedlings. *Egyptian J. Biol.*, 10, 20–27.

Jala, S. and Goyal, D. (2006). Fly-ash as a soil ameliorant for improving crop production – A review. *Bioresour. Technol.*, 97, 1136–1147.

Jalapathi, R. L., Venkatachari, A., Sundara Rao, W.V.B. and Raj Reddy, K. (1977). *Curr. Sci.*, 46(2), 50.

Jayasinghearachchi, H.S. and Seneviratne, G. (2004). A bradyrhizobial-*Penicillium* spp. biofilm with nitrogenase activity improves N_2 fixing symbiosis of soybean, *Biol. Fertil. Soils*, 40, 432–434.

Jeffries, D.L., Klopatek, J.M., Link, S.O. and Bolton, H. Jr. (1992). Acetylene reduction by cryptogamic crusts from a black brush community as related to resaturation and dehydration. *Soil Biol. Biochem.*, 24, 1101–1105.

Jha, K.K., Ali, M.A., Singh, R. and Bhattacharya, P.B. (1965). Increasing rice production through the inoculation of *Tolypothrix tenuis*, a nitrogen fixing blue-green alga. *J. Indian Soc. Soil Sci.*, 13, 161–166.

Jones, K. and Stewart, W.D.P. (1969). Nitrogen turnover in marine and brackish habitats. III. The production of extracellular nitrogen by *Calothrix scopularaum. Journal of the Murine Biological Association, U.K.*, 49, 475–488.

Jones, K. and Wilson, R.E. (1978). The fate of nitrogen fixed by a free-living blue-green alga. *Ecol. Bull.*, 26, 158–163.

Kannaiyan, S. (1999). *Bioresources technology for sustainable agriculture*. Associated Publishing Company.

Karthikeyan, N., Prasanna, R., Sood, A., Jaiswal, P., Nayak, S. and Kaushik, B.D. (2009). Physiological characterization and electron microscopic investigations of cyanobacteria associated with wheat rhizosphere. *Folia Microbiologica*, 54, 43–51.

Karthikeyan, N., Prasanna, R., Lata, N. and Kaushik, B.D. (2007). Evaluating the potential of plant growth promoting cyanobacteria as inoculants for wheat. *European Journal of Soil Biology*, 43, 23–30.

Kaushik, B.D. (2005). Blue-green algae (Cyanobacteria) in Agriculture. *In:* Advances in Microbiology at IARI 1961–2004. B.D. Kaushik (ed.). Mounto Publishing House, New Delhi. pp. 19–34.

Kaushik, B.D. (1985). Effect of native algal flora on nutritional and physicochemical properties of sodic soils. *Acta Botanica Indica*, 13, 143–147.

Kaushik, B.D. (1994). Algalization of rice in salt affected soils. *Ann. Agril. Res.*, 15, 105–106.

Kaushik, B.D. (1998). Use of cyanobacterial biofertilizer in rice cultivation: A Technology Improvement. *In*: Cyanobacterial Biotechnology. G. Subramanian, B.D. Kaushik, G.S. Venkataraman (Eds). Science Publs. Inc. USA. pp. 211–222.

Kaushik, B.D. and Prasanna, R. (1998). Improved cyanobacterial biofertilizer production and N-saving in rice cultivation. pp.145–155. *In:* D.B. Sahoo. and S.Z. Quasim(eds), Sustainable Aquaculture, P.P.H. Publications, New Delhi.

Khan, Z.U.M., Tahmida Begum, Z.N., Mandal, R. and Hossain, M.Z. (1994). Cyanobacteria in rice soils. *World J Microbiol Biotechnol.*, 10, 296–298.

Khush, G.S. (2005). What it will take to feed 5.0 billion rice consumers in 2030. *Plant Mol. Biol.*, 59, 1–6.

Kikuchi, E. and Kurihara, Y. (1977). *In vitro* studies on the effects of tubicids on the biological, chemical and physical characteristics of submerged ricefield soil and overlying water. *Oikos*, 29, 348–356.

Kopteva, Z.H.P. (1970). Biosynthesis of thiamine, riboflavin and Vitamin B_{12} by some blue-green algae. *Mikrobiol. Zh (Kiev).*, 32, 429–433.

Kundu, D.K. and Ladha, J.K. (1995). Efficient management of soil and biologically fixed N_2 in intensively cultivated rice fields. *Soil Biol. Biochem.*, 27, 431–439.

Lee, K.K. and Watanabe, I. (1977). Problems of the acetylene reduction technique applied to water-saturated paddy soils. *Appl. Environ. Microbiol.*, 34, 654–660.

Lee, K.K., Castro, T. and Yoshida, T. (1977). Nitrogen fixation throughout growth and varietal differences in nitrogen fixation by the rhizosphere of rice planted in pots. *Plant and Soil*, 48, 613–619.

Ley, S.N. (1959). The effect of nitrogen fixing blue-green alga on the yield of rice plant. *Acta Hydrob. Sinica*, 4, 440–444.

Lightfoot, C., Roger, P.A., Cagauan, A.G. and DelaCraz, C.R. (1990). A fish crop may improve rice yields and rice fields. *Naga*, 13, 12–13.

Lin, X., Zhou, W., Zhu, D., Chen, H. and Zhang, Y. (2006). Nitrogen accumulation, remobilization and partitioning in rice (*Oryza sativa* L.) under an improved irrigation practice. *Field Crops Res.*, 96, 448–454.

Lowendorf, H.S. (1980). Biological nitrogen fixation in flooded rice. Agronomy Paper No. 1305. Agronomy Dept, Cornell Univ. Ithaca, NY.

Mac Rae, I.C. and Castro, T.F. (1967). Nitrogen fixation in some tropical rice soils. *Soil Sci.*, 103, 277–280.

Mah Thien-Fah, C. and O'Toole George, A. (2001). Mechanisms of biofilm resistance to antimicrobial agents. *Trends Microbiol.*, 9(1), 34–39.

Mahajan, A., Choudhary, A.K., Jaggi, R.C. and Dogra, R.K. (2003). Importance of biofertilizers in sustainable agriculture. *Farmer's Forum*, 3, 17–19.

Mandal, B., Vlek, P.L.G. and Mandal, L.N. (1999). Beneficial effects of blue-green algae and *Azolla*, excluding supplying nitrogen, on wetland rice fields: A review. *Biol. Fertil. Soils*, 28, 329–342.

Manjunath, M., Prasanna, R., Sharma, P., Nain, L. and Singh, R. (2011). Developing PGPR consortia using novel genera *Providencia* and *Alcaligenes* along with cyanobacteria for wheat. *Archives of Agronomy and Soil Science*, 57, 873–887.

Mateo, P., Bonilla, I., Fernandez-Valiente, E. and Sanchez-Maseo, E. (1986). Essentiality of boron for dinitrogen fixation in *Anabaena* sp. PCC7119. *Plant Physiol.*, 81, 430–433.

Matsuguchi, T. (1979). Factors affecting heterotrophic nitrogen fixation in submerged rice soils. *In:* Nitrogen and Rice. International Rice Research Institute, Los Banos, Philippines. pp. 207–222.

Mayland, H.F. and McIntosh, T.H. (1966). Availability of biologically fixed atmospheric nitrogen-15 to higher plants. *Nature*, 209, 421–422.

Meharg, A.A. (2004). Arsenic in rice-understanding a new disaster for South-East Asia. *Trends Plant Sci.*, 9, 415–417.

Miam, M.H. and Stewart, W.D.P. (1985). Fate of nitrogen applied as *Azolla* and cyanobacteria (cyanobacteria) in water-logged rice soils as ^{15}N tracer study. *Plant Soil*, 83, 363–370.

Miflin, B. J., & Lea, P. J. (1976). The pathway of nitrogen assimilation in plants. *Phytochemistry*, 15(6), 873–885.

Misra, H.S. and Tuli, R. (1994). Nitrogen fixation by *Plectonema boryanum* has a photosystem II independent component. *Microbiology*, 140, 971–976.

Misra, H.S. and Tuli, R. (2000). Differential expression of photosynthesis and nitrogen fixation genes in the cyanobacterium *Plectonema boryanum. Plant Physiology*, 122, 731–736.

Mittra, B.N., Karmakar, S., Swain, D.K. and Ghosh, B.C. (2005). Flyash – A potential source of soil amendment and a component of integrated plant nutrient supply system. *Fuel*, 84, 1447–1451.

Molisch, H. (1926). *Pflanzenbiologie in Japan*. G. Fischer.

Molope, M.B., Page, E.R. and Grieve, I.C. (1985). A comparison of soil aggregate stability test using soils with contrasting cultivation histories. *Commun. Soil Sci. Plant Anal.*, 16, 315–322.

Nain, L., Rana, A., Joshi, M., Jadhav, S.D., Kumar, D., Shivay, Y.S., Paul, S. and Prasanna, R. (2010). Evaluation of synergistic effects of bacterial and cyanobacterial strains as biofertilizers for wheat. *Plant Soil*, 331, 217–230.

Nayak, S. and Prasanna, R. (2007). Soil pH and its role in cyanobacterial abundance and diversity in rice field soils. *Applied Ecol. Environ. Res.*, 5(2), 103–113.

Nayak, S., Prasanna, R., Dominic, T.K. and Singh, P.K. (2001). Floristic abundance and relative distribution of different cyanobacterial genera in rice field soil at different crop growth stages. *Phykos*, 40(1and2), 15–22.

Nayak, S., Prasanna, R., Prasanna, B.M. and Sahoo, D.B. (2007). Analyzing diversity among Indian isolates of *Anabaena* (Nostocales, Cyanophyta) using morphological, physiological and biochemical characters. *World Journal of Microbiology and Biotechnology*, 23, 1575–1584.

Nayak, S., Prasanna, R., Prasanna, B.M. and Sahoo, D.B. (2009). Genotypic and phenotypic diversity of *Anabaena* isolates from diverse rice agro-ecologies of India. *J. Basic Microbiol.*, 49, 165–177.

Nayak, S., Prasanna, R., Pabby, A., Dominic, T.K. and Singh, P.K. (2004). Effect of Cyanobacteria – *Azolla* biofertilizers on nitrogen fixation and chlorophyll accumulation at different depths in soil cores. *Biol. Fertil. Soils*, 40, 67–72.

Nguyen, M.L., Haynes, R.J. and Goh, K.M. (1995). Nutrient budgets and status in three pairs of conventional and alternative and mixed cropping farms in Canterbury, New Zealand. *Agriculture Ecosystems and Environment*, 52, 149–162.

Norman, R.J., Guindo, D., Wells, B.R. and Wilson, C.E. (1992). Seasonal accumulation and partitioning of nitrogen-15 in rice. *Soil Sci. Soc. Amer. J.*, 56, 1521–1527.

Okuda, A. and Yamaguchi, M. (1952). Algae and atmospheric nitrogen fixation in paddy soils. II. Relation between the growth of cyanobacteria and physical or chemical properties of soil and effect of soil treatments and inoculation on the nitrogen fixation. *Mem. Res. Inst. Food Sci. Kyoto Univ.*, 4, 1–11.

Olkarinen, M. (1996). Biological soil amelioration as the basis of sustainable agriculture and forestry. *Biol Fertil Soils*, 22, 342–344.

Osa-Afiana, L.O. and Alexander, M. (1981). Factors affecting predation by a microcrustacean (*Cypris* sp.) on nitrogen fixing blue-green algae. *Soil Biol. Biochem.*, 13, 27–32.

Pabbi, S., Prasanna, R., Dhar, D.W. and Singh, P.K. (2000). *In:* Biofertilizers. *Cyanobacteria and Azolla*. Singh, P.K., Dhar, D.W., Pabbi, S., Prasanna, R. and Arora, A. (eds). Venus Printers and Publishers, New Delhi. pp. 83–99.

Pandey, V.N., Gupta, R.N. and Srivastava, A.K. (1993). Effect of Cyanobacteria-Biofertilizer supplement on growth and yield of rice (*Oryza sativa*) variety Saket-4. *Nat. Acad. Sci. Letters*, 16(11 and12), 281–284.

Patel, S.R., Srivastava, S.K. and Chandravanshi, B.R. (1984). *Research News*, 9(3), 25.

Patrick, W.H. Jr. (1982). Nitrogen transformations in submerged soils. *In:* Nitrogen in Agricultural Soils. Agronomy Monograph No. 22, pp. 449–365. American Agronomy Society. Madison.

Pearce, J., Leach, C.K. and Carr, N.G. (1969). The incomplete tricarboxylic acid cycle in the blue green alga *Anabaena variabilis*. *J. Gen. Microbiol.*, 55, 371–378.

Peters, G.A. and Mayne, B.C. (1974). *Azolla-Anabaena azollae* relationship II Localization of nitrogenase activity as assayed by acetylene reduction. *Plant Physiology*, 53, 119–126.

Prasanna, R., Sharma, B.K., Sharma, R.K. and Kaushik, B.D. (1998). Standardization of growth parameters and formulation of medium for cyanobacterial biofertilizer strains. *Indian Journal of Microbiology*, 38, 211–215.

Prasanna, R. and Kaushik, B.D. (1994). Physiological and molecular genetic aspects of nitrogen fixation in non-heterocystous cyanobacteria. *Indian Journal of Experimental Biol*ogy, 32, 248–251.

Prasanna, R. and Kaushik, B.D. (1995). Nitrogen fixation and *nif* gene organization in branched heterocystous cyanobacteria: Variation for the presence of *xisA*. *Folia Microbiologica*, 40(2), 176–180.

Prasanna, R., & Kaushik, B. D. (2006). Cyanobacteria in soil health and sustainable agriculture. *Health and Environment*, 3, 91–105.

Prasanna, R. and Nayak, S. (2007). Influence of diverse rice soil ecologies on cyanobacterial diversity and abundance. *Wetlands Ecol. Managmt.*, 15, 127–134.

Prasanna, R., Babu, S., Rana, A., Kabi, S.R., Chaudhary, V., Gupta, V., Kumar, A., Shivay, Y.S., Nain, L. and Pal, R.K. (2013a). Evaluating the establishment and agronomic proficiency of cyanobacterial consortia as organic options in wheat-rice cropping sequence. *Experimental Agriculture*, 49, 416–434.

Prasanna, R., Chaudhary, V., Gupta, V., Babu, S., Kumar, A., Shivay, Y.S. and Nain, L. (2013b). Cyanobacteria mediated plant growth promotion and bioprotection against *Fusarium* wilt in tomato. *European Journal of Plant Pathology*, 13, 337–353.

Prasanna, R., Jaiswal, P., Singh, Y.V. and Singh, P.K. (2008). Influence of biofertilizers and organic amendments on nitrogenase activity and phototrophic biomass of soil under wheat. *Acta Agron. Hung.*, 56(2) , 149–159.

Prasanna, R., Gupta, V., Natarajan, C. and Chaudhary, V. (2010). Allele mining for chitosanases and microcystin-like compounds in *Anabaena* strains. *World Journal of Microbiology and Biotechnology*, 26, 717–724.

Prasanna, R., Jaiswal, P., Nayak, S., Sood, A. and Kaushik, B.D. (2009a). Cyanobacterial diversity in the rhizosphere of rice and its ecological significance. *Indian Journal of Microbiology*, 49, 89–97.

Prasanna, R., Joshi, M., Rana, A., Shivay, Y.S. and Nain, L. (2012). Influence of co-inoculation of bacteria– cyanobacteria on crop yield and C-N sequestration in soil under rice crop. *World Journal of Microbiology and Biotechnology*, 28, 1223–1235.

Prasanna, R., Lata, N., Tripathi, R., Gupta, V., Middha, S., Joshi, M., Ancha, R. and Kaushik, B.D. (2008). Evaluation of fungicidal activity of extracellular filtrates of cyanobacteria – Possible role of hydrolytic enzymes. *Journal of Basic Microbiology*, 48, 186–194.

Prasanna, R., Lata, N., Pandey, A.K. and Nayak, S. (2010). Exploring the ecological significance of microbial diversity and networking in the rice ecosystem. *In:* Soil Biology and Agriculture in the Tropics. Patrice Dion (Ed.), Soil Biology series. Springer. pp. 257–298.

Prasanna, R., Nain, L., Ancha, R., Shrikrishna, J., Joshi, M. and Kaushik, B.D. (2009b). Rhizosphere dynamics of inoculated cyanobacteria and their growth-promoting role in rice crop. *Egyptian J. Biol.*, 11, 26–36.

Prasanna, R., Pattnayak, S., Sugitha, T.C.K., Nain, L. and Saxena, A.K. (2011). Development of cyanobacterium based biofilms and their in vitro evaluation for agriculturally useful traits. *Folia Microbiologica*, 56, 49–58.

Prasanna, R., Tripathi, U., Dominic, T.K., Singh, A.K., Yadav, A.K. and Singh, P.K. (2003). An improvised technique for measurement of nitrogen fixation by blue-green algae and *Azolla* using moist soil core from rice fields. *Expt. Agric.*, 39, 145–150.

Prasanna, R., Triveni, S., Bidyarani, N., Babu, S., Yadav, K., Adak, A., Khetarpal, S., Pal, M., Shivay, Y.S. and Saxena, A.K. (2013c). Evaluating the efficacy of cyanobacterial formulations and biofilmed inoculants for leguminous crops. *Archives of Agronomy and Soil Science*, 60, 349–366.

Querijero-Palacpac, N.M., Milagrosa, M. and Boussiba, S. (1990). Mass cultivation of the nitrogen-fixing cyanobacterium *Gloeotrichia natans,* indigenous to rice-fields. *J. Appl. Phycol.*, 2, 319–325.

Rai, U.N., Pandey, K., Sinha, S., Singh, A., Saxena, R., & Gupta, D.K. (2004). Revegetating fly ash landfills with Prosopis juliflora L.: impact of different amendments and Rhizobium inoculation. *Environment International*, 30(3), 293–300.

Rajaram, H. and Apte, S.K. (2003). Heat-shock response and its contribution to thermotolerance of the nitrogen-fixing cyanobacterium *Anabaena* sp. strain L-31. *Arch Microbiol*, 179, 423–429.

Rajaram, H. and Apte, S.K. (2008). Nitrogen status and heat-stress dependent differential expression of the *cpn60* chaperonin gene influences thermotolerance in the cyanobacterium *Anabaena. Microbiology*, 154, 317–325.

Rajaram, H., Ballal, A.D., Apte, S.K., Weigert, T. and Schumann, W. (2001). Cloning and characterization of the major *groESL* operon from a nitrogen-fixing cyanobacterium *Anabaena* sp. strain L-31. *Biochim Biophys Acta*, 1519, 143–146.

Rakshit, S.K., Loan, N.T. and Johnsen, S. (1999). Fate of blue green algae in the food web of flooded rice field ecosystems. *Biol. Fertil. Soils*, 29, 141–145.

Rana, A., Joshi, M., Prasanna, R., Shivay, R.S. and Nain, L. (2012). Biofortification of wheat through inoculation of plant growth promoting rhizobacteria and cyanobacteria. *Eur J Soil Biol*, 118–126.

Rao, D. L. N., & Burns, R. G. (1990). Use of blue-green algae and bryophyte biomass as a source of nitrogen for oil-seed rape. *Biology and fertility of soils*, 10(1), 61–64.

Rathore, A.L., Chipde, S.J. and Pal, A.R. (1995). Direct and residual effect of bio-organic and inorganic fertilizers in rice-wheat cropping system. *Indian J. Agron.*, 40, 14–19.

Rausch, T., & Wachter, A. (2005). Sulfur metabolism: a versatile platform for launching defence operations. *Trends in plant science*, 10(10), 503–509.

Rautaray, S.K., Ghosh, B.C. and Mittra, B.N. (2003). Effect of fly ash, organic wastes and chemical fertilizers on yield, nutrient uptake, heavy metal content and residual fertility in rice-mustard cropping system under acid lateritic soils. *Bioresour. Technol.*, 90, 275–283.

Reddy, B.R., Apte, S.K. and Thomas, J. (1989). Enhancement of cyanobacterial salt tolerance by combined nitrogen. *Plant Physiol*, 89, 204–210.

Relwani, L.L. (1963). Role of blue-green algae on paddy yield. *Curr. Sci.*, 32, 417–418.

Reynaud, P.A. and Metting, B. (1988). Colonization potential of cyanobacteria on temperate irrigated soils of Washington State, U.S.A. *Biol. Agri. Hort.*, 5, 197–208.

Reynault, J., Sasson, A., Pearson, H.W. and Stewart, W.D.P. (1975). Nitrogen fixing algae in Morocco. *In:* Nitrogen fixation by Free Living Microorganisms. Stewart, W.D.P. (ed.). Cambridge Univ. Press, London. pp. 229–248.

Rice, W.A. and Paul, E.A. (1971). The acetylene reduction assay for measuring nitrogen fixation in waterlogged soil. *Can. J. Microbiol.*, 17, 1049–1056.

Roger, P.A. and Kulasooriya, S.A. (1980). Cyanobacteria and rice. International Rice Research Institute, Manila.
Roger, P.A. (1991). Reconsidering the utilization of cyanobacteria in wetland rice cultivation. *In:* Biological N_2 fixation associated with rice production. Dutta, S.K. and Sloger, C. (eds). Oxford and IBH, New Delhi. pp. 119–141.
Roger, P. A. (1996). *Biology and management of the floodwater ecosystem in rice fields*. Int. Rice Res. Inst.
Roger, P.A. and Ladha, J.K. (1992). Biological N_2 fixation in wetland rice fields: Estimation and contribution to nitrogen balance. *Plant Soil*, 141, 41–55.
Roger, P.A. and Reynaud, P.A. (1979). Ecology of blue-green algae in paddy fields. *In:* Nitrogen and Rice, IRRI, Los Banos. pp. 289–309.
Roger, P.A. and Reynaud, P.A. (1982). Free living blue-green algae in tropical soils. *In:* Microbiology of tropical soils and plant productivity. Dommergues, Y. and Diem, H. (eds). Martinus Nijhoff Press, The Hague. pp. 147–168.
Roger, P.A. and Watanabe, I. (1986). Technologies for utilizing biological nitrogen fixation in lowland rice: Potentialities, current usage and limiting factors. *Fer. Res.*, 27, 470–477.
Roger, P.A., Grant, I.F., Reddy, P.M. and Watanabe, I. (1987). The photosynthetic aquatic biomass in wetland rice fields and its effect on nitrogen dynamics. *In:* Efficiency of nitrogen fertilizers for rice. IRRI, Manila. pp. 43–68.
Roger, P.A., Reddy, P.M. and Remulla-Jimenez, R. (1988). Photodependent acetylene reducing activity (ARA) in ricefields under various fertilizer and biofertilizer management. *In:* Nitrogen fixation: Hundred years after. Bothe, H., De Bruijn,, F.J. and Newton, W.E. (eds). Gustav Fischer, Stuttgart, New York. pp. 827.
Roger, P.A., Zimmerman, W.J. and Lumpkin, T.A. (1993). Microbiological management of wetland rice fields. *In:* Soil microbiological ecology-application in agricultural and ecological management. Baline Metting, F. Jr. (ed.). Marcel Dekker, New York, Basel Hong Kong. pp. 417–455.
Rogers, S.L. and Burns, R.G. (1994). Changes in aggregate stability, nutrient status, indigenous microbial populations and seedling emergence following inoculation of soil with *Nostoc muscorum. Biol. Fertil. Soils*, 18, 209–215.
Rowell, P., Enticott, S. and Stewart, W.D.P. (1977). Glutamine synthetase and nitrogenase activity in the blue green alga *Anabaena cylindrica. New Phytologist*, 79, 41–54.
Roychoudhury, P., Pillai, G.R., Pandey, S.L., Krishnamurti, G.S.R. and Venkataraman, G.S. (1983). Effect of blue-green algae on aggregate stability and rice yield under different irrigation and nitrogen levels. *Soil Till. Res.*, 3, 61–66.
Saha, K.C. and Mandal, L.N. (1980). A green house study on the effect of inoculation of blue-green algae in an alluvial soil treated with P and Mo on the yield of rice and changes in the N content of soil. *Plant Soil*, 57, 23–30.
Saha, K.C. and Mandal, L.N. (1979). Effect of algal growth on the availability of P, Fe and Mn in rice soils. *Plant Soil*, 52, 139–149.
Saha, K.C., Panigrahi, B.C. and Singh, P.K. (1982). Cyanobacteria or *Azolla* addition on the nitrogen and phosphorus availability and redox potential of a flooded rice soil. *Soil Biol. Biochem.*, 14 (1), 23–26.
Saleque, M.A., Abedin, M.J., Bhuiyan, N.I., Zaman, S.K. and Panaullah, G.M. (2004). Long-term effects of inorganic and organic fertilizer sources on yield and nutrient accumulation of lowland rice. *Field Crops Res.*, 86, 53–65.
Santra, S.C. (1991). Rice field cyanobacteria and its utilization prospect as bio-fertilizer in West Bengal, India. *Proc. Nat. Symp.* on cyanobacterial nitrogen fixation, New Delhi. pp. 385–389.
Santra, S.C. (1993). Biology of rice-fields. *In: Cyanobacteria*. Daya Publishing House, Delhi. pp. 184.
Schramm, J.R. (1914). The relation of certain species of grass-green algae to elementary nitrogen. *Ann. Mo. Bot. Gdn*,, 1, 157–184.
Seneviratne, G., Thilakaratne, R. M. M. S., Jayasekara, A. P. D. A., Seneviratne, K. A. C. N., Padmathilake, K. R. E., & De Silva, M. S. D. L. (2009). Developing beneficial microbial biofilms on roots of non legumes: A novel biofertilizing technique. In *Microbial strategies for crop improvement* (pp. 51–62). Springer Berlin Heidelberg.

Sevilleja, R.C., Cagauan, A.G. (1992). Rice-fish Asia working group (CLSU-FAC component) program report. 1st July to 31st December 1991). Central Luzon State University, Nuera Ecija, Philippines.

Shanmugasundaram, S.S. (1996) Consolidated Report, Mission mode project on technology development and demonstration of algal biofertilizer, MKU, Madurai, India.

Sherr, B.F., Sherr, E.B. and Berman, T. (1983). Grazing, growth and ammonium excretion rates of a heterotrophic microflagellate fed with four species of bacteria. *Appl. Environ. Microbiol.*, 45, 1196–1201.

Sikka, R. and Kansal, B.D. (1995). Effect of fly-ash application on yield and nutrient composition of rice, wheat and on pH and available nutrient status of soils. *Bioresour. Technol.*, 51, 199–203.

Singh, A.L. and Singh, P.K. (1987). Nitrogen fixation and balance studies of rice soil. *Biol. Fert. Soils*, 4, 15–19.

Singh, J.S., Pandey, V.C. and Singh, D.P. (2011). Efficient soil microorganisms: A new dimension for sustainable agriculture and environmental development. *Agric Ecosyst Environ*, 140, 339–353.

Singh, K.N. and Singh, R.P. (1992). Farming system research and development-A holistic approach for sustainable agriculture growth. *In:* Proceedings of XII National Symposium on Resource Management for Sustained Crop Production held in Rajasthan Agriculture Univerisity, Bikaner during Feb. 25–28, 391–398.

Singh, P.K. (1978). Nitrogen economy of rice soils in relation to nitrogen fixation by blue-green algae and *Azolla. In:* National Symposium on increasing rice yields in Kharif. CRRI., Cuttack.

Singh, P.K. (1985). Nitrogen fixation by blue-green algae in paddy fields. *In:* Rice Research in India, ICAR, New Delhi. pp. 344–362.

Singh, P.K. (1988). Biofertilization of rice crop. *In:* Biofertilizers: potentialities and problems. Sen, S.P. and Palit, P. (eds). Plant Physiology Forum, Calcutta. pp. 109–114.

Singh, P.K. (1989). Use of *Azolla* in Asian Agriculture. *Applied Agricultural Research*, 4, 149–161.

Singh, P.K. and Bisoyi, P.K. (1989). Blue-green algae in rice fields. *Phykos*, 28, 181–195.

Singh, P.K., Panigrahi, B.C. and Satapathy, K.B. (1981). Comparative efficiency of *Azolla,* blue-green algae, other organic manures in relation to N and P availability in a flooded rice soil. *Palnt Soil.*, 62, 35–44.

Singh, R.N. (1961). Role of blue-green algae in nitrogen economy of Indian agriculture. ICAR, New Delhi.

Sinha, R.K., Valani, D., Chauhan, K. and Agarwal, S. (2014). Embarking on a second green revolution for sustainable agriculture by vermiculture biotechnology using earthworms: Reviving the dreams of Sir Charles Darwin. *Int J Agric Health Saf*, 1, 50–64.

Smith, V.H. and Crews, T. (2014). Applying ecological principles of crop cultivation in large scale algal biomass production. *Algal Res*, 4, 23–34.

Sprent, J.I. and Sprent, P. (1990). Nitrogen fixing organisms—Pure and applied aspects. Chapman and Hall, London.

Stewart, P.S. (2002). Mechanisms of antibiotic resistance in bacterial biofilms. *Int. J. Med. Microbiol.*, 292, 107–113.

Stewart, W.D.P. (1967). Transfer of biologically fixed nitrogen in a sand dune slack region. *Nature*, 214, 603–604.

Stewart, W.D.P. and Lex, M. (1970). Nitrogenase activity in *Plectonema boryanum* 594. *Arch. Microbiol.*, 75, 250–260.

Stewart, W.D.P., Rowell, P., Ladha, J.K. and Sanpio, M.J.A. (1979). Blue-green algae (cyanobacteria)—Some aspects related to their sources of fixer nitrogen in paddy soils. *In: Nitrogen and Rice*, IRRI, Los Banos. pp. 263–285.

Subramanyan, R. and Manna, G.B. (1966). Relative response of the rice plant to blue-green algae and ammonium sulphate in bulk trials. *Curr. Sci.*, 35, 482–483.

Sundara Rao, W.V.B., Goyal, S.K. and Venkataraman, G.S. (1963). Effect of inoculation of *Aulosira fertilissima* on rice plants. *Curr. Sci.*, 32, 366–367.

Swaminathan, M.S. (1982). Biotechnology research and third world agriculture. *Science*, 218(4576), 967–972.
Swarnalakshmi, K., Prasanna, R., Kumar, A., Pattnaik, S.C., Chakravarty, K.C., Shivay, S.A., Singh, B.R. and Saxena, A.K. (2013). Evaluating the influence of novel cyanobacterial biofilmed biofertilizers on soil fertility and plant fertility and plant nutrition in wheat. *Eur J Soil Biol*, 55, 107.
Swarnalakshmi, K., Dhar, D.W. and Singh, P.K. (2006). Cyanobacteria: A potential biofertilizer for sustainable rice cultivation. *Proc. Indian Natn. Sci. Acad.*, 72, 135–143.
Swarnalakshmi, K., Dhar, D.W. and Singh, P.K. (2007a). Evaluation of blue green algal inoculation on specific soil parameters. *Acta Agronomica Hungarica*, 55(3), 307–313.
Swarnalakshmi, K., Dhar, D.W. and Singh, P.K. (2007b). Potential of Cyanobacteria on C-N content and yield attributes of rice crop under phytotronic condition. *Indian Journal of Ecology*, 34(1), 47–49.
Thomas, J. (1997). Biological nitrogen fixation. *Nucl India*, 15, 2–8.
Thomas, J. and Apte, S.K. (1984). Sodium requirement and metabolism in nitrogen-fixing cyanobacteria. *J Biosci*, 6, 771–794.
Thomas, J. (1970). Absence of the pigments of photosystem II of photosynthesis in heterocysts of blue-green algae. *Nature*, 228, 181–183.
Thomas, J., Wolk, C.P., Shaffer, P.W., Austin, S.M. and Galonsky, A. (1975). The initial organic products of fixation of ^{15}N-labelled nitrogen gas by the blue green alga *Anabaena cylindrica. Biochem. Biophys. Res. Commun.*, 67, 501–507.
Tien, C.J. (2002). Biosorption of metal ions by fresh water algae with different surface characteristics. *Process Biochem.*, 38, 605–613.
Tirol, A.C., Roger, P.A. and Watanabe, I. (1982). Fate of nitrogen from blue green alga in a flooded rice soil. *Soil Sci. Plant Nutr.*, 28, 559–569.
Tisdall, J. M., & Oades, J. (1982). Organic matter and water-stable aggregates in soils. *Journal of soil science*, 33(2), 141–163.
Tripathi, R. D., Dwivedi, S., Shukla, M. K., Mishra, S., Srivastava, S., Singh, R., … & Gupta, D. K. (2008). Role of blue green algae biofertilizer in ameliorating the nitrogen demand and fly-ash stress to the growth and yield of rice (Oryza sativa L.) plants. *Chemosphere*, 70(10), 1919–1929.
Tripathi, R.D., Srivastava, S., Mishra, S., Singh, N., Tuli, R., Gupta, D.K. and Maathuis, F.J.M. (2007). Arsenic hazards: Strategies for tolerance and remediation by plants. *Trends Biotechnol.*, 25, 158–165.
Tripathi, R.D., Vajpayee, P., Singh, N., Rai, U.N., Kumar, A., Ali, M.B., Kumar, B. and Yunus, M. (2004). Efficacy of various amendments for amelioration of flyash toxicity: Growth performance and metal composition of *Cassia siamea* Lamk. *Chemosphere*, 54, 1581–1588.
Tupik, N.D. (1973). Study of the content of group B Vitamins in cells of some blue-green algae in dependence on the culture age. *Ukr. Bot. Zh.*, 30, 636–639.
Venkataraman, G.S. and Shanmugasundaram, S. (1992). Algal Biofertilizer Technology for Rice. Madurai Kamraj University, Madurai, India. pp. 1–24.
Venkataraman, G.S. (1961). The role of blue-green algae in agriculture. *Sci. Cult.*, 27, 9–13.
Venkataraman, G.S. (1972). Algal biofertilizers and Rice cultivation. Today and Tomorrow Printers, Faridabad, Haryana, India.
Venkataraman, G.S. (1975). The role of cyanobacteria in tropical rice cultivation. *In:* Nitrogen fixation by Free living Microorganisms. Stewart, W.D.P. (ed.). Cambridge Univ. Press. pp. 207–218.
Venkataraman, G.S. (1979). Algal inoculation of rice fields. *In:* Nitrogen and Rice. IRRI, Los Banos. pp. 311–321.
Venkataraman, G.S. and Goyal, S.K. (1968). Influence of blue-green algal inoculation on the crop yield of rice plants. *Soil Sci. Plant Nutr.*, 14(6), 249–251.
Venkataraman, G.S. and Goyal, S.K. (1969). Influence of blue-green algae on the high yielding paddy variety IR 8. *Sci. Culture*, 35, 58.

Venkataraman, G.S. and Neelakantan, S. (1967). Effect of extracellular constituents of the nitrogen fixing blue-green algae *Cylindrospermum muscicola* on the root growth of rice seedlings. *J. Gen. Appl. Microbiol.*, 13, 53–61.

Venkataraman, G.S. (1981). Cyanobacteria for rice production. FAO Soils Bulletin No. 46.

Ventura, W. and Watanabe, I. (1983). ^{15}N dilution technique of assessing the contribution of nitrogen fixation to rice plant. *Soil Sci. Plant Nutr.*, 29, 123–131.

Viswanath, B. (1932). Society of Biological Chemists, India, Annual publication.

Watanabe, A. (1959). Distribution of N_2-fixing blue-green algae in various areas of South and East Asia. *J. Gen. Appl. Microbiol.*, 5, 21–29.

Watanabe, A. (1965). Studies on the blue-green algae as green manure in Japan. *Proc. Nat. Acad. Sci. India*, 35A, 361–369.

Watanabe, A. and Kiyohara, T. (1960). Decomposition of blue-green algae as affected by the action of soil bacteria. *J. Gen. Appl. Microbiol.*, 5, 175–179.

Watanabe, A. and Yamamoto, Y. (1971). Algal nitrogen fixation in the tropics. *Plant Soil* (Spl. vol), 403–413.

Watanabe, A., Nishigaki, S. and Konishi, C. (1951). Effect of nitrogen fixing cyanobacteria on the growth of rice plant. *Nature*, 168, 748–749.

Watanabe, I. and Cholitkul, W. (1979). Field studies on nitrogen fixation in paddy soils. *In:* Nitrogen and rice, IIRI, Los Banos, Philippines. pp. 223–239.

Watanabe, I. (1986). Nitrogen fixation by non-legumes in tropical agriculture with special reference to wetland rice. *Plant and Soil*, 90, 343–357.

Watanabe, I. and Roger, P.A. (1984). Nitrogen fixation in wetland rice fields. *In:* Subba Rao, N.S. (ed.). Current developments in biological nitrogen fixation. Oxford & IBH, New Delhi. pp. 237–276.

Watanabe, I., Bai, K.Z., Berja, N.S., Espinas, L.R., Ho, O. and Subudhi, R.P.R. (1981). The *Azolla-Anabaena* complex and its use in rice culture. *IRRI Research Paper Series No.* 69.

Watanabe, I., Lee, K.K., Alimagno, B.V., Sato, M., Del Rosario, D.C. and De Guzman, M.R. (1977). Biological nitrogen fixation in paddy field studies by *in situ* acetylene-reduction assays. *IRRI Research Paper Series* No., 3, 1–16.

Webb, J.S., Givskov, M. and Kjelleberg, S. (2003). Bacterial biofilms: Prokaryotic adventures in multicellularity. *Curr. Opin. Microbiol.*, 6, 578–585.

Whitton, B. A., & Potts, M. (2012). Introduction to the cyanobacteria. In *Ecology of Cyanobacteria II* (pp. 1–13). Springer Netherlands.

Wilson, C.E., Norman, R.J. and Wells, B.R. (1989). Seasonal uptake patterns of fertilizer nitrogen applied in split applications to rice. *Soil Sci. Soc. Amer. J.*, 53, 1884–1887.

Wilson, J.T., Greenes, S. and Alexander, M. (1980). Effect of microcrustaceans on blue-green algae in flooded soil. *Soil Biol. Biochem.*, 12, 237–240.

Winter, G. (1935). Uber die Assimilation der Luftstickstoffs durch endopbytische Blaualgen. *Beitr. Biol. Pfl.*, 23, 295–335.

Wolk, C.P., Thomas, J., Shaffer, P.W., Austin, S.M. and Galonsky, A. (1976). Pathway of nitrogen metabolism after fixation of ^{15}N-labelled nitrogen gas by the cyanobacterium, *Anabaena cylindrica. J. Biol. Chem.*, 251, 5027–5234.

Wyatt, J.T. and Silvey, J.K.G. (1969). Nitrogen fixation by *Gloeocapsa. Science*, 165, 908–909.

Yadav, D.N., Tiwari, T.N. and Singh, B.B. (1988). *In:* Seminar on 'Problem of soil and reclamation technology'. N.D. University of Agric. and Technology, Faizabad.

Yamaguchi, M. (1976). Nitrogen fixation by micro-organisms in paddy soils in relation to their fertility. *In:* The fertility of paddy soils and fertilizer applications for rice. ASPAC, Taiwan, Republic of China. pp. 60–75.

Yanni, Y.G. (1992). Fertilizer responses to nitrogen and cyanobacteria in presence of insecticides. *Soil Biol. Biochem.*, 24, 1085–1088.

Yoshida, T. and Ancajes, R. R. (1971). Nitrogen-fixing activity in upland and flooded rice fields. *Soil Sci. Soc. America J.*, 37, 42–46.

Yoshida, T. and Rinaudo, G. (1982). Heterotrophic N, fixation in paddy soils. *In:* Microbiology of Tropical soils and Plant Productivity. Dommergues, Y.R. and Diem, H.G. (Eds). Martinus-Nijhoff, The Hague. pp. 75–107.

Chapter 17
Role of Microalgae in Microbial Fuel Cell

Soumya Pandit and Debabrata Das

17.1 Introduction

The reliance on conventional fossil fuels has resulted in the imminent energy catastrophe with the combined challenge of global warming and the depletion of these energy reserves (Nayak et al., 2013). Our earth today is facing many environmental problems, ranging from pollution, global warming due to the accumulation of CO_2 in the atmosphere, depletion of natural energy sources like coal and petroleum and the increasing need for sustainable energy sources (Bentley, 2002). Research on renewable methods for producing energy has received utmost attention in last few years. Standing in such a situation, the use of microalgae to convert CO_2 into potential biomass coupled with their ability to produce oxygen gas, assumes strategic importance (Popp et al., 2014). Significant research is being carried out in this field to exploit this ability of microalgae and integrate it with microbial fuel cells. This integration becomes especially favourable considering the fact that the phototrophic organisms act as in-situ generators of oxygen which facilitate the reaction in cathode chamber of the MFC. Further, microalgae also effectively removes phosphorous and nitrogen from the wastewater which might not be possible solely by the MFCs (Rozendal et al., 2008). The use of phototrophic organisms in MFCs leads to the development of photosynthetic microbial fuel cells or PMFCs (Rosenbaum et al., 2010).

S. Pandit (✉) • D. Das
Department of Biotechnology, Indian Institute of Technology, Kharagpur 721302, India
e-mail: sounipf@gmail.com

D. Das (ed.), *Algal Biorefinery: An Integrated Approach*,
DOI 10.1007/978-3-319-22813-6_17

17.2 General Concepts

17.2.1 Microbial Fuel Cell Technology

Microbial fuel cells are power generating bioelectrochemical systems (BESs) based on exploitation of biocatalytic reactions with electroactive microbes (Malvankar and Lovley, 2012). This is a rapidly evolving technique for generation of bioelectricity. Typically an MFC consists of two chambers: a cathodic chamber and an anodic chamber, with an ion exchange membrane (IEM) separating them. Electrons and protons are generated at the anode by biocatalysis through substrate oxidation. The electrons are transferred to the anode surface by electron mediators, or shuttles by direct membrane associated electron transfer or by nano-wires produced by bacteria, or by other means. The electrons then flow to the cathode by an external circuit resulting in a current flow from the cathode to the anode (Fig. 17.1). The protons move through the IEM into the cathodic chamber, where along with the electrons they reduce oxygen into water (Logan, 2008). This oxygen may be obtained by pumping air into the cathodic chamber (in case of double chambered MFC) or by the direct contact of the cathode with air (single chambered MFC).

There is inflow and outflow of various gases in the MFC compartments. The dissolved oxygen (DO) concentration is a major factor determining the power production of the MFC (Pandit et al., 2012b). After conducting experimental studies it was found that the optimum DO concentration was around 6.6 mg/l. When DO concentration is below this optimum level, the current generation of the MFC is limited by oxygen. When glucose, acetate or wastewater is used as substrate, the outflow gas is CO_2. Alkaline conditions inside the cathode, due to the accumulation of hydroxide

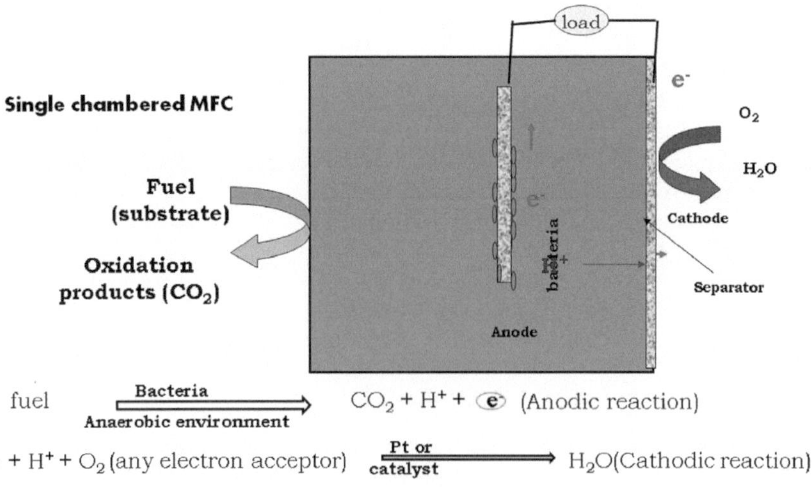

Fig. 17.1 Schematic configuration of single chambered MFC.

ions, increases absorption of CO_2 from the anode. These hydroxyl ions result from oxygen reduction in the cathode (Khilari et al., 2013). Oxygen delivery and carbon dioxide accumulation are presently the major limitations in the practical application of MFCs (Gajda et al., 2013).

17.2.2 Microalgal Biomass Culturing

Algae help in generation of biochemical energy by conversion of solar energy. It accounts for more than half of the photosynthetic output on earth. With their high growth rates and high CO_2 fixation rates (can reach up to 6.24 kg/m^3/day), microalgae are being taken into serious consideration for bioenergy production (Skjånes et al., 2007). The photosynthetic process (a biological redox reaction carried out by algae where they use solar energy to produce carbohydrates and oxygen, as well as additional compounds) might be used for energy generation or synthesis of more complex substances. Biological method of CO_2 sequestration using green algae and cyanobacteria is considered as most promising method (Markou et al., 2014). The use of algae for CO_2 sequestration has several advantages e.g. green algae has several applications: mitigating CO_2 to reduce global warming, wastewater treatment, production of biofuels, biofertilizer and other important products like food products, antibiotics and pigments (Kumar et al., 2013).

Green algae and cyanobacteria are the vast group of both facultative photoautotrophic and photoheterotrophic microorganisms. Their uniqueness from other microorganisms is the presence of chlorophyll and having photosynthetic ability in a single algal cell, therefore allowing easy operation for biomass generation, and effective genetic and metabolic research in a much shorter time period than conventional plants. Algae cultivation is carried out in illuminated environments which might be natural or artificial. They grow autotrophically using CO_2 as a carbon source and light as an energy source. Some algae can also substitute CO_2 fixation with available carbon sources in the media, thus showing heterotrophic growth. Also, the two growth modes might be combined in a mixotrophic growth mode, with photosynthesis and respiratory metabolism operating simultaneously. This results in parallel assimilation of CO_2 and organic carbon. The major shortage for use of microalgae is the need for designing specially illuminated autotrophic photobioreactors PBRs (Olivieri et al., 2014). Heterotrophic growth mode helps in overcoming this bottleneck. Also, some algae under heterotrophic mode have higher biomass, growth rates, ATP production, nitrogen content and lipid content than in autotrophic mode. On the contrary, there is a limited number of heterotrophic microalgal species. Also, energy expenses increase by supplementing an organic substrate, thereby, increasing the chances of growth inhibition by surplus organic substrate and failure to produce light induced metabolites. Moreover, heterotrophic mode of cultivation is more prone to contamination and competition from other microorganisms.

Green algae contains both chlorophyll *a* and *b* types, mainly α and/or β- carotene and γ-carotene of carotenoids, and starch and oils as storage products. Light requirement of the typical algae is lower than the higher plants. Two moles of water dissociates into four moles of protons and electrons and one mole of oxygen molecule (Eq. 17.1).

$$2H_2O \rightarrow 4H^+ + 4e^- + O_2 \tag{17.1}$$

17.3 Integrating Microalgae and MFC

For capturing energy in the form of bioelectricity using photosynthetic metabolism in an MFC system, the photosynthetic organisms employ charge separation and release electrons and protons, throughout flow chain reactions. This process is imitating the lagooning technique which is a usually used natural treatment process. During this treatment, the synergistic effects between heterotrophic microorganisms and algae are taking place, in which oxygen is produced by algae using solar energy and metabolizing nutrients and bicarbonates produced by heterotrophic microorganisms metabolizing organic matter and oxygen. Through this concatenation, the conventional aeration systems could be replaced by the more environmental and economic sustainable photosynthetic one (Kruzic and Kreissl, 2009). Thus, it is assumed that electricity can be produced by the growing cathodic algae in MFC, where oxygen is produced by the photosynthesis process (Juang et al., 2012).

This assumption has been applied in MFC by utilizing a substrate which is oxidized at the anode and employing an algal biocathode which acts as a biological electron acceptor simultaneously reducing CO_2 to biomass (Singh and Singh, 2014). Generally, for electron shuttling in cathode chamber, a mediator is used. The electrons flowing from the anode side are transported to the catholyte and reduce the oxidized state mediator, which then penetrates the algal cell store. These shuttled electrons are consumed by growing algal cells, in their metabolic pathways to convert CO_2 in to oxygen and biomass (Parlevliet and Moheimani, 2014). The oxidized mediator is released by the cells to the media and the cycle is reiterated where the mediator is once again reduced by the electrons in the catholyte (Powell et al., 2009). During illumination, the biochemical reactions that take place at the anode and cathode are demonstrated by Eqs (17.2) and (17.3), respectively (Zhou et al., 2012a, b).

$$C_6H_{12}O_6 + 6H_2O \xrightarrow{oxidation} 6CO_2 + 24H^+ + 24e^- \tag{17.2}$$

$$6CO_2 + 12H^+ + 12e^- \xrightarrow{light} -C_6H_{12}O_6 - (\text{biomass}) + 3O_2 \tag{17.3}$$

When sunlight falls on algae, they utilize it to carry out photosynthesis, use CO_2 to generate organic matter and biomass, and simultaneously consume oxygen during dark phase and obtain energy by oxidizing the formerly produced organic matter (Eq. 17.4) (González del Campo et al., 2013).

$$C_6H_{12}O_2 + 6O_2 \rightarrow 6CO_2 + 6H_2O \quad (17.4)$$

Conversely, some photosynthetic cyanobacteria could be operated as a bioanode catalyst, in which the formation of biofilm maintains the electrochemical potential. For example, *Spirulina platensis* could be used as a bioanode catalyst; it doesn't need a mediator for accepting the generated electron (Fu et al., 2010).

17.4 Configurations of Photosynthetic MFC

Solar energy is one of the major sustainable energy resources. Advanced approaches to convert solar energy into bioelectricity developed with bioelectrochemical systems (BESs) made in the last ten years, in which many of these photovoltaic devices have the ability to separate photosynthetic energy and heterotrophic dark electricity production in the absence of artificial mediators. Photosynthetic MFCs are the one which involve an anode or cathode, with a biofilm enclosing photosynthetic microorganisms, in which photosynthesis is carried out and as an outcome they act as electron donors and also as producers of organic metabolites. Removal of carbon dioxide by this integrated PMFC is another additional benefit (De Schamphelaire and Verstraete, 2009). Configuration of such PMFCs is the main task in order to increase the power density and obtain long term performance so as to get a cost effective system. Four different configurations of PMFCs are schematized and detailed in the following sections.

17.4.1 Coupled PMFC

In this concatenated PMFC, bioanodic MFC is connected to a PBR, in which CO_2 is pumped directly from the MFC to the PBR. This configuration functions in the absence of ion exchange membrane which simplifies its structure and makes it cost-effective to scale up. A photosynthetic microbial cathodic half-cell, using the *Chlorella vulgaris* microalgae as the direct electron acceptor, was developed earlier (Powell et al., 2009) where the half-cell was concatenated to a fermentative yeast anode, creating a complete coupled MFC. This design was tested into an existing bioethanol plant to create coupled MFC with the existing industrial yeast bioreactor acting as anodic half cells. This twin benefit integrated system is used to produce power for the existing bioethanol plant, and simultaneously metabolize CO_2 emissions, from bioethanol production, through photosynthesis by the microalgae growth in the cathodic PBR half-cell. Furthermore, the biodiesel is produced as energy by-product during the growth of microalgae (Powell and Hill, 2009). To achieve all these benefits, a synthetic chemical mediator should be supplemented to the anode chamber to allow electron shuttling between yeast cells and the electrode. The cathodic half-cell was aerated with feed-air containing 10 % v/v CO_2 which is sparged directly into the cell culture, and illuminated by sunlight to facilitate photosynthesis by microalgae in the PBR.

Similar concept was demonstrated by joining a glass PBR to MFC to form the PMFC (Strik et al., 2008a). The illuminated PBR was used for the preliminary start-up of algal growth. A sparger was used to pump air inside the reactor, whilst the MFC has dual electrodes separated by a cation exchange membrane. Jiang et al. proposed upflow MFC and PBR coupled system for bioelectricity generation and wastewater treatment (Fig. 17.2) (Jiang et al., 2013). The upflow MFC mainly comprised a plastic cylinder with carbon fibre brush electrodes, and glasswool/bead layers separator between anode and cathode compartments. An external column PBR was coupled with the upflow MFC, in which the effluent from the cathode compartment of the up flow MFC was continuously pumped into the column PBR. Continuous illumination was provided to the microalgal culture and a purge of CO_2 (effluent of MFC) and air mixture gas, to help it grow.

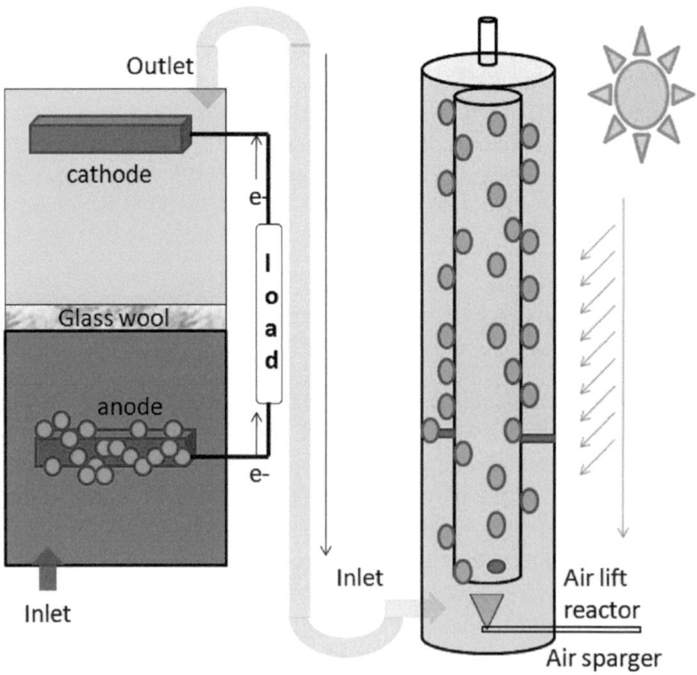

Fig. 17.2 Schematic diagram of upflow type MFC integrated with PBR.

17.4.2 Single Chamber PMFC

S. platensis has the ability to directly shuttle electrons to the electrode, without the requirement of mediators. A membrane-less single chamber PMFC design was employed with photosynthetic bioanode (Lin et al., 2013). In this design, there's a simple scheme that allows the direct attachment of microalgae to the anode and it was suggested to construct an electrical power generator (Fig. 17.1). Some blue–green microalgae were used as a biocatalyst for electricity production. The required electro potential can be created by anodic biofilm, where the respiration reaction in the dark and the photosynthetic reaction in light can generate current (Fig. 17.3).

Mixed culture of bacterial and microalgal cells can be used to improve the efficiency of PMFC as a synergistic approach (Nishio et al., 2013) in which the microalgal cells have the potential to use organic materials like acetate, which is assimilated by the bacterial cells as a substrate for electricity generation (Fig. 17.4). This design can be used as a MFC and portable bio-battery, with the application of both light and dark conditions. The photosynthetic reaction is activated by illumination which is considered as a reversible process for recharging the MFC to operate for longer times.

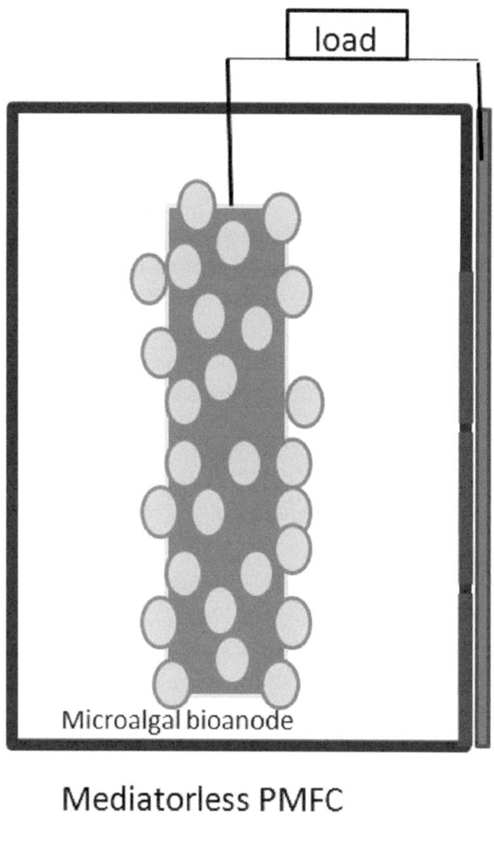

Fig. 17.3 Schematic configuration of MFC with microalgae catalysed bioanode.

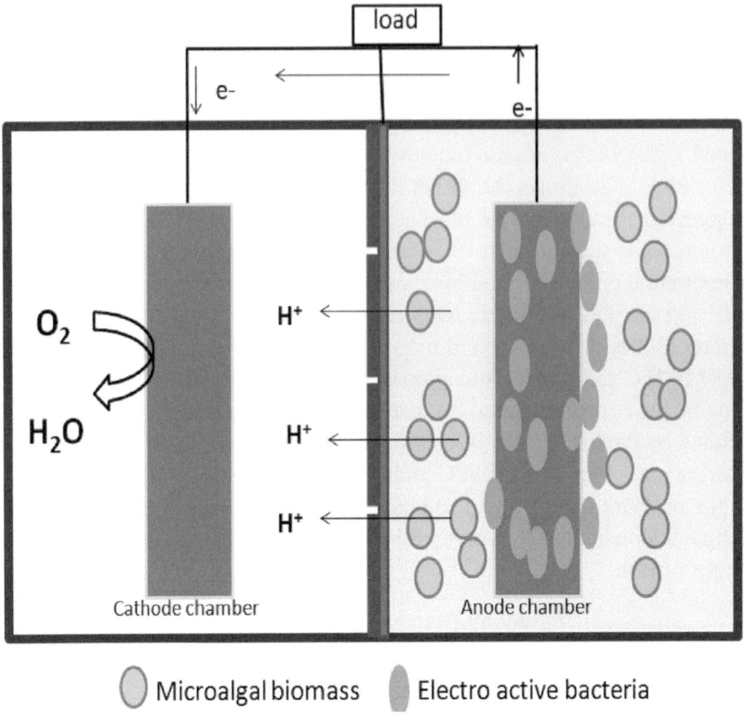

Fig. 17.4 Schematic configuration of co-culture (electro active bacteria and microalgae cells) catalysed design.

Another type of the single chamber PMFC is photo-biological fuel cell in which a PEM is sandwiched between anode and cathode. Microalgae was used to form anodic biofilm with atmospheric CO_2 as a carbon source in mixotrophic nutritional mode which increases the practicality of the association between autotrophic and heterotrophic metabolism in the same system by facilitating the conditions in which all types of CO_2 (atmospheric and organic) could be consumed (Gajda et al., 2013).

17.4.3 Dual Chambers PMFC

The algal photosynthesis is used as the source of oxygen in the cathodic chamber using a dual chambered PMFC (Lobato et al., 2013). This setup consists of two chambers separated by an ion exchange membrane (IEM). Usually, the inoculum for the anodic compartment is the activated sludge from a wastewater treatment plant, which is covered during operation to exclude light and thus to avoid the growth of algae (Rodrigo et al., 2009). On the other hand, the cathode compartment contains a culture of microalgae, which is illuminated for 12 h a day. The anodophilic bacteria produce CO_2, which is transferred to the cathode compartment in some of these configurations so that it can be utilized by the microalgae through the photosynthesis process. This can be achieved by creating a vent at the top of each chamber which is connected by a tube with a funnel-shaped gas collector placed in the anode side in order to simplify the piping of the produced CO_2 into the cathode for microalgal biomass production via photosynthesis (Wang et al., 2010).

An alternate method is the utilization of microalgae as a bioanodic catalyst within a dual chamber PMFC separated by PEM with a chemical cathodic catalyst (Raman and Lan, 2012). Overall, the dual chamber setup can be efficiently started up by a direct three-stage process involving the separate production of bacteria and microalgae cultures, subsequently the substitution of the mechanical aeration system by the microalgae culture and finally a shift in the light dosage from the continuous input to the dynamic light/dark regime.

17.4.4 Photosynthetic Sediment MFC (PSMFC)

Energy can be produced through the naturally existing differences in potential (De Schamphelaire et al., 2008) by using a configuration in which an anode is buried in sediment and a cathode submerged in the water laying on top of the sediment. Such configuration is called sediment microbial fuel cell (SMFC) or benthic MFC. This energy is exploited through the reducing power of microorganisms in the sediment, directly created by the oxidation of organic molecules or the redox reactions of inorganic reduced complexes i.e. sulphur and the cathodic reaction of the SMFC includes the reduction of electron acceptors like the dissolved oxygen in water (Fig. 17.5).

A configuration which incorporates microalgae in SMFC was proposed, in which biogenic compartment was added, replacing the cathodic compartment (Jeon et al., 2012). The CO_2 produced by anodic bacterial activity is consumed by algal cells, and the O_2 produced by the algae is consumed by the PSMFC's cathode compartment for current generation. This PSMFC generally consists of an anode placed in the middle of a sediment layer which is covered with sand and a cathode compartment which is filled with microalgal culture medium. The PSMFC is normally operated in the presence of a light source for photosynthesis to take place.

17.5 Power Generation by PMFC

Normally, microalgae can be cultivated in the anode or cathode compartments of the PMFC. Anodic microalgae have the ability to assimilate substrate, generating electrons and transferring these electrons directly to the anode without the aid of a shuttling mediator; hence they can be used to generate current directly. The role of cathodic microalgae is different, where they are used as biological oxygenators

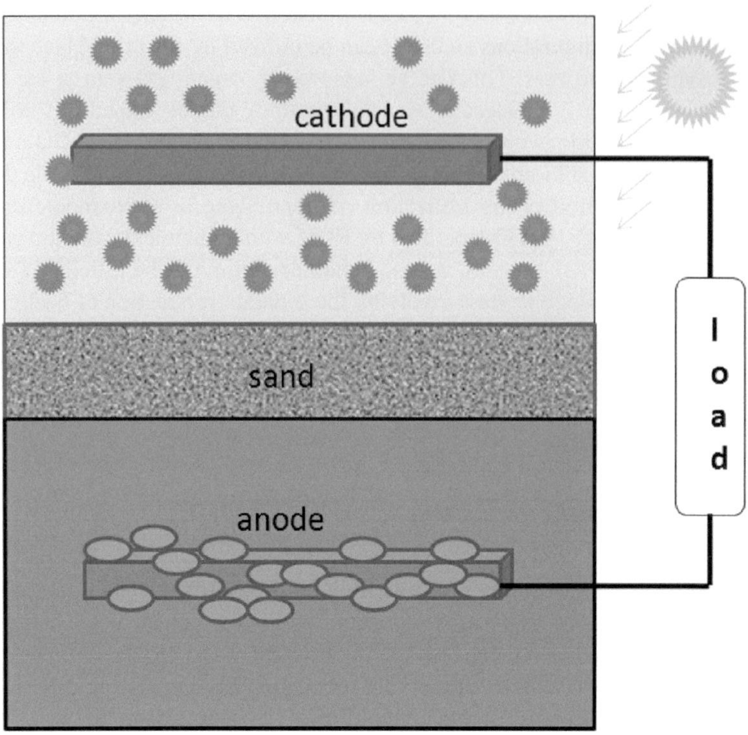

Fig. 17.5 Schematic configuration of photosynthetic sediment MFC (PSMFC).

instead of the mechanical ones which add up to the total cost of generated energy of MFC. A number of trials were performed in order to obtain the maximum benefits of microalgal MFC either in anode or cathode compartments.

17.5.1 PMFC with Anode Catalyzed Microalgae

Microalgae develop a biofilm on the anode of PMFC which has the ability to assimilate substrate generating electrons, which are then transported to the anode either directly or via a mediator.

17.5.1.1 Photosynthetic Bacteria at the Anode with Mediators

During the period 1995–2005, extensive e research was carried out in the field of mediator-based PMFCs. Electrons cannot be transferred from the normal microbial electron transport systems to the electrode due to the non-conductive nature of the cell surface structures. Mediators are typically redox molecules (e.g. ubiquinones, dyes and metal complexes) that can form reversible redox couples, are stable in both oxidized and reduced form, are not biologically degraded and are not toxic towards the microbial consortium. Electrochemical mediators are, therefore, employed to render electron transfer from the microbial cells to the electrode. It may be noted that the overall efficiency of the electron transfer mediators also depends on many other parameters, and in particular on the electrochemical rate constant of the mediator re-oxidation, which depends on the electrode material (Fig. 17.6).

Cyanobacteria species such as *Anabaena* and *Synechocystis* were identified to function as biocatalysts with HNQ (2-Hydroxy-1,4-napthaquinone) as a mediator (Yagishita et al., 1999). The artificial redox mediator helped in shuttling electrons from the microalgae to the anode. Power generation increases in the dark phase during which oxidation of intracellular carbon sources (glycogen) occurs, and electrons are recovered with BES (Yagishita et al., 1998). On the other hand, power production was limited by oxygen production during light phase reaction. The bottleneck for these early PMFCs was that the mediators used were unsustainable and not

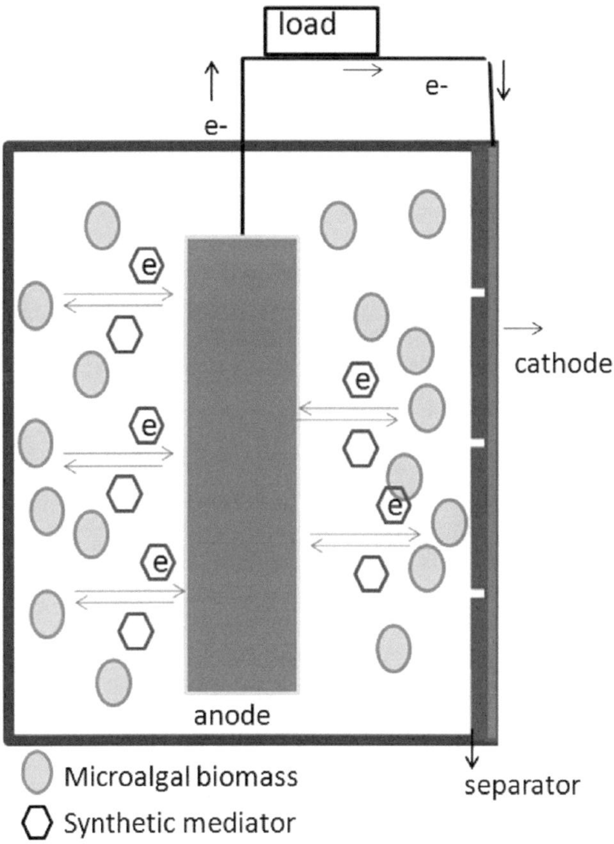

Fig. 17.6 Schematic configuration of PSMFC with chemical mediator in anode chamber.

environment-friendly. Hence, their popularity has been drastically reduced, and become obsolete.

17.5.1.2 Hydrogen Generating Photosynthetic Bacteria with an Electrocatalytic Anode

The general idea of using hydrogen-generating photosynthetic bacteria along with catalyst loaded anode is biohydrogen production photosynthetic bacteria followed by *in situ* oxidation of hydrogen on the electrocatalytic surface of anode. A direct biophotolysis of H_2 production is a biological process which utilizes solar and photosynthetic systems similar to plants which convert water into chemical energy. With hydrogen gas recovery, it is advantageous to generate electric power from biohydrogen (Fig. 17.7).

To reap the advantages of maintaining a very low partial pressure of hydrogen, the photosynthetic hydrogen gas production units are coupled with *in situ* hydrogen oxidation by an electro catalytic conversion step. In this process, originally tested in 1964, H_2/H^+ serves as a natural electron mediator between the microbial metabo-

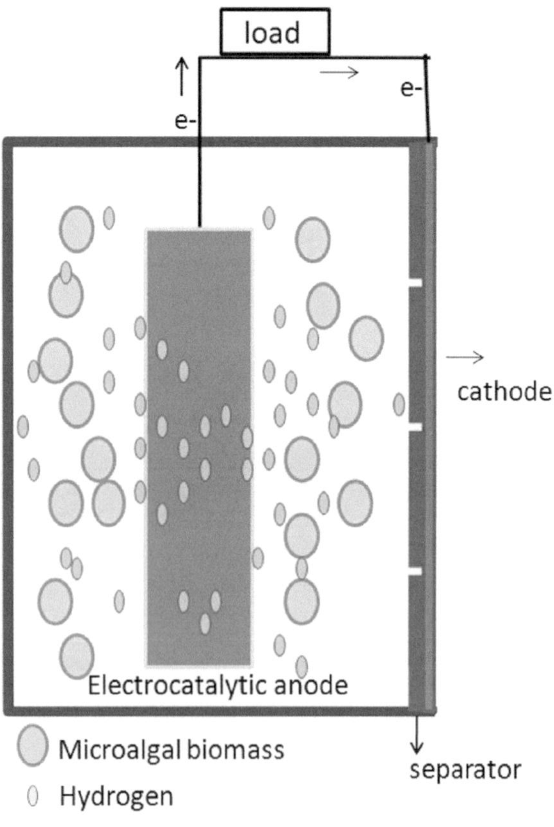

Fig. 17.7 Schematic diagram of PBR (for biohydrogen generation) with electrocatalytic anode (for in situ current production).

lism and the anode. It was found that only after a period of dark followed by illumination, microalgae acquire ability to produce H_2 because dark period creates anaerobic environment. In a recent study involving the green alga *Chlamydomonas reinhardtii*, it was found that *in situ* hydrogen removal (keeping partial pressure of hydrogen very low) is beneficial for increasing hydrogen production. Sulphur deprivation on green algae, *C. reinhardtii,* was found to create anaerobic conditions and hydrogen evolution by biophotolysis under light using photosynthetic pathway (Melis, 2002). Studies have also been done on anoxygenic photosynthetic bacteria, producing H_2 in a photo fermentation process, coupled with electro catalytic electrodes for immediate removal of hydrogen (Cho et al., 2008). Further, direct dependence of power production on photosynthetic activity has been determined using *Rhodobactre sphaeroides* (3 W/m^3 in light vs. 0.008 W/m^3 in dark). Hydrogen was produced by splitting of water, which can sustain for longer period of time. Hydrogen production in a sulphur-deprived condition is the result of two different electrons transfer pathways: the residual photosynthesis (PSII-dependent pathway, resulting from water photolysis) and simultaneously with a PSII-independent pathway that uses catabolism of endogenous starch reserves as an additional source of electrons. A combined process of H_2 generation and *in situ* oxidation of the hydrogen to produce electricity could be a cheaper process than a two-stage process of hydrogen collection followed by the oxidation of hydrogen. Nevertheless, the hydrogen yield and rate of hydrogen production are very low; process is complex, and complicated. The most important requirement for an electro catalytic PMFC is an electro catalyst that is stable and cost efficient. Typical H_2/H^+ exchange membrane fuel cell catalysts are prone to poisoning and inactivation under dirty microbial conditions (Nießen and U.S, 2004). Though the platinum catalyst typically used can be protected using conductive polymers, cheaper non-noble metal electro catalysts, like tungsten carbide (WC), can be a more promising substitute. However, even with these catalysts, stability issues should be overcome before it becomes possible to use them practically on a large scale (Rosenbaum et al., 2007).

17.5.1.3 Direct Electron Transfer Between Photosynthetic Bacteria and Electrodes

The transfer of electrons from the microorganism to the anode without the presence of any artificial redox mediators, electro catalytic electrodes or heterophillic bacteria is known as Direct Electron Transfer (DET). It still remains unclear whether this process actually takes place in nature (Rabaey et al., 2009). Till date, publications which identified DET at the anode also included the use of electro catalysts, like Pt or polyaniline, which are catalytically active towards H_2 generated by the photosynthetic algae. Again, independent from the BES, potentially conductive microbial nano-wires have been discovered for some cyanobacteria like *Synechocystis* sp. PCC 6803, which suggest that DET might be possible in nature. Synechocystis PCC-6803 was used as exoelectrogen to generate anode potential (Fig. 17.3). Electrically conductive nano-wires were observed in these cyanobacteria at anode chamber under excess light and CO_2 limiting condition. More interestingly, even bicarbonate reduction using light in a photo-bio-cathode was shown as a DET process since the autotrophic microbes, in the absence of any organic matter, did not produce oxygen, neither were any electro catalysts present, nor did flushing the cathode to wash away soluble redox mediators disrupt the current (Gorby et al., 2006). This anoxygenic photosynthetic process would use CO_2 as the e^- acceptor with the cathode as the e^- donor, mimicking iron (II)–mineral oxidation as described for photoautotrophic bacteria. However, for this process, an unidentified mixed culture was used as inoculum, which indicates presence of other autotrophic bacteria that might have played a role. Further research is, therefore, required to confirm DET process occurrence in pure bacterial cultures.

Some microalgal species have the ability to shuttle electrons directly to the anode without the need of mediator, e.g. *S. platensis* and based on this ability, it was examined by several authors for current generation using a single chamber membraneless and mediator-free PMFC. Different electrode types were used in these studies, i.e., gold and platinum (Fu et al., 2010). Fu et al. observed power density of 1.64 mW/m, which was amplified by Lin et al., (2013) to reach 10 mW/m^2. The amplification was probably due to the usage of gold anode and graphite cathode (Lin et al., 2013). In this type of PMFC, *S. platensis* is carrying out the photosynthesis reaction to produce oxygen as a by-product while exposed to light. Then it acts as an oxidant and possibly inhibits anodic oxidation reactions, reducing the generated electrical power. The PMFC's electricity generation gets influenced by the chlorophyll content of microalgal film where lower voltage is obtained in light conditions as compared to dark conditions (Cho et al., 2008). This negative influence of light on the voltage is reversed with other PMFC configurations in which light intensity increases the output voltage (Cao et al., 2008).

In order to achieve a sustainable system which is able to sustain on CO_2 and concurrently obtain a value added biomass in a biorefinery concept, membrane was included in the same PMFC configuration and operated with the mixed culture of microalgae inoculated in the anode side (Venkata Subhash et al., 2013). Low power density was obtained (0.004 mW/m^2), almost certainly because of the presence of different strains in the mixed culture with different abilities toward direct electron shuttling to the anode. The reason could also be the mixotrophic mode that was applied using both the atmospheric CO_2 and wastewater as carbon sources. Dissolved oxygen produced during the photosynthesis process was found to be the major restrictive factor towards dropping performance. In addition, electrons installed at the electrode in light condition were higher as compared to that installed in dark condition under oxygenic environment (Chandra et al., 2012). Light source and intensity also plays a major role in chlorophyll development, stomata opening and photosynthesis in phototrophic cells (here microalgae). Lan et al. (2013) studied the influence of light on the PMFC power density in terms of light source and intensity, as they can significantly affect the chlorophyll development, stomata opening and photosynthesis process in microalgal cells. So, the *Chlamydomonas reinhardtii* was used as a bioanode catalyst in PMFC and was illuminated with monochromatic red and blue lights with different intensities. The observed PMFC power density was directly proportional to light intensity, with the superiority of red light to blue one, with maximum value of 13 mW/m^2. In another study, the same microalgal strain was used to optimize the electrode distance within two chambers PMFC with graphite electrodes and the application of dynamic light/dark regime (Raman and Lan, 2012). Maximum power density of 0.82 mW/m^2 was achieved at electrode distance of 14.7 cm, with a substantial reduction in the internal resistance.

17.5.1.4 Synergism Between Phototrophic Microorganisms and Mixed Heterotrophic Bacteria in Sediments

Photosynthetic processes help in accumulating organic matter which subsequently undergoes oxidation by heterotrophic microbial catalysts and consequently electricity is produced by the PMFCs. Photosynthetic producers and heterotrophic consumers form various synergistic relationships in the eco system, often along with diverse anode-respiring bacterial species. In recent years, researchers investigated CO_2 fixation by photosynthesis combined with heterotrophic electricity generation with three types of systems.

In saline and freshwater sediments, algae and some bacteria, like cyanobacteria, are capable of supplying organic matter (e.g. excreted polysaccharides) to heterophillic bacteria through photosynthesis, hence maintaining synergistic communities in ecosystems like microbial mats (Strik et al., 2008b). The same synergistic relationship was examined in freshwater sediment- type PMFC for generating electricity. The PMFC when containing microbial community under illumination produced current continuously. Similar to early studies, electricity generation in this case also showed inverse relationship with illumination. The magnitude of current increased in the absence of light and decreased when light was provided. Current production decreases under continuous illumination possibly due to accumulation of oxygen. Studies were also conducted using marine microbiota a sediment-type PMFCs with microbial anode and cathode (Kaku et al., 2008). Photosynthetic microorganisms here in the overlaying water produced O_2, for reduction by the cathode, and organic matter, which is used as a carbon source in the anode in the anaerobic sediment. This forms a self-maintaining synergistic BES, consuming light and producing electricity. Other than in the previous studies, light dependent current generation was observed in this work, because the system depended on O_2 production at the cathode.

17.5.1.5 Ex situ Photosynthesis Coupled with Mixed Heterotrophic Bacteria at a Dark Anode

Externally generated biomass formed by photosynthesis can also be added to heterotrophic MFCs to generate electricity (Fig. 17.8).

Algae can be used as anodic fuel to generate electricity in these *ex situ* PMFCs. The energy value of algal biomass comes about because its carbon molecules con-

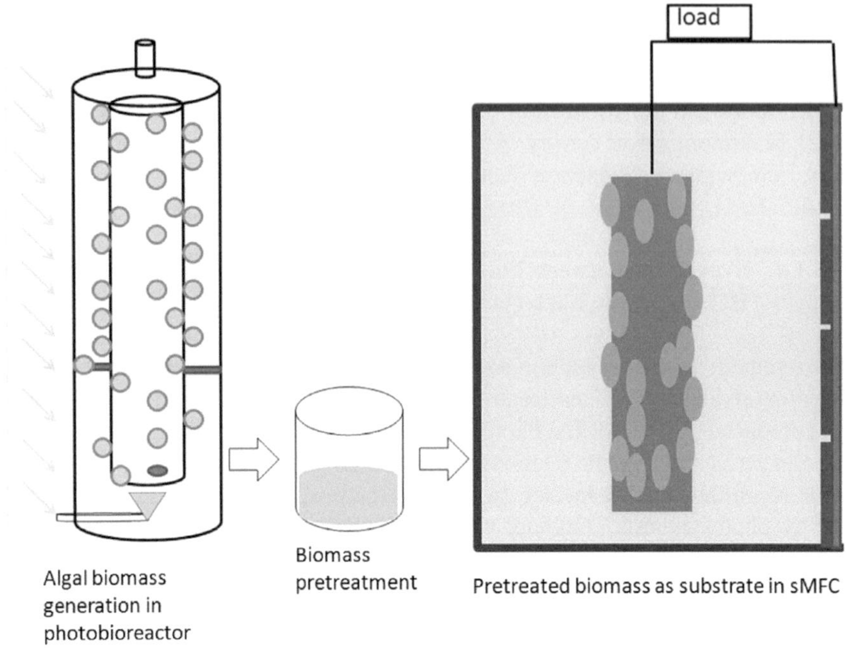

Fig. 17.8 Flow chart of algal biomass generation in airlift PBR followed by pre-treatment and its utilization in MFC as substrate.

tain high-energy electrons. Biofuels from algae are the "third generation" of biofuel feedstock as they can potentially address most of the concerns about first- and second-generation fuels. Algal biomass is rich in carbohydrates, protein and lipids though its concentration varies depending upon strain to strain and on various physico-chemical parameters influencing the metabolic pathway. One of the most attractive features of microalgal biomass production is the potential to fix CO_2 from atmosphere. Between 1.6 and 2 g of CO_2 is captured for every gram of microalgae biomass produced. The microalgal cell wall contains significant amount of the cellulose and hemicelluloses for which pre-treatment is required to break down cell structure and disrupt crystalline structure of cellulose for the accessibility of the cellulose prior to its use as substrate for biofuel generation. This pretreated substrate can be converted into anaerobic production of electricity (Lakaniemi et al., 2012a). Electro active bacteria (EAB) can channel the electrons and their energy to generate bioelectricity in MFCs, society's most widely useful energy form—directly without combustion. These systems require separate PBRs for optimal algal growth (no shading by electrode) and less complicated dark MFC system for optimal generation of electricity (Lakaniemi et al., 2012b). In the first study, dried algae powder was added to the MFC and desirable results were obtained. In the second study, the MFC and PBR were connected in series. At this point it is of special interest to note that there are limitations of feeding complex organic matter (like algae cells) to a mixed heterotrophic bacterial community in an MFC. This limitation arises in the form of very low columbic efficiencies, of only about 2.8 %. *Chlorella vulgaris* (2500 mg COD/L), a phytoplankton containing more than 50 % protein and *Ulva lactuca* (2500 mg COD/L), a macrophyte containing about 60 % carbohydrates, was used in MFCs to produce electricity and their performance was compared. Maximum power densities obtained was 980 mW/m^2 for *C. vulgaris* and 760 mW/m^2 for *U. lactuca* (Velasquez-Orta et al. 2009).

A better modified system would be a PBR along with immobilised cyanobacteria, to generate easily degradable metabolic products, in series with a dark MFC to increase the columbic efficiency of the system, since higher efficiency levels are reached by carboxylic acids than complex materials (Rabaey et al. 2009).

17.5.2 PMFC with Microalgae at the Cathode

It is possible to generate O_2 by photosynthesis both *in situ* and *ex situ*, by recirculating the solution from the PBR to the PMFC cathode. The aim of this is to provide the terminal e^- acceptor oxygen without aeration (Logan, 2008). The concept evolved during the very early development of BES. In the study, the PMFC was provided with O_2 at the cathode by marine algae. In more recent research work, the alga *Chlorella vulgaris* was grown at the cathode and it was claimed that there was mediated electron transfer with the cathode since an artificial mediator was added. However, it is also believed that the oxygen also played a very important role in the transfer process. The artificial redox mediator used might have transferred electrons directly from the non-catalytic cathode to the O_2 produced by photosynthesis activity of the algae (Powell et al., 2009). In another recent study, *in situ* O_2 generation in a PMFC using an undefined mixed culture was used to reverse the anode and cathode during the dark and light phases respectively. It might be strategic to have small amounts of O_2 at the anode since it would provide advantageous energetic profiles under micro-aerobic conditions. Also, as mentioned earlier, photosynthetic O_2 generated by bacteria at the cathode on top of the anaerobic sediment, during the light phase, might be advantageous with regard to current generation in sediment type PMFCs (Strik et al., 2010).

17.5.2.1 Microbial Carbon Capture Cell (MCC)

Carbon dioxide (CO_2) was also demonstrated as the major gaseous end product when either glucose or acetate was used as substrate and in an MFC with real wastewater. In MCC, with light illumination, the microalgae in the cathode chamber can utilize CO_2 from the anode as carbon source for photosynthesis and produce oxygen, which is electron acceptor for electricity generation. Further, CO_2 in exhausted gas from different industries can be converted to useful biomass with photosynthetic creature, achieving simultaneous electricity generation, CO_2 sequestration, wastewater treatment and biomass production. Flue gas which contains large amounts of CO_2 could provide suitable buffering capacity and high NOx-containing flue gas could be used for power augmentation since nitrate can act as potential electron acceptor.

Using microalgae as bio-cathodes in PMFCs help in replacing the mechanical aeration methods. This reduced the cost of the process and is hence more sustainable as well as economic. The algae cultivated in the cathode can also reduce the CO_2 generated from bacterial metabolism and respiration. CO_2 could be used as electron acceptor in cathode. It is already reported that the use of bicarbonate buffer resulted in decrease in the internal resistance and an increased power density (Fan et al., 2007) and continuous addition of CO_2 to cathodes maintained sustainable catholyte pH and improved the anolyte pH, alkalinity and conductivity. Studies have been conducted with *C. vulgaris* to study CO_2 capture in the cathodic chamber. At 10 % v/v CO_2 concentration, the maximum cell growth of 3.6 mg l^{-1} h^{-1} was obtained which generated a power density of 2.7 mW m^{-2}. Use of algae in MFCs help to convert them into complete microbial systems, reducing the need for mechanical energy and hence reducing the net cost of the process. This configuration consists of anodic and cathodic chambers separated by a PEM, and wastewater as an anolyte. The cathode can be illuminated by different light regimes, for 12–24 h, and the importance of light for the working process can be determined. The cell voltage as well as the DO decreases in the dark phase reaction, the acclamation phase power density becoming around 13.5 mW/m^2. Wang et al. (2010) developed another configuration of the PMFCs called the microbial carbon capture cell (MCC), which employs the ability of the algae *C. vulgaris* to reduce CO_2 emissions. The CO_2 generated in the anode was removed by the microalgae, while the soluble inorganic carbon was transformed into algal biomass. The output voltage of the MCC (610 mV) was comparable to that of an MFC (630 mV), generating a maximum power density of 5.6 W/m^3 (Wang et al., 2010).

From these values it was confirmed that the algae are capable of producing enough oxygen for optimum operation of the MFCs. However, the electricity generated was not constant due to fluctuations in the dissolved oxygen concentration, which is dependent on the illumination in the chamber. It was also discovered that the cathode polarization resistance is higher than that of the anode due to which the cathodic reaction is the limiting factor in this configuration. To reduce this resistance, immobilized *C. vulgaris* is used in the cathode in MCC, the configuration simultaneously generating electricity, treating wastewater and producing biodiesel. 85 % COD removal was achieved with a power density of 2485.35 mW/m^3. Compared to suspended cells, the immobilized cells had about 58 % more columbic efficiency (Zhou et al., 2012b). Cao et al. (2009) demonstrated that in the presence of light, biocathode could be used to directly reduce bicarbonate. When the biocathode was used in an MFC, the maximum power density obtained was 15-fold larger than that produced using a plain carbon cathode. This indicated the possibility of direct electron transfer between a cathode and microorganisms for the fixation of carbon dioxide in biomass, while at the same time allowing the generation of electricity from biodegradable organic matter. Pandit et al. (2012a) investigated the performance of the MCC with *Anabaena* sparged with CO_2–air mixture as compared with that of a conventional cathode sparged with air only (Fig. 17.9).

The power densities achieved were 33.3 % higher for *Anabaena* sparged with a CO_2–air mixture, as compared to air sparging only. The experimental results suggested that flue gas which contains large amounts of CO_2 might provide suitable buffering capacity and high NOx-containing flue gas could be used for power augmentation since nitrate can act as potential electron acceptor. Power generation in an MCC depends on both light and bicarbonate utilization.

To determine the applicability of *C. vulgaris* for PMFC operation without an oxidant, studies were conducted with organic-rich sediments in the anode. The CO_2 generation increase is independent of the generated current and this in turn inhibits CH_4 production (Jin et al., 2011). This reflects the PSMFC capability to provide microalgae biomass producing method by the oxidation of organics based on current generation. A mixed culture of *Chlorella* and *Phormidium* can be used to study the effect of current intensity on power generation by PMFCs (Juang et al., 2012). High light power should be avoided if oxygen in the cathodic chamber is produced by algal photosynthesis. Syntrophic interactions between the bacteria and the algae were employed to make a photo-solar cell, using mixed cultures of *C. reinhardtii* and the iron reducing bacteria *Geobacter sulferreducens*. The maximum power density was 41 mW/m^2. The algae produced formate in the absence of light, which was oxidised by the bacteria to generate current.

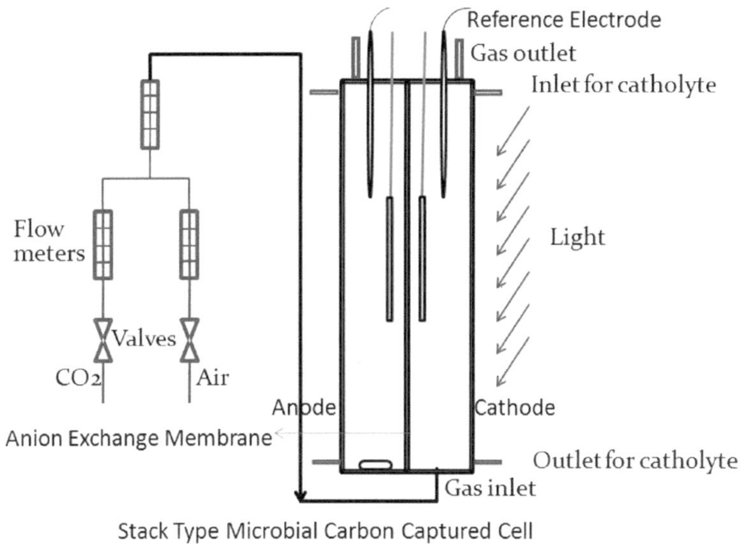

Fig. 17.9 Schematic design of the MCC apparatus with CO_2 sparging system.

17.5.3 *PMFC with Integrated PBR*

Another possible configuration, as already mentioned, is formed by connecting a membrane-less up-flow MFC to a PBR containing the microalgae. This can be used for treating wastewater as well as for generating electrical energy. Wastewater is fed into the MFC to reduce COD, phosphorous, and nitrogen and to produce electricity. The effluent leaving the cathode then enters the PBR to reduce remaining phosphorous and nitrogen by the microalgae. The maximum power density obtained in this case was 481 mW/m^2, along with 78 % COD removal. A modified configuration, with a membrane introduced, was capable of producing electricity regularly for 100 days with a maximum power density of 110 mW/m^2. This MFC can continuously produce electricity and algal biomass along with treatment of wastewater.

17.6 Bottleneck of PMFCs

Microalgae are capable of producing high cell density with high reaction velocity and resistance to hazardous materials. However, effective supply of CO_2 becomes a problem in cases involving high density biomass. MFCs with bio-cathodes need to be improved further as great over potentials are expressed, resulting in energy loss. Resistance of charge transfer is also a problem when accompanied by open air bio-cathodes, though the oxygen mass transfer is reduced.

Power generation rapidly decreases with the increase in the thickness of the biofilm at the cathode, on both of electrode and current collector. Also, shortening of the start-up time for biocathodes affects the output significantly. The cells involving biocathodes are very sensitive to pH change, which affect the electrostatic charges, hydrophobicity cell morphology and sizes of cells. In spite of using buffer to control the pH change, it still remains a major problem.

Another important concern is the role of carbon source. Its limitation affects the performance of the MFCs. Autotrophic growth mode is slower than heterotrophic mode, and this affects the overall time of the process.

Molecular diffusion and electro osmosis are other challenges limiting the high performance of MFCs, occurring due to the crossover of organic materials from the anode to the cathode through the separating membrane. The crossover of substances reduces cathodic potential and alters its surface leading to poisoning of the catalyst, which affects the columbic efficiency of the system. The cathode is turned irreversibly to an aerobic heterotrophic biofilm due to the high COD content inflow and oxygen supply is restricted to the cathodic biofilm.

New MFC configurations are required to be designed to adjust the supply of CO_2, light and nutrients to the microalgae. Overall, biologically catalysed cathodes generate lower power as compared to chemically catalysed ones, which are the main challenge facing microalgal cathodes. In dual chambered MFC, cathode chamber becomes alkaline after long term operation, as a result operating voltage drops

(Pandit et al., 2012b). For MCC, this pH splitting or imbalance can be minimized through CO_2 sparging (Pandit et al., 2012a). However cations dissolved in medium have been found to negatively affect the photosynthetic process. Algal carries net negative charge in their surface and hence are a potential adsorber of polyvalent cations present in the surrounding medium. Adsorption of polyvalent cations on the surface of algae causes morphological changes or can replace/block the prosthetic metal atoms in the active site of relevant enzymes leading to photosynthesis inhibition. Light intensity at the depth of dense algal suspension is greatly reduced because of the absorption and scattering of light. Attenuation of light intensity is dependent of its wavelength, cell concentration, penetration distance of light and the geometry of photobiocathode (Lan et al., 2013).

17.7 Perspective

There was a boom in experimental research work on photosynthesis coupled with MFCs. PMFC with artificial mediator or electrocatalytic anode has become obsolete due to very slow kinetics and non- sustainability. Currently, power generation using pretreated algal biomass has become a popular topic. However, the method applied for hydrolysis or pre-treatment must be cost effective, easy to handle, energy efficient, maximum yield of fermentable sugars before its use as substrate in MFC. Glycerol after biodiesel extraction and other carbohydrate-rich part of carbon dioxide sequestered can be utilized for bioenergy recovery. Apart from this, sediment PMFC is emerging as promising means which has ability to provide electric current in remote places. It utilizes substrate from excreted organic matter from cyanobacteria or dead microalgal biomass and produces enough power to drive environmental biosensors.

DET in photobiocathode has also drawn maximum attention in recent past. The effect of small changes in the environmental conditions, like presence or absence of O_2, changes in the pH, dynamics of the microbial communities, and conductivity changes, on the current generation in MFCs, will be hard to differentiate from DET. Another indication of DET might be the increased electrochemical signals from the light sensitive and electrochemically active mediators, like quinone, upon illumination with sunlight. Thus, all these possibilities have to be taken into consideration and eliminated before a general acceptance emerges regarding the occurrence of DET process in physical systems. If this is further verified to be correct, then a full PMFC with, for example, the promising *in situ* hydrogen oxidation in the photo-anode and anoxygenic photosynthesis in the photo-biocathode becomes a real possibility. Photobiocathode appears to be promising as the power generation of algal containing biocathode is comparable with abiotic one. Nevertheless unlike abiotic cathode photobiocathode is sustainable for long-term operation. Moreover, CO_2 sequestered biomass can be harvested from the photobiocathode and utilized for biofuel and fertilizer generation. The entire process is environmentally benign and economically viable. Though investigations of energy recovery from wastewater using MCCs have been found promising, these have been limited to laboratory-scale experiments and relatively optimal operational conditions. At current stage, the most likely limiting factors for successful scale up of this technology are the high fabrication cost and large internal resistance of MCC. Therefore, significant optimization of various parameters for the operation of MCCs will be required for most of the applications.

References

Bentley, R.W. (2002). Global oil & gas depletion: An overview. *Energy Policy*, 30(3), 189–205.

Cao, X., Huang, X., Boon, N., Liang, P. and Fan, M. (2008). Electricity generation by an enriched phototrophic consortium in a microbial fuel cell. *Electrochemistry Communications*, 10(9), 1392–1395.

Cao, X., Huang, X., Liang, P., Boon, N., Fan, M., Zhang, L. and Zhang, X. (2009). A completely anoxic microbial fuel cell using a photo-biocathode for cathodic carbon dioxide reduction. *Energy & Environmental Science*, 2(5), 498–501.

Chandra, R., Venkata Subhash, G. and Venkata Mohan, S. (2012). Mixotrophic operation of photo-bioelectrocatalytic fuel cell under anoxygenic microenvironment enhances the light dependent bioelectrogenic activity. *Bioresource Technology*, 109, 46–56.

Cho, Y.K., Donohue, T.J., Tejedor, I., Anderson, M.A., McMahon, K.D. and Noguera, D.R. (2008). Development of a solar-powered microbial fuel cell. *Journal of Applied Microbiology*, 104(3), 640–650.

De Schamphelaire, L. and Verstraete, W. (2009). Revival of the biological sunlight-to-biogas energy conversion system. *Biotechnology and Bioengineering*, 103(2), 296–304.

De Schamphelaire, L., Rabaey, K., Boeckx, P., Boon, N. and Verstraete, W. (2008). Outlook for benefits of sediment microbial fuel cells with two bio-electrodes. *Microbial Biotechnology*, 1(6), 446–462.

Fan, Y., Hu, H., & Liu, H. (2007). Sustainable power generation in microbial fuel cells using bicarbonate buffer and proton transfer mechanisms. *Environmental science & technology*, 41(23), 8154–8158.

Fu, C.-C., Hung, T.-C., Wu, W.-T., Wen, T.-C. and Su, C.-H. (2010). Current and voltage responses in instant photosynthetic microbial cells with Spirulina platensis. *Biochemical Engineering Journal*, 52(2–3), 175–180.

Fu, C.-C., Su, C.-H., Hung, T.-C., Hsieh, C.-H., Suryani, D. and Wu, W.-T. (2009). Effects of biomass weight and light intensity on the performance of photosynthetic microbial fuel cells with Spirulina platensis. *Bioresource Technology*, 100(18), 4183–4186.

Gajda, I., Greenman, J., Melhuish, C. and Ieropoulos, I. (2013). Photosynthetic cathodes for Microbial Fuel Cells. *International Journal of Hydrogen Energy*, 38(26), 11559–11564.

González del Campo, A., Cañizares, P., Rodrigo, M.A., Fernández, F.J. and Lobato, J. (2013). Microbial fuel cell with an algae-assisted cathode: A preliminary assessment. *Journal of Power Sources*, 242, 638–645.

Gorby, Y.A., Yanina, S., McLean, J.S., Rosso, K.M., Moyles, D., Dohnalkova, A., Beveridge, T.J., Chang, I.S., Kim, B.H., Kim, K.S., Culley, D.E., Reed, S.B., Romine, M.F., Saffarini, D.A., Hill, E.A., Shi, L., Elias, D.A., Kennedy, D.W., Pinchuk, G., Watanabe, K., Ishii, S., Logan, B., Nealson, K.H. and Fredrickson, J.K. (2006). Electrically conductive bacterial nanowires produced by *Shewanella oneidensis* strain MR-1 and other microorganisms. *Proceedings of the National Academy of Sciences*, 103(30), 11358–11363.

Jeon, H.J., Seo, K., Lee, S.H., Yang, Y.-H., Kumaran, R.S., Kim, S., Hong, S.W., Choi, Y.S. and Kim, H.J. (2012). Production of algal biomass (*Chlorella vulgaris*) using sediment microbial fuel cells. *Bioresource Technology*, 109, 308–311.

Jiang, H., Luo, S., Shi, X., Dai, M. and Guo, R. (2013). A system combining microbial fuel cell with photobioreactor for continuous domestic wastewater treatment and bioelectricity generation. *Journal of Central South University*, 20(2), 488–494.

Jin, J., Yang, L., Chan, S.M.N., Luan, T., Li, Y. and Tam, N.F.Y. (2011). Effect of nutrients on the biodegradation of tributyltin (TBT) by alginate immobilized microalga, Chlorella vulgaris, in natural river water. *Journal of Hazardous Materials*, 185(2–3), 1582–1586.

J. Nießen, U.S. (2004). Flourinated Polyanilines as Superior Materials for Electrocatalytic Anodes in Bacterial Batteries. *Electrochemistry Communications*, 6, 571–575.

Juang, D.F., Lee, C.H. and Hsueh, S.C. (2012). Comparison of electrogenic capabilities of microbial fuel cell with different light power on algae grown cathode. *Bioresource Technology*, 123, 23–29.

Kaku, N., Yonezawa, N., Kodama, Y. and Watanabe, K. (2008). Plant/microbe cooperation for electricity generation in a rice paddy field. *Applied Microbiology and Biotechnology*, 79(1), 43–49.

Khilari, S., Pandit, S., Ghangrekar, M.M., Das, D. and Pradhan, D. (2013). Graphene supported α-MnO_2 nanotubes as a cathode catalyst for improved power generation and wastewater treatment in single-chambered microbial fuel cells. *RSC Advances*, 3(21), 7902–7911.

Kruzic, A.P. and Kreissl, J.F. (2009). Natural Treatment and On-site Systems. *Water Environment Research*, 81(10), 1346–1360.

Kumar, K., Roy, S. and Das, D. (2013). Continuous mode of carbon dioxide sequestration by C. sorokiniana and subsequent use of its biomass for hydrogen production by E. cloacae IIT-BT 08. *Bioresource Technology*, 145, 116–122.

Lakaniemi, A.-M., Tuovinen, O.H. and Puhakka, J.A. (2012a). Anaerobic conversion of microalgal biomass to sustainable energy carriers—A review. *Bioresource Technology*, 135, 222–231.

Lakaniemi, A.-M., Tuovinen, O.H. and Puhakka, J.A. (2012b). Production of Electricity and Butanol from Microalgal Biomass in Microbial Fuel Cells. *BioEnergy Research*, 5(2), 481–491.

Lan, J.C.-W., Raman, K., Huang, C.-M. and Chang, C.-M. (2013). The impact of monochromatic blue and red LED light upon performance of photo microbial fuel cells (PMFCs) using Chlamydomonas reinhardtii transformation F5 as biocatalyst. *Biochemical Engineering Journal*, 78, 39–43.

Lin, C.-C., Wei, C.-H., Chen, C.-I., Shieh, C.-J. and Liu, Y.-C. (2013). Characteristics of the photosynthesis microbial fuel cell with a Spirulina platensis biofilm. *Bioresource Technology*, 135, 640–643.

Lobato, J., González del Campo, A., Fernández, F.J., Cañizares, P. and Rodrigo, M.A. (2013). Lagooning microbial fuel cells: A first approach by coupling electricity-producing microorganisms and algae. *Applied Energy*, 110, 220–226.

Logan, B.E. (2008). Microbial Fuel Cells. Wiley-Interscience.

Malvankar, N.S. and Lovley, D.R. (2012). Microbial Nanowires: A New Paradigm for Biological Electron Transfer and Bioelectronics. *ChemSusChem*, 5(6), 1039–1046.

Markou, G., Vandamme, D. and Muylaert, K. (2014). Microalgal and cyanobacterial cultivation: The supply of nutrients. *Water Research*, 65, 186–202.

Melis, A. (2002). Green alga hydrogen production: progress, challenges and prospects. *International Journal of Hydrogen Energy*, 27(11), 1217–1228.

Nayak, B.K., Pandit, S. and Das, D. (2013). Biohydrogen. *In*: Kennes, C. and Veiga, ría C. (eds). Air Pollution Prevention and Control. Chapter 15, 345–381. John Wiley & Sons, Ltd.

Nishio, K., Hashimoto, K. and Watanabe, K. (2013). Light/electricity conversion by defined cocultures of Chlamydomonas and Geobacter. *Journal of Bioscience and Bioengineering*, 115(4), 412–417.

Olivieri, G., Salatino, P. and Marzocchella, A. (2014). Advances in photobioreactors for intensive microalgal production: Configurations, operating strategies and applications. *Journal of Chemical Technology and Biotechnology*, 89(2), 178–195.

Pandit, S., Nayak, B.K. and Das, D. (2012a). Microbial carbon capture cell using cyanobacteria for simultaneous power generation, carbon dioxide sequestration and wastewater treatment. *Bioresource technology*, 107, 97–102.

Pandit, S., Ghosh, S., Ghangrekar, M.M. and Das, D. (2012b). Performance of an anion exchange membrane in association with cathodic parameters in a dual chamber microbial fuel cell. *International Journal of Hydrogen Energy*, 37(11), 9383–9392.

Parlevliet, D. and Moheimani, N.R. (2014). Efficient conversion of solar energy to biomass and electricity. *Aquatic Biosystems*, 10, 4.

Popp, J., Lakner, Z., Harangi-Rákos, M. and Fári, M. (2014). The effect of bioenergy expansion: Food, energy, and environment. *Renewable and Sustainable Energy Reviews*, 32, 559–578.

Powell, E.E. and Hill, G.A. (2009). Economic assessment of an integrated bioethanol–biodiesel–microbial fuel cell facility utilizing yeast and photosynthetic algae. *Chemical Engineering Research and Design*, 87(9), 1340–1348.

Powell, E.E., Mapiour, M.L., Evitts, R.W. and Hill, G.A. (2009). Growth kinetics of Chlorella vulgaris and its use as a cathodic half cell. *Bioresource Technology*, 100(1), 269–274.

Rabaey, K., Angenent, L. and Schroder, U. (2009). Bioelectrochemical Systems: From Extracellular Electron Transfer to Biotechnological Application. IWA Publishing.

Raman, K. and Lan, J.C.-W. (2012). Performance and kinetic study of photo microbial fuel cells (PMFCs) with different electrode distances. *Applied Energy*, 100, 100–105.

Reimers, C.E., Stecher, H.A., Westall, J.C., Alleau, Y., Howell, K.A., Soule, L., White, H.K. and Girguis, P.R. (2007). Substrate Degradation Kinetics, Microbial Diversity, and Current Efficiency of Microbial Fuel Cells Supplied with Marine Plankton. *Applied and Environmental Microbiology*, 73(21), 7029–7040.

Rodrigo, M.A., Cañizares, P., García, H., Linares, J.J. and Lobato, J. (2009). Study of the acclimation stage and of the effect of the biodegradability on the performance of a microbial fuel cell. *Bioresource Technology*, 100(20), 4704–4710.

Rosenbaum, M., He, Z. and Angenent, L.T. (2010). Light energy to bioelectricity: photosynthetic microbial fuel cells. *Current Opinion in Biotechnology*, 21(3), 259–264.

Rosenbaum, M., Zhao, F., Quaas, M., Wulff, H., Schröder, U. and Scholz, F. (2007). Evaluation of catalytic properties of tungsten carbide for the anode of microbial fuel cells. *Applied Catalysis B: Environmental*, 74(3–4), 261–269.

Rozendal, R.A., Hamelers, H.V.M., Rabaey, K., Keller, J. and Buisman, C.J.N. (2008). Towards practical implementation of bioelectrochemical wastewater treatment. *Trends in Biotechnology*, 26(8), 450–459.

Singh, S.P. and Singh, P. (2014). Effect of CO_2 concentration on algal growth: A review. *Renewable and Sustainable Energy Reviews*, 38, 172–179.

Skjånes, K., Lindblad, P. and Muller, J. (2007). BioCO_2—A multidisciplinary, biological approach using solar energy to capture CO_2 while producing H_2 and high value products. *Biomolecular Engineering*, 24(4), 405–413.

Strik, D.P.B.T.B., Hamelers, H. V.M. and Buisman, C.J.N. (2010). Solar Energy Powered Microbial Fuel Cell with a Reversible Bioelectrode. *Environmental Science & Technology*, 44(1), 532–537.

Strik, D.P.B.T.B., Terlouw, H., Hamelers, H.V.M. and Buisman, C.J.N. (2008a). Renewable sustainable biocatalyzed electricity production in a photosynthetic algal microbial fuel cell (PAMFC). *Applied Microbiology and Biotechnology*, 81(4), 659–668.

Strik, D.P.B.T.B., Hamelers (Bert), H.V.M., Snel, J.F.H. and Buisman, C.J.N. (2008b. Green electricity production with living plants and bacteria in a fuel cell. *International Journal of Energy Research*, 32(9), 870–876.

Velasquez-Orta, S.B., Curtis, T.P. and Logan, B.E. (2009). Energy from algae using microbial fuel cells. *Biotechnology and Bioengineering*, 103(6), 1068–1076.

Venkata Subhash, G., Chandra, R. and Venkata Mohan, S. (2013). Microalgae mediated bioelectrocatalytic fuel cell facilitates bioelectricity generation through oxygenic photomixotrophic mechanism. *Bioresource Technology*, 136, 644–653.

Wang, X., Feng, Y., Liu, J., Lee, H., Li, C., Li, N. and Ren, N. (2010). Sequestration of CO_2 discharged from anode by algal cathode in microbial carbon capture cells (MCCs). *Biosensors and Bioelectronics*, 25(12), 2639–2643.

Yagishita, T., Sawayama, S., Tsukahara, K.-I. and Ogi, T. (1998). Performance of photosynthetic electrochemical cells using immobilized Anabaena variabilis M-3 in discharge/culture cycles. *Journal of Fermentation and Bioengineering*, 85(5), 546–549.

Yagishita, T., Sawayama, S., Tsukahara, K. and Ogi, T. (1999). Effects of glucose addition and light on current outputs in photosynthetic electrochemical cells using Synechocystis sp. PCC6714. *Journal of Bioscience and Bioengineering*, 88(2), 210–214.

Zhou, M., He, H., Jin, T. and Wang, H. (2012a). Power generation enhancement in novel microbial carbon capture cells with immobilized Chlorella vulgaris. *Journal of Power Sources*, 214, 216–219.

Zhou, M., He, H., Jin, T. and Wang, H. (2012b). Power generation enhancement in novel microbial carbon capture cells with immobilized Chlorella vulgaris. *Journal of Power Sources*, 214, 216–219.

Chapter 18
Remediation of Domestic and Industrial Effluents Using Algae

G. Gera, S.N. Yewalkar, and S.N. Nene

18.1 Introduction

In the environment, microbial activity is considered as one of the most important mechanisms for the abatement of water-borne pollutants. In the natural and anthropogenic environment the wide range of contaminates are not eliminated by single specie but by the complex interaction of a mixed microbial population performing complementary reactions. This principle is very much applicable for the treatment of the industrial and domestic wastewater, which has excess of nitrogenous compounds (N) and phosphates (P). N and P along with various organic pollutants, if not properly treated, would create a devastating impact on natural aquatic ecosystems. Among the many other disturbing impacts, most prevalent is the phenomenon of eutrophication, which is the accumulation of high levels of organic matter and the decomposing organisms, which deplete the oxygen in water, and causing the death of other organisms, such as fish. The excess nutrients in the aquatic ecosystem support the growth of various phytoplanktons. It not only spoils the water quality but also adversely affects the whole aquatic ecosystem. This chapter will address the possibility of effectively utilizing a natural microbial flora/consortium, enriched with the rapidly growing algae for the remediation of polluted water bodies.

G. Gera • S.N. Yewalkar • S.N. Nene (✉)
Biochemical Engineering Unit, Chemical Engineering & Process Development Division, National Chemical Laboratory, Dr. Homi Bhabha Road, Pune 411008, India
e-mail: sanjay.nene@gmail.com

D. Das (ed.), *Algal Biorefinery: An Integrated Approach*,
DOI 10.1007/978-3-319-22813-6_18

18.2 Potential of Microalgae for Treating the Domestic and Industrial Effluents

Nutrient removal is becoming an important priority for wastewater treatment plants due to its detrimental impact on water bodies receiving this effluent. Dairy wastewater, swine manure or wastewater from piggery, food processing industries (fish processing, slaughter house waste) and agro-industrial waste are rich in nitrates, phosphate and other organic content. The current biological methods of nutrient removal make use of aerobic and anaerobic/anoxic methods, that are inadequate for effective removal of these nutrients (Singh and Thomas, 2012). These constituents are often the nutrients required for the growth of green microalgae. Thus soluble nitrates and phosphate get removed from the waste when treated with green microalgae. At this point it is important to make a distinction between algal blooms and microalgae. Algal blooms are multicellular phytoplankton known to naturally produce bio-toxins and grow as visible patches in water bodies. They are known to harm the health of plants, animals and adversely affect the environment. Periodic mechanical removal of these algal blooms is one of the solutions to avoid the dangerous side effect on the ecosystem. The other option is maximum removal of nutrients from the waters. Microalgae, on the contrary, are typically found in freshwater and marine systems. They are unicellular species which exist individually, or in chains or groups, are known to fix light photosynthetically and act as feed for rotifers, fish and other aquatic life forms.

This biological method of removal of the phosphates and nitrates from the wastewater has many advantages over waste treatment process currently used in terms of

1. Effective uptake of N and P if sufficient solar energy is available
2. Simultaneous production of O_2 and consumption of CO_2 in presence of light
3. No requirement for the supply of extra organic or inorganic nutrients
4. Providing oxygenated water due to activity of algae
5. Less sludge accumulation
6. Absence of generation of secondary pollutants; ecologically, a safe technique
7. Generation of microalgal biomass which can be utilized for feedstock, fertilizer, biogas and biofuel (Xin et al., 2010; Wang et al., 2013; Woertz et al., 2009a, b).

Algal cake generated in the wastewater treatment and its utilization for the biodiesel production is very attractive option for reducing energy, nutrients, fresh water cost as well as reducing green house gas emission (Olguin, 2012).

18.2.1 Direct Use of Algae

This algal system has more potential for the waste treatment in tropical and subtropical regions. Table 18.1 summarizes the various algae used for the treatment of the domestic and industrial effluents and their effectiveness in removal of phosphate and nitrates. Growth of bacteria along with these algae must be acknowledged for

Table 18.1 Algae growing in the wastewater by exploiting the nutrients present in the wastewater

Wastewater source	Algae used	%N	%P	Time (days)	Reference
Secondary treated sewage	*Phormidium laminosum*	48.7	99.7	2	Sawayama et al. (1998)
Municipal wastewater	*Chlorella* sp.	58.85	97.1	24	Wang et al. (2013)
Domestic wastewater	*Chlorella* sp.	36	12.5	3	Singh and Thomas (2012)
	C. vulgaris	47.35	18.8	3	
	Scenedesmus quadricauda	42.8	31.3	3	
	S. dimorphus	45.1	25		
Agroindustrial waste	*C. vulgaris*	60	100	9	Gonzfilez et al. (1997)
Piggery waste	*Euglena viridis*	56.77	60	8	Godos et al. (2010)
	C. sorokiniana	78.57	45	8	
Municipal wastewater	*C. vulgaris*	81.6	92.8	10	Lau et al. (1995)
Simulated wastewater	*C. vulgaris* immobilized	100 %	93.9	1	Tam and Wong (2000)
Municipal wastewater	Mixed algal sp.	96 %	99 %	15	Woertz et al. (2009a, b)
Piggery wastewater	*S. quadricauda*	92.2 %	75 %	19	Gantar et al. (1991)

their symbiotic relation and enhancing the remediation of the wastewater (Sriram and Seenivasan, 2012). These symbiotic bacteria utilize complex organic matter present in the wastewater as a carbon source and oxygen produced during algal photosynthesis. The bacterial growth enhances COD reduction of the wastewater. Though some algae are reported to degrade some toxic phenolic compounds, the bacterial symbiotic partner has the lion's share in degrading the toxicants present in industrial wastewater (Pinto et al., 2003).

18.2.2 Photosynthetic Aeration

Mechanical agitation system or aerators are generally employed in the wastewater treatment process for providing sufficient aeration required for bacterial degradation of biological matter. However, it is the most energy intensive step as compared to any other process parameter. Brandi et al. (2000) and Sánchez-Monedero et al. (2008) have suggested the need for a submerged oxygenation system instead of mechanical aeration as the later system was found to be less environment friendly, due to high rate of aerosol dispersion around the tanks containing microorganisms. This was likely to promote air-borne transmission of pathogenic bacteria and fungi with an adverse bearing on human health (Li et al., 2013). The algal growth in the wastewater treatment tank provides a significant part of the oxygen (by photosynthesis) required by aerobic bacteria to breakdown the organic matter present in the wastewater and lowering the BOD level. Bacteria would, on the other hand, generate CO_2 required by algae as carbon source for growth thereby establishing a symbiotic relationship between the two. Oxygen generated by microalgae promotes the bacterial biodegradation of recalcitrant pollutants like phenolics, organic solvents, and polycyclic aromatic hydrocarbons (PAH's). Therefore, it is especially advantageous as the recalcitrant and toxic compounds can be easily degraded aerobically than the anaerobic process (Munoz and Guieysse, 2006).

In eutrophic lakes dissolved oxygen concentration and pH increase are attributed to algae and this results in the inactivation of the faecal coliforms as well as *E. coli.* This is particularly beneficial to communities in developing countries who use raw untreated water from lakes and other freshwater sources (Ansa et al., 2011). The algae *Chlorella sorokiniana* in combination with various pollutant specific bacteria e.g. *Ralstonia basilensis* for salicylate, *Acinetobacter haemolyticus* for phenol, *Pseudomonas migulae* and *Sphingomonas yanoikuyae* for phenanthrene were able to degrade the specified pollutants completely by photosynthetic aeration without any external aeration mechanism (Borde et al., 2003). The same consortium of *R. basilensis* and *C. sorokiniana* was reported to remove sodium salicylate at 87 mg L^{-1} h^{-1} in a continuous closed photobioreactor with the 77 mg O_2 L^{-1} h^{-1} of oxygenation capacity which is very much close to that of the mechanical aerator (Muñoz et al., 2004). In this manner, the synergistic relationship between algae and bacteria is efficient for the treatment of industrial waste by minimizing the cost of aeration, which is a major cost centre in the wastewater treatment by conventional methods.

18.3 Microbial Selection

The first step in wastewater treatment is to 'know the wastewater' and characterization of physico-chemical parameters of wastewater. This is followed by selection of the proper algal specie for the wastewater treatment. Growth of genera Scenedesmus and Chlorella are commonly observed in the water contaminated with domestic sewage in warmer climates (Jalal et al., 2011). Water site rich in silicates supports growth of diatoms (Brzezinski et al., 1998). Analysis of metal polluted rivers, receiving waste from a paper mill, containing 20–80 ppm chromium showed growth of *Oscillatoria*, *Phormidium*, *Scenedesmus* and *Pandorina* (Cervantes et al., 2001). In short, remediation of water occurs naturally by selective algae growing in the wastewaters; their growth being controlled by the nutrients present in the water and environmental factors like temperature and light. Considering this point, one should select the algae/mixture of the algae for the treatment, based on the organic and inorganic components present in the wastewater, ambient temperature and light conditions of the natural climatic ecosphere. Proper selection of algae will help in faster (less retention time), effective (maximum removal of the nutrients, toxic components) treatment of the wastewater and generate more biomass. This will make the process cost effective. Isolating the algae/consortia growing in the various wastewaters naturally is very relevant field of research to select the proper algae for the wastewater remediation.

18.3.1 Microbial Tolerance

Though wastewater nutrients support algal growth, it also has several compounds that are potentially toxic to the microalgae, such as heavy metals and other recalcitrants, especially in the industrial wastewater. Heavy metals like Cr are potential inhibitors of photosynthesis; at higher concentration they affect the cell machinery responsible for oxygen evolution notably the electron transport chain of the resistant algal culture (Yewalkar et al., 2013). This definitely hampers the growth rate of the algae. Phenolic compounds are one of the major industrial water pollutants that adversely affect the growth of algae and bacteria.

Piggery wastewater and swine manure supports the growth of algae after dilution. The growth rate of *Scenedesmus* sp. was increased by three-fold after supplementing the medium with 3 % (v/v) fermented swine urine (Kim et al., 2007). Gogdo et al. (2010) found that *C. sorokiniana* and *E. viridis* were found to grow well in the four to eight times diluted piggery wastewater and were able to remove P and N (*E. viridis* removed approximately 60 % N and P; *C. sorokiniana* was able to remove 80 % N and 50 % P). However, *S. obliquus* was unable to sustain continuous growth while *S. platensis* growth was completely inhibited in the eight times diluted swine manure. This growth inhibition was due to its sensitivity towards the high concentration of NH_3 found in the piggery wastewater. The microalgae species studied by them showed different degrees of intolerance to NH_3. The use of NH_3 tolerant microalgae may improve the biodegradation of piggery wastewater.

The tolerant algal species can be obtained by genetic manipulation, cell acclimation to progressively higher pollutant concentration or isolating the algae from the site of contamination where the prevailing microorganisms have already been exposed to the contaminates.

18.3.2 Microbial Interaction

Heterotrophic bacteria play a significant role in aquatic ecosystems. They are not only decomposers in the aqueous phase but also convert dissolved organic carbon into particulate organic carbon (White et al., 1991). These bacteria grow symbiotically in wastewater. Microalgae provide O_2 essential for the aerobic bacteria, which aerobically degrade the organic pollutants. The algae consume the CO_2 generated by bacterial respiration. However, due to algae growth the pH and DOC of the wastewater was found to increase. This may allow selective bacteria to grow. The exo-polysaccharides synthesized by the algae were found to support the growth of microorganisms. This microbial interaction is an essential requirement in the efficient treatment of wastewater treatment.

18.3.3 Microbial Growth Rate

The microbial growth rate of algae depends on the concentration of nutrient, and inhibitory substances. Bacterial growth rate is several orders of magnitude higher than that of algal. This difference in growth rate may lead to an imbalance in population of bacteria and algae. However, the availability of oxygen to bacteria and carbon dioxide to algae often acts as a rate-limiting step. Mouget et al., (1995) found increased growth rate of *S. bicellularis* and *Chlorella* when co-cultured with *Pseudomonas diminuta* and *P. vesicularis*. As the wastewater is highly complex and variable with respect to the chemical composition as compared to the synthetic media, growth rate of the algae may differ accordingly. It is also clear that bacteria enhance the algal growth by utilizing the photosynthetic oxygen evolved by algae, which would otherwise have caused them some deleterious effects due to photo-oxidation.

18.3.4 Microalgae Predominance

Those algae which have a higher adaptability to the environment in terms of faster rates of P and N assimilation, shifts in pH and levels of DOC, will have a faster growth rate in wastewater. The faster growing algae only predominate in wastewater. In uncontrolled environments it is difficult to maintain specific microalgal cultures, especially in case of raw sewage it is difficult in both open ponds and closed photobioreactor to maintain any selectivity in the algal flora because the sewage being treated may have some of its own algal flora that will grow along with the inoculated algal culture. It is our observation that many a times inoculating the domestic sewage with a selected algal specie, is often completely displaced by filamentous blue green algae or *Chlorella* at the end of the treatment.

18.3.5 Inoculation and Selection

The rate of removal of the nitrates and phosphates from the wastewater depends on the rate of algal growth. The rate of removal of phosphates and nitrates can be accelerated by increasing the initial cell density of inoculum (Lau et al., 1995). Inoculation with immobilized algal for an improved degree of removal of a pollutant is another approach (Tam and Wong et al., 2000). However, both these approaches are difficult to implement at a large scale. At larger scale, preparation of an algal inoculum can be done using either fresh water for development of the algae or raw sewage or activated sludge (Munoz and Guieysse, 2006).

18.4 Waste Stabilization Ponds

For the wastewater treatment stabilization ponds are widely employed in which wastewater is treated by natural processes using consortium of algae and bacteria. These waste stabilization ponds have proven to be one of the most stable, cost-effective and easy to operate process for the treatment of domestic and industrial wastewater treatment (Kayombo et al., 2000). These ponds can be effectively employed in the tropical and sub-tropical countries having moderate to high temperature, as energy from the sun is the sole requirement for its operation. Apart from the low energy requirement, it is easily manageable and requires simple maintenance in terms of cleaning and surveillance. Removal of pathogens is one of the key advantages of using waste stabilization ponds (Mara, 1987). Contrary to this, closed photobioreactor systems are more energy intensive, complex to operate and require sophisticated materials of construction. However, PBRs are more flexible to operate as compared to the open pond systems, being more amenable to changes in physiological and biological nature of algal species and the improved mass and heat transfer, mixing and reduction in photo-inhibition and photo-oxidation. The closed photobioreactors are found as either flat plate, tubular, vertical column or helical photobioreactor systems. They have lower footprint and better control over some parameters like the circulation speed, agitation, temperature, dissolved O_2 and CO_2 etc. which make them more beneficial for the treatment of wastewater using algae.

Stabilization ponds are the shallow man-made basins for the treatment of wastewater by using a natural process. They are also known as oxidation ponds or sewage lagoons or redox ponds and are widely used for secondary wastewater treatment. The first reported stabilization pond for the treatment of wastewater was Mitchell lake of an average depth of about 1.4 m in the city of San Antonio, Texas, in 1901 (Gloyna, WHO report 1971). These are widely used in the tropical and sub-tropical countries where the sunlight is available in natural abundance. WSP are very suitable for the developing countries like India where sunlight is in abundance and operating cost is very low. Removal of pathogens by the combined treatment with algae and bacteria makes it more attractive for treating wastewater. Temperature, evaporation rate, waste flow and receiving BOD are the important considerations while designing the WSP. West Bengal is the only state in India where the WSP are installed, as Calcutta is one of the biggest wastewater fed-fisheries in the world. The four places in West Bengal where these WSP are located at Titagarh, Panihati, Bally (North Howrah) and on the outskirts of Nabadwip. The history of Calcutta East wastewater fed-fisheries is 80 years old and is still in place, providing employment to some 4000 people (Mara, 1997). Stabilization ponds can be classified as anaerobic pond, facultative pond and maturation pond which are employed in series or in parallel, depending upon the type of waste effluent, for effective wastewater treatment.

18.4.1 Anaerobic Pond

Anaerobic sewage ponds are generally designed on the basis of volumetric BOD loading rates (g/m^{3}d). In order to maintain anaerobic conditions and prevent odour emissions, the value of volumetric BOD loading rate should lie within 100–400 g m^{-3}d^{-1}. The performance of anaerobic pond significantly depends upon its temperature as the BOD volumetric rate increases with the rise in temperature (Mara, 1997). They generally have negligible dissolved oxygen and contain no algae. Anaerobic ponds are commonly 1–5 m deep and the bacterial consortium in the pond is very sensitive to the pH < 6. BOD removal up to 80 % can be achieved by using anaerobic ponds in series with the HRT of 1–2 days (Rao, 1972). The main advantage of an anaerobic pond is that it not only stabilizes the organic matter of wastewater but also has a lower land area requirement as compared to facultative ponds.

18.4.2 Facultative Pond

Facultative pond or lagoons (1.5–2 m deep) are designed based on the surface BOD loading (kg ha^{-1} d^{-1}) and act as primary and secondary facultative ponds. A primary facultative pond operates as a primary clarifier for the wastewater received from the anaerobic pond. Aerobic degradation of organic matter by the bacteria is prominent in the primary facultative pond. The secondary facultative pond receives this particle-free wastewater. Raw wastewater is simultaneously treated in anaerobic and primary facultative ponds along with the secondary facultative pond. Facultative ponds need clear particle-free water for deep light penetration to promote algal photosynthesis and decomposition of organic matter. A facultative pond consists of three layers during its operation namely aerobic zone, facultative zone and anaerobic zone. In the uppermost aerobic zone mixing is promoted by wind and photosynthetic oxidation by microalgae takes place increasing the dissolved oxygen, which prevents release of odourous gases like H_2S. Organic wastes are decomposed to acids and alcohols which are further degraded into CO_2, NH_3, H_2S and CH_4 in the facultative zone. The algal bacteria consortia present in the facultative zone help in nutrient and BOD depletion respectively.

In the secondary facultative pond the remaining BOD is lowered by heterotrophic bacteria namely *Pseudomonas*, *Flavobacterium*, *Archromobacter* and *Alcaligenes* sp. (Abdel-Raouf et al., 2012). The bacterial and microalgal growth is supported by the symbiotic relationship that exists between the two viz. oxidation of BOD by bacteria using the O_2 released by the photosynthetic growth of algae which in turn uses the CO_2 released by the bacteria. The dissolved oxygen concentration in the facultative pond increases after sunrise as a result of photosynthetic activity of algal culture and gets depleted at sunset. Increase in CO_2 released by the bacteria often leads to increase in the pH of the waste up to 8–9. At such an alkaline pH fecal coli from bacteria are killed thereby improving the quality of wastewater treatment. Several weeks (2–3 weeks) are required for the treatment of waste in the facultative pond which makes it less efficient. Moreover the growth of rotifers and protozoas which feed on algae may lead to depletion in dissolved oxygen level giving rise to bad odour due to anaerobic decomposition of organic matter. Cyanobacterial growth in the pond too reduces the light penetration thereby leading to the death of selective algal strain and inhibiting the organic matter decomposition.

18.4.3 Maturation Pond

Pathogen removal is the most important task performed by maturation ponds from the wastewater received from facultative ponds and are known as polishing ponds. Maturation ponds are generally 0.8–1.5 m deep where a pond depth of 1.0 m is common. Their main purpose is to provide high quality effluent by getting rid of pathogens from the facultative pond. For designing the maturation ponds residence time, temperature, pH and light intensity are the major parameters of importance. A light intensity over wavelengths of 425–700 nm are reported to cause death of most faecal bacteria, with the exception of *Vibrio cholerea* (Mara and Pearson, 1986). Maturation ponds are not designed to lower BOD significantly as compared to facultative and anaerobic ponds, where 90 % lowering of BOD takes place. In short, due to lower initial organic loading concentration BOD removal is depressed. Effective nutrient removal can also be achieved if operated in conjugation with the algae and/or fish breeding (Tilley et al., 2014). Maturation ponds are strictly used when treated wastewater is required for irrigation and should contain faecal coliform bacteria <1000 per 100 ml according to WHO guidelines.

In the waste stabilization ponds, helminthic eggs and cysts are removed during the process of sedimentation whereas *V. cholerae* is reported to be killed by the presence of low sulphide concentration in anaerobic pond (Mara, 1987). Waste stabilization ponds are more reliable and cost effective as they can be constructed using the relatively cheap local materials. If designed correctly it gives consistent and high quality treated effluent, which can be reused, in the aquaculture and for crop irrigation. A poor design may lead to odour emission in case of anaerobic ponds. Therefore, there is a need for expert supervision for periodical removal of sludge and its disposal (Fig. 18.1).

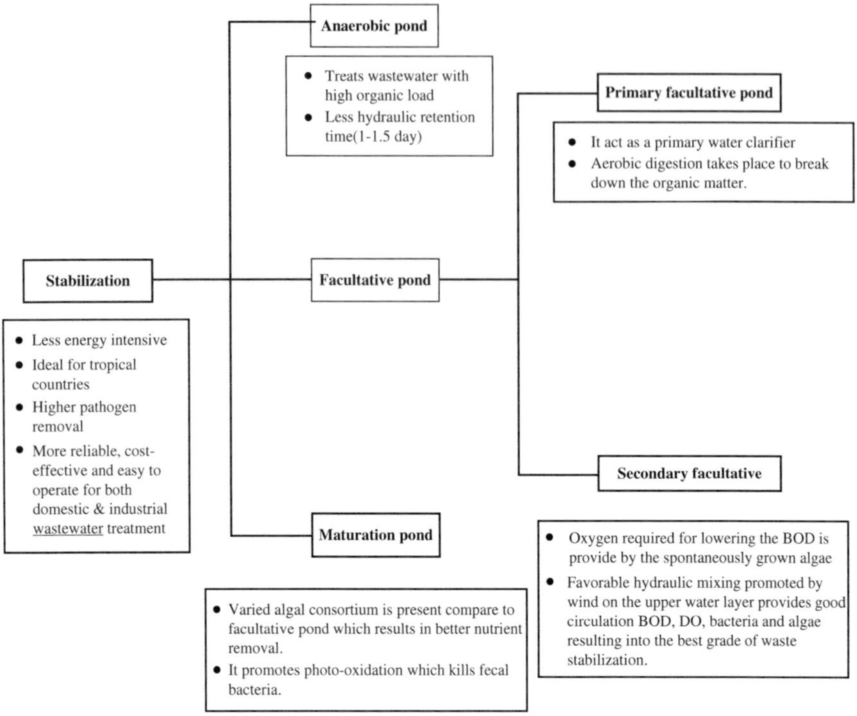

Fig. 18.1 Flow diagram of strategies involved in wastewater treatment.

18.5 Photobioreactor Design

Microalgae rich in lipid and carbohydrate are selected to provide an alternative source of renewable energy to meet the increasing fuel demands all over the world. This biological source of energy was largely independent of expensive raw materials and used solar energy as the primary source of energy for conversion of CO_2 to carbohydrate and lipid by photosynthesis (Schenk et al., 2008; Mata et al., 2009; Tang et al., 2012). Design of photobioreactors for cultivation of algae has received much attention in the past two decades, with the realization that more engineering efforts are needed for designing these low-energy consuming and high throughput photobioreactor system for the microalgae cultivation. Currently for the microalgae cultivation, open and closed photobioreactor systems are employed.

18.5.1 Open Pond System

Open systems comprise the open raceway pond and waste stabilization pond. Open raceway ponds were in use from the late 1950's and was studied extensively for growing algae (David et al., 1953). Open raceway ponds are the shallow ponds fabricated from cheap materials including plastic cladding, designed for the cultivation of algae. They are generally 0.3–0.6 m deep and 0.8–1 m wide having a paddle wheel for effective water circulation and mixing. Open raceway ponds are also termed as high rate algal ponds because of their effectiveness in treating the wastewater effluent at a HRT of 8–10 d. 90–95 % of nutrient and 80 % of COD removal is reported from the fish farm waste and domestic wastewater by using high rate algal pond in 10 d HRT (Posadas et al., 2014). These systems are relatively economical and are used commercially for growing algae as a food, feed and fuel source. Some of the disadvantages associated with open raceway pond include less control over algal culture conditions, bigger footprints, lower light penetration and poor productivity. Much research is needed to minimize the water evaporation losses by optimizing the operating conditions especially turbulence in the raceway pond.

High rate algal ponds can treat upto 35 g BOD $m^{-2}d^{-1}$ (175 g BOD $m^{-3}d^{-1}$ in a 0.2 m deep pond) as compared to 5–10 g BOD m^{-2} d^{-1} (5–10 g BOD $m^{-3}d^{-1}$ in a 1 m

Fig. 18.2 Open raceway pond (Courtesy: National chemical laboratory).

Table 18.2 Comparison of different closed photobioreactors

Type of closed photobioreactor	Advantages	Disadvantages
Flat plate PBR	Provides large surface area, good light path, good biomass productivities.	Scale-up requires many compartments and support materials.
	Low oxygen buildup	Difficult to control temperature.
		Hydrodynamic stress to algal strain with some degree of wall growth.
		Gradients of pH, dissolved oxygen and CO_2 build up.
Tubular PBR	Low shear stress on tubes with good biomass productivities, good mass transfer and good mixing.	Larger foot prints.
	Low shear stress because of good mixing.	Wall growth which affects light penetration.
		Decrease in illumination surface area on scale-up.
		Requires sophisticated material.
		Comparatively expensive.
LED PBR	Rapid light utilization rate	Low illumination surface area.
		Heat generation and wall growth.

deep pond) in a waste stabilization pond (Munoz and Guieysse, 2012). Stringent water discharge norms into the water bodies led to polishing of the treated effluent by intermittent use of sand filters in combination with phytoremediation using reeds (Racault and Boutin, 2005) (Fig. 18.2).

18.5.2 Closed Photobioreactor

Contrary to the open pond system a closed photobioreactor is not open to the environment, but is enclosed in a loop where the section exposed to light could be tubes, cylinders, channels or flat plates made of transparent materials like glass or polymers (polycarbonate, acrylates, etc.) (Tables 18.2 and 18.3). It provides a controlled environment for the growth of microalgae in terms of pH, mixing, light intensity, culture density and temperature. As it often promotes the monoculture of microalgae, higher biomass productivity can be achieved for a longer period of time as compared to an open pond reactor; besides the contaminants (rotifers and protozoans) are restricted because of the controlled environment. Tubular photobioreactors are the most commonly used system for the microalgal cultivation. It has been

Table 18.3 Pollutant removal by using algal photobioreactor

Type of PBR	Type of waste	Volume of the reactor	Pollutant removal achieved	Type of reactor material	Algal strain	Light intensity	Bio-mass concentration	Residence Time	References
LED PBR	Artificial wastewater	35 litre	62 % total Kjeldahl nitrogen	Plexiglass	*Scenedesmus* sp.	30 μmol/m²/s	600 mg/L	10 days	Carlo et al. (2011)
			50 % total phosphate						
Semi closed loop diesel PBR	Domestic wastewater	NA	NA	NA	*Pediastrum algae* sp.	Sulphur lamp	NA	NA	Frank. US patent (2012)
Glass bottle reactor	Swine wastewater	9 litre	NH_3-N = 45 %	Glass	*Chlorella vulgaris*	NA	NA	96 hr	Mezzari et al. (1987)
			>60 % COD reduction						
Continuous stirred tank PBR	Domestic wastewater	5 litre	NO_3 = 43–54 %	Glass	*Chlorella* sp.	4000 ± 300 lux	NA	2–3 days	Singh and Thomas (2012)
			NO_2 = 83–95 %		*Chlorella vulgaris*				
			PO_4 = 70–92 %		*Scenedesmus qudriquada*				
					Scenedesmus dimorphous				
High rate algal ponds (sloping)	Acidic effluent from alginate industry	720 litre	NA	NA	*Chroococcus turgidus*	Sunlight 55–60 klux	NA	4 days	Sivasubramaniam et al. (2009)
Pilot sloping pond	Soft drink industry wastewater	NA	71 % hardness reduction	RCC	*Chlorococcum humicola*	Sunlight	NA	7 days	Sivasubramanian et al. (2012a, b)
			63–74 % hardness reduction		*Chlorella Conglomerata*				
	Textile dyeing waste effluent	NA	87.42 % hardness reduction		*Chlorococcum humicola*	Sunlight	NA	7 days	
HRAP	Leather processing industry chemical effluent	NA	• TKN = 74.28 %	NA	*Chlorella vulgaris*	Sunlight	NA	5 days	Hanumantha Rao et al. (2011)
			• COD = 54 %						
			• BOD = 49.60 %						
			• TP = 99.9 %						

(contiuned)

Table 18.3 (continued)

Type of PBR	Type of waste	Volume of the reactor	Pollutant removal achieved	Type of reactor material	Algal strain	Light intensity	Bio-mass concentration	Residence Time	References
Pilot slope tank	Oil drilling effluent	NA	Sulphur = 95.8 % BOD = 93.20 % TDS = 80.79 %	NA	*Chlorococcum humicola*	Sunlight	NA	10 days	Sivasubramanian et al. (2012a, b)
Pilot slope tank	Detergent industry effluent	1500 L	• 63.79 % reduction in total hardness • BOD = 13.95 % • COD = 13.20 %	NA	*Chlorococcum humicola*	15–75 klux of sunlight	NA	Continuous process	Sivasubramaniam et al. (2011)
Tubular PBR	Secondary treated wastewater	110 L	NH_3-N = >90 %	Low density polyethylene	Mixed culture dominated by *Desmodesmus* sp.	NA	20 gm/m^{2}d	20 min.	Wiley et al. (2013)
Batch closed system	Domestic wastewater	NA	Total nitrogen = 55.26 % Total inorganic carbon ≈ 92 %	Glass	*Chlorella vulgaris*	100.8 μmol/m^{2}s	≈130 mg/L	216 hr.	Kim et al. (2010)
Shake flask	Partially treated wastewater	50 ml	Total nitrogen = 83.1 % Total phosphate = 91.2 % COD = 77.8 %	Glass	*Chlorella vulgaris*	Flourescent light	200 mg/L	10-14 days	Al-Mamun et al. (2012)
Shake flask	Municipal wastewater	500 ml	Total phosphate = 90 % Total nitrogen = 17.04–58.85 %	Glass	*Chlorella* sp.	60 μmol/m^{2}s	0.278 g/L (dw)	24 days	Wang et al. (2012)
NA	Municipal wastewater	NA	NH_4-N = 82.4 % Total phosphate = 90.6 % COD = 83 % (maximum removal)	NA	*Chlorella* sp.	NA	NA	9–10 days	Wang et al. (2010)

Algal turf scrubber raceway pond	Dairy manure effluent	3500 L	Total nitrogen and phosphates = 50–80 %	Cement concrete	Consortium of *Rhizoclonium hieroglyphicum* (the most abundant species), *Microspora willeana Lagerh.*, *Ulothrix ozonata*, *Rhizoclonium hieroglyphicum* and *Oedogonium* sp.	NA	11–14 g dw/m^{2}d	4–12 days	Mulbry et al. (2008)
Glass jar	Pulp and paper industry wastewater	1 L	COD = 58 %	Glass	*Chlorella* and *Diatom* sp.	3.4–5.8 klx	40 mg/L	5–10 days	Tarlan et al. (2002)
			Colour = 84 %						
			AOX = 80 %						
Glass bottles	Settled and diluted piggery waste	1 L	COD = 88.0 %	Glass	*Chlorella vulgaris*	Solar radiation = 0.60–0.75 W/cm^2	NA	190 h	Travieso et al. (2006)
Glass tank photobioreactor	Piggery wastewater	3.5 L	TOC = 42–55 %	Glass	*Scenedesmus obliquus*	10,000 lux	NA	300–500 hrs.	de Godos et al. (2010)
			NH_4 = 21–39 %		*Chlorella sorokiniana*				
					Spirulina platensis				
					Euglena viridis and consortium of two microalgae sp.				
Photobioreactor	Dairy effluent	1 L	Total phosphates = 97.1 %	Polycarbonate	Immobilized *chlorella pyrenoidosa*	55 μmol/m^{2}s	NA	4 days	Yadavalli and Heggers (2013)
			BOD = 87.83 %						
			COD = 84.55 %						
			NH_4^+-N = 98.28 % (calculated)						
Flat bottom cylindrical containers	Artificial wastewater	0.2 m^3	Total nitrogen = 47 % (winter), 79 % (summer)	NA	*Scenedesmus obliquus*	18.1–215 μmol/m^{2}s (winter)	66.15 mg/L	3 days	Gómez-Villa et al. (2005)
			Total phosphates = 45 % (winter), 73 % (summer)			143–2215 μmol/m^{2}s (summer)			

(contiuned)

Table 18.3 (continued)

Type of PBR	Type of waste	Volume of the reactor	Pollutant removal achieved	Type of reactor material	Algal strain	Light intensity	Bio-mass concentration	Residence Time	References
High rate algal pond	Fish farm and domestic wastewater	180 L	COD = 77 ± 9 %	Flexible PVC	*Microalgae consortium*	5435 Wh $m^{-2}d^{-1}$	5 g/m^2d	10–20 days	Posadas et al. (2015)
			Total Kjeldahl nitrogen = 83 ± 10 %						
			Total phosphorus = 94 ± 6 %						
Hollow fibre membrane photobioreactor	Synthetic wastewater	500 ml	NO_3 = 63 %	Glass	*Spirulina platensis*	9000 lux	2131 mg/L	12 days	Kumar et al. (2010)
Bubble column reactor	Piggery wastewater	500 ml	Total nitrate = 628 mg/L	Glass	*Botryococcus braunii*	100 μE m^{-2} s^{-1}	7.5 g/L	12 days	Jin (2003)
			Total phosphate = 90–95 %						
High rate algal pond	Domestic wastewater	100 ha	Ammonia = >98 %	NA	Mixed microalgae culture	NA	20 g/m^2d	7–13 days	Woertz et al. (2009)
			Phosphorus = >96 %						
Shake flask	Process water from hydrothermal carbonization unit	150 ml	Total nitrogen = 45.5–59.9 %	Glass	*Chlorella vulgaris*	100 μE m^{-2} s^{-1}	0.013–0.054 g $L^{-1}d^{-1}$	5 days	Du et al. (2012)
			Total phosphate = 85.8–94.6 %						
			COD = 50.0–60.9 %						
Shake flask	Toxic aromatic pollutant	21–25 ml	Pollutant removal = >85 %	Glass	*Chlorella sorokiniana*	2500 lux	NA	48–144 hrs.	Borde et al. (2003)
Shake flask	Mixed domestic and industrial wastewater	1 L	BOD = 89.29 %	Glass	*Cyanobacteria* sp.: (*Anabaena oryzae*, *Anabaena variabilis*) and *Tolypothrix ceytonica*	NA	NA	7 days	El-Bestawy (2008)
			COD = 73.68 %						
			FOG = 94.63 %						
			TSS = 38.84 %						
			TDS = 64.37 %						

RCC = Reinforced cement concrete, FOG = Fats oil and grease, TDS = Total dissolved solids, TSS = Total suspended solids

Fig. 18.3 Closed photobioreactors 3A: Tubular photobioreactor, 3B: Flat plate photobioreactor.

reported that trading of algal biomass is expensive (up to 5000 €/t) and is used mostly for animal feed and not for production of biofuel (Lehr and Posten, 2009). Cultivating microalgae in wastewater for lipid production provides an alternative for cost reduction of biomass along with concomitant phycoremediation at no extra cost for nutrient supplementation. However, using closed photobioreactor for wastewater treatment makes it relatively expensive as they are difficult to operate and scale up as compared to open raceway pond. Also supplying air rich in carbon dioxide is not economical unless a suitable source (e.g. stack gases) is available in the immediate vicinity. In order to achieve high microalgal biomass productivity along with the metabolites with a by-product value, Habiba et al. (2012) suggested the need for innovative research strategies for designing and developing the photobioreactors in combination with the genetically engineered strains. Table 18.2 gives an overview of the advantages and disadvantages associated with the different designs of closed photobioreactor (Fig. 18.3).

18.5.3 Mixing

A favourable degree of mixing plays a pivotal role for the growth of microalgae and subsequent reduction of nutrient level, good mass transfer and sufficient light exposure for the photosynthetic activity. Waste stabilization ponds generally do not require a mixing mechanism as the wind (surface aeration) plays a major role; however, at times it may lead to the formation of anaerobic zones and increases the mass transfer limitations. Optimal mixing is one of the abiotic factor that greatly affects the pollutant removal, as too much of mixing may lead to shear stress of the algal cells leading to their damage and thereby increasing the operational cost (Su et al., 2011). It has been also reported that the highest productivity of the algal culture of *Chlorella* sp. was achieved in a Plexiglass bowl at 300 rpm due to the centrifugal action of a rapid mixing device named as Algraton. Three such Algratons having 1 m in diameter are claimed to sustain a human diet of 1600 calories per day from the algal biomass produced (Oswald, 2003).

Paddle wheels are the most commonly used mixing device in the open raceway ponds designated as one of the low cost instrument for efficient mixing (Oswald and Gotaas, 1957). However, many sophisticated designs in the form of swirl vanes were developed even for the smaller tubular photobioreactor, which provides mixing by causing helical flow for circulating the culture (Wiley, 2013). Such type of mixing has reduced settling of the microalgae by 86 % of the total biomass. But such kind of mixing instruments add up to the cost of the closed photobioreactor, whereas paddle wheels are the most efficient mechanism in open ponds and HRAP as they are profitable in terms of mixing as well as energy cost.

18.5.4 Biomass Harvesting

Algal biomass harvesting is proving to be the barrier for the algal technology to proliferate economically for energy production as compared to the conventional methods (Rawat et al., 2011). Commercial production of algae for human consumption on an industrial scale started in the early 1960's in Japan and extensive research was conducted in the US, Germany and other countries for use as a food supplement. Growing *Chlorella* sp. on a commercial scale as a food source was considered seriously as early as 1947–48 (Burlew, 1953; Borowitzka, 1999). Unfortunately, most of the algal species, which grow at faster rates, are unicellular and hence, pose problem for harvesting. Some of the desirable characteristics of algae for the cost effective harvesting are discussed by Borowitzka (1999) in Table 18.4.

Sedimentation, filtration, centrifugation, flocculation, sonication and flotation are various methods of choice for algal biomass harvesting and can be broadly classified into two important steps viz. dewatering and drying. The most critical engineering features related to reduction of the cost of the whole process will be covered in the area of biomass harvesting. In general practice, flocculation followed

Table 18.4 Characteristics of algal biomass for efficient harvesting

Desirable characteristics	Advantages	Disadvantages
Growth in extreme environment	Reduces problems with competing species and predators	Only limited number of species are available and some extreme environments are difficult to maintain on a large scale (e.g. Cold)
Rapid growth rate	Provides competitive advantage over competing species and predators; reduces pond area required	Growth rate is usually inversely proportional to the cell size; i.e. fast growing cells are usually very small
Large cell size, colonial or filamentous morphology	Reduces harvesting costs	Large cells usually grow slower than small cells
Wide tolerance of environmental conditions	Less control of culture conditions required for reliable culture	–
Tolerance of shear force	Allows cheaper pumping and mixing methods to be used	–
High cell content of the product	Higher value of biomass	Products are usually secondary metabolites, high concentrations mean slower growth

by gravity sedimentation is the most commonly used method for wastewater treatment as there are large volumes to treat and low biomass concentration (Grima et al., 2003). Multivalent cations and cationic polymers are generally added as a flocculent in the broth to reduce or neutralize the negative surface charge in order to agglomerate the microalgal cells in suspension. The flocculent should be inexpensive, non-toxic and effective in low concentration (as observed for chitosan). Alum is the most widely used multivalent metal salt in wastewater treatment process especially for flocculating *Scenedesmus* and *Chlorella* sp. (Golueke and Oswald, 1965).

According to Jorquera et al. (2010), the energy requirement for algal biomass harvesting can be significantly reduced if microalgae can be concentrated about 30–50 times by coagulation-flocculation followed by gravity sedimentation prior to the dewatering stage. Centrifugation provides good results by compact thickening of the algal slurry but it is one of the most energy intensive processes as compared to three other techniques studied by Sim et al. (1988) for harvesting microalgae namely chemical flocculation followed by floatation and continuous filtration with a fine weave belt filter. Recently Evodos has come up with an improved centrifuge with spiral internals that has been extensively tested on harvesting algae. Separation of algae using electrical energy is also a very effective process as algae carry a net negative surface charge and can be concentrated in an electric field. The major advantage of this process is that there is no addition of chemicals but its high power requirement and electrode cost makes it unsuitable for large scale harvesting of algae (Uduman et al., 2010).

Commercial mechanical harvesters are in practice to yield algal biomass designed to separate algae from a continuous moving belt through vacuum before it passes through the culture broth (Shepherd, 2012). One more continuous belt harvester system being used by Algaventure systems, Inc. is based on a capillary extraction mechanism. In this technique the primary belt is in contact with a secondary belt made of superabsorbent polymer material on which the water is absorbed and the dried biomass is collected on the primary belt while water is drained out from secondary belt by compression before it comes in contact with the primary belt (Youngs and Cook, 2010; Christenson and Sims, 2011). In a patent application filed by Mendez et al. (2009) for Sapphire Energy it describes a genetically modified algae to facilitate controlled flocculation and easier harvesting. The algae is modified to express a ligand or receptor for encouraging flocculation. Such genetically modified algal species including cyanobacteria have been used for production of ethanol, where conventional cell harvesting is avoided, by enclosing the growing culture in a greenhouse where the water vapour and ethanol condense on the ceiling and gets collected in a channel (Woods et al., 2010). Preference of algal harvesting method largely depends upon the nature of the product. For instance, if algal biomass is meant to be used as food or feed supplement then flocculants to be used should be compatible with the food standards and should not cause toxic effects. Certain applications do not require harvesting such as in aquaculture where the mussels, cladocerans, rotifiers, shell fish or fish are directly fed with algae. This process is termed as biological filtration and the feeders are termed as filter feeders which also help in effluent removal along with biomass generation (Guterstam and Etnier, 1991). A company, Live Fuel Inc. utilizes fish as a means of harvesting algae by the planktivorous fish tilapia for oil and fishmeal (Wu et al., 2010a). Immobilized algae are considered to be the best option for continuous production and harvesting of algal biomass; however, very little information is available on its use on large scale (Christenson and Sims 2011). Extensive research is needed in the area of biomass harvesting which can reduce the overall cost of algal production.

18.5.5 Biomass Concentration

Production of algal biomass using wastewater effluent has been in use since the late 1950's and can be defined as the energy stored as new biomass per unit of light absorbed. In addition to the light intensity, oxygen accumulation and shear stress are the limiting factors for the microalgae productivity in the photobioreactor design (Janssen et al., 2002). Due to the abundance of nutrients like nitrogen, phosphorus and carbon found in wastewater, it is an enriched media for biomass production along with the phycoremediation. Although unicellular algae grow at a faster rate with high accumulation of nutrients but they are difficult to harvest and increases the biomass harvesting cost (Pittman, 2011). High rate algal ponds are best suited for the algal biomass production using wastewater with minimum environmental impact, but the algal species needs to have tolerances to seasonal and diurnal variations in outdoor conditions. At times they form aggregates which increase the ease of harvesting especially if they have high accumulation of lipids and other valuable products (Park et al., 2010). However, shading effects occur when the algal biomass concentration increases resulting in reduction of available oxygen required by bacteria as the algae enter the dark respiration phase. The amount of algal biomass concentration obtained in open photobioreactors is almost the same as that in closed photobioreactors e.g. 20 g m^{-2} d^{-1} of maximum biomass concentration is reported in high rate algal pond treating domestic wastewater as well as in tubular photobioreactor using the secondary treated wastewater (see Table 18.3). Also, the algal turf scrubber raceway pond gives 11–14 g m^{-2} d^{-1} of algal biomass by using the dairy manure effluent (Mulbry et al., 2008). In high rate algal ponds, algae are susceptible to consumption by herbivorous protozoa and zooplankton (e.g. rotifers and cladocerans) which can reduce the biomass concentration and cause the pond to crash in a matter of days (Park et al., 2011).

Sun dried, spray dried or the compressed tablets are the common form of microalgal biomass product available in the market from the microalgal biotechnology. *Chlorella* and *Spirulina* dominate the microalgal market with an average production of 2000 and 3000 t/yr respectively and other products are limited to very few taxa viz. *Dunaliella*, *Nostoc* and *Aphanizomenon* (Pulz and Gross, 2004). Genetically transformed microalgae are providing a new ray of hope for the mass production of high value products as the microalgae lacks cell differentiation and exhibit a much simpler system for genetic manipulation as compared to higher plants.

18.5.6 Surface/Volume Ratio

For obtaining superlative algal growth, maximum sunlight exposed surface to volume ratio is one of the key parameter for designing the photobioreactor. A comparison of surface to volume ratios in different types of photobioreactor is given by Pietro Carlozzi (2003) in which he reports helical type of photobioreactor to have maximum surface to volume ratio. Large surface to volume ratio photobioreactor offers maximum light utilization and greatly improves the productivity. Higher illuminated surface to volume ratios of horizontal or inclined tubular and flat plate photobioreactor makes them ideally suited to wastewater bioremediation; however, they cannot be used for axenic culture as sterilizing the reactor medium is difficult (Lee, 2001).

Immobilizing algae for better algal productivity has gained much attention recently as immobilized algal surface provides higher surface to volume ratio for light illumination (Burke, 2013; Ozkan et al., 2012). Rotating photobioreactors with the algal culture immobilized on the surface is emerging as an improved method for the nutrient removal from wastewater and prevention of eutrophication. These rotating photobioreactors are capable of providing high surface to volume ratios thereby increasing the gas transfer, light illumination, biomass productivity and making harvesting inexpensive (Burke, 2013). Maximum biomass yield of 24.94 ± 2.07 g m^{-2}

of *B. braunii* on dry basis is obtained in algal biofilm photobioreactor with significant reduction in energy and water requirement for cultivation (Ozkan et al., 2012). Higher biomass productivity can be achieved by immobilizing algae as compared to suspended culture if surface area is large (Middlebrooks et al., 1974). Tubular photobioreactors provide maximum surface area to sunlight exposure and hence, are more suitable to outdoor mass cultivation (Brennan and Owende (2009). Munoz and Guieysse (2006) demonstrated the action of algal biofilm on the pollutant removal in the vertical flat bed and horizontal algal turf reactor in terms of surface to volume ratio. Here, a vertical flat bed photobioreactor with biofilm attached to the reactor wall was illuminated from the side where the most active algae were not directly exposed to the pollutant bulk liquid. Bacteria consumed the pollutant and its concentration decreased through the biofilm. The most active algae were rejuvenated and the depleted oxygen concentration in the biofilm restored. In contrast a horizontal algal turf reactor containing microalgae when directly illuminated by light and exposed to pollutant water resulted in pollutant toxicity to the biofilm. A maximum degradation rate of 295 mg BOD l^{-1} h^{-1} was achieved in an algal turf reactor (Muñoz et al., 2009).

In order to get maximum surface to volume ratio for higher illuminated surface area, wide land area is required especially for wastewater treatment using microalgae. In this scenario, waste stabilization ponds are more economical and easy for maintenance. Algal-bacterial processes suitable for treatment of 60–4000 m^3 d^{-1} of wastewater with the load of 30–1800 kg d^{-1} can be established depending on the local land prices (Munoz and Guieysse, 2006). As the artificial lighting adds load to the cost of wastewater treatment in algal-bacterial photobioreactor, sunlight is the sole and freely available power source for the oxygenation. Hence, surface to volume ratio plays a key role in designing the photobioreactor, which incorporates the factors such as the biomass concentration, the hydrodynamic regime and the incident light intensity. Along these lines, open raceway ponds and waste stabilization pond provides maximum surface exposure and are best suited for phycoremediation.

18.5.7 Hydraulic Retention Time (HRT)

One of the most important factors that governs the successful and cost efficient wastewater treatment process is the hydraulic retention time. It was always suggested to keep the HRT long enough such that dilution rate should not exceed the maximum algal growth rate μ_{max} and thereby prevents wash out effects during wastewater treatment. However, longer HRTs are not preferred as it may result into slower growth rate due to nutrient limitation and increased internal shading. HRT varies greatly with the seasonal changes, being higher in winter as compared to summers, especially, in case of outdoor waste effluent treatment using open raceway ponds and high rate algal ponds. Nutrient removal was very much affected by the seasonal factor as reported by Gómez-Villa et al. (2005), where 45 % reduction in phosphorus was achieved in winter as compared to 73 % reduction obtained during summer with the *Scenedesmus obliquus* cultivated outdoors in artificial wastewater. Dissolved nitrogen concentrations of 53 % and 21 % of their initial values in winter and summer respectively, were used. Usually 2–6 d of HRT is recommended for the outdoor wastewater treatment system comprising waste stabilization ponds and raceway ponds (Mara and Pearson, 1986). However, similar HRTs are observed in case of closed photobioreactor process (Singh and Thomas, 2012). At a shorter HRT of 98 h and 183 h 98 % elimination of phosphorus and almost 100 % removal of ammonium was achieved in a stirred tank photobioreactor at 25 °C for urban wastewater using *Scenedesmus obliquus* (Martínez et al., 2000). Almost quantitative removal of the pollutant was achieved with shorter HRTs if the algal photobioreactor system was combined with a membrane bioreactor for polishing and safe disposal of treated waste effluent (Singh and Thomas, 2012; Wiley et al., 2013).

18.6 Influence of Environmental Parameters

Algal growth and nutrient uptake in the wastewater is dependent on abiotic and biotic factors. Abiotic factors include physical parameters like pH, light intensity, temperature, colour/opacity of the wastewater, and chemical parameters like concentration of macro and micro nutrients in the wastewater. Biotic factors include the initial density of the algal cells, presence of zooplanktons, algal pathogens etc. (Lau et al., 1995; Park et al., 2011; Grobbelaar, 2009).

18.6.1 Chemical Abiotic Factors in the Waste/Nutrients

Microalgae require energy source in the form of light energy for autotrophic growth or organic carbon for the heterotrophic growth. The mixotrophic algae growing in the microbial consortia of the wastewater may compete with the auxotrophs, as they are able to withstand the fluctuations in the type and concentration of the nutrients. The growth of algae in the wastewater is also determined by the concentration of the micro and macronutrients present in the wastewater. The common nutrients present in the wastewater are: phosphorus (that occurs as organic phosphates), nitrogen (that occurs as ammonia (NH_4^+), nitrate (NO_3^-), nitrite (NO_2^-) and urea ($CO(NH_2)_2$) and carbon.

18.6.1.1 Phosphorus

Phosphorus is one of the important macronutrients required by algae to grow as it participates in formation of vital organic molecules. RNA is the most abundant phosphorus-containing molecule followed by other nucleic acids (DNA), phospholipids and ATP. The concentration of nitrogen and phosphorous present in the water is considered to be a fundamental factor that directly influences algal growth kinetics. Total phosphorus concentration in weak untreated domestic wastewater is around 4 mg L^{-1}, in medium untreated domestic wastewater it is around 8 mg L^{-1} and strong untreated domestic wastewater has 15 mg L^{-1} total phosphorous (Rawat et al., 2011).

The micro alga have tendency to store the phosphorus in the cell in the form of polyphosphate (PPB) granules when they are growing in the environment with excess phosphate concentration. These reserves can be sufficient to prolong the growth in phosphate deficiency in the surrounding medium (Shivkumar et al., 2012; Larsdotter, 2006; Markou et al., 2012).

18.6.1.2 Nitrogen

Nitrogen is among the most important macro nutrients and a growth limiting nutrient. The concentration of nitrogen at which cell growth gets inhibited depends upon the culture conditions and the algal species (Arumugam et al., 2013). Nitrogen can be utilized as NO_3, NO_2 or NH_4 and also as N_2. Some of the nitrogen fixing cyanobacteria like *Anabaena*, *Spirulina* and *Oscilatoria* can use N_2 diazotrophically. They fix the atmospheric nitrogen into ammonia by the enzyme nitrogenase (Sawayama et al., 1998). Total nitrogen concentration in the weak untreated domestic wastewater is approximately 20 mg L^{-1}, in medium untreated domestic wastewater around 40 mg L^{-1} and strong untreated domestic wastewater shows 85 mg L^{-1} total nitrogen (Rawat et al., 2011).

Algae remove nitrogen from environment less efficiently than phosphate. It is observed that complete phosphate removal from the growth medium is possible at any N/P ratio. However, removal of nitrogen from the waste is dependent on N/P ratio. *Scenedesmus* sp. required N/P ratio in the range of 5:1 to 20:1 for maximum nitrogen removal efficiency (Xin et al., 2010). Kapdan and Aslan (2008) reported the optimum N/P ratio for *Chlorella vulgaris* was 8:1. Nitrogen starvation negatively affects the PSII photosynthetic system, causing decrease in chlorophyll and carotenoid content. The flow of photosynthetically fixed carbon is diverted to an accumulation of lipids or carbohydrates rather than for the synthesis of proteins.

18.6.1.3 Carbon

Carbon is an essential requirement for growth and can be taken up in either organic or inorganic form. Dissolved CO_2 provide carbon to the algae for the biomass production when algae grow phototrophically. Most algae utilize dissolved inorganic and organic carbon in wastewater, while heterotrophic or mixotrophic algae tend to use only organic carbon.

Inorganic carbon species utilized by algae are CO_2 and HCO_3. Intracellular carbon is in the form of HCO_3, which get converted to CO_2 by enzyme carbonic anhydrase (Larsdotter, 2006). CO_2 dissolved in water forms a weak acid/base buffer system, namely biocarbonate/carbonate buffer system. This is one of the most important buffer system present in natural water. Cyanobacteria and chlorophyceae can grow with upto 18 % dissolved CO_2 in the cultivation medium (Matkou and Georgakakis, 2011). Besides this some algae utilize organic carbon present in the wastewater in the form of sugar, organic acids, glycerol, acetate heterotrophically. However, the ability to grow heterotropically or mixotropically on one or more carbon substrates is species dependent (Muhling et al., 2005). Total organic carbon (TOC) concentration in the weak untreated domestic wastewater is approximately 80 mg L^{-1}, in medium untreated domestic wastewater it is around 160 mg L^{-1} and strong untreated domestic wastewater shows 290 mg L^{-1} total TOC (Rawat et al., 2011).

Carbon: nitrogen (C:N) and carbon:phosphorus (C:P) ratios in domestic sewage (C:N 3.5:1; C:P 20:1) and dairy lagoon water (C:N 3:1; C:P 10:1) are low compared to typical ratios required by rapidly growing algal biomass (C:N 6:1; C:P 48:1). This dearth of carbon limits growth of the algae and results in incomplete removal of the nutrients from the wastewater (Woertz et al. 2009a, b). CO_2 is most costly nutrient required for the cultivation of the microalgae (Borowitzka, 1992). Hence the system that couples a waste CO_2 source can reduce the cultivation cost and mitigate the CO_2 (Yewalkar et al., 2011).

18.6.2 Physical Abiotic Factors

The physical abiotic factors, which control the growth of algae in the wastewater, are temperature, pH and opacity (for light penetration) of the wastewater.

18.6.2.1 Temperature

Temperature of the wastewater is an important parameter. It determines or controls gas solubility (O_2 and CO_2), pH, ionic equilibrium of the wastewater etc. Most of the algal species are able to carry out photosynthesis and growth over a wide range of temperature from 15 °C to 30 °C. Increase of temperature in this range increases the growth rate of the algae till it reaches an optimum value, beyond which it causes cell growth inhibition. At lower temperatures algae show photo-inhibition at higher light intensity. The warmer climates of tropical and sub-tropical countries support the outdoor cultivation of the algae in wastewater. In these regions growth of algae in the outdoor photobioreactors are controlled by the seasonal fluctuations of temperature. In temperate zones and countries situated above 40°N latitude (with the exclusion of France, Italy, Belgium, Russia, Germany, Ukraine, Turkmenistan), climatic conditions are unsuitable for outdoor algal cultivation (Zittelli et al., 1996). In such conditions closed photobioreactor with temperature control remains the only possible option.

18.6.2.2 Light

Different algal species show a variable response to increase and decrease in light intensity in outdoor open ponds. Algae show photo-acclimatization by synthesizing or degrading the active components of its photosynthetic machinery. Whereas at sub-saturating light intensity, chlorophyll pigments and the photosynthetic reaction centre proteins D1 and D2 increase, an over-saturated light intensity causes phot-oxidation in algal cells, leading to degradation of the photosynthetic pigments and protein.

Algae use many techniques to remain at the surface of the water and thereby catch the maximum light intensity. These include synthesis of gas vacuoles, accumulation of the fat and synthesis of the mucilage, which help reduce the density of the algal cell and prevent it from sinking. However, not all algae are able to float. Many of them sink and are unable to get light. To avoid this, one simple option is to keep the depth of the photobioreactor as low as possible.

Light shielding occurs when cell density increases considerably. If the wastewater has many suspended particles, they also cause shielding. The colour and transparency of the wastewater also controls light penetration (Larsdotter, 2006; Shivkumar et al., 2012).

18.6.2.3 pH

pH of the wastewater affects many biochemical process which control algal growth and metabolism. In the photosynthetic algae, CO_2 assimilation causes pH of the wastewater to rise above 10. pH of the wastewater also decides which inorganic species of the carbon will get fixed during the photosynthesis. pH can increase to 11 if CO_2 is limiting and inorganic carbonate has been used as the source of inorganic carbon. Nitrogen assimilation by algae is also affected by pH. If ammonia is used as nitrogen source, then the pH of the medium turns acidic (as low as 3); however, use of nitrates raises pH of the medium to an alkaline range. Hence pH regulates not only algal growth but also the nitrogen removal efficiency. Optimal pH for the algal growth is 8. However many algae can grow in more alkaline conditions (Park et al., 2011).

18.6.3 Biotic Factors

A number of biotic factors also determine the mode of algal growth, extent of algal growth and alga cell density.

18.6.3.1 Cell Density

The nutrient removal efficiency from the wastewater with algal system is directly related to the cell mass or number of active cells. More the number of active algal cells, rate of nutrient removal will be faster and retention time of the wastewater will be reduced. However, very dense algal cell culture (more than 1×10^7 cells/ml) results in self-shielding. Because of self-shielding the algal cells may shift to a mixotrophic or heterotrophic mode of nutrition. This can be avoided by providing proper mixing (Lau et al., 1995). Immobilizing algae (Tam and Wong, 2000; Burke, 2013; Ozkan et al., 2012) can also provide a higher cell density.

18.6.3.2 Zooplanktons Grazers and Predators

In the treatment of wastewater in open pond reactor by algal, one cannot avoid grazers. In this ‘artificial uncontrolled ecosystem’, grazers form the primary consumers of algae. The herbivorous protozoa and zooplanktons (like rotifers and cladocerans) can reduce algal concentration to a very low level within just a few days, causing ponds to crash. If cell density of the rotifers and cladocerans exceeds 10^2 L^{-1} they can reduce the algal population by 90 % within two days. Daphnia was responsible for 99 % reduction of chlorophyll in open pond (Park et al., 2011). Zooplanktons grazing may be controlled through physical treatments like filtration, centrifugation and low dissolved oxygen/high carbon loading. Chemical treatments to control the grazers are by application of chemicals, invertebrate hormone mimics, increased pH, and free ammonia addition.

The parasitic fungi *Chytidium* and some algae attacking virus may grow along with the algal culture and spoil the food chain completely. Control of such organisms is vital for effective waste treatment. Fungal parasites and grazers are the most ubiquitous biotic drivers of decimation of the algal community. Some of the most effective methods to control the growth of zooplanktons are by reducing aeration, reducing retention time and adjustment of pH to 11.

18.6.4 Dissolved Oxygen Concentration (DOC)

Dissolved oxygen in the raw wastewater is very low. Many conventional systems used energy intensive aerations systems to increase the dissolved oxygen for the oxidation of the organic matter. In case of algal wastewater remediation, algae generate the oxygen during the photosynthesis. This photosynthetic generated oxygen is responsible for the oxidation/biodegradation of organic material. Algae like *C. sorokiniana* and *E. viridis* proved their potential to treat the piggery waste efficiently with this photosynthetic oxygenation (Godos et al., 2010). In high rate oxidation ponds, intense photosynthesis increased DOC by 200 % during day time. High dissolved oxygen levels in water have negative impact on algal growth (Park et al., 2011). Fortunately O_2 saturation cannot cause harm during the biodegradation process. The oxygen generated by the algae is utilized by the heterotrophic, symbiotic bacteria. The DOC levels remain low during the degradation of all organic pollutants. However, after depletion of the organic pollutants the DOC levels raised rapidly (Munoz and Guieysse, 2006).

18.7 Future Prospects

Wastewater treatment becomes economical with the introduction of algal biomass in the consortium of waste utilizing organisms. This sustainable approach can further influence the lowering of cost of treating wastewater. Practically, if suitable methods of algal biomass separation/harvesting are developed, algal biomass can provide byproduct credit. The algal biomass generated during wastewater treatment can be utilized for energy production or as a bio-fertilizer, but may not be suitable for food, animal feed, nor nutritional components as it is grown on wastewater. The algal biomass generated on the industrial wastewater may not be useful as a bio-fertilizer in case of accumulation of heavy metal in the algal biomass (Munoz and Guieysse, 2006).

18.7.1 Potential Uses of Algal-bacterial Biomass

The best use of algal-bacterial biomass generated from the wastewater treatment is for the energy production in various ways as listed below:

Biodiesel: If the algae grown on the waste has high amount of lipids, then lipids can be extracted. The extracted algal oil (raw microalgal lipid) after transesterification gets converted to renewable, non-toxic, biodegradable biodiesel.

Thermal decomposition of the algal biomass can give different types of energy fuels depending on temperature used during the conversion process.

Gasification: Partial oxidation of the biomass at 800–1000 °C produces syngas (combustible mixture of carbon dioxide, hydrogen gas, nitrogen and methane). Syngas can be directly used as fuel.

Thermo chemical liquefaction: The wet biomass is subjected to a thermal treatment (300–350 °C) at high pressure (50–200 atm) in presence of catalyst to produce a bio-oil.

Pyrolysis: Thermal conversion is carried out in the presence or absence of catalyst and in absence of air/oxygen. Pyrolysis of biomass produces charcoal, condensable organic liquids (acetic acid, acetone, methanol) and non-condensable gases. Pyrolysis of algal biomass was found to produce higher quality bio-oil than lignocellulosic compounds.

Combustion: Direct combustion or burning of biomass in presence of air for conversion to energy in the form of hot gases has been practiced. The conversion efficiency of the algal biomass to energy is more favourable than that of coal; however, it requires dry biomass containing low amount of water (<50 %) (Rawat et al., 2011; Olguin, 2012).

18.7.2 Combining Wastewater Treatment with CO_2 Mitigation

Algae can utilize the P and N in the wastewater and generate biomass. On the same lines CO_2 from the flue gas generated from heavy industries (cement, petroleum, power plants and oil industries) can be utilized to improve the growth rate and lipid content of algae. This approach is also known as "CO_2 mitigation". In large-scale cultivation of algae, to avoid the CO_2 limitation, concentrated CO_2 sparging is essential to improve the growth rate and lipid accumulation. The CO_2 contained in a typical flue gas is an ideal source of C for algal photosynthesis. However, the flue gas in addition to CO_2 also has SO_2 and NOx, which inhibit the growth of algae drastically. This inhibition could be overcome by buffering the water with $CaCO_3$. It may be necessary to pretreat the flue gas for desulphurization or for removal of inhibitory substances. The other limitations of CO_2 mitigation using algal biomass is the low specific rate of consumption of CO_2 and the land requirement for constructing very large raceway ponds. Several scores of hectares of land would be required for building algal photobioreactors/cultivation ponds and pretreatment of flue gas (McGinn et al., 2011; Yewalkar et al., 2011). The problem can be partially solved by separating CO_2 from the flue gas by membrane separation followed by compression and transportation to the site of cultivation or piping the flue gas directly from the stacks to the site of cultivation. However, this would substantially add to variable costs and add to the complexity of algal biomass production.

18.8 Conclusion

The treatment of the wastewater with algae-bacteria system is efficient for removal of phosphorous and nitrogen to an acceptable level. It generates algal biomass as a byproduct with marginal added cost for nutrients and water. However, harvesting of algal biomass remains a major challenge. Prevention of predators, especially in wastewater streams will be major cause of worry. This sustainable approach needs further development and refining of existing techniques and processes for cost effective production of algal biomass and its separation and harvesting. It is clear, however, that algae are here to stay in the bioremediation of wastewater.

References

Abdel-Raouf, N., Al-Homaidan, A. A., & Ibraheem, I. B. M. (2012). Microalgae and wastewater treatment. *Saudi Journal of Biological Sciences, 19*(3), 257–275.

Al-Hadabi, H., Al-Balushi, T., Al-Aisri Sahar Talebi, S. and Al-Barwani, T. (2012). A critical review of wastewater treatment in photobioreactors for improving microalgae growth. Proceedings of the World Congress on Engineering.

Al-Mamun, A., Amid, A., Karim, I. A., Rashid, S., & Shah, M. (2012). Phytoremediation of partially treated wastewater by Chlorella vulgaris.

Al-Shayji, Y.A., Puskas, K. and Al-Daher, R. (1994). Production and separation of algae in a high-rate pond system. *Environment International*, 20 (4), 541–550.

An, J.Y., Sim, S.J. Lee, J.S. and Kim, B.W. (2003). Hydrocarbon production from secondarily treated piggery wastewater by the green alga *Botryococcus braunii. Journal of Applied Phycology*, 15, 185–191.

Ansa, E.D.O., Luberding, H.J., Ampofo, J.A. and Gijzen, H.J. (2011). The role of algae in the removal of *Escherichia coli* in a tropical eutrophic lake. *Ecological Engineering*, 37, 317–324.

Aragon-Bustos, A., Padilla-Borja, R. and Ros de Ursinos-Fiestas, J.A. (1992). Experimental study of recovery of algae cultured in effluents from the anaerobic biological treatment of urban wastewaters. *Resources, Conservation and Recycling*, 6, 293–302.

Arnold, M. (2013). Sustainable algal biomass products by cultivation in wastewater flows. VTT Technology Julkaisija-Utgivare Publication147.

Arumugam, M., Agrawal, A., Arya, C.M. and Ahmed, Z. (2013). Influence of nitrogen sources on biomass productivity of microalgae *Scenedcdesmus bijugatus. Bioresource Technology*, 131, 246–249.

Becker, E.W. (1983). Limitations of heavy metal removal from wastewater by means of algae. *Water Res.*, 17(4), 459–466.

Becker, E.W. and Venkatraman, L.V. (1984). Production and utilization of blue-green alga *Spirulina* in India. *Biomass*, 4, 105–125.

Benemann, J.R. (1979). Production of nitrogen fertilizer with nitrogen-fixing blue-green algae. *Enzyme Microbial Technology*, 1, 141–229.

Borde, X., Guieysse, B., Delgado, O., Muñoz, R., Hatti-Kaul, R., Nugier-Chauvin, C., Patin, H. and Mattiason, B.(2003). Synergistic relationships in algal–bacterial microcosms for the treatment of aromatic pollutants. *Bioresource Technology*, 86, 293–300.

Borowitzka, M.A (1999). Commercial production of microalgae: Ponds, tanks, and fermenters. *Progress in Industrial Microbiology*, 35, 313–321.

Borowitzka, M. (1992). Algal biotechnology products and processes – Matching science and economics. *Journal of Applied Phycology*, 4, 267–279.

Borowitzka, M.A. (1998). Limits to Growth: Wastewater treatment with algae. Biotechnology Intelligence Unit, 203–226.

Brandi, G., Sisti, M. and Amagliani, G. (2000). Evaluation of environmental impact of microbial aerosols generated by wastewater treatment plants utilizing different aeration systems. *Journal of Applied Microbiology*, 88, 845–852.

Brennan, L. and Owende, P. (2009). Biofuels from microalgae: A review of technologies for production, processing and extractions of biofuels and co-products. *Renewable and Sustainable Energy Reviews*, 805, 1–21.

Brzezinski, M.H. Villareal, T.A. and Lipschultz, F. (1998). Silica production and contribution of diatoms to new and primary production in the central north pacific. *Marine Ecology Progress Series*, 167: 89–104.

Burke D.A. (2013). The rotating photo-bioreactor as a method for removing soluble nutrients from waters. American Society of Agricultural and Biological Engineers publications.

Burlew, J.S. (1953). Algal culture from laboratory to pilot plant. Carnegie Institution of Washington publication 600.

Cervantes, C., Campos-Garcia, J., Devars, S., Gutierrez-Corona, F., Loza-Tavera. H., Torres-Guzman, J.C. and Moreno-Sanchez, R. (2001). Interactions of chromium with microorganisms and plants. *FEMS Microbiol. Rev.*, 25, 335–347.

Chaumont, D. (1993). Biotechnology of algal biomass production: A review of systems for outdoor mass cultures. *Journal of Applied Phycology*, 5, 593–604.

Chen, C., Yeh K., Aisyah, R., Lee, D. and Chang, J. (2011). Cultivation, photobioreactor design and harvesting of microalgae for biodiesel production: A critical review. *Bioresource Technology*, 102(1), 71–81.

Cheng, L., Zhang, L., Chen, H. and Gao, C. (2006). Carbon dioxide removal from air by microalgae cultured in a membrane-photobioreactor. *Separation and Purification Technology*, 50, 324–329.

Chisti, Y. (2007). Research review paper: Biodiesel from microalgae. *Biotechnology Advances*, 25, 294–306.

Christenson, L. and Sims, R. (2011). Research review paper: Production and harvesting of microalgae for wastewater treatment, biofuels and bioproducts. *Biotechnology Advances*, 29, 686–702.

Cohen, Z. (1999). Chemicals from microalgae. CRC Press Taylor & Francis group, 387–406.

Degen, J., Uebele, A., Retze, A., Schmid-Staiger, U. and Trösch, W. (2001). A novel airlift photobioreactor with baffles for improved light utilization through the flashing light effect. *Journal of Biotechnology*, 92, 89–94.

Du, Z., Hu, B., Shi, A., Ma, X., Cheng, Y., Chen, P., Liu, Y., Lin, X and Ruan, R. (2012). Cultivation of microalga *Chlorella vulgaris* using recycled aqueous phase nutrients from hydrothermal carbonization process. *Bioresource Technology*, 126, 354–357.

El-Bestawy, E. (2008). Treatment of mixed domestic-industrial wastewater using cyanobacteria. *Journal of Industrial Microbiology and Biotechnology*, 35, 1503–1516.

Eroğlu, E., Gündüz, U., Yücel, M., Türker, L. and Eroğlu, I. (2004). Photobiological hydrogen production by using olive mill wastewater as a sole substrate source. *International Journal of Hydrogen Energy*, 29, 163–171.

Frank, D.L. (2012). Semi-closed loop alga-diesel fuel photobioreactor using wastewater. United States patent US8183032.

Gantar, M., Obreht, Z. and Dalmacija, B. (1991). Nutrient removal and algal succession during the growth of *Spirulina platensis* and *Scenedesmus quadricauda* on swine wastewater. *Bioresource Technology*, 36, 167–171.

Gloyna, E.F. (1971). Waste stabilization ponds. World Health Organization monograph series no. 60.

Godos, I., Vargas, V., Blanco, S., González, M., Soto, R., Garcia-Encína, P., Becares, E. and Muñoz, R. (2010). A comparative evaluation of microalgae for the degradation of piggery wastewater under photosynthetic oxygenation. *Bioresource Technology*, 101, 5150–5158.

Golueke, C.G. and Oswald, W.J. (1965). Harvesting and processing sewage grown planktonic algae. *Journal of Water Pollution Control Fed*, 37, 471–498.

Gómez-Villa, H., Voltolina, D., Nieves, M. and Piña, P. (2005). Biomass production and nutrient budget in outdoor cultures of *Scenedemus obliquus* (Chlorophyceae) in artificial wastewater, under the winter and summer conditions of Mazátlan, Sinaloa, Mexico. *Vie Et Milieu*, 55, 121–126.

Gonzalez, L.E., Cañizers, R.A. and Baena, S. (1997). Efficiency of ammonia and phosphorus removal from a colombian agroindustrial wastewater by the microalgae *Chlorella vulgaris* and *Scenedesmus dimorphus. Bioresource Technology*, 60, 259–262

Gonzfilez, L.E., Cafiizares, R.O. and Baena, S. (1997). 'Efficiency of ammonia and phosphorus removal from a colombian agroindustrial wastewater by the microalgae Chlorella vulgaris and Scenedesmus dimorphus', Bioresource Technology, 40, 2016–2022.

Grima, E., Acién Fernández, F., Camacho, F. and Chisti, Y. (1999). Photobioreactors: Light regime, mass transfer, and scale up. *Journal of Biotechnology*, 70, 231–247.

Grima, E., Belarbi, E., Acién, Fernández, F., Medina, A. and Chisti, Y. (2003). Recovery of microalgal biomass and metabolites: Process options and economics. *Biotechnology Advances*, 20, 491–515.

Grobbelaar, U. J. (2009). Factors governing algal growth in photobioreactors: The "open" versus "closed" debate. *Journal of Applied Phycology,* 21, 489–492.

Guterstam, B., & Etnier, C. (1991). Ecological engineering for wastewater treatment: theoretical foundations and practical realities. In *Ecological engineering for wastewater treatment. Proceedings of the International Conference held at Stensund Folk College, Sweden, March 24–28, 1991.* (pp. 38–54). Bokskogen.

Habiba, A. H., Al-Balushi, T., & Talebi, S. (2012). A Critical Review of Wastewater Treatment in Photobioreactors for Improving Microalgae Growth. In *Proceedings of the World Congress on Engineering* (Vol. 3).

Hanumantha Rao, P., Ranjith Kumar, R., Raghavan, B. G., Subramanian, V. V., & Sivasubramanian, V. (2011). Application of phycoremediation technology in the treatment of wastewater from a leather-processing chemical manufacturing facility. *Water Sa*, 37(1), 07–14.

Hirata, S., Hayashitani, M., Taya, M. and Tone, S. (1996). Carbon dioxide fixation in batch culture of *Chlorella* sp. using a photobioreactor with a sunlight-collection device. *Journal of Fermentation and Bioengineering*, 81, 470–472.

Hoffman, J.P. (1998). Mini review: Wastewater treatment with suspended and non-suspended algae. *Journal of Phycology*, 34, 757–763.

Hu, Q., Guterman, H. and Richmond, A. (1996). A flat inclined photobioreactor for outdoor mass cultivation of photoautotrophs. *Biotechnology and Bioengineering*, 51, 51–60.

Ignacio de Godos, Vargas, V.A., Blanco, S., García Gonházalez, M.C., Soto, R., García-Encina, P.A., Becares, E. and Muñoz, R. (2010). A comparative evaluation of microalgae for the degradation of piggery wastewater under photosynthetic oxygenation. *Bioresource Technology*, 101, 5150–5158.

Jalal, K.C.A., Alam, Z., Matin, W.A., Kamaruzzaman, B.Y., Akbar, J. and Hossain, T. (2011). Removal of nitrates and phosphate from municipal wastewater sludge by *Chlorella vulgaris, Spirulina platensis* and *Scenedsmus quadricauda. IIUM Engineering Journal*, 12, 125–132.

Janssen, M., Tramper, J., Mur, L. and Wijffels, R. (2002). Enclosed outdoor photobioreactors: Light regime, photosynthetic efficiency, scale-up and future prospects. *Biotechnology and Bioengineering*, 81, 193–210.

Jin, Y., Ju, J., Kim, J., Lee, S., Kim, J. Y., Park, S. H., … & Suh, H. (2003). Design, synthesis, and electroluminescent property of CN-poly (dihexylfluorenevinylene) for LEDs. *Macromolecules*, 36(19), 6970–6975.

Jorquera, O., Kiperstok, A., Sales, E.A., Embirucu, M. and Ghirardi, M.L. (2010). Comparative energy life-cycle analyses of microalgal biomass production in open ponds and photobioreactors. *Bioresource Technology*, 101, 1406–1413.

Kapdan, I.K. and Aslan, S. (2008). Application of the Stover–Kincannon kinetic model to nitrogen removal by Chlorella vulgaris in continuously operated immobilized photobioreactor system. *Journal of Chemical Technology and Biotechnology*, 83, 998–1005.

Kayombo, S., Mbwette, T.S.A., Mayo, A., Katima, J.H.Y. and Jørgenson, S. (2000). Modelling diurnal variation of dissolved oxygen in waste stabilization ponds. *Ecological Modeling*, 127, 21–31.

Kebede-Westhead, E., Pizarro, C. and Mulbry, W. (2006). Treatment of swine manure effluent using fresh water algae: Production, nutrient recovery, and elemental composition of algal biomass at four effluent loading rates. *Journal of Applied Phycology*, 18, 41–46.

Kim, K.M., Park, W.J., Park, S.C., Kim, J. S., Jeune, H.K., Chang, U.M. and Acreman J. (2007). Enhanced production of Scenedesmus Spp. (green microalgae) using a new medium containing fermented swine wastewater. *Bioresource Technology*, 98, 2220–2228.

Kim, J., Lingaraju, B. P., Rheaume, R., Lee, J. Y., & Siddiqui, K. F. (2010). Removal of ammonia from wastewater effluent by Chlorella vulgaris. *Tsinghua Science & Technology*, 15(4), 391–396.

Kumar, A., Yuan, X., Sahu, A., Ergas, S., Langenhovine, H. and Dewulf, J. (2010). A hollow fiber membrane photo-bioreactor for CO_2 sequestration from combustion gas coupled with wastewater treatment: A process engineering approach. *Journal of Chemical Technology and Biotechnology*, 85, 387–394.
Kumar, K., Dasgupta, C., Nayak, B., Lindblad, P. and Das, D. (2011). Review: Development of suitable photobioreactors for CO_2 sequestration addressing global warming using green algae and cyanobacteria. *Bioresource Technology*, 102, 4945–4953.
Larsdotter, K. (2006). Wastewater treatment with microalgae-a literature review. Vatten, 62(1), 31.
Lau, S. P., Tam, F.Y.N. and Wong, S.Y. (1995). Effect of algal density on nutrient removal from primary settled wastewater. *Environmental Pollution*, 89, 59–66.
Lee, Y. (2001). Microalgal mass culture systems and methods: Their limitation and potential. *Journal of Applied Phycology*, 13, 307–315.
Lehr, F. and Posten, C. (2009). Closed photo-bioreactors as tools for biofuel production. *Current Opinion in Biotechnology*, 20, 280–285.
Li, Y., Zhang, H., Qiu, X., Zhang, Y. and Wang, H. (2013). Dispersion and risk assessment of bacterial aerosols emitted from rotating-brush aerator during summer in a wastewater treatment plant of Xi'an, China. *Aerosol and Air Quality Research*, 13, 1807–1814.
Lundquist, T., Woertz, I., Quinn, N.W.T. and Benemann, J. (2010). A realistic technology and engineering assessment of algae biofuel production. Energy Biosciences Institute, University of California.
Mallick, N. (2002). Biotechnological potential of immobilized algae for wastewater N, P and metal removal: A review. *Biometals*, 15, 377–390.
Mara, D. (1987). Waste stabilization ponds: problems and controversies. *Water Quality International*, (1).
Mara, D. (1997). Design Manual for waste stabilization ponds in India. University Print Services, Leeds.
Mara, D. (2009). Waste stabilization ponds: Past, present and future. *Desalination and Water Treatment*, 4, 85–88.
Mara, D. and Pearson, H. (1986). Waste stabilization pond research: Experimental methods and data analysis. Regional research seminar on waste stabilization ponds.
Markou, G., Angelidaki, I. and Georgakakis, D. (2012). Microalgal carbohydrates: An overview of the factors influencing carbohydrate production, and of main Bioconversion technologies for the production of biofuels., *Appl Microbiol Biotechnol.*, 96, 631–645.
Martínez, E., Sánchez, S., Jiménez, M., El Yousfi, F. and Muñoz, L. (2000). Nitrogen and phosphorus removal from urban wastewater by the microalga *Scenedesmus obliquus*. *Bioresource Technology*, 73, 263–272.
Mata, T., Martins, A. and Caetano, N. (2009). Microalgae for biodiesel production and other applications: A review. *Renewable and Sustainable Energy Reviews*, 757, 1–16.
Matkou, G. and Georgakakis, D. (2011). Cultivation of filamentous cyanobacteria (blue-green algae) in agroindustrial waste and wastewater: A review. *Applied Energy*, 88, 3389–3401.
McGinn, J.P., Dickinson, E. K., Bhatti, S., Jean-Claude, F., Guiot, R.S. and O'Leary, J.B.S. (2011). Integration of microalgae cultivation with industrial waste remediation for biofuel and bioenergy production: Opportunities and limitation. *Photosynth. Res.*, 109, 231–247.
Mendez, M., Behnke, C., Poon, Y. and Lee, P. (2009). Induction of flocculation in photosynthetic Organisms. Patent cooperation treaty application no. WO 2009158658.
Merchuk, J.C., Ronen, M., Giris, S. and Arad (Malis), S. (1998). Light/Dark cycles in the growth of the red microalga *Porphyridium* sp. *Biotechnology and Bioengineering*, 59, 705–713.
Mezzari, M. P., da Silva, M. L., Viancelli, A., Ibelli, A. M., Kunz, A., & Soares, H. M. (1987, October). BACTERIA-MICROALGAE INTERACTIONS DURING NITRIFICATION/DENITRIFICATION PROCESSES IN A PHOTOBIOREACTOR TREATING SWINE WASTEWATER. In Embrapa Suínos e Aves-Artigo em anais de congresso (ALICE). In: SIMPÓSIO INTERNACIONAL SOBRE GERENCIAMENTO DE RESÍDUOS AGROPECUÁRIOS E AGROINDUSTRIAIS, 3., 2013, São Pedro, SP. Anais… São Pedro, SP: SBERA, 1987.

Middlebrooks, J., Porcella, B., Gearheart, A., Marshall, R., Reynolds, H. and Grenney J. (1974). Techniques for algae removal from wastewater stabilization ponds. *Journal of Water Pollution Control Federation*, 46, 2676–2695.

Miyamoto, K., Wable, O. and Benemann, J. (1988). Vertical tubular reactor for microalgae cultivation. *Biotechnology Letters*, 10, 703–708.

Monedero–Sánchez, M., Aguilar, M., Fenoll, R. and Roig, A. (2008).Effect of the aeration system on the levels of airborne microorganisms generated at wastewater treatment plants. *Water Research*, 42, 3739–3744.

Morweiser, M., Kruse, O., Hankamer, B. and Posten, C. (2010). Developments and perspectives of photobioreactors for biofuel production. *Applied Microbiology and Biotechnology*, 87, 1291–1301.

Mouget, J-L., Dakhama, A., Lavoie, C.M. and da la Noiie, J. (1995). Algal growth enhancement by bacteria: Is consumption of photosynthetic oxygen involved? *FEMS Microbiology Ecology*, 18, 35–43.

Muhling, M., Belay, A. and Whitton A.B. (2005). Screening Arthrospira (*Spirulina*) stains for heterotrophy. *J Applied Phycology*, 17, 129–135.

Mulbry, W., Kondrad, S., Pizzaro, C. and Westhead-Kebede, E. (2008). Treatment of dairy manure effluent using freshwater algae: Algal productivity and recovery of manure nutrients using pilot-scale algal turf scrubbers. *Bioresource Technology*, 99, 8137–8142.

Munoz, R., & Guieysse, B. (2006). Algal–bacterial processes for the treatment of hazardous contaminants: a review. *Water research, 40*(15), 2799–2815.

Muñoz, R., Köllner, C. and Guiyesse, B. (2009). Biofilm photobioreactors for the treatment of industrial wastewaters. *Journal of Hazardous Materials*, 161(1), 29–34.

Muñoz, R., Köllner, C., Guieysse, B. and Mattiasson, B. (2004). Photosynthetically oxygenated salicylate biodegradation in a continuous stirred tank photobioreactor. *Biotechnology and Bioengineering*, 87, 797–803.

Ogbonna, J. and Tanaka, H. (2000). Light requirement and photosynthetic cell cultivation: Development of processes for efficient light utilization in photobioreactors. *Journal of Applied Phycology*, 12, 207–218.

Olguin, E. (2012). Dual purpose microalgae–bacteria-based systems that treat wastewater and produce biodiesel and chemical products within a Biorefinery. *Biotechnology Advances*, 30, 1031–1046.

Oswald, W. (2003). My sixty years in algology. *Journal of Applied Phycology*, 15, 99–106.

Oswald, W. J., & Gotaas, H. B. (1957). Photosynthesis in sewage treatment. *Trans. Am. Soc. Civ. Eng, 122*, 73–105.

Oswald, W., Asce, A., Gotaas, H. and Asce, M. (1955). Photosynthesis in sewage treatment. Proceedings in Transactions for American Society of Civil Engineers.

Ozkan, A., Kinney, K., Katz, L. and Berberoglu, H. (2012). Reduction of water and energy requirement of algae cultivation using an algae biofilm photobioreactor. *Bioresource Technology*, 114, 542–548.

Park, J.B.K., Craggs, R. and Shilton, A. (2011). Wastewater treatment high rate algal ponds for biofuel production. *Bioresource Technology*, 102, 35–42.

Park, J., Jin, H. F., Lim, B. R., Park, K. Y., & Lee, K. (2010). Ammonia removal from anaerobic digestion effluent of livestock waste using green alga Scenedesmus sp. *Bioresource technology, 101*(22), 8649–8657.

Pearson, H., Mara, D., Mills, S. and Smallman, D. (1987). Physico-chemical parameters influencing faecal bacterial survival in waste stabilization ponds. *Water Science and Technology*, 19, 145–152.

Pinto, G., Pollio, A., Previtera, L., Stanzionel, M. and Temussi F. (2003). Removal of low molecular weight phenols from olive oil mill wastewater using microalgae. *Biotechnology Letters*, 25, 1657–1659.

Pittman, J. K. (2011). Vacuolar Ca 2+ uptake. *Cell calcium, 50*(2), 139–146.

Pittman, J., Dean, A. and Osundeko, O. (2011). The potential of sustainable algal biofuel production using wastewater resources. Bioresource Technology Special Issue: Biofuels – II: *Algal Biofuels and Microbial Fuel Cells*, 102 , 17–25.

Posadas, E., Bochon, S., Coca, M., García-González, M. C., García-Encina, P. A., & Muñoz, R. (2014). Microalgae-based agro-industrial wastewater treatment: a preliminary screening of biodegradability. *Journal of Applied Phycology, 26*(6), 2335–2345.

Posadas, E., Muñoz, A., Gonzinález, M., Muñoz, R. and García-Encina, P. (2015). A case study of a pilot high rate algal pond for the treatment of fish farm and domestic wastewater. *Journal of Chemical Technology and Biotechnology*. (in press)

Pozza, C., Schmuck, S. and Mietzel, T. (2011). A novel photobioreactor with internal illumination using Plexiglas rods to spread the light and LED as a source of light for wastewater treatment using microalgae. World Congress on Water, Climate and Energy. Dublin, Ireland.

Pulz, O. (2001). Photobioreactors: Production systems for phototrophic microorganisms. *Applied Microbiology and Biotechnology*, 57, 287–293.

Pulz, O. and Gross, W. (2004).Valuable products from biotechnology of microalgae. *Applied Microbiology and Biotechnology*, 65, 635–648.

Racault, Y. and Boutin, C. (2005). Waste stabilisation ponds in France: State-of-the-art and recent trends. *Water Science & Technology*, 51, 1–9.

Rao, P., Kumar, R., Raghavan, B., Subramanian, V. and Sivasubramanian, V. (2011). Application of phycoremediation of wastewater from a leather-processing chemical manufacturing facility. *Water SA*, 37, 7–14.

Rao, S. (1972). A low cost waste treatment method for the disposal of distillery waste (spent wash). *Water Research*, 6, 1275–1282.

Rawat, I., Kumar, R., Mutanda, T. and Bux, F. (2011). Dual role of microalgae: Phycoremediation of wastewater and biomass production for sustainable biofuels production. *Applied Energy*, 88, 3411–3424.

Rosenberg, J., Oyler, G., Wilkinson, L., and Betenbaugh, M. (2008). A green light for engineered algae: Redirecting metabolism to fuel a biotechnology revolution. *Current Opinion in Biotechnology*, 19, 430–436.

Sánchez-Monedero, M. A., Aguilar, M. I., Fenoll, R., & Roig, A. (2008). Effect of the aeration system on the levels of airborne microorganisms generated at wastewater treatment plants. *Water Research, 42*(14), 3739–3744.

Sawayama, S., Rao, K.K. and Hall O.D. (1998). Nitrate and phosphate ion removal from water by *Phormidium laminosum* immobilized on hollow fibres in a photo-bioreactor. *Applied Microbiology and Biotechnol*, 49, 463–468.

Schenk, P., Skye, R., Hall, T., Stephens, E., Marx, U., Mussgnug J., Posten, C., Kruse, O. and Hankamer, B. (2008). Second generation biofuels: High-efficiency microalgae for biodiesel production. *Bioenergy Research*, 1, 20–43.

Shepherd, S. (2012). System for harvesting algae in continuous fermentation. United States patent no. US 0226061.

Shivkumar, G., Xu, J., Thompson, R. W., Yang, Y., Randol-Smith, P. and Weathers, P.J. (2012). Integrated green algal technology for the bioremediation and biofuel. *Bioresource Technol*, 107, 1–9.

Sim, T., Goh, A. and Becker, E. (1988). Comparison of Centrifugation, dissolved air floatation and drum filtration techniques for harvesting sewage-growing algae. *Biomass*, 16, 51–62.

Singh, G. and Thomas, P. (2012). Nutrient removal from membrane bioreactor permeate using microalgae and in a microalgae membrane photoreactor. *Bioresource Technology*, 117, 80–85.

Sivasubramanian, V. and Muthukumaran, M. (2012). Large scale phycoremediation of oil drilling effluent. *Journal of Algal Biomass Utilization*, 3, 5–17.

Sivasubramanian, V., Sankaran, B., Murali, R. and Subramanian, V. (2011). Micro algal technology to correct pH and reduce sludge in an acidic effluent from a detergent industry. *Journal of Algal Biomass Utilization*, 2, 95–102.

Sivasubramanian, V., Subramanian, V. and Muthukumaran, M. (2012). Phycoremediation of effluent from a soft drink manufacturing industry with a special emphasis on nutrient removal—A laboratory study. *Journal of Algal Biomass Utilization*, 3, 21–29.

Sivasubramanian, V., Subramanian, V. Muthukumaran, M. and Murali, R. (2012). Algal Technology for effective reduction of total hardness in wastewater and industrial effluents. *Phykos*, 42(1), 51–58.

Sivasubramanian, V., Subramanian, V., Raghavan, B. and Ranjithkumar, R. (2009). Large scale phycoremediation of acidic effluent from an alginate industry. *Science Asia*, 35, 220–226.

Sriram, S. and Seenivasan, R. (2012). Microalgae cultivation in wastewater for nutrient removal. *Journal of Algal Biomass Utilization*, 2, 9–13.

Su, Y., Mennerich, A. and Urban, B. (2011). Municipal wastewater treatment and biomass accumulation with a wastewater-born and settleable algal-bacterial culture. *Water Research*, 45, 3351–3358.

Tam, N. F. Y., & Wong, Y. S. (2000). Spatial variation of heavy metals in surface sediments of Hong Kong mangrove swamps. *Environmental Pollution, 110*(2), 195–205.

Tang, H., Chen, M., Simon Ng, K. Y., & Salley, S. O. (2012). Continuous microalgae cultivation in a photobioreactor. *Biotechnology and bioengineering, 109*(10), 2468–2474.

Tarlan, E., Dilek, F. and Yetis, U. (2002). Effectiveness of algae in the treatment of a wood-based pulp and paper industry wastewater. *Bioresource Technology*, 84, 1–5.

Tilley, E., Ulrich, L., Lüthi, C., Reymond, P. and Zurbrügg, C. (2014). Compendium of sanitation systems and technologies. 2nd revised edition, Swiss Federal Institute of Aquatic Science and Technology (Eawag), Duebendorf, Switzerland.

Tilzer, M. (1983). The importance of fractional light absorption by photosynthetic pigments for phytoplankton productivity in lake Constance. *American Society of Limnology and Oceanography*, 28, 833–846.

Travieso, L., Benítez, F., Sánchez, E., Borja, R., Martín, A., & Colmenarejo, M. F. (2006). Batch mixed culture of Chlorella vulgaris using settled and diluted piggery waste. *Ecological Engineering, 28*(2), 158–165.

Tredici, M. and Zittelli, G. (1997). Efficiency of sunlight utilization: Tubular versus flat photobioreactors. *Biotechnology and Bioengineering*, 57, 187–197.

Uduman, N., Qi, Y., Danquah, M., Forde, G. and Hoadley, A. (2010). Dewatering of microalgal cultures: A major bottleneck to algae-based fuels. *Journal of Renewable and Sustainable Energy*, 2, 12701–12715.

Ugwu, C., Aoyagi, H. and Uchiyama, H. (2008). Photobioreactors for mass cultivation of algae. *Bioresource Technology*, 99, 4021–4028.

Vunjak-Novakovic, G., Kim, Y., Wu, X., Berzin, I. and Merchuk, J. (2005). Air-lift bioreactors for algal growth on flue gas: Mathematical modeling and pilot-plant studies. *Industrial and Engineering Chemistry Research*, 44, 6154–6163.

Wang, H., Xiong, H., Hui, Z., & Zeng, X. (2012). Mixotrophic cultivation of Chlorella pyrenoidosa with diluted primary piggery wastewater to produce lipids. *Bioresource Technology, 104*, 215–220.

Wang, C., Yu, X., Lv, H. and Yang, J. (2013). Nitrogen and phosphorus removal from municipal wastewater by the green alga *Chlorella* sp., *Journal of Environmental Biology*, 34, 421–425.

Wang, L., Min, M., Li, Y., Chen, P., Chen, Y., Liu, Y., Wang, Y. and Ruan, R. (2009). Cultivation of green algae *Chlorella* sp. in different wastewaters from municipal wastewater treatment plant. *Applied Biochemistry and Biotechnology*, 162, 1174–1186.

Wang, L., Min, M., Li, Y., Chen, P., Chen, Y., Liu, Y., … & Ruan, R. (2010). Cultivation of green algae Chlorella sp. in different wastewaters from municipal wastewater treatment plant. *Applied biochemistry and biotechnology, 162*(4), 1174–1186.

White, P.A., Kalff, J., Rasmussen, B.J. and Gasol, M.J. (1991). Effect of temperature and algal biomass on bacterial production and specific growth rate in fresh water and marine habitats. *Microb. Ecol.*, 21, 99–118.

Wiley, P. E. (2013). Microalgae cultivation using offshore membrane enclosures for growing algae (OMEGA).

Wiley, P., Harris, L., Reinsch, S., Tozzi, S., Embaye, T., Clark, K., McKuin, B., Kolber, Z., Adams, R., Kagawa, H., Richardson, T., Malinoswki, J., Beal, C., Claxton, M., Geiger, E., Rask, J., Campbell, J. and Trent, J. (2013). Microalgae cultivation using offshore membrane enclosures for growing algae (OMEGA). *Journal of Sustainable Bioenergy Systems*, 3, 18–32.

Woertz, I., Feffer, A., Lundquist, T. and Nelson, Y. (2009). Algae grown on dairy and municipal wastewater for simultaneous nutrient removal and lipid production for biofuel feedstock. *Journal of Environmental Engineering*, 135, 1115–1122.

Woertz, I., Fulton, L. and Lundquist, T. (2009). Nutrient removal and green house gas abatement with CO_2 supplemented algal high rate ponds. Paper written for Water Environment Federation WEFTEC, 7924–7936.

Woods, R., Legere, E., Moll, B., Unamunzaga, C. and Mantecon, E. (2010). Closed photobioreactor system for continued daily in situ production, separation, collection, and removal of ethanol from genetically enhanced photosynthetic organisms. United States Patent no. US 7682821.

Wu, B., Stephen, D., Morgenthaler, G. and Jones, D. (2010a). Systems and methods for producing biofuels from algae. United States Patent no. 0077654.

Wu, B., Stephen, D., Morgenthaler, G. and Jones, D. (2010b). Systems and methods for producing biofuels from algae. United States Patent no. US 20100077654.

Xin, L., Hong-Ying, H., Ke, G. and Ying-Xue, S. (2010). Effect of different nitrogen and phosphorus concentrations on the growth, nutrient uptake, and lipid accumulation of fresh water microalga *Scenedesmus* sp. *Bioresource Technology*, 101, 5494–5600.

Yadavalli, R. and Heggers, G. (2013). Two stage treatment of dairy effluent using immobilized *Chlorella pyrenoidosa. Journal of Environmental Health Science and Engineering*, 11(36), 1–6.

Yewalkar, N.S., Dhumal, N.K. and Sainis, K.J. (2013). Effect of chromate on photosynthesis in Cr(VI)–resistant *Chlorella. Phytosynthetica*, 51, 565–573.

Yewalkar, S., Li, B., Posarac, D. and Duff, S.(2011). Potential for CO_2 fixation by *Chlorella pyrenoidosa* grown in oil sands tailings water. *Energy and Fuels*, 25, 1900–1905.

Youngs, R. and Cook, J. (2010). Method and apparatus for separating particles from a liquid. United States Patent no. US 20100224574.

Zhang, Y., Zhou, X., Su, H., Chen, X., Zhang, C. and Li, J. (2011). Photobioreactor for carbon dioxide mitigation in wastewater. United States Patent no. US 0159581.

Zittelli, G.C., Tomasello, V., Pinzani, E. and Tredici, R.M. (1996). Outdoor cultivation of Arthrospira platensis during autumn and winter in temperate climates. *Journal of Applied Phycology*, 8, 293–301.

Chapter 19
Bioremediation of Toxic Metals Using Algae

Panchali Bhattacharya, Nabanita Chakraborty, and Ruma Pal

19.1 Introduction

The rapidly growing population and expanding technological activities have accelerated the rate of addition of numerous poisonous pollutants especially the metal ions to the surrounding environment. These pollutants become deleterious due to their mobilization, transport and deposition in the various aquatic as well as terrestrial ecosystems. The cyanobacteria and algae (commonly called together 'Algae') constitute the most ancient groups of autotrophic microorganisms and are invariably affected by the presence of metal ions in the environment (Whitton, 1970). Algae are the organisms which can resist the metal toxicity by biochemical, chemical and physical mechanisms resulting in cell surface adsorption, metabolism dependent accumulation and precipitation (Gadd, 1988). They instantly interact with metal pollutants differently at cellular level showing different responses and tolerance mechanisms, termed as 'algae-metal interactions'—which is the basis of phytoremediation process.

Metals and metalloids can be characterized based on their toxicity level towards biological organisms (Gadd, 1993). One of the most common toxic heavy metal is lead, causing severe damage to living organisms and arsenic, another toxic metalloid, ranking 28th in abundance on the earth's crust, is widely encountered in the environment and severely damages metabolic pathways in different organisms from prokaryotes to human beings. Special emphasis would be given in this review on phycoremediation (removal of toxic metals by algae) of these two toxic elements and their probable mechanisms.

To understand the phycoremediation process by algae, it is important to investigate the responses of particular alga to particular metal and its tolerance or sensitivity,

P. Bhattacharya • N. Chakraborty • R. Pal (✉)
Department of Botany, University of Calcutta,
35, Ballygunge Circular Road, Kolkata 700019, India
e-mail: rpalcu@rediffmail.com

D. Das (ed.), *Algal Biorefinery: An Integrated Approach*,
DOI 10.1007/978-3-319-22813-6_19

together with the metal uptake capacity. The metal ions are either actively accumulated by living cells or adsorbed by dead biomass of algae, may also be chelated with the extracted metabolites, polysaccharides or other constituents of the cell surfaces. Overall the metal uptake by the algal biomass is considered to be quite complex phenomenon, influenced by several physico-chemical processes of the living or dead cells together with the external factors (Gadd, 1992). Overall there are two processes: (i) more rapid metabolism independent process or adsorption and, (ii) comparatively slower metabolism dependent process or uptake. In the first category, the metal ions remain adsorbed on the cell surface ligands and in other process or active absorption, transportation of metal ions through the cell membrane into the cytoplasm occurred (Bates et al., 1982; Mehta and Gaur, 2005). Thus the ability of cyanobacteria and microalgae to sorb metal pollutants from the surrounding environment at a higher concentration is a fascinating phenomenon which may easily be compared to that of other chemical sorbents. Therefore, they are highly suitable for use in the treatment of metal contaminated industrial or other effluents.

Almost countless reports are there regarding toxic metal removal by algae, few are mentioned here. As much as 98–100 % metal (Cd, Pb, Mn) removal efficiency have been recorded by many marine and fresh water algal genera (Esteves et al., 2000; Ele-Sheekh et al., 2005; Cervantes et al., 2001) and it seems clear that different species of algae accumulate metal in different degrees (Jordanova et al., 1999). On the other hand, in the natural ecosystem, biomagnification of toxic metals by algal genera may affect the entire food chain. Therefore, it is necessary to study the metal sorption capacity of algal genera for using them in bioremediation purpose on one hand and for pollution risk assessment in ecological niche, on the other. A number of reviews have been published by several authors giving detail account of metal sorption by algae, their tolerance mechanisms and biotechnology (Whitton et al., 1981; Genter, 1996; Mehta and Gaur, 2005). Whitton (1984) reviewed the metal accumulation by algae including biomonitoring of metals from natural population, assay of metal composition from algal population in laboratory condition, algal adaptation to elevated level of metals and effect of different metals on species and community composition. Rai et al. (1981) reviewed in detail the tolerance mechanisms of different algal genera giving special emphasis on metal binding on cell surface, exudation of metal complex ligands, efflux of metal ions and sequestration by phytochelatins and metallothioneins intracellularly. Mehta and Gaur (2005) critically reviewed the metal accumulation process including data analysis, mode of action, factors affecting metal sorption, regulation and reuse of metals in detail. A comprehensive review on microbial arsenic resistance systems have been elaborately illustrated by Mukhopadhyay et al. (2002), which stated about the global geocycling of arsenic, nature of arsenic resistant genes, arsenate reductase families, arsenite oxidation and methylation processes by microbes.

Many algae exhibit tolerance to high concentration of toxic metals showing higher level of biosorption (De Filippis and Pallaghy, 1994). The tolerance limit differs for different metals for same or different algal genera (Whitton, 1970; Foster, 1982; Takamura et al., 1989; Agrawal and Kumar, 1975; Harding and Whitton, 1976; Say et al., 1977; Whitton, 1980). Metal tolerance of algae is the result of different physiological activities like cell surface adsorption, secretion of extracellular ligands for metal complexation, exclusion of metals and sequestration by phytochelatins, sequestration of ROS by stress enzymes and other chelators etc. (Rai et al., 1981; Whitton, 1970; Mehta and Gaur, 2005). Among the metals used for these studies, most commonly used metals are cadmium, chromium (Cr II, III and VI), copper, nickel and zinc (De Carvalho et al., 1995; Chong and Volesky, 1996; Sandau et al., 1996; Roux, 1998; Singh et al., 1998; Zhou et al., 1998; Lau et al., 1999; Chong et al., 2000; Mehta and Gaur, 2001a, b, c; Yin et al., 2001; Cossich et al., 2002; Mehta et al., 2002a, b; Chaisuksant, 2003; Chojnacka et al., 2004; Feng and Aldrich, 2004; Hashim and Chu, 2004; Lee et al., 2004; Sheng et al., 2004b; Chojnacka et al., 2005; Gardea-Torresdey et al., 2005; Vijayaraghavan et al., 2005). Other metals tested are aluminium, cobalt, iron, mercury and lead (Ting et al., 1995; Matheickal and Yu, 1996; Sandau et al., 1996; Gardea-Torresdey et al., 1998; Ozer et al., 1999; Carrilho and Gilbert, 2000; Klimmek et al., 2001; Feng and Aldrich, 2004; Lee et al., 2004; Chojnacka et al., 2004; Prasher et al., 2004; Mahapatra and Gupta, 2005; Vijayaraghavan et al., 2005). Among the precious metals Au and Ag have been tried by a few authors for bioaccumulation study and to estimate the safe and toxic concentration also (Steele and Thursby, 1983; Green et al. 1986; Ting et al., 1995; Niu and Volesky, 2000; Lengke et al. 2006a, b). Our laboratory has published a series of papers on gold accumulation and recovery by algal genera of different groups like cyanobacteria, chlorophyta, diatoms, etc., associated with nanogold production or reduction of Au^{3+} to Au^0 (Nayak et al., 2006; Chakraborty et al., 2006; 2009; Parial et al., 2012).

Cyanobacterial members are quite efficient in metal removal process. Therefore, many authors used several cyanobacterial strains for bioaccumulation studies. Most successfully used taxa are *Oscillatoria*, *Anabaena*, *Spirulina*, *Lyngbya*, *Synechococcus* PCC 7942, *Synechocystis*, *Microcystis* etc. (Sandau et al., 1996; Gardea-Torresdey et al., 1998; Pradhan et al., 1998; Singh et al., 1998; Ahuza et al., 1999; Donmez et al., 1999; Klimmek et al., 2001; Chojnacka et al., 2004; Chojnacka et al., 2005; Mahapatra and Gupta, 2005). Among the chlorophycean members Chlorella vulgaris and a few other species like *Scenedesmus*, *Selenastrum* and *Cladophora* are most commonly used for bioaccumulation studies for Cd, Cu, Ni, Zn and Au (Keeney et al., 1976; Sandau et al., 1996; Donmez et al., 1999; Lau et al., 1999; Chong et al., 2000; Mehta and Gaur, 2001a, b, c; Mehta et al., 2002a, b). A large number of seaweeds have been employed for metal removal process like *Laminaria*, *Sargassum*, *Ulva*, *Ceramium*, *Ecklonia*, *Fucus*, *Gigartina*, *Padina*, *Ascophyllum* and *Palmaria* for Au, Co, Cd, Cu and Pb accumulation study (De Carvalho et al., 1995; Sandau et al., 1996; Yu and Kaewsarn, 1999; Niu and Volesky, 2000; Yin et al., 2001; Ofer et al., 2003; Sheng et al., 2004b; Feng and Aldrich, 2004; Hashim and Chu, 2004; Lee et al., 2004; Prasher et al., 2004; Vijayaraghavan et al., 2005). Zinc accumulation by Macrocystis (999.50 mg g^{-1}) indicated almost 100 % accumulation (Pradhan et al., 1998). For lead, maximum accumulation observed was 349.09 mg g^{-1} by *Laminaria japonica* (Lee et al., 2004) and that of nickel was 437.98 mg g^{-1} by *Chlorella vulgaris* (Mehta et al., 2002a, b). The seaweed genus *Ascophyllum* accumulated 129.9 mg g^{-1} Cr (III) (Kratochvil and

Volesky, 1998), whereas 146.12 mg g^{-1} cadmium sorption was noticed by *Laminaria japonica* (Yin et al., 2001). A few authors studied the role of alginate and fucoidan present in the cell wall of brown algae in metal binding process (Davis et al., 2003). Haug (1967) reported different degrees of binding capacity of various metals by alginic acid extracted from *Laminaia digitata* in a descending series: Pb^{2+} -Cu^{2+} -Cd^{2+} -Ba^{2+} -Sr^{2+} -Ca^{2+} -Co^{2+} -Ni^{2+} -Mn^{2+} -Mg^{2+}.

Availability of lead on the earth's surface is quite high (5–25 mg kg^{-1}), evolving from rocks, being released into the environment as gases during volcanic activity and associated with natural mobilisation into the environment (Goldberg and Gross, 1971). Chow (1968) reported that lead content of lakes and rivers varies between 1 and 10 μg L^{-1}. As lead is one of the major heavy metals and is a potent environmental pollutant, it gained considerable importance in environmental research.

In contrast to higher plants comparatively less data are available regarding Pb uptake, transport and detoxification process in algal systems. A few reports are available regarding bioaccumulation of Pb by different algal genera. Lead content of red snow alga *Chlamydomonas* from Greenland and Spitspergen (Fjerdingstad et al., 1974) were estimated employing proton induced X-ray spectrometry and found higher Pb content in Spitspergen (42.1 μg g^{-1}) sample than that of Greenland (13.2 μg g^{-1}). The high content of Pb in *Chlamydomonas* indicates the Pb contaminated environment due to more industrialisation in Spitspergen. Harding and Whitton (1978) reported the Zn and Pb content of *Nitella flexilis* from Pb contaminated reservoir polluted by mining activities. The seaweed genus *Laminaria japonica* and cyanobacterial member *Lyngbya taylori* showed comparatively high amount of Pb accumulation, which was 349.09 mg g^{-1} and 304.56 mg g^{-1} (DW) respectively (Lee et al., 2004; Klimmek et al., 2001). Another seaweed genus *Ecklonia* accumulated 243 to 281 mg g^{-1} of lead (Matheickal and Yu, 1996; Feng and Aldrich 2004). But other genera like *Spirulina*, *Schizomeris*, *Synechococcus*, *Chlorella* and *Palmaria* showed 0.01 to 65.47 mg g^{-1} of Pb accumulation (Chojnacka et al., 2004; Sandau et al., 1996; Ozer et al., 1999; Prasher et al., 2004). Holan and Volesky (1994) reported Pb accumulation as much as 1.1 to 1.3 mmol g^{-1} in phaeophycean genera like *Ascophyllum*, *Sargassum* and *Fucus*.

Arsenic, the other toxic metalloid, is widely spread in different layers of earth's crust, with a concentration range from 0.1 to more than 1000 ppm (mg kg^{-1}) in soil, 50–400 ppm in atmospheric dust, up to 2.6 ppb in seawater, and up to 0.4 ppb in fresh water (Mukhopadhyay et al., 2002). In many countries, arsenic contamination in ground water have been reported like Bangladesh, India, China, Taiwan etc. and investigated by several authors (Dhar et al., 1997; Biswas et al. 1998; Mandal et al., 1996, 1997; Liangfang and Jianghong, 1994; Chen et al., 1995; Tondel et al., 1999). The permissible limit for drinking water is only 0.01 mg L^{-1}, as designated by the World Health Organization (WHO). The national standard of arsenic concentration for drinking water in Bangladesh and India is 0.05 mg L^{-1}, which is much higher than the WHO standard limit. The highest arsenic concentration has been recorded as 0.9 mg L^{-1} in Nadia district of West Bengal, India which is 90 times than the WHO standard limit and almost 2 mg L^{-1} in Bangladesh (Chakraborti, 1999; Tondel et al., 1999).

Different valency states of arsenic like, −3, 0, +3 and +5 are present in nature. Among them, arsenite [As(III)] is the dominant form under reducing conditions whereas in oxygenated environments, arsenate [As(V)] is the stable form. As is the predominant form in soil and in groundwater and in submerged soil condition, the predominant form is arsenite. Methylated As are present in agricultural land, where microorganisms based conversion from inorganic arsenic to organic forms are reported including different forms of arsenic, like momomethyl arsinic acid (MMAA) and dimethyl arsinic acid (DMMA) (Takamatsu et al., 1982). However, the main source of arsenic on the Earth's surface is the igneous activity, i.e., formed during volcanic eruption. A schematic diagram for conversion is given in Fig. 19.1.

In marine ecosystem, the flora and fauna like phytoplanktons, macroalgae, crustaceans, mollusks and larger fishes, being continuously exposed to arsenic pollution, convert arsenate to MMA, DMA or other forms of organic storage. They either store these compounds or secret into the environment (Knowles and Benson, 1983; Frankenberger, 2001). Many algal species are reported to accumulate organoarseni-

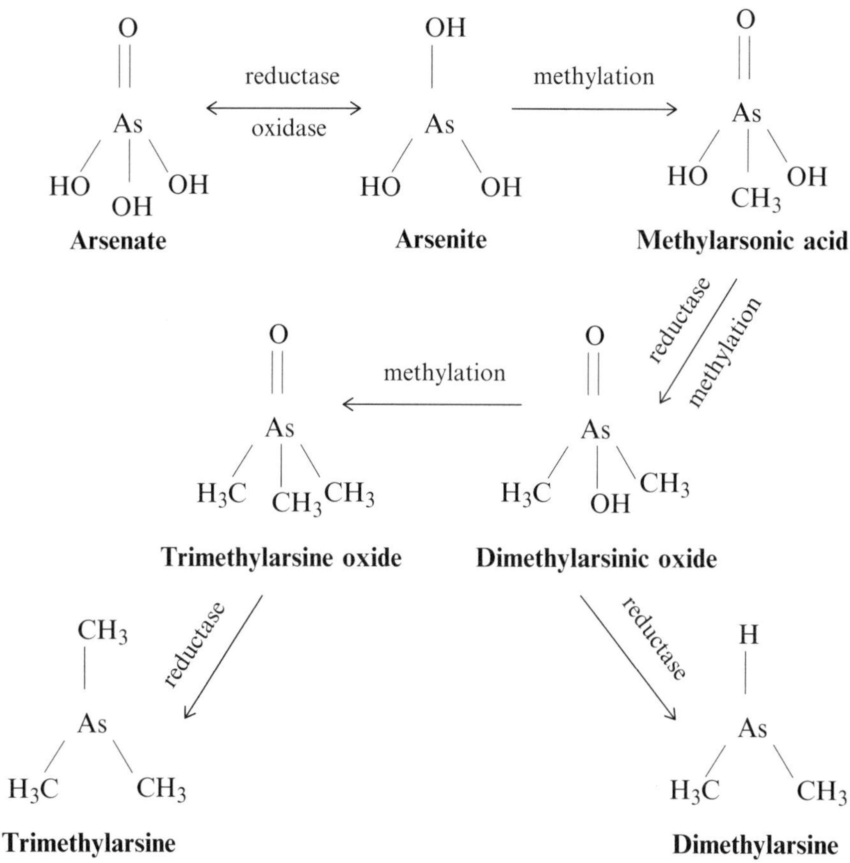

Fig. 19.1 Schematic diagram showing different forms of arsenic.

cal compounds like water-soluble arsenosugars (i.e. dimethylarsenosugars) and lipid-soluble compounds (arsenolipids). Fish and marine invertebrates store 99 % of accumulated arsenic in the form of arsenobetaine, which are passed through the food chain by phytoplanktons (Klumpp 1980). Arsenobetaine in turn is degraded by microbial metabolism in coastal seawater sediments to methylarsonic acid and inorganic arsenic. This way biological cycling of arsenic occurs in the marine system (Mukhopadhyay et al., 2002). Arsenic is metabolized by the human body quite differently from food and water depending on the chemical species administered (National Research Council, 1999; Eisler, 1994) and also in the animal species (Aposhian, 1997; Vahter, 2000; Mitchell et al., 2000). As a result of accumulation and bio-transformation processes, the organic arsenic concentration varies from 1–100 mg kg^{-1} in algae and marine animals (Cullen and Nelson 1993). Among the nontoxic organic forms arsenocholine (AsC), arsenobetaine (AsB) and arsenosugars are the major biosynthetic products in marine animals (Gailer et al., 1995; Larsen, 1995) and as a result trace amount of MMA and DMA are sometimes detected in seafood products.

Arsenic toxicity in algae and their tolerance limits have been reported by many authors. Algae accumulate and transform arsenate because of its similarity (analogous) to the essential and often growth-limiting nutrient PO_4 [(PO-$(OH)_3$]. Arsenic induced growth inhibition tests were performed in *Chlamydomonas reinhardii* with 100 mg L^{-1} As(V) (Jurewicz and Buikema, 1980). But no significant effect on growth of the freshwater diatom *Asterionella formosa* was recorded, when exposed to 160 μg L^{-1} As(V) (Conway, 1978). Sanders (1979b) exposed the diatom Skeletonema costatum to organic and inorganic arsenic and found that DMA had no significant effect on carbon uptake, and additions of phosphate to the media reduced the arsenate toxicity. In case of other taxa also growth were not affected in high concentrations of arsenite or arsenate, like *Tetraselmis chui* and *Hymenomonas carterae* (Bottino et al., 1978), Dunaliella sp. (Yamaoka et al., 1988), *Chlorella vulgaris* (Maeda et al., 1985) in 100 to 2000 mg L^{-1} As(V). Therefore, it can be inferred that arsenic does not affect the growth of some algae and cyanobacteria at even high concentrations. The pH level of experimental media also affects the toxicity of arsenate as observed by Michnowicz and Weaks (1984) in *Selenastrum capricornutum*, where growth enhancement is recorded at a higher pH. There are many reports regarding arsenic toxicity in diatoms. Hollibaugh et al. (1980) studied the toxicity of arsenic to *Thalassiosira aestevalis*.

Different studies have been done from time to time to understand the arsenic accumulating capacity of different cyanobacterial and algal population in natural and experimental conditions (Imamul Huq et al., 2005; Shamsuddoha et al., 2006). Maeda et al. (1987) exposed the cyanobacterium (*Nostoc* sp.) to 1 and 10 mg As(V) L^{-1} for 32 d and found 32 and 77 mg As kg^{-1} of dry cell weight with no significant effect on growth. Methylation and excretion of As by arsenic resistant genus Phormidium has been reported by Maeda et al. (2004). They also reported increased growth rate of algal biomass up to 100 mg g^{-1} As in growth medium. *Skeletonema costatum* was found to increase their arsenic concentrations by 40 % (Sanders and Windom, 1980) and other phytoplanktons accumulate arsenic from 5.7 to

17.7 mg kg^{-1} (dry weight) when cultured for 48–96 h at 25 μg L^{-1} As(V) (Sanders et al., 1989). A study by Maeda et al. (1992) found that arsenate was actively accumulated when the cells were exposed during the early exponential phase of *Chlorella*. The other unicellular green alga *Dunaliella salina* accumulated more arsenic at higher nitrogen concentrations (Yamaoka et al., 1992). Reuther (1992) observed that when arsenate was added to a freshwater model ecosystem, it was readily accumulated by plankton with arsenic residues of 37–47 mg kg^{-1} (dry weight) at 5 μg L^{-1} As(V) exposure and >200 mg kg^{-1} at 50 μg L^{-1} As(V) after 65 d exposures. Arsenic also induced changes in cellular metabolites. Accumulation of inorganic As increased the beta carotene and fatty acids (C18:1 and C18:3) and water extractable carbohydrate content in the cells of *D. salina* (Yamaoka et al. 1992).

Marine macro-algae like *Ascophyllum* and *Fucus* are known to accumulate As and selenium showing concentration factors of 1000 to 10000, compared to their environment (Lunde, 1970; Klumpp and Peterson, 1979). Several authors detected arseno-sugars using anion exchange HPLC or ICP-MS or by other methods from different genera of sea weeds like *Porphyra*, *Fucus*, *Sargassum*, *Ceramuim*, *Padina*, *Enteromorpha*, *Ulva*, *Eichlonia* etc. (McSheehy and Szpunar, 2000; Šlejkovec et al., 2006; Edmonds and Francesconi, 1981; Madsen et al., 2000). Many reports have illustrated the extraction and separation of arsenosugar species from marine algae. *Laminaria japonica* (brown algae), *Fucus serratus* and *Porphyra* (red algae) were found to contain four arsenosugars and methylarsonicals (Karthikeyan and Hirata, 2003). In environmental waters several algae influence the speciation of arsenic also (Bottino et al., 1978; Conway, 1978). Green alga *Chlorella* has been reported to reduce arsenate to arsenite (Knauer and Hemond, 2000). Garcia-Salgado et al. (2006) identified As(V) in *Hizikia* (46 ± 2 mg g^{-1}), *Sargassum* (38 ± 2 mg g^{-1}) and *Chlorella* (9 ± 1 mg g^{-1}) samples and DMA in *Chlorella* (13 ± 1 mg g^{-1}). Several authors have reported different species of arsenic like As(V), As(III), arsenobetaine, arsenocholine, arsenosugars, tri-MeOH-ribose, glycerol triethylated arsenoriboside; DMAE, dimethylarsenoyl ethanol; MA, methyl arsonate; DMA, dimethyl arsinate; TETRA, tetramethylarsonium ion; TMAO, trimethylarsine oxide in marine macroalgae, *Laminaria*, *Sargassum*, *Undaria*, *Hizika*, *Pelvetia*, *Myelophycus*, *Ceramium*, *Gelidium*, *Cystoseira*, *Enteromorpha*, *Fucus*, *Padina*, *Polisyphonia*, *Ulva*, *Chladophora*, *Chlorella*, *Eucheuma* etc. (Meier et al., 2005; Garcia-Salgado et al., 2006; Hirata and Toshimitsu, 2007; Šlejkovec et al., 2006; Rubio et al., 2010).

19.2 Mechanism of Phycoremediation

Algal cell walls from different groups with varied chemical nature have played important role in metal sorption process. Generally carboxylic group of cell wall polysaccharide play a predominant role in metal uptake by cyanobacteria and eukaryotic algae (Chojnacka et al., 2005). The other functional group like sulphonate, amino and hydroxyl groups in adsorption of various metal ions have also been reported (Mehta and Gaur, 2005). Thiol group also plays an important role in sorption of metals like Cd at lower pH (Sheng et al., 2004a). Metal sorption by brown algae like *Ascophyllum*, *Sargassum* etc. is high due to their alginate content (Davis et al., 2004).

A number of metals are reported to bind to intracellular polyphosphate granules of algal cells. Therefore, the metals remain in bound form in presence of high concentration of PO_4^{3-}. Reports are available regarding Zn and Cd binding to polyphosphate granules (Bates et al., 1985; Walsh and Hunter, 1992). X-ray microanalysis revealed that polyphosphate bodies in *Chlorella* treated with Al, Fe, Cu and Zn (Wong et al., 1994) and in *Anabaena cylindrica* exposed to lead (Swift and Forciniti, 1997).

Algal enzyme systems are active to combat the metal stress. Some enzymes take active part in metal detoxification process by reducing the toxic metal to non-toxic form or by quenching the ROS (Hassan and Scandalios, 1990; Rice-Evans et al., 1996; Fridovich, 1997; Asada, 1999). In *Chlorella*, mercuric chloride and phenyle mercuric acetate is reduced to metallic, volatile mercury by NADPH or NADH (Ben-Bassat and Mayer, 1977).

Algal extracellular polysaccharides (EPS) are potential compounds for metal removal process due to the presence of different metal chelating ligands and are used in different biotechnological purposes. Uronic acid and $COOH^-$ group present in EPS of cyanophycean algae and $COOH^-$ and SO_4^- of heteropolysaccharides of green algal genera serve as prime metal chelating components. Oxidation of Myo-inositol is the key step in the formation of plant extracellular polymeric substances like gum, mucilage, glycoprotein etc. (Loewus and Loewus, 1983). From this point of view excess production of algal EPS in metal stressed condition can be attributed to oxidizing capacity of the ROS produced within the cells.

Several cyanophycean and chlorophycean algal species have been reported to produce copper-complexing polysaccharidic ligands. Pistocchi et al. (1997) reported higher extracellular carbohydrate production in *Cylindrotheca fusiformis* than *Gymnodinium* sp. with increased toxic Cu concentration of 0.2 to 0.5 ppm after 12 to 16 days of growth. McKnight and Morel (1979, 1980) detected strong copper complexing chelate in culture of cyanophytes at stationary growth phase. Results show that cyanobacteria complexes are generally stronger than those of eukaryotic ones. Nordi et al. (2005) investigated the potentiality of EPS produced by *Anabaena spiroides* in binding Mn(II), Cu(II), Pb(II) and Hg(II) and successfully used for bioremoval process. The algal cell may show the metal resistance employing different biochemical pathways as shown in Fig. 19.2.

19.3 Phytoremediation and Oxidative Stress

Nonessential metals in sub lethal concentrations trigger oxidative stress to the plants leading to the formation of persistent reactive oxygen species (ROS), which damages the cell organelles and disturbs the cellular metabolism. Enzymatic

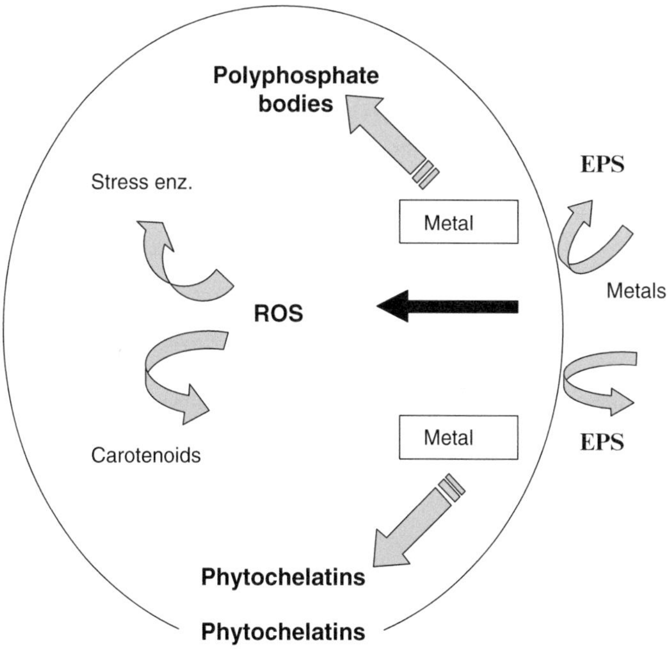

Fig. 19.2 Pathways of metal sequestration in the algal cell (arrows indicate metal sequestration pathways).

components like superoxide dismutase (SOD), catalase (CAT), glutathione peroxidase (GPX), ascorbate peroxidase (APX) and glutathione reductase (GR) as well as non-enzymatic molecules such as ascorbic acid, cystein, reduced glutathione, α-tocopherol, hydroquinone, carotenoids and polyamines form the anti-oxidative systems of the cell suppressing the ROS level, therefore, reducing the metal toxicity. This is the main basis for phycoremediation of toxic metals. In metal exposure, oxidative stress results from reduced antioxidant enzymatic defense, or low molecular mass like glutathione, α-tocopherol, ascorbate and/or an increase in production of reactive oxygen species (ROS) (Mallick and Mohan, 2000; Okamoto and Colepicolo, 2001; Pinto et al., 2003). The internal ROS level exceeds the tolerance level for inducing oxidation of lipids, proteins and nucleic acids toxicity results (Halliwell and Gulleridge, 1999). Under severe metal pollution, not only excessive oxidation occurs, but the efficiency of anti-oxidative defence is greatly altered. In non-resistant plants these mechanisms are weak. Algae generally respond at molecular level quickly to combat the toxicity (Reed and Gadd, 1990; Rodriguez-Ariza et al., 1991; Holovská et al., 1996; Okamoto et al., 1996; Pinto et al., 2003).

One important major non-protein thiol-glutathione plays main role in scheming the organism's antioxidant defense mechanism; especially the reduced thiols are reported to recycle the antioxidants, vitamin E and vitamin C (Constantinescu et al., 1993). It also acts as a reductant in the highly oxidizing environment of photosynthetic cells in plants (Alscher, 1989; Noctor and Foyer, 1998). The SH group of GSH can be used to reduce peroxides derived as by-products of metabolism which in turn may lead to enhance the peroxidation of membrane lipids and loss of cell viability (De Vos, 1992; Pinto et al., 2003). Therefore, consumption of cellular GSH in scavenging reactive species and free radicals produced in metal stressed condition is obvious. Reduction in GSH amount can also be seen in case of enhanced activity of H_2O_2 removing ascorbate oxidase (Apx) system. Actually, transfer of signal for translation of Apx transcript is inhibited by high levels of GSH (Karpinski et al., 1997). Under conditions of oxidative stress when Apx activity is required, GSH levels would be depleted to reduce dehydroascorbate (DHA) to ascorbate (AA) by means of DHA reductase. There are reports on correlation of acute metal stress and decreased reduced glutathione pool as found in treated cells in *Gonyaulax polyedra* exposed to the toxic metal Pb^{2+} (Okamoto et al., 1999). Nagalakshmi and Prasad (2001) observed progressive depletion of GSH content with increasing concentration of Cu.

The harmful H_2O_2 is removed catalytically by catalase (CAT) and ascorbate peroxidase (APX). Catalytic breakdown of H_2O_2 to H_2O and O_2 is induced by CAT and this occurs in peroxisomes (Halliwell and Gulleridge, 1999), but in cyanobacteria catalase activity is located in cytosol (Regelsberger et al., 1999). Catalysis of H_2O_2 is done differently by APX, which removes it by using it to oxidize ascorbate, producing mono-dehydro-ascorbate (MDHA) and H_2O. There are at least three distinct isoenzymes of APX: thylakoid bound, stromal and cytosolic APX which are restricted to higher plants, algae and some cyanobacteria (Mittler and Zilinskas, 1993). These enzymes counteracting H_2O_2 exposure, is part of the integrated net of strategies that make the redox status of algal cells (Barros et al., 2003). An increase in peroxidase activity is regarded as a reliable indicator of stress from toxicity of heavy metals/metalloids, which may cause disruption of the plasma membrane by lipid peroxidation and the ROS production (Macfarlane and Burchett, 2001).

Dismutation of O_2^- by SOD produces H_2O_2, a weak oxidizing agent that can cross the cell membrane easier than $O_2.^-$ and possesses a steady state high concentration (Chance et al., 1979). A 7-fold increase in catalase activity (90.7 nmol min^{-1} mg^{-1} proteins) was observed by Loretto et al. (2005) in *Scytosiphon lomentaria* inhabiting copper enriched coastal environments indicating involvement of catalase in buffering oxidative stress *in vivo*. The SOD generally catalyses superoxide anion radicals produced in different compartments of plant cells to H_2O_2. On the other hand, transition heavy metals (e.g. Cu, Fe) catalyse the formation of •OH radicals from O_2^- (superoxide) in the nonenzymatic Fenton reaction. The protective function of CAT is limited due to its localization mainly in peroxisomes. Ascorbate (ASC) is known as a major primary antioxidant, reacting directly with •OH, superoxide (O_2^-) and singlet •O_2 (Buettner and Jurkiewicz, 1996). Ascorbate peroxidase (APX) and glutathione reductase (GR) are vital constituents of the ascorbate-glutathione pathway which are required to scavenge H_2O_2 and to maintain the redox state of the cell (Asada, 1992). Under oxidative stress increased GR activity could be required to supply reduced glutathione (GSH) to the ascorbate-glutathione cycle.

Flavonoids are found in higher plants and brown algae and are directly linked with scavenging •OH, ONOOH (peroxynitrous acid) and HOCl (hypochlorous acid) in order to inhibit lipid peroxidation. Since flavonoids bind to metal ions, the scavenging efficiency of flavonoids is directly proportional to the number of hydroxyl groups (Rimbach et al., 2003).

Under unfavourable environmental conditions, among all amino acids, proline is accumulated rapidly and more frequently. It acts as a powerful secondary antioxidant reducing the oxidized form of α-tocopherol (Buettner and Jurkiewicz, 1996).

Lipid peroxidation of membranes is an indicator of oxidative damage, which is caused by free radicals and hydroperoxides (Smirnoff, 1993). It involves oxidative degradation of polyunsaturated fatty acyl residues of membranes (Girotti, 1990). A reduced level of saturated fatty acids and high levels of unsaturated fatty acids of membranes in several plant species are brought about by metal ions through lipid peroxidation (Halliwell and Gulleridge, 1999). These results suggest that decreased activities of antioxidant enzymes could result in an increased level of lipid peroxidation, thus contributing to damage of cell membranes leading to cell death (Blum and Ebercon, 1981; Marcum, 1998; Abernethy et al., 1989).

Enhanced ROS level generally induce antioxidant synthesis in algal cultures, depending on the duration and severity of the stress applied (Okamoto et al., 1996). Generally the type and duration of exposure to metal/metalloid ions, either acute or chronic, alter the level of antioxidants and create different oxidative status (Okamoto and Colepicolo, 2001). Therefore, higher level of cellular antioxidants could allow cells to combat chronic stress, whereas, sometimes a sudden generation of high levels of ROS over a short period can surpass the total antioxidant capacity resulting toxicity. Antioxidant capacities of GSH, NADPH and ascorbate are likely to occur first in acute stress, which lowers the GSH pool and depletes NADPH levels.

In contrast to information pertaining to antioxidative defense in microorganisms to the effects of metals on bacteria, fungi, dinoflagellates and diatoms, very little is known about antioxidant defense system in other algae by arsenic stress. Pandey et al. (2012) studied the upregulations and downregulations in antioxidant system of *Anabaena* sp. PCC 7120 exposed to arsenic and found that an up-regulation of CAT, peroxiredoxin (Prx), thioredoxin (Trx) and oxidoreductase, and also an appreciable induction in phytochelatin content, GST activity and transcripts of phytochelatin synthase, arsenate reductase and arsenite efflux genes—*asr*1102, *alr*1097 which echoed their role in As sequestration and shielding of the organism from As toxicity. They established that, up-regulation in metabolic and antioxidative defense proteins, phytochelatin and GST together with the *ars* genes play a central role in detoxification and survival of *Anabaena* under As stress. *In vitro* study was done by Zutshi et al. (2014) to determine the toxic effect of sodium arsenate (0–100 mM) on an aquatic cyanobacteria *Hapalosiphon fontinalis*-339. At this level they found that MDA production was enhanced that probably resulted in decreased growth of the test organism.

Accumulation of enzymatic and non-enzymatic substance e.g. SOD, CAT, APX activities and proline, total glutathione showed an efficient antioxidative potential mechanism in *Hapalosiphon fontinalis*-339. Srivastava et al. (2009) provided comprehensive information on arsenic induced oxidative stress and changes in antioxidative defense system of *Anabaena doliolum*. They concluded that the cyanobacterium may survive better in As(V) than As(III) contaminated fields because of its low toxicity and pronounced induction of antioxidative defense system. The present authors already reported arsenic induced changes in stress enzymes and other stress related compounds in *Phormidium laminosum* (Bhattacharya and Pal, 2011). Therefore, metals and metalloids lead to the activation of a defense mechanism inside the cells in the form of antioxidants, which lead to the reduction of toxicity. Thus it is important to study the biochemical modulations of the antioxidant defense system of cyanobacteria in Arsenic stress and to fully understand its potential as a suitable material for bioremediation process.

19.4 Genetic Tolerance to Arsenic

Rosen and his group (1994, 1997, 2002) did an extensive study on biochemistry of As detoxification and its genetic control in *E. coli*. In prokaryotic system, several authors opined the fact that Pit and Pst are the two PO_4^- transporters and both of them catalyse AsO_4^- uptake, where the Pit system appears to be predominant (Willsky and Malamy, 1980). In the eukaryote system, like *Saccharomyces cerevisiae* different PO_4^- transporters participate in AsO_4^- accumulation (Yompakdee et al., 1996). Sanders et al. (1997) recorded the glycerol facilitator of *E. coli* which transports both As III and Sb III—the trivalent metalloid transported as Glp F being a member of the aquaporin superfamily. Aqua-glyceroporins transport neutral organic solutes like glycerol and urea. The Fps 1p, the homologue of Glp F, has recently been discovered as the route of uptake for arsenite in *S. cerevisiae* (Wysocki et al., 2001). Liu et al. (2002) have also shown that mammalian aqua-glyceroporin catalyses uptake of trivalent metalloids. The genes responsible for As transport and detoxification have also been characterized by several authors (Liu et al., 2002).

It is known that the ancient environment was not oxidizing and As(III) was the most dominant form, therefore, early organisms have evolved with a detoxification mechanism of As(III), mainly the extrusion system. According to Dey and Rosen (1995), bacteria show two basic mechanisms of arsenite extrusion—one is with carrier protein, where energy is supplied by the membrane potential of the cell and the other by an AsO_3^- translocating ATPase. For arsenate reduction process, three independently evolved families of arsenate reductase enzymes have been recognized whose sequences also have been identified as a product of ars operon (Mukhopadhyay et al., 2002). Cytosolic arsenite is also detoxified by removal process (Rosen, 1999). Cole et al. (1994) reported that members of multidrug resistance associated protein (MRP) is responsible for AsO_3^- resistance in eukaryotic As extrusion systems. Not much is however known about extrusion mechanisms in algae. Some work has been done by the present group which is illustrated in Fig. 19.3.

As mentioned earlier, three different families of arsenate reductase are reported in different organisms. The product of the *arsC* gene from the *E. coli* plasmid R773 was reported to be the first family of arsenate reductases. Several Gram-negative bacteria harbour this enzyme, which uses glutaredoxin as a source of reducing equivalents. The other two types of pI258 from *Staphylococcus aureus* and *Bacillus subtilis* differ significantly from *E. coli*. In these cases, instead of glutaredoxin, the

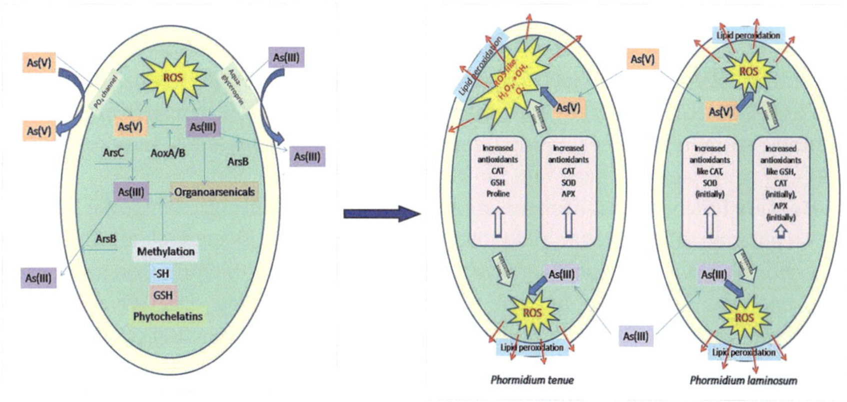

Fig. 19.3 Arsenic toxicity and defence mechanism in cyanobacteria Phormidium spp. (as studied by our group).

arsenate reductase is related to low-molecular-weight protein tyrosine phosphatases and uses thioredoxin as the source of reducing equivalents. On the other hand another family of arsenate reductase from *Saccharomyces cerevisiae*, represented by the Acr2p enzyme, is also similar to a protein phosphatases which contains CDC25a. Generally these *arsC* are responsible for reduction to the more toxic form arsenite, than exported by the carrier protein ArsB. The *E. coli* plasmid R773 and *Staphylococcus aureus* pI258 bearing *arsB* gene show similar activity encoding an integral membrane protein that expel arsenite. When ArsB interacts with ArsA, an arsenite-stimulated ATPase, proteins can also function as an arsenite pumps. The ARR3 protein from *S. cerevisiae* (formerly ACR3) and the ArsB gene of the *B. subtilis* of *ars* operon are considered as second family of arsenite carriers. In addition, another protein ArsH from gene *arsH* has been found to be essential for resistance to arsenite and arsenate both in *Yersinia enterocolitica* and *Acidothiobacillus ferroxidans* (Lopez-Maury et al., 2003). In some plasmid-determined systems of Gram-negative bacteria, the arsenic efflux pump consists of a two-component ATPase complex. The *arsA* gene product is a soluble ATPase subunit (Rosen et al., 1999), which physically associates with an integral membrane protein, the product of the *arsB* gene (Tisa and Rosen, 1990; Gladysheva et al., 1994). In most chromosomal arsenic resistance systems of Gram negative bacteria and the plasmids and chromosomes of Gram-positive bacteria, though adjacent *arsB* and *arsC* genes are found, but there is no *arsA* gene (Silver et al., 2001).

Eukaryotes such as *Saccharomyces cerevisiae*, the arsenic resistance gene cluster is similar to that of bacteria (Bobrowicz et al., 1997). Here, three adjoining genes remain in cluster, ARR1, ARR2 and ARR3. The first gene, ARR1, appears to produce a yeast transcriptional regulator and its disruption leads to hypersensitivity to arsenite and arsenate. Mukhopadhyay et al. (2000) found functional yeast ARR2 arsenate reductase gene in *E. coli.* Galperin et al. (1998) recorded Arr3p (the protein product of ARR3) as a member of a family of arsenite carrier efflux bacterial and archaeal members and is unrelated to the larger family of ArsB proteins found in many bacterial *ars* operons (including in *E. coli* and *Staphylococcus aureus*). According to them they evolved as a result of convergent evolution. In addition to the three ARR gene products, another yeast protein, Ycf1p, which is an ABC ATPase, also contributes to resistance to As(III) and Sb(III), being located in the vacuolar membrane and by pumping glutathione adducts, $As(GS)_3$ and presumably $Sb(GS)_3$ from the cytoplasm into the vacuole.

But very little work has been reported in cyanobacteria and algae where the function and regulation of these *ars* genes are still unknown (Cervantes et al., 2006). In *Synechocystis* PCC 6803 an arsenic and antimony resistance operon reported showing *arsC*-encoding a putative arsenate reductase, *arsB*-encoding a putative arsenite-antimonite carrier and *arsH*-encoding a protein of unknown function (Lopez-Maury et al., 2003). It was reported that *arsC* mutants were sensitive only to arsenate, while *arsB* mutant strains were sensitive to arsenite, arsenate, and antimonite. They also observed that purified recombinant ArsR protein bound to the *arsBHC* promoter-operator region and dissociated in the presence of Sb(III) or As(III) but not in the presence of As(V), suggesting that trivalent metalloids are the true inducers of the system.

Proteomics study in combination with morphological, physiological and biochemical variables have been employed by Pandey et al. (2012) in arsenic treated cyanobacterium *Anabaena* sp. PCC7120 to unravel its survival strategies. In this study it was revealed that 13 were novel (hypothetical) ones out of total 45 differentially expressed proteins. They also proposed hypothetical model which explains the interaction of metabolic proteins associated with the survival of *Anabaena* sp. PCC7120 under As stress.

19.5 Model Developed for Bioremoval of Metal/Metalloids

Various conventional physical, chemical and biological methods have so far been practiced to remove pollutants from industrial wastewater. Sometimes these methods are costly due to large chemical requirements and excessive sludge production and are with operational difficulties. There is, therefore, always a requirement for a low cost simple to operate system for treating industrial wastewaters. Use of activated algae-reactor is in well practice over the past few years (McGriff and McKinney, 1972; McShan et al., 1974; Lee et al., 1980).

A few models have already been proposed for metal removal process and use of 'algae-pond' being a popular and widespread technology for last few decades (McGriff and McKinney, 1972; McShan et al., 1974; Lee et al., 1980). A continuous flow system consisting of three rectangular algae reactors, connected in series was designed by Aziz and Ng (1993) for removing organic and synthetic pollutants along with metals from industrial wastewater. Experimental data obtained suggested that activated algae-reactor was successfully able to remove organic pollutants, colour, nutrients and toxic metals from wastewater in a cost effective manner. Bender et al. (1994) also used glass column packed with cyanobacterial for removing metals like zinc and manganese from contaminated water. Within 3 h retention time 96 % Zn and 86 % Mn could be removed by the column. Boi-filter, AlgaSORB-

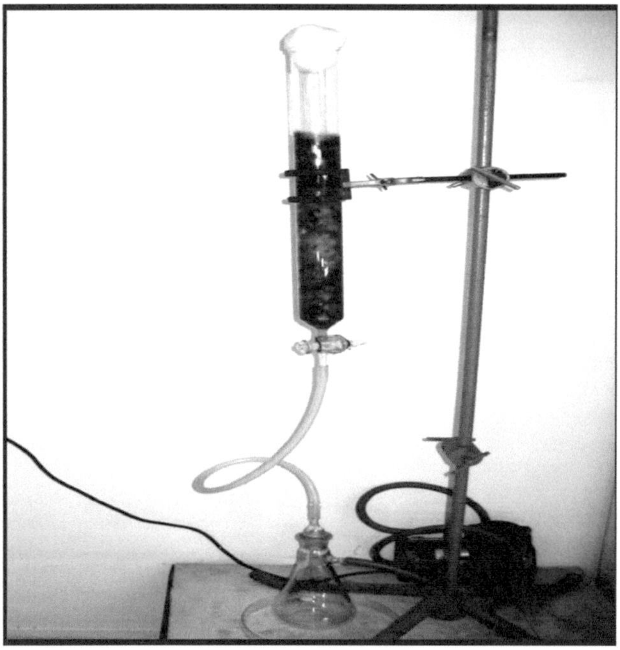

Fig. 19.4 Column type algae based Biofilter for metal removal from contaminated water showing a glass column with live algal mat enmeshed with glass wool, collection flask and pump.

scy (*Scytonema*-dimethyl-formamide slurry) over a polymer-modified silica gel, was suitable for 100 % As(III) removal (Prasad et al., 2006). The present group already reported 95.8 % removal of the Pb from 5 mg L^{-1} Pb solution using *L. majuscula* as bioreagent (Fig. 19.4; Chakraborty et al., 2011).

References

Abernethy, R.H., Thiel, D.S., Peterson, N.S. and Helm, K. (1989). Thermotolerance is developmentally dependent in germinating wheat seed. *Plant Physiol.*, 89, 569–576.

Agrawal, M. and Kumar, H.D. (1975). Response of *Chlorella* to mercury pollution. *Ind J Ecol.*, 21, 94–98.

Ahuza, P., Gupta, R. and Saxena, R.K. (1999). Zn biosorption by *Oscillatoria anguistissima. Process Biochem*, 34, 77–85.

Alscher, R.G. (1989). Biosynthesis and antioxidant function of glutathione in plants. *Physiologia plantarum*, 77, 457–464.

Aposhian, H.V. (1997). Enzymatic Methylation of Arsenic Species and Other New Approaches to Arsenic Toxicity. *Ann Rev Pharmacol Toxicol*, 37, 397–419.

Asada, K. (1992). Ascorbate Peroxidase: A hydrogen peroxide scavenging enzyme in plants. *Physiol Plant*, 85, 235–241.

Asada, K. (1999). The water-water cycle in chloroplasts: Scavenging of active oxygens and dissipation of excess photons. *Ann Rev Plant Physiol Plant Mol Biol*, 50, 601–639.

Aziz, M.A. and Ng, W.J. (1993). Industrial wastewater treatment using an activated algae-reactor. *Water Sci Technol*, 28, 71–76.

Barros, M.P., Granbom, M., Colepicolo, P. and Pedersén, M. (2003). Temporal mismatch between induction of superoxide dismutase and ascorbate peroxidase correlates with high H_2O_2 concentration in seawater from clofibrate-treated red algae *Kappaphycus alvarezii. Arch Biochem Biophys*, 420, 161–168.

Bates, S.S., Tessier, A., Campbell, P.G.C. and Buffle, J. (1982). Zinc adsorption and transport by *Chlamydomonas variabilis* and *Scenedesmus subspicatus* (Chlorophyceae) grown in semicontinuous culture. *J Phycol*, 18, 521–529.

Bates, S.S., Tessier, A., Campbell, P.G.C. and Letourneau, M. (1985). Zinc-phosphorus interactions and variation in zinc accumulation during growth of *Chlamydomonas variabilis* (Chlorophyceae) in batch culture. *Can J Fish Aquat Sci*, 32, 86–94.

Ben-Bassat, D. and Mayer, A.M. (1977). Reduction of mercury chloride by *Chlorella*: Evidence for a reducing factor. *Physiologia Plantarum*, 40(3), 157–162.

Bender, J., Gould, J.P., Vatchaapijarn, Y., Young, J.S. and Philips, P. (1994). Removal of zinc and manganese from contaminated water with cyanobacteria mats. *Water Environ Res*, 66, 679.

Bhattacharya, P. and Pal, R. (2011). Response of algae to arsenic toxicity. *J Appl Phycol*, 23, 293–299.

Biswas, B.K., Dhar, R.K., Samanta, G., Mandal, B.K., Chakraborti, D., Faruk, I., Islam, K.S., Choudhury, M.M., Islam, A. and Roy, S. (1998). Detailed study report of Samta, one of the arsenic-affected villages of Jessore District. *Bangladesh Curr Sci*, 74, 134–145.

Blum, A. and Ebercon, A. (1981). Cell membrane stability as a measure of drought and heat tolerance in wheat. *Crop Sci*, 21, 43–47.

Bobrowicz, P., Wysocki, R., Owsianik, G., Go!eau, A. and Ulaszewski, S. (1997). Isolation of three contiguous genes, ACR1, ACR2 and ACR3, involved in resistance to arsenic compounds in the yeast *Saccharomyces cerevisiae. Yeast*, 13, 819–828.

Bottino, N.R., Newman, R.D., Cox, E.R., Stockton, R., Hoban, M., Zingaro, R.A. and Irgolic, K.J. (1978). The effects of arsenate and arsenite on the growth and morphology of the marine unicellular algae *Tetraselmis chui* (Chlorophyta) and *Hymenomonas carterae* (Chrysophyta). *J Exp Mar Biol Ecol*, 33(2), 153–168.

Buettner, G.R. and Jurkiewicz, B.A. (1996). Catalytic metals, ascorbate and free radicals: Combinations to avoid. *Radiat Res*, 145, 532–541.

Carrilho, E.N. and Gilbert, T.R. (2000). Assessing metal sorption on the marine alga *Pilayella littoralis. J Environ Mon*, 2, 410–415.

Cervantes, C., Campos-García, J., Devars, S., Gutiérrez-Corona, F., Loza-Tavera, H., Torres-Guzmá, J.C. and Moreno-Sánchez, R. (2001). Interactions of chromium with microorganisms and plants. *FEMS Microbiol Rev*, 25(3), 335–347.

Cervantes, C., Ji, G., Ramirez, J.L. and Silver, S. (2006). Resistance to arsenic compounds in microorganisms. *FEMS Microbiol Rev*, 15(4), 355–367.

Chaisuksant, Y. (2003). Biosorption of Cd (II) and Cu (II) by pretreated biomass of marine alga *Gracilaria fisheri. Environ Technol*, 24, 1501–1508.

Chakraborti, D. (1999). 29 Dec 1999 message from Dipankar Chakraborti. Arsenic Crisis Information Centre. Retrieved from www.bicn.com/acic. (Last updated 24.09.2002).

Chakraborty, N., Banerjee, A., Lahiri, S., Panda, A., Ghosh, A.N. and Pal, R. (2009). Biorecovery of gold using cyanobacteria and eukaryotic alga with special reference to nanogold formation – A novel phenomenon. *J Appl Phycol*, 21, 145–152.

Chakraborty, N., Banerjee, A. and Pal, R. (2011). Accumulation of lead by free and immobilized cyanobacteria with special reference to accumulation factor and recovery. *Biores Tech*, 102(5), 4191–4195.

Chakraborty, N., Pal, R., Ramaswami, A., Nayak, D. and Lahiri, S. (2006). Diatom: A potential bio-accumulator of gold. *J Radioanal Nucl Chem*, 270, 645–649.

Chance, B., Sies, H. and Boveris A. (1979). Hydroperoxide metabolism in mammalian organs. *Physiol Rev*, 59, 527–605.

Chen, S.L., Yeh, S.J., Yang, M.H. and Lin, T.H. (1995). Trace-element concentration and arsenic speciation in the well water of a Taiwan area with endemic blackfoot disease. *Biol Trans Elem Res*, 48, 263–274.
Chojnacka, K., Chojnacki, A. and Gorecka, H. (2004). Stress metal removal by *Spirulina* sp. from copper smelter and refinery effluent. *Hydrometallurgy*, 73, 147–153.
Chojnacka, K., Chojnacki, A. and Gorecka, H. (2005). Biosorption of Cr^{3+}, Cd^{2+} and Cu^{2+} ions by blue-green algae *Spirulina* sp.: Kinetics, equilibrium and the mechanism of the process. *Chemosphere*, 59, 75–84.
Chong, A.M.Y., Wong, Y.S. and Tam, N.F.Y. (2000). Performance of different microbial species in removing nickel and zinc from industrial wastewater. *Chemosphere*, 41, 251–257.
Chong, K.H. and Volesky, B. (1996). Metal biosorption equilibria in a ternary system. *Biotechnol Bioeng*, 49, 629–638.
Chow, T.J. (1968). Isotope analysis of sea-water by mass spectrometry. *J Water Pollut Control Fed*, 40, 399–411.
Cole, S.P., Sparks, K.E., Fraser, K., Loe, D.W., Grant, C.E., Wilson, G.M. and Deeley, R.G. (1994). Pharmacological characterization of multidrug resistant MRP-transfected human tumor cells. *Cancer Res*, 54(22), 5902–5910.
Constantinescu, A., Han, D. and Packer, L. (1993). Vitamin E recycling in human erythrocyte membranes. *J Biol Chem*, 268(15), 10906–10913.
Conway, H.L. (1978). Sorption of Arsenic and Cadmium and Their Effects on Growth, Micronutrient Utilization, and Photosynthetic Pigment Composition of *Asterionella formosa*. *J Fisher Res Board Can*, 35, 286–294.
Cossich, E.S., Tavares, C.R.G. and Ravagnani, T.M.K. (2002). Biosorption of chromium (III) by *Sargassum* sp. *Biomass Electronic J Biotechnol*, 5, 133–140.
Cullen, W.R. and Nelson, J.C. (1993). The biotransformation of monomethylarsonate and dimethyl arsenate into arsenobetaine in sea water and mussels. *Appl Organomet Chem*, 7, 319–327.
Davis, T.A., Ali, F.E.C., Giannitti, E., Volesky, B. and Mucci, A. (2004). Cadmium biosorption by *S. fluitans*: Treatment, resilience and uptake relative to other *Sargassum spp.* and brown algae. *Water Qual Res J*, 39(3), 185–191.
Davis, T.A., Llanes, F., Volesky, B., Pulido, G.D., Mccook, A. and Mucci, A. (2003). ^{1}H-NMR Study of Na Alginates Extracted from *Sargassum* spp. in Relation to Metal Biosorption. *Appl Biochem Biotechnol*, 110, 75–90.
De Carvalho, R.P., Chong, K.H. and Volesky, B. (1995). Evaluation of the Cd, Cu and Zn biosorption in two-metal systems using algal biosorbent. *Biotechnol Prog*, 11, 39–44.
De Filippis, L.F. and Pallaghy, C.K. (1994). Heavy metals: Sources and biological effects. *In:* Algae and water pollution. Rai, L.C., Gaur, J.P., Soeder, C.J. (Eds), E Schweizerbartsche Verlagsbuchhandlung (Nägele u. Obermiller) Stuttgart. pp. 31–77.
De Vos, C.H.R. (1992). Glutathione depletion due to copper induced phytochelatin, synthesis causes oxidative stress in *Silene cucubalus*. *Pant Physiol*, 98, 853–858.
Dey, S. and Rosen, B.P. (1995). Dual mode of energy coupling by the oxyanion translocating ArsB protein. *J Bacteriol*, 177, 385–389.
Dhar, R.K., Biswas, B.K., Samanta, G., Mandal, B.K., Chakraborti, D., Roy, S., Jafar, A., Islam, A., Ara, G. and Kabir, S. (1997). Groundwater arsenic calamity in Bangladesh. *Curr Sci*, 73, 48–59.
Donmez, G.C., Aksu, Z., Ozturk, A. and Kutsal, T. (1999). A comparative study on heavy metal biosorption characteristics of some algae. *Process Biochem*, 34, 885–892.
Edmonds, J.S. and Francesconi, K.A. (1981). Arseno-sugars from brown kelp (*Ecklonia radiata*) as intermediates in cycling of arsenic in a marine ecosystem. *Nature*, 289, 602–604.
Eisler, R. (1994). A review of arsenic hazards to plants and animals with emphasis on fishery and wildlife resources. *In:* Arsenic Exposure and Health. Chappel, W.R., Abernathy, C.O. and Cothern, C.R. (Eds), Sci Technol Lett, Northwood. Pp. 185–259.

Ele-Sheekh, M.M., El-Shouny, W.A., Osman, M.F.H. and El-Gammal, W.E. (2005). Growth and heavy metals removal affinity of *Nostoc muscorum* and *Anabaena subcylindrica* in sewage and industrial wastewater effluent. *Environ Toxicol Pharmacol*, 19, 357–365.

Esteves, A.J.P., Valdman, E. and Leite, S.G.F. (2000). Repeated removal of cadmium and zinc from an industrial effluent by waste biomass *Sargassum* sp. *Biotechnol Lett*, 22, 499–502.

Feng, D. and Aldrich, C. (2004). Adsorption of heavy metals by biomaterials derived from marine alga *Ecklonia maxima*. *Hydrometallurgy*, 73, 1–10.

Fjerdingstad, E., Kemp, K., Fjerdingstad, E. and Vanggaard, L. (1974). Chemical analyses of red "snow" from East-Greenland with remarks on *Chlamydomonas nivalis* (Bau.) Wille. *Arch Hydrobiol*, 73, 70–83.

Foster, P.L. (1982). Metal resistance of chlorophyta from rivers polluted by heavy metals. *Freshwater Biol*, 12, 41–61.

Frankenberger, W.T. Jr. (Ed.) (2001). Environmental Chemistry of Arsenic. Marcel Dekker, New York.

Fridovich, I. (1997). Superoxide anion radical, superoxide dismutase and related matters. *J Biol Chem*, 250, 18515–18517.

Gadd, G.M. (1988). Accumulation of metals by microorganisms and algae. *In:* Biotechnology. Rehm, H.J. (Ed.), VCH, Weinheim. pp. 401–434.

Gadd, G.M. (1992). Microbial control of heavy metal pollution. *In:* Fry, J.C., Gadd, G.M., Herbert, R.A., Jones, C.W. and Watson-Craik, I.A. (Eds). Microbial control of pollution, Cambridge University Press.

Gadd, G.M. (1993). Interactions of fungi with toxic metals. *New Phytol*, 124, 25–60.

Gailer, J., Francesconi, K.A., Edmonds, J.S. and Irgolic, K.J. (1995). Metabolism of arsenic compounds by the blue mussel Mytiilus edulis after accumulation from seawater spiked arsenic compounds. *Appl Organanomet Chem*, 9, 341–355.

Galperin, M.Y., Walker, D.R. and Koonin, E.V. (1998). Analogous enzymes: Independent inventions in enzyme evolution. *Genome Res*, 8, 779–790.

Garcia-Salgado, S., Quijano, M.A. and Bonilla, J. (2006). Determination of soluble toxic arsenic species in alga samples by microwave-assisted extraction and high performance liquid chromatography-hydride generation-inductively coupled plasma-atomic emission spectrometry. *J Chromatogr A*, 1129, 54–60.

Gardea-Torresdey, J.L., Peralta-Videa, J.R., Rosa, G. and Parsons, J.G. (2005). Phytoremediation of heavy metals and study of the metal coordination by X-ray absorption spectroscopy. *Coord Chem Rev*, 249, 1797–1810.

Gardea-Torresdey, J.L., Tiemann, K.J., Gamez, G., Dokken, K. and Yacaman, M.J. (1998). Innovative technology to recover gold (III) from aqueous solutions by using *Medicago sativa* (alfalfa). *In:* Proceedings of the 1998 Annual Conference on Hazardous Waste Research. Erickson, L.E. and Rankin, M.M. (Eds), Kansas State Univ, Manhattan, KS. pp. 122–133.

Genter, R.B. (1996). Ecotoxicology of inorganic chemical stress on algae. *In:* Algal ecology – Freshwater benthic ecosystems. Stevenson, R.J., Bothwell, M.L. and Lowe, R.L. (Eds) Academic Press. California. pp. 403–468.

Girotti, A.W. (1990). Photodynamic lipid peroxidation in biological systems. *Photochem Photobiol*, 51(4), 497–509.

Gladysheva, T.B., Oden, K.L. and Rosen, B.P. (1994). Properties of the arsenate reductase of plasmid R773. *Biochemistry*, 33, 7288–7293.

Goldberg, E.D. and Gross, M.G. (1971). Man's impact on terrestrial and oceanic ecosystems. Matthews, W.H., Smith, F.E. and Goldberg, E.D. (Eds). MIT Press, Cambridge, Massachusetts, Part V.

Green, B., Hosea, M., Mcpherson, R., Henzl, M., Alexander, M.D. and Darnall, D.W. (1986). Interaction of gold (I) and gold (III) complexes with algal biomass. *Environ Sci Technol*, 20, 627–632.

Halliwell, B. and Gulleridge, J.M.C. (1999). Free radicals in biology and medicine. Oxford University Press, 3, 936.

Harding, J.P.C. and Whitton, B.A. (1976). Resistance to zinc of *Stigeoclonium tenue* in the field and the laboratory. *Br Phycol J*, 11, 417–426.
Harding, J.P.C. and Whitton, B.A. (1978). Zinc, cadmium and lead in water, sediments and submerged plants of the Derwent Reservoir, Northern England. *Wat Res*, 12, 307–316.
Hashim, M.A. and Chu, K.M. (2004). Biosorption of cadmium by brown, red and green seaweeds. *Chem Eng J*, 97, 249–255.
Hassan, H.M. and Scandalios, J.M. (1990). Superoxide dismutases in aerobic organisms. *In:* Stress responses in plants: Adaptation and acclimatation mechanisms. Alscher, R.G. and Cumming, J.R. (Eds). Wiley-Liss, New York. pp. 175–199.
Haug, A. (1967). The affinity of some divalent metals to different types of alginates. *Acta Chem Eng J*, 70, 115–124.
Hirata, S. and Toshimitsu, H. (2007). Determination of arsenic species and arsenosugars in marine samples by HPLC-ICP-MS. *Appl Organomet Chem*, 21(6), 447–454.
Holan, Z.R. and Volesky, B. (1994). Biosorption of lead and nickel by biomass of marine algae. *Biotechnol Bioeng*, 41, 819–825.
Hollibaugh, J.T., Seibert, D.L.R. and Thomas, W.H. (1980). A comparison of the acute toxicities of ten heavy metals to phytoplankton from Saanich Inlet, BC, Canada. *Estuar Coast Mar Sci*, 10, 93–105.
Holovská, K., Lenártová, V., Pedrajas, J.R., Peinado, J., Lopéz-Barea, J., Rosival, I. and Legáth, J. (1996). Superoxide dismutase, glutathione peroxidase and glutathione reductase in sheep organs. *Comp Biochem Physiol*, 115(B), 451–456.
Imamul Huq, S.M., Bulbul, A., Choudhury, M.S., Alam, S. and Kawai, S. (2005). Arsenic bioaccumulation in a green alga and its subsequent recycling in soil of Bangladesh. *In:* Natural Arsenic in Groundwater: Occurrence, Remediation and Management. Bundschuh, J. and Bhattacharya, P. (Eds), Balkema, New York. pp. 119–124.
Jordanova, A., Strezov, A., Ayranov, M., Petkov, N. and Stoilova, T. (1999). Heavy metal assessment in algae, sediments and water from the Bulgarian Black Sea Coast. *Water Sci Tech*, 39, 207–212.
Jurewicz, S., & Buikema, A. L. (1980). Effects of arsenate on algae, Daphnia, and mosquito fish. *Va J Sci*, *31*(124), 124.
Karpinski, S., Escobar, C., Karpinska, B., Creissen, G. and Mullineaux, P. (1997). Photosynthetic electron transport regulates the expression of cytosolic ascorbate peroxidase genes in Arabidopsis during excess light stress. *Plant Cell*, 9, 627–640.
Karthikeyan, S. and Hirata, S. (2003). Arsenic speciation in environmental samples. *Anal Lett*, 36, 2355–2366.
Keeney, W.L., Breck, W.G., Van Loon, G.W. and Page, J.A. (1976). The determination of trace metals in *Cladophora glomerata – C. glomerata* as a potential biological monitor. *Wat Res*, 10(11), 981–984.
Klimmek, S., Stan, H.J., Wilke, A., Bunke, G. and Buchholz, R. (2001). Comparative analysis of the biosorption of cadmium, lead, nickel and zinc by algae. *Environ Sci Technol*, 35, 4283–4288.
Klumpp, D.W. (1980). Accumulation of Arsenic from Water and Food by *Littorina littoralis* and *Nucella lapillus*. *Mar Biol*, 58, 265–274.
Klumpp, D.W. and Peterson, P.J. (1979). Arsenic and other trace elements in the waters and organisms of an estuary in SW England. *Environ Poll*, 19(1), 11–20.
Knauer, K. and Hemond, H. (2000). Accumulation and Reduction of Arsenate by the freshwater green Alga *Chlorella* sp. (Chlorophyta). *J Phycol*, 36, 506–509.
Knowles, F.C. and Benson, A.A. (1983). The biochemistry of arsenic. *Trends Biochem Sci*, 8, 178–180.
Kratochvil, D. and Volesky, B. (1998). Advances in biosorption of heavy metals. *TIBTECH*, 16, 291–300.
Larsen, E.H. (1995). Speciation of dimethylarsinyl-riboside derivatives (arsenosugars) in marine reference materials by HPLC-ICP-MS. *Fresenius' J Anal Chem*, 352(6), 582–588.

Lau, P.S., Lee, H.Y., Tsang, C.C.K., Tam, N.F.Y. and Wong, Y.S. (1999). Effect of metal interference, pH and temperature on Cu and Ni biosorption by *Chlorella vulgaris* and *Chlorella miniata*. *Environ. Technol.*, 20, 953–961.

Lee, B.Y., Lee, K.W., McGarry, M.G. and Graham, M. (1980). Overview of wastewater treatment and resource recovery. IDRC Report. Workshop on High-Rate Algae-Ponds. Singapore.

Lee, H.S., Suh, J.H., Kim, B.L. and Yoon, T. (2004). Effect of aluminium in two-metal biosorption by an algal biosorbent. *Minerals Eng*, 17, 487–493.

Lengke, M.F., Fleet, M.E. and Southam, G. (2006a). Morphology of gold nanoparticles synthesized by filamentous cyanobacteria from gold (III)-chloride complexes. *Langmuir*, 22, 2780–2787.

Lengke, M.F., Ravel, B, Fleet, M.E., Wanger, G., Gordon, R.A. and Southam, G. (2006). Mechanisms of gold bioaccumulation by filamentous cyanobacteria from gold (III)-chloride complex. *Envir Sci Technol*, 40, 6304–6309.

Lengke, M.F. and Southam, G. (2006). Bioaccumulation of gold by sulfate-reducing bacteria cultured in the presence of gold (I)-thiosulfate complex. *Geochem Cosmochim Acta*, 70, 3646–3661.

Liangfang, W. and Jianghong, H. (1994). Chronic arsenism from drinking water in some areas of Xinjiang, China. *In:* Arsenic in the Environment, Part II: Human Health and Ecosystem effects. Nriagu, J.O. (Ed.). John Wiley & Sons Inc., New York. pp. 159–172.

Liu, J. and Rosen, B.P. (1997). Ligand interactions of the ArsC arsenate reductase. *J Biol Chem*, 272(34), 21084–21089.

Liu, J., Zheng, B., Aposhian, H.V., Zhou, Y., Chen, M.L., Zhang, A. and Waalkes, M.P. (2002). Chronic arsenic poisoning from burning high-arsenic-containing coal in Guizhou, China. *Environ Health Perspect*, 110, 119–122.

Lo´pez-Maury, L., Florencio, F.J. and Reyes, J.C. (2003). Arsenic Sensing and Resistance System in the Cyanobacterium Synechocystis sp. Strain PCC 6803. *J Bacteriol*, 185(18), 5363–5371.

Loewus, F.A. and Loewus, M.W. (1983). Myo-inositol: Its biosynthesis and metabolism. *Annu Rev Plant Physiol*, 34, 137–161.

Loretto, C., Moenne, A. and Correa, J.A. (2005). Antioxidant responses in Scytosiphon lomentaria (Phaeophyceae) inhabiting copper-enriched coastal environments. *J Phycol*, 41, 1184–1195.

Lunde, G. (1970). Analysis of arsenic and selenium in marine raw materials. *J Sci Fd Agric*, 21, 242–247.

Macfarlane, G.R. and Burchett, M.D. (2001). Photosynthetic pigments and peroxidase activity as indicators of heavy metal stress in the grey mangrove, Avicennia marina (Forsk.). *Vier Mar Pollut Bull*, 42, 233–240.

Madsen, A.D., Goessler, W., Pedersen, S.N. and Francesconi, K.A. (2000). Characterization of an algal extract by HPLC-ICP-MS and LC-electrospray MS for use in arsenosugar speciation studies. *J Anal At Spectrom*, 15, 657–662.

Maeda, H., Hori, S., Ohizumi, H., Segawa, T., Kakehi, Y., Ogawa, O. and Kakizuka, A. (2004). Effective treatment of advanced solid tumors by the combination of arsenic trioxide and L-buthionine-sulfoximine. *Cell Death Differ*, 11, 737–746.

Maeda, S., Kumar, K., Maeda, M., Higashi, S. and Takashita, T. (1987). Bioaccumulation of arsenic by fresh water algae (Nostoc sp.) and the application to the removal of inorganic arsenic from an aqueous phase. *Appl Organomet Chem*, 1, 363–370.

Maeda, S., Kusadome, K., Arima, H., Ohki, A., Naka, K. (1992). Biomethylation of arsenic and its excretion by the alga Chlorella vulgaris. *Appl Organomet Chem*, 6(4), 407–413.

Maeda, S., Nakashima, S., Takeshita, T. and Higashi, S. (1985). Bioaccumulation of Arsenic by freshwater algae and the application to the removal of Inorganic arsenic from an aqueous phase, Part II, by Chlorella vulgaris isolated from Arsenic Polluted environment. *Separation Sc Tech*, 20(2&3), 153–161.

Mahapatra, H. and Gupta, R. (2005). Concurrent sorption of Zn (II), Cu (II) and Co (II) by Oscillatoria angustissima as a function of pH in binary and ternary metal solutions. *Bioresource Technol.*, 96, 1387–1398.

Mallick, N. and Mohan, F.H. (2000). Reactive oxygen species: Response of algal cells. *J Plant Physiol*, 157, 183–193.

Mandal, B.K., Roy Chowdhury, T., Samanta, G., Basu, G.K., Chowdhury, P.P., Chanda, C.R., Lodh, D., Karan, N.K. and Dhar, R.K. (1996). Arsenic in groundwater in seven districts of West Bengal, India: The biggest arsenic calamity in the world. *Curr Sci*, 70, 976–986.

Mandal, B.K., Roy Chowdhury, T., Samanta, G., Basu, G.K., Chowdhury, P.P., Chanda, C.R., Lodh, D., Karan, N.K., Dhar, R.K. and Tamili, D.K. (1997). In reply to "chronic arsenic toxicity in West Bengal". *Curr Sci*, 72, 114–117.

Marcum, K.B. (1998). Cell membrane thermostability and whole-plant heat tolerance of Kentucky bluegrass. *Crop Science*, 38,1214–1218.

Matheickal, J.T. and Yu Q. (1996). Biosorption of lead from aqueous solutions by marine alga Ecklonia radiate. *Water Sci Technol*, 34, 1–7.

McGriff, E.C. and McKinney, R.E. (1972). The removal of nutrients and organics by activated algae. *Wat Res*, 6, 1155–1168.

McKnight, D.M. and Morel, F.M.F. (1979). Release of weak and strong copper complexing agents by algae. *Limnol Oceanogr*, 24, 823–837.

McKnight, D.M. and Morel F.M.M. (1980). Copper complexation by siderophores from filamentous blue-green-algae. *Limnol Oceanogr*, 25, 62.

McShan, M., Trieff, N.M. and Grajger, D. (1974). Biological treatment of waste-waters using algae and artemia. *J Wat Poll Control Fed*, 46, 1742–1750.

McSheehy, S. and Szpunar, J. (2000). Speciation of Arsenic in edible algae by bidimensional size exclusion anion exchange HPLC with dual ICP-MS and Electrospray Ms/MS detection. *J Annal At Spectrom*, 15, 79–87.

Mehta, S.K. and Gaur, J.P. (2001a). Characterization and optimization of Ni and Cu sorption from aqueous solution by *Chlorella vulgaris. Ecol Eng*, 18, 1–13.

Mehta, S.K. and Gaur, J.P. (2001b). Concurrent sorption of Ni^{2+} and Cu^{2+} by *Chlorella vulgaris* from a binary metal solution. *Appl Microbiol Biotechnol*, 55, 379–382.

Mehta, S.K. and Gaur, J.P. (2001c). Removal of Ni and Cu from single binary metal solutions by free and immobilized Chlorella vulgaris. *Eur J Protistol*, 37, 261–271.

Mehta, S.K. and Gaur, J.P. (2005). Use of algae for removing heavy metal ions from wastewater: Progress and prospects. *Crit Rev Biotech*, 25, 113–152.

Mehta, S.K., Singh, A. and Gaur, J.P. (2002). Kinetics of adsorption and uptake of Cu^{2+} by *Chlorella vulgaris*: Influence of pH, temperature, culture age and cations. J Environ Sci Health, Part A 37, 399–414.

Mehta, S.K., Tripathi, B.N. and Gaur, J.P. (2002). Enhanced sorption of Cu^{2+} and Ni^{2+} by acid-pretreated *Chlorella vulgaris* from single and binary metal solutions. *J Appl Phycol*, 14, 267–273.

Meier, J., Kienzl, N., Goessler, W. and Francesconi, K.A. (2005). The occurrence of thioarsenosugars in some samples of marine algae. *Environ Chem*, 2(4), 304–307.

Michnowicz, C. J., & Weaks, T. E. (1984). Effects of pH on toxicity of As, Cr, Cu, Ni and Zn to Selenastrum capricornutum Printz. *Hydrobiologia*, *118*(3), 299–305.

Mitchell, R.D., Ayala-Fierro, F. and Carter, D.E. (2000). Systemic indicators of inorganic arsenic toxicity in four animal species. *J Toxicol Environ Health*, 59, 119–134.

Mittler, R. and Zilinskas, B.A. (1993). Detection of ascorbate peroxidase activity in native gels by inhibition of the ascorbate-dependent reduction of nitroblue tetrazolium. *Anal Biochem*, 212, 540–546.

Mukhopadhyay, R., Rosen, B.P., Phung, L.T. and Silver, S. (2002). Microbial arsenic: From geocycles to genes and enzymes. *FEMS Microbiol Rev*, 26, 311–325.

Mukhopadhyay, R. and Rosen, B.P. (2002). Arsenate reductases in prokaryotes and eukaryotes. *Environ Health Perspect*, 110 Suppl (5), 745–748.

Mukhopadhyay, R., Shi, J. and Rosen, B.P. (2000). Purification and characterization of ACR2p, the Saccharomyces cerevisiae arsenate reductase. *J Biol Chem*, 275, 21149–21157.

Nagalakshmi, N. and Prasad, M.N.V. (2001). Responses of glutathione cycle enzymes and glutathione metabolism to copper stress in Scenedesmus bijugatus. *Plant Sci*, 160, 291–299.
National Research Council (1999). Arsenic in Drinking Water. National Academy Press, Washington DC.
Nayak, D., Nag, M., Banerjee, S., Pal, R., Laskar, S. and Lahiri, S. (2006). Preconcentration of 198Au in a green alga, Rhizoclonium. *J Radioanal Nucl Chem*, 268, 337–340.
Niu, H. and Volesky, B. (2000). Gold-cyanide biosorption with L-cysteine. *J Chem Technol Biotechnol*, 75, 436–442.
Noctor, G. and Foyer, C.H. (1998). Ascorbate and glutathione: Keeping active oxygen under control. *Ann Rev Plant Physiol Plant Mol Biol*, 49, 249–279.
Nordi, C.S.F., Vieira, A.A.H. and Nascimanto, O.R. (2005). The metal binding capacity of Anabaena spiroides extracellular polysaccharide: An EPR study. *Process Biochem*, 40, 2215–2224.
Ofer, R., Yerachmiel, A. and Shmuel, Y. (2003). Marine microalgae as biosorbent for cadmium and nickel in water. *Water Environ Res*, 75, 246–253.
Okamoto, O.K., Asano, C.S., Aidar, E. and Colepicolo, P. (1996). Effects of cadmium on growth and superoxide dismutase activity of the marine microalga Tetraselmis gracilis. *J Phycol*, 32, 74–79.
Okamoto, O.K. and Colepicolo, P. (2001). Circadian protection against reactive oxygen species involves daily synthesis of manganese and iron-containing superoxide dismutase in Gonyaulax polyedra. *Biol Rhythm Res*, 32, 439–448.
Okamoto, O.K., Shao, L., Hastings, J.W. and Colepicolo, P. (1999). Acute and chronic effects of toxic metals on viability, encystment and bioluminescence in the dianoflagellate Gonyaulax polyedra. Comp Biochem Physiol, C 123, 75–83.
Ozer, A., Ozer, D. and Ekiz, H.I. (1999). Application of Freundlich and Langmuir models to multistage purification process to remove heavy metal ions by using Schizomeria. *Process Biochemistry*, 34(9), 919–927.
Pandey, S., Rai, R. and Rai, L.C. (2012). Proteomics combines morphological, physiological and biochemical attributes to unravel the survival strategy of Anabaena sp. PCC7120 under arsenic stress. *J Proteomics*, 75, 921–937.
Parial, D., Patra, H.K., Dasgupta, A.K. and Pal, R. (2012). Screening of different algae for green synthesis of gold nanoparticles. *Eur J Phycol*, 47, 22–29.
Pinto, E., Sigaud-Kutner, T.C.S., Leitao, M.A.S., Okamoto, O.K., Morse, D. and Colepicolo, P. (2003). Heavy metal-induced oxidative stress in algae. *J Phycol*, 39, 1–11.
Pistocchi, R.F., Balboni, G.V. and Boni. L. (1997). Copper toxicity and carbohydrate production in the microalgae Cylindrotheca fusiformis and Gymnodinium sp. *Eur J Phycol*, 32, 125–132.
Pradhan, S., Singh, S., Rai, L.C. and Parker, D.L. (1998). Evaluation of metal biosorption efficiency of laboratory-grown Microcystis under various environmental conditions. *J Microb Biotechnol*, 8, 53–60.
Prasad, B.N. and Godward, M.B.E. (1968). Cytology of irradiated Mougeotia sp. *Nucleus*, 11, 43–49.
Prasad, B. B., Banerjee, S., & Lakshmi, D. (2006). An AlgaSORB column for the quantitative sorption of arsenic (III) from water samples. *Water quality research journal of Canada*, *41*(2), 190–197.
Prasher, S.O., Beaugeard, M., Hawari, J., Bera, P., Patel, R.M. and Kim, S.H. (2004). Biosorption of heavy metals by red algae (*Palmaria plamata*). *Environ Technol*, 25, 1097–1106.
Rai, L.C., Gaur, J.P. and Kumar, H.D. (1981). Phycology and heavy-metal pollution. *Biol. Rev Cambridge Philos Soc*, 56, 99–151.
Reed, R.H. and Gadd, G.M. (1990). Metal tolerance in eukaryotic and prokaryotic algae. *In:* Heavy metal tolerance in plants: Evolutionary aspects. Shaw, A.J. (Ed.). CRC Press, Boca Raton. pp. 1105–1118.
Regelsberger, G., Obinger, C., Zoder, R., Altmann, F. and Peschek, G.A. (1999). Purification and characterization of a hydroperoxidase from the cyanobacterium Synechocystis PCC 6803: Identification of its gene by peptide mass mapping using matrix assisted laser desorption ionization time-of-flight mass spectrometry. *FEMS Microbiol Lett*, 170, 1–12.

Reuther, R. (1992). Arsenic introduced into a littoral freshwater model ecosystem. *Sci Total Environ*, 115, 219–237.

Rice-Evans, C.A., Miller, N.J. and Paganga, G. (1996). Structure-antioxidant activity relationships of flavonoids and phenolic acids. *Free Radic Biol Med*, 20, 933–956.

Rimbach, G., De Pascual-Teresa, S., Ewins, B.A., Matsugo, S., Uchida, Y., Minihane, A.M., Turner, R., VafeiAdou, K. and Weinberg, P.D. (2003). Antioxidant and free radical scavenging activity of isoflavone metabolites. *Xenobiotica*, 33(9), 913–925.

Rodriguez-Ariza, A., Dourado, G., Navas, J.I., Pueyo, C. and Lopez-Barea, J. (1991). Pro-mutagen activation by fish liver as a biomarker of littoral pollution. *Environ Mol Mutagen*, 24, 116–123.

Rosen, B.P., Bhattacharjee, H., Zhou, T.Q. and Walmsley, A.R. (1999). Mechanism of the ArsA ATPase. *Biochim Biophys Acta*, 1461, 207–215.

Rosen, B.P. (1999a). Families of arsenic transporters. *Trends Microbiol*, 7, 207–212.

Roux, J.C. (1998). Use of dead recovered biomass to adsorb heavy metals from wastewaters. Water Quality Intl, June 21–26. Vancouver, Canada.

Rubio, R., Ruíz Chancho, M.J. and López-Sánchez, J.F. (2010). Sample pre-treatment and extraction methods that are crucial to arsenic speciation in algae and aquatic plants. TrAC Trends Anal Chem, 29(1), 53–69.

Sandau, E., Sandau, P. and Pulz, O. (1996). Heavy metal sorption by marine algae and algal by-products. *Acta Biotechnol* 16, 227–235.

Sanders, J.G. (1979). Effects of arsenic speciation and phosphate concentration on arsenic inhibition of *Skeletonema costatum* (Bacillariophyceae). *J Phycol*, 15(4), 424–428.

Sanders, J.G. and Windom, H.L. (1980). The uptake and reduction of arsenic species by marine algae. *Estuar cstlmar Sci*, 10, 555–567.

Sanders, J.G, Osman, R.W. and Riedel, G.F. (1989). Pathways of arsenic uptake and incorporation in estuarine phytoplankton and the filter-feeding invertebrates *Eurytemora affinis*, *Balanusim provisus* and *Crassostrea virginica. Mar Biol*, 103, 319–325.

Sanders, O.I., Rensing, C., Kuroda, M., Mitra, B. and Rosen, B.P. (1997). Antimonite is accumulated by the glycerol facilitator GlpF in *Escherichia coli. J Bacteriol*, 179, 3365–3367.

Say, P.J., Diaz, B.M. and Whitton, B.A. (1977). Influence of zinc on lotic plants. I: Tolerance of Hormidium sp. to zinc. *Freshwater Biol*, 7, 357–376.

Shamsuddoha, A.S.M., Bulbul, A. and Imamul Huq, S.M. (2006). Accumulation of arsenic in green algae and its subsequent transfer to the soil–plant system. *Bangladesh J Microbiol*, 22(2), 148–151.

Sheng, P.X., Tan, L.H., Chen, J.P. and Ting, Y.P. (2004b). Biosorption performance of two brown marine algae for removal of chromium and cadmium. *J Dispersion Sci Technol*, 25, 679–686.

Sheng, P.X., Ting, Y.P., Chen, J.P. and Hong, L. (2004a). Separation of lead, copper, cadmium, zinc and nickel by marine algal biomass: Characterization of biosorption capacity and investigation of mechanisms. *J Colloid Interf Sci*, 275, 131–141.

Silver, S., Phung, L.T. and Rosen, B.P. (2001). Arsenic metabolism: Resistance, reduction and oxidation. *In:* Environmental Chemistry of Arsenic. Frankenberger, W.T. Jr. (Ed.). Marcel Dekker, New York. pp. 247–272.

Singh, S., Pradhan, S. and Rai, L.C. (1998). Comparative assessment of Fe^{3+} and Cu^{2+} biosorption by field and laboratory grown Microcystis. *Process Biochem*, 33, 495–504.

Šlejkovec, Z., Kápolna, E., Ipolyi, I. and van Elteren, J.T. (2006). Arsenosugars and other arsenic compounds in littoral zone algae from the Adriatic Sea. *Chemosphere*, 63(7), 1098–1105.

Smirnoff, N. (1993). The role of active Oxygen in the response of plants to water deficit and dessication. *New Phytol*, 125, 27–58.

Srivastava, A.K., Bhargava, P., Thapar, R. and Rai, L.C. (2009). Differential response of antioxidative defense system of Anabaena doliolum under arsenite and arsenate stress. *J Basic Microbiol*, 49, S63–S72.

Steele, R.L. and Thursby, G.B. (1983). A toxicity test using life stages of Champia parvula (Rhodophyta). *In:* Aquatic toxicology and hazard assessment: Sixth symposium. Bishop, W.E., Cardwell, R.D., Heidolph, B.B. (Eds). ASTM STP 802, American Society for Testing Materials, Philadelphia, pp. 73–89.

Swift, D.T. and Forciniti, D. (1997). Accumulation of lead by Anabaena cylindrical: Mathematical modeling and an energy dispersive X-Ray study. *Biotechnol Bioeng*, 55(2), 408–418.

Takamatsu, T., Aoki, H. and Yoshida, T. (1982). Determination of arsenate, arsenite, monomethylarsonate, and dimethylarsinate in soil polluted with arsenic. *Soil Sci*, 133, 239–246.

Takamura, N., Kasai, F. and Watanabe, M.M. (1989). Effects of Cu, Cd and Zn on photosynthesis of freshwater benthic algae. *J Appl Phycol*, 1, 39–52.

Ting, Y.P., Teo, W.K. and Soh, C.Y. (1995). Gold uptake by Chlorella vulgaris. *J Appl Phycol*, 7, 97–100.

Tisa, L.S. and Rosen, B.P. (1990). Molecular characterization of an anion pump: The ArsB protein is the membrane anchor for the ArsA protein. *J Biol Chem*, 265, 190–194.

Tondel, M., Rahman, M., Magnuson, A., Chowdhury, I.A., Faruquee, M.H., Ahmad, S.A. (1999). The relationship of arsenic levels in drinking water and the prevalence rate of skin lesions in Bangladesh. *Environ Health Perspect*, 107, 727–729.

Vahter, M. (2000). Genetic polymorphism in the biotransformation of inorganic arsenic and its role in toxicity. *Toxicol Lett*, 209, 112–113.

Vijayaraghavan, K., Jegan, J., Palanivenu, K. and Velan, M. (2005). Biosorption of copper, cobalt and nickel by marine green alga Ulva reticulata in a packed column. *Chemosphere*, 60, 419–426.

Walsh, R.S. and Hunter, K.A. (1992). Influence of phosphorus storage on the uptake of cadmium by the marine alga Macrocystis pyrifera. *Limnol Oceanogr*, 37(7), 1361–1369.

Whitton, B.A. (1980). Zinc and plants in rivers and streams. *In:* Zinc in the environment. Part (II): Health effects. Nriagu, J.O. (Ed.). John Wiley, New York. pp. 364–400.

Whitton, B.A. (1984). Algae as monitors of heavy metals in freshwaters. *In:* Algae as Ecological Indicators. Shubert, L.E. (Ed.). Academic Press. New York.

Whitton, B.A., Say, P.J. and Wehr, J.D. (1981). Use of plants to monitor heavy metals in rivers. *In:* Heavy metals in northern England: Environmental and biological aspects. Say, P.J. and Whitton, B.A. (Eds). Botany Department, University of Durham. pp. 135–145.

Whitton, B.A. (1970). Toxicity of heavy metals to Chlorophyta from running waters. *Arch Mikrobiol*, 72, 353–360.

Willsky, G.R. and Malamy, M.H. (1980). Effect of arsenate on inorganic phosphate transport in Escherichia coli. *J Bacteriol*, 144, 366–374.

Wong, S.L, Nakamoto, L. and Wainwright, J.F. (1994). Identification of toxic metals in affected algal cells in assay of wastewaters. *J Appl Phycol*, 6, 405–414.

Wysocki, R., Chéry, C.C., Wawrzycka, D., van Hulle, M., Cornelis, R., Thevelein, J.M. and Tamás, M.J. (2001). The glycerol channel Fps1p mediates the uptake of arsenite and antimonite in Saccharomyces cerevisiae. *Mol Microbiol*, 40(6), 1391–1401.

Yamaoka, Y. and Takimura, O. (1986). Marine algae resistant to inorganic arsenic. *O Agric Biol Chem*, 50, 185–186.

Yamaoka, Y., Takimura, O., & Fuse, H. (1988). Environmental factors relating to arsenic accumulation by Dunaliella sp. *Applied organometallic chemistry*, *2*(4), 359–364.

Yamaoka, Y., Takimura, O., Fuse, H., & Kamimura, K. (1992). Effects of arsenic on the organic component of the alga Dunaliella salina. *Applied organometallic chemistry*, *6*(4), 357–362.

Yin, P., Yu, Q., Lin Z. and Kaewsarn, P. (2001). Biosorption and desorption of cadmium (II) by biomass of *Laminaria japonica. Environ Technol*, 22, 509–514.

Yompakdee, C., Ogawa, N., Harashima, S. and Oshima, Y. (1996b). A putative membrane protein, Pho88p, involved in inorganic phosphate transport in *Saccharomyces cerevisiae. Mol Gen Genet*, 251, 580–590.

Yu, Q. and Kaewsarn, P. (1999). Fixed-bed study of copper (II) removal from aqueous solutions by marine alga Durvillaea potatorum. *Environ Technol*, 20, 1005–1008.

Zhou, J.L., Huang, P.L. and Lin, R.G. (1998). Sorption and desorption of Cu and Cd by macroalgae. *Environ Pollut*, 101, 67–75.

Zutshi, S., Bano, F., Ningthoujam, M., Habib, K. and Fatma, T. (2014). Metabolic Adaptations to Arsenic-Induced Oxidative Stress in *Hapalosiphon fontinalis*-339. *Int J Innov Res Sci Engg Tech*, 3(2), 9386–9394.

Index

D. Das (ed.), *Algal Biorefinery: An Integrated Approach*,
DOI 10.1007/978-3-319-22813-6

MIX
Papier aus verantwortungsvollen Quellen
Paper from responsible sources
FSC® C105338

If you have any concerns about our products,
you can contact us on
ProductSafety@springernature.com

In case Publisher is established outside the EU,
the EU authorized representative is:
Springer Nature Customer Service Center GmbH
Europaplatz 3, 69115 Heidelberg, Germany

Printed by Libri Plureos GmbH
in Hamburg, Germany